NO-TILLAGE AND SURFACE-TILLAGE AGRICULTURE

NO-TILLAGE AND SURFACE-TILLAGE AGRICULTURE

The Tillage Revolution

Edited by

MILTON A. SPRAGUE

Professor, Emeritus Rutgers,
The State University of New Jersey
New Brunswick, New Jersey

GLOVER B. TRIPLETT

Professor of Agronomy
Mississippi State University
Mississippi State, Mississippi

A WILEY-INTERSCIENCE PUBLICATION

JOHN WILEY & SONS

New York Chichester Brisbane Toronto Singapore

Library of Congress Cataloging in Publication Data:

Main entry under title:

No-tillage and surface-tillage agriculture.

"A Wiley-Interscience publication."
Includes bibliographies and index.
1. No-tillage. 2. Tillage. I. Sprague, Milton A.
II. Triplett, Glover B. (Glover Brown), 1930–
III. Title: Surface-tillage agriculture.

S604.N62 1986 631.5′8 85-26587
ISBN 0-471-88410-3

Printed in the United States of America

10 9 8 7 6 5 4 3 2 1

CONTRIBUTORS

John N. All
Department of Entomology
University of Georgia
Athens, Georgia

V. Allan Bandel
Department of Agronomy
University of Maryland
College Park, Maryland

Michael G. Boosalis
Department of Plant Pathology
University of Nebraska
Lincoln, Nebraska

James E. Box
Southern Piedmont Conservation Research Center
USDA-ARS
Watkinsville, Georgia

Pierre Crosson
Resources for the Future
Washington, D.C.

Benjamin L. Doupnik
Department of Plant Pathology
University of Nebraska
South Central Extension and Research Center
Clay Center, Nebraska

Michael Duffy
Economic Research Service
U.S. Department of Agriculture
Washington, D.C.
Present address:
Department of Economics
Iowa State University
Ames, Iowa

Richard H. Fox
Department of Agronomy
Pennsylvania State University
University Park, Pennsylvania

Donald R. Griffith
Department of Agronomy
Purdue University
West Lafayette, Indiana

Michael Hanthorn
Economic Research Service
U.S. Department of Agriculture
Washington, D.C.

Rattan Lal
International Institute of Tropical Agriculture
Ibadan, Nigeria

Jerry V. Mannering
Department of Agronomy
Purdue University
West Lafayette, Indiana

Gerald J. Musick
Department of Entomology
University of Arkansas
Fayetteville, Arkansas

Charles J. Scifres
Department of Range Science
Texas A&M University
College Station, Texas

Milton A. Sprague
Department of Soils and Crops
Rutgers University
New Brunswick, New Jersey

Grant W. Thomas
Department of Agronomy
University of Kentucky
Lexington, Kentucky

Ray I. Throckmorton
357 Harris Ave.
Clarendon Hills, Illinois

Glover B. Triplett
Department of Agronomy
Mississippi State University
Mississippi State, Mississippi
Formerly: Ohio
State University/
OARDC,
Wooster, Ohio

Robert W. Van Keuren
Department of Agronomy
Ohio State University/Ohio
Agricultural Research and
Development Center
Wooster, Ohio

John E. Watkins
Department of Plant Pathology
University of Nebraska
Lincoln, Nebraska

Gail A. Wicks
University of Nebraska
West Central Research and
Extension Center
North Platte, Nebraska

A. Douglas Worsham
Department of Crop Science
North Carolina State University
Raleigh, North Carolina

PREFACE

The nation was alerted to the urgent need for better farming methods on May 11, 1934, when dust clouds darkened eastern skies over the nation's capitol. It was not the first soil erosion observed, but it was by far the most dramatic, and its magnitude had an impressive impact on events that followed. Several years of drought and depressed prices had severely weakened the country's agriculture, and soil loss was out of control. Conservation became the first call to action, followed by decades of research to preserve the soil and to restore, expand, and extend the agricultural economy. Tillage intensity needed to be greatly reduced and improved germplasm introduced. A new agriculture evolved.

During the brief period of only a quarter century, a broadly based expansion of research led to a new and revolutionary change in tillage management. It was preceded by several decades of isolated, trial and error attempts to prepare a suitable seedbed by replacing the plow with shallow surface tillage. A new generation of agricultural scientists concentrated their efforts on land use, soil improvement, plant breeding, pest control, and crop management technology. The first objective led to retention of plant residue as a mulch cover on the soil surface as protection from wind and rain. The second emphasized less tillage for weed control and water conservation. A third was improved mineral nutrition. Beginning with the forages and small grains, mostly on the steeper slopes and western plains, plowing and repeated cultivation gradually evolved into greatly reduced tillage systems designed for use with most of the major row crops. Many advantages beyond that of soil stability were exhibited, including more efficient use of labor, fuel, land, fertilizers, pest control and water.

The introduction of herbicides, following World War II, provided the new tool with which to replace excessive tillage. Chemical energy was substituted for tractor energy to reduce competition from weeds during and after seedling establishment. New environments, arising from major reductions in tillage, not only provided new solutions (such as soil and water conservation) to old problems, but also gave rise to new problems and opportunities. Percolation rates changed, and mineral fertilizers moved differently in soils; crop rooting patterns changed with less evaporation loss from the soil surface, and insect and disease cycles were altered with higher humidities near the undisturbed soil surface. Many scientific disciplines were involved to cope with crop safety, mineral fertilization, seed placement, and pest control in different geographic regions where different cultivars, soils, cropping systems, and economics prevailed. Many of the variables, being collectively linked, needed to be integrated in new ways. Reports of these investigations are found in a myriad of science journals, bulletins, and circulars, foreign and domestic. From the vast array of available information, this book endeavors to impart order and understanding of the principles involved, advantages to be gained, and pitfalls to be avoided.

The concept is that of a drastically reduced tillage agriculture, free of weed competition, with a dead mulch of plant material on an undisturbed soil surface. Attempts are made to identify the differences, similarities, and complexities among the integrated systems that have evolved and to coordinate the adjustments necessary for them to best fit the sites and situations for which they are intended. The objectives are to devise new and more favorable environments in which crops can flourish, at the same time rebuilding the natural resource base on which crop production depends. In some instances problems have been only partially solved. However, rationale is available of the progress that has been made, and the direction reduced tillage will likely follow in the future.

This book consists of four major parts, with the intent to portray the basic principles within which surface-tillage and no-tillage management systems function:

1. Environmental resources and their management including soil, climate, mulch–moisture relationships, mineral nutrition, fertilizer, and seed placement.
2. Characteristics of the major cropping systems determined by species adaptation, site situations, equipment, and economics in different geographic regions.
3. Pest control management, including weed competition, diseases, insects, and other predators.
4. Economic and summary evaluations of the factors that motivate, support, and sustain new surface-tillage systems for a permanent agriculture.

Within this framework the editors have attempted to assemble and coordinate the experience and judgments of 22 authors as experts representing the major disciplines involved. The text is international in scope, although much of the research has been conducted in only a sampling of certain areas and regions. The book is intended for professional scientists, students, advisors, industry, administrators, and innovative farmers plus others involved in leadership roles to strengthen and add permanence to agricultures throughout the world. Its strength arises from a careful selection of experienced specialists most closely associated with the research and development of this tillage revolution.

The metric system of weights and measures has been used in the interest of uniformity. A simplified conversion table is provided in Appendix B. The common and chemical names of herbicides mentioned in the text are listed in Appendix A. The terms "cultivar" and "variety" are used interchangeably.

We wish to express our sincere appreciation to the authors of each chapter for their willing efforts to coordinate their contributions into the whole work so that it might present an organized appraisal of a complex subject. Reviewers' comments have been particularly helpful with respect to many of the chapters. These reviewers include: Drs. S. R. Race, J. L. Peterson, J. A. Meade, R. D. Ilnicki, R. L. Flannery, and J. C. F. Tedrow of Rutgers University; R. J. Aldrich, USDA, and A. F. Wiese, of the Missouri and Texas Agricultural Experiment Stations, respectively.

We are deeply grateful to John Hannon and Chester Wells, station editors of the New Jersey and Mississippi Agricultural Experiment Stations, respectively, for their suggestions in the preparation of certain parts of the manuscript.

Special thanks are due Katherine Eckardt, Department of Soils and Crops, Rutgers University, for her efficient and patient help in typing and proofreading letters and manuscripts far beyond the call of secretarial duty, and to Elizabeth Murdock, Patricia Ryan and Cynthia Gourley for their willing assistance.

Appreciation is extended to the New Jersey Agricultural Experiment Station and to the Mississippi Agricultural and Forestry Experiment Station for the opportunity they provided us in accomplishing our goals.

Above all, we are grateful for the encouragement of our families, who persevered during our efforts to complete this volume.

MILTON A. SPRAGUE
GLOVER B. TRIPLETT

New Brunswick, New Jersey
Mississippi State, Mississippi
March 1986

CONTENTS

NO-TILLAGE AND SURFACE-TILLAGE AGRICULTURE

1

OVERVIEW

M. A. SPRAGUE
Professor Emeritus
Department of Soils and Crops
Rutgers University
New Brunswick, New Jersey

INTRODUCTION

The world has witnessed a long succession of changes as its agriculture progressed from a nomadic harvest of natural products to the sophisticated cropping systems of the twentieth century. Most of these changes have been interrelated and have evolved stepwise, each made possible by those changes that preceded it. All were a result of peoples' observation of biological principles at work, their basic needs, their imagination, and the courage to test their theories of cause and effect. Their accumulative effects have produced major improvements in yield, quality and the efficiency of labor management. Without this progress in farming methods, tools, materials, and crops, mankind could not have achieved the standards of living it enjoys today. In practice this stepwise growth, motivated by urgent need and good leadership, has drastically enhanced mankind's ability to feed itself.

Some of the changes have had a dramatic impact on the rise and fall of civilizations; other changes have been minor, insignificant, and soon superseded. Some changes have contributed to improving the ability of a crop to capture the sun's energy and convert it efficiently into useful products; other changes have consisted of a newly designed hoe, plow, or planter. Much of the progress has been crop-oriented and soil-related, yet many of the developments have been at the expense of soil and water resources. Most have improved efficiency in the use of natural resources for short-term gain. Some also have improved the long-term environments in which crops grow. Only during the past century and a half have improvements in agronomy materially recharged the soil's capacity to produce. Many major elements are now being put into proper order to provide the long-awaited permanently productive agriculture. The most recent is the concept of alternatives to excessive use of tillage.

GOAL: A PERMANENT AGRICULTURE

Civilizations long have sought a permanently productive agriculture as essential to their survival. In recent decades the need has become more urgent as populations approach the carrying capacity of our globe. Mankind's early history consisted primarily of a search for new land, new crops, new tools, and more power with which to use them. Throughout much of mankind's existence, tillage of the soil was the dominate activity. As its intensity increased it brought with it a matching crescendo of soil deterioration, which continues unchecked in much of the world. Since the turn of the century, fertilizers have been applied sufficiently to have a major impact. Together with better cultivars, more effective pest control, and greatly expanded irrigation, some agricultures have increased total production many times. Still, mismanaged soils in some regions have been so depleted of the requirements for growth that it has become impossible for people to feed themselves. Most of the losses have been related to tillage management.

For thousands of years new lands have been available on which to expand and build new agricultures. As new land was occupied, increasingly more intensive cultivation relieved the soil of its accumulated natural reservoir of fertility and moisture. The slow natural soil-building processes were reversed as excessive tillage led to loss of topsoil, structure, and organic matter—the very life in the soil so essential for high production. Crop growth, initially vigorous, gradually waned, leveling off at a low plateau.

Carter and Dale (1974) characterized mankind as having "marched across the earth leaving deserts as his footprints." Their studies of many civilizations that have existed in the last 6000 years show that a civilization has remained viable in one location for an average of from 40 to 60 generations (1000 to 1500 years) before declining. They identified more than 30 civilizations that followed this sequence. The Fertile Crescent, Carthage, Syria, and Rome are examples of civilizations that rose and then degenerated as their agricultures weakened. Some nations today are following a similar pattern. One notably lasted much longer, that of the lower Nile River, where the soils were replenished annually with mineral nutrients, organic matter, and moisture through regular flooding with waters from the rain forests of central Africa. Most of the wars throughout history have been a search for more land, new land. Interruptions in this sequence frequently have given rise to famine. The world now demands that we vigorously seek alternative solutions.

TOWARD LESS TILLAGE

Major agricultural changes have been more frequent during the past century as the impact of science has been brought to bear on farm production. Each successive new discovery forms a base for further research as more efficient techniques are superimposed on new and improved germplasm. The objective of the recent thrust toward less tillage has been to limit mechanical disturbance of the soil to that required for seed placement. Its purpose is to control soil erosion, enhance crop performance, and use energy more efficiently. In principle, it introduces an alternate means of vegetation control as a replacement for plowing and surface cultivation during seedbed preparation and subsequent growth of the crop.

Eliminating repeated physical manipulation of the soil layer is a radical departure from long-established conventional practice. Therefore, other well-established cultural practices must be reevaluated and adjusted to compensate for the new problems that arise. A soil profile remains undisturbed from year to year. A dead and decaying mulch of crop residues covers the surface. Microclimates are greatly altered, affecting temperatures and evaporation rates. Fertilizer materials are no longer plowed down, but are placed on the surface to percolate into the root zone. Soil microflora and fauna reproduce in different sequences. The new and different environments have been shown to influence crop growth patterns both directly and indirectly in many ways.

Over the short term a reduction in tillage may be expected to lower fuel, machinery, and labor costs per unit of land. However, new problems and costs must be assessed along with adjusted returns in order to establish a new balance. Encouragingly, more precise control over many environmental stresses affecting both crops and pests appears possible. Freedom from tillage restrictions suggests opportunity for land use adjustments. Because the environment can be more precisely controlled, more nearly optimum performance of the crop can be anticipated. The principal objective becomes exceedingly complex: that of creating an environment as ideal as possible for the crop during each stage in its life cycle.

PRODUCTION PRESSURES

As the twentieth century comes to a close, new lands are no longer available in great abundance. The settling of our globe approaches completion. Many undeveloped areas in the tropics will be costly and difficult to bring into production, but offer some promise for certain crops. The cold and forbidding taiga and tundra hold little promise other than for timber and grazing lands. The greatest opportunity for increasing world food supplies lies in restoring productivity where it has been lost. Population growth rates are at an all time high, and the deserts are expanding rapidly under increased pressure to produce. The *United Nations Yearbook* (1978) indicates worldwide only 11.2% of the approximately 13.1 billion hectares of land is currently considered arable land (Table 1.1). Another 23.4% of the total land area is in permanent grassland, 31.2% is in forest, and 34.2% is unusable for agriculture. Dudal (1976) contends that arable land could be doubled providing modern technology could be applied to relieve mineral, drought, and excess water stresses.

The *Global 2000 Report* (Barney, 1980) indicated that even allowing for the beneficial effects of improved technology, the world's population by the year 2000 may be within a few generations of reaching the planet's carrying capacity. Projections place the population in excess of 6.3 billion in the year 2000, up 1.7% per year from 4.1 billion in 1975. Projecting further, approximately 10.8 billion people can be expected to need a productive agriculture by 2025. It is sobering to note that only 4 of 12 groups of nations produced grain in excess of per capita consumption in 1970 and 3 in 1975. The projected grain deficit in the

TABLE 1.1. World Summary of Land Use, 1977

		Land Use (million hectares)				
		Agricultural Land				
	Total Land Area	Total	Arable Land and Land in Permanent Crops	Permanent Grasslands	Forest Land	Other Lands[a]
Africa	2,965	1,007	209	798	637	1,321
North America	2,136	621	267	354	718	797
Canada	912	67	44	23	326	529
United States	913	428	187	241	290	194
Mexico	193	98	23	75	71	24
Central America	108	28	13	15	31	50
South America	1,754	551	108	443	921	282
Europe	473	229	142	87	155	89
Asia (excluding USSR)	2,677	996	458	538	571	1,110
USSR	2,227	606	232	374	920	701
Oceania	843	512	47	465	155	176
Australia	762	496	45	451	107	159
New Zealand	27	14	0.4	14	7	6
World	13,074	4,520	1,462	3,058	4,077	4,477

[a]Cities, roads, swamps, lakes, rivers, ice and snow, permafrost, desert.

SOURCE: *FAO United Nations Yearbook*, 1978.

less-developed countries by the year 2000 is placed at 14.6 kilograms per capita, up from 10.8 kilograms in 1970. More recent appraisals differ in detail but confirm the trend.

Either population trends must be reversed or production accelerated if civilizations are to prosper. Two major problems related to production are involved, deterioration of the soil and efficiency in the use of available water. Conservation alone of soil and water resources is insufficient. A system of recharging these resources is necessary on a continuing basis to gain the productive stability required. The new tillage systems discussed in this volume offer much encouragement.

ANCIENT TILLAGE SYSTEMS

In ancient times tillage involved tools that merely scratched a shallow furrow in the soil. With enough passes over a field, the weeds were killed and the soil was exposed, smoothed, and pulverized. If seed were to germinate and grow, seed–soil contact was needed. Greek and Roman writers were emphatic in expressing need for a "crumbly" condition of the soil and advocated "repeated plowing and cross plowing up to 12 times during the year prior to sowing" (Fussell, 1965). The light-textured soils of the semiarid Mediterranean regions were easily worked, and frequent tillage was able to keep them weed-free. However, excessive tillage took its toll.

When people moved to the wetter climates of northern Europe, they found heavier clay soils and more perennial weeds that required a deeper furrow to subdue the vegetation by complete coverage. Turning over a furrow required more power than oxen could provide (Jope, 1956). The faster, stronger horse could not be used as a draft animal until invention of the horse collar during the Middle Ages, for, without an improved harness, as the horse lowered his head to bring the strength of his hindquarters to bear, he choked himself.

In 1733 Jethro Tull, an English land-owner, held the philosophy that the more and the deeper the tillage the better. Tull considered tillage as a substitute for manure, since "by deeper tillage we can enlarge our field . . . though the external surface of it be confined within narrow bounds." According to his theory, repeated deep plowing brought more soil within reach of the plant, and thoroughly pulverized soil was necessary to make the particles fine enough to be "insumed" by the roots. Tillage "pulverized the soil by attrition," whereas "dung by fermentation." Tull devoted most of his life to improving plows and other tillage implements since he considered "every plant is earth and the growth and true increase of a plant is the addition of more earth."

Designs of many moldboard plows were introduced during the eighteenth and nineteenth centuries in attempts to plow deeper and bury unwanted vegetation. New weeds were killed with surface cultivation. The optimum shape to turn the furrow was worked out mathematically by Thomas Jefferson in 1798 (Jope, 1956). Now people were able to attack the soil more intensely than ever before. Breaking the virgin sods of the United States' midcontinent is an example of a new area opened for crop production too rapidly. It became national agricultural policy to encourage settlement of the midcontinent as quickly as possible, and the new moldboard plow became the key tool in this endeavor.

OPENING MIDDLE AMERICA

The year 1862 was a memorable one. The transcontinental railroad in the United States was nearly complete. The Morrill Act (Land Grant System) and the establishment of the U.S. Department of Agriculture represented a firm federal commitment to broad training and support of an expanding national agriculture. The year 1862 also saw passage of the Homestead Act, whereby 160 acres were given to any citizen willing to work the land. Not only was the frontier opened for growing and shipping crops and livestock by rail east and west, but within 20 years open-range grazing would feel competition from a new pattern of development, that of a walk-plow agriculture of grains, mostly wheat (*Triticum vulgare* L.). No longer would the interior of the United States consist of frontiersmen and Indians with a few domestic settlers around a trading post. Given free land, new equipment, transportation and eager markets, the region was to experience an enthusiastic boom (Voigt, 1976). With the Civil War finished there came a massive import of human activity intent on conquering the land, first with open-range cattle and sheep, then with the plow. As with most things done too rapidly, there were costly mistakes.

Corn (*Zea mays* L.), cotton (*Gossipium* spp.), hay, and milk dominated the higher-rainfall areas to the east, and the bread grains moved to the better suited climates and soils farther west. So intense was the effort, from the Canadian Prairies through the Dakotas to Texas, that the reserves of moisture in the surface layers of soil were soon exhausted. In many low rainfall areas, dryland plowing and surface tillage continued throughout one year to conserve precipitation from the first season to provide moisture for a crop of grain the second. Fall plowing, to catch and hold more snow in winter's high winds, became commonplace. Fall plantings of winter grains were grazed. Such frequent and extended soil exposure caused serious problems.

It was not only the plow that exhausted the range. After the Civil War, Texas was bursting with cattle. Hordes of men recently out of uniform were eager to seek new lives in the rapidly opening lands beyond the Mississippi River (Voigt, 1976). Transportation improved to supply the demands of people from coast to coast, and settling of the West took a remarkably short time considering the vastness of the land area. Cattle numbers in the 17 western range states are placed at 7.9 million head in 1870, 12.9 million in 1880, and 21.6 million by 1886 (Silcox, 1936). Overcrowding bared the soil and encouraged the spread of unpalatable weedy species. Daniel Freeman was granted the first homestead near Beatrice, Nebraska, in 1869, and by 1900, 242,908 farm units encompassing 98.3 million acres (39.8 million ha) were listed. Crop farming was developing within the same territory and on land first used for grazing. In 1910 there were 373,337 farm units on 110.9 million acres (44.9 million ha). The first big irrigation dam (the Cheesman) was completed in 1905; the Shoshone was completed in 1910. By 1930 farm units reached 775,748 on over 392 million acres (158.8 million ha), 19 million (7.7 million ha) of which were irrigated; animal units exceeded 32 million (Silcox, 1936). In less than six decades an intensive agriculture was introduced on a large scale to the semiarid range.

Crash programs are frequently inadequately planned and wasteful. Excessively heavy grazing pressures and unwise use of the plow and disk took a severe toll. One of the nation's greatest natural resources, the native sod itself, fast disappeared. Without a live grass cover, much of the soil resource beneath was exposed, lost, and destroyed. The water ran off or evaporated instead of penetrating to recharge the soil reservoir. Soil lost both its structure and organic matter; salinity expanded. Unprotected soil surfaces drifted in a region where wind velocities were high. Cattlemen were looking so intently at the economic and social opportunities that they did not note the destruction of the country's three greatest natural resources: the sod, the soil, and the water. The United States was rapidly following the sequence to disaster so clearly described by Carter and Dale (1974).

The concept of less tillage began early in the twentieth century and slowly gained momentum as its merits were revealed. Initial efforts consisted of several isolated attempts to control erosion by restricting use of the moldboard plow. Cotton (1910) in New York slowed water runoff and reduced gully erosion by surface-disking of hilly land to prepare a seedbed for pasture. Jardin (1913) in Kansas advocated a surface residue mulch by disking grain stubble to prevent soil from blowing. Such changes in tillage for seedbed preparation constituted a major departure from conventional plowing at that time. It was a trashy seedbed, and more seed was needed.

UNCONTROLLED EROSION

In the United States during the 1920s and 1930s Hugh H. Bennett called out to the country to recognize the urgency of saving the remaining topsoil. His central focus was on conserving soil. His efforts were enhanced by dramatic events of the time, which had been in the making for decades as side effects of a too rapidly expanding agriculture by an uninformed, unguided nation. Throughout much of the eastern humid region of the country huge gullies had come to dominate the landscape as heavy rainfalls rushed over bare soils. A local church fell into a gulch 300 ft (95 m) deep near Province, Georgia (Brink, 1951). Large areas of farmland were abandoned as their productive capacity diminished. Moorhouse (1909), in the first volume of the *Agronomy Journal*, reported "land has been so disfigured by the rush of water that no attempt is made to use those parts of the farm once they have been under tillage." In the drier plains states from Texas to Canada, strong winds raised giant dust storms from soils plowed for grain before and after World War I. Three such storms out of the southwest in 1934 were estimated to carry eastward over 300 million t (282.8 million mt) of topsoil each, equivalent to 150,000 acres (60,750 ha) of topsoil 12 in. (30.5 cm) deep. Dust settled on decks of ships hundreds of miles off the Atlantic Coast (Brink, 1951). During a single storm in May of 1936, 1 ft (30.5 cm) of brown snow fell in northern Wisconsin.

Bennett (1939) waged war nationwide on gullies and sheet erosion. His techniques included contour terraces, grass waterways, shelter belts, and strip cropping. His purpose was soil conservation, particularly to slow water runoff and to shield the soil surface from wind and rain. With words and pictures he portrayed the growing desolation that followed soil loss. The covered fence rows, abandoned farmsteads, foreclosed mortgages, and long unemployment rosters were all testament to the urgency of saving soil that future generations might prosper. On one occasion Bennett personally carried a Morgan long-plow to the Senate Office Building while a Senate investigating committee awaited his testimony (Brink, 1951). Here was "smoking gun" evidence of the culprit responsible for scarring the landscape. Soil erosion had become a national menace!

Hugh Bennett's dramatic efforts resulted in establishment of the Soil Erosion Service in 1933 in the Department of Interior, which two years later became the Soil Conservation Service in the Department of Agriculture. During the next quarter-century, conservation expanded to involve the entire United States and was spreading abroad. Several developments encouraged acceptance of the program. The Civilian Conservation Corps was established during the 1930s to provide meaningful employment for idle youth who needed constructive work to do. Government cost incentives to farmers encouraged adoption. Surplus grain production after World War II reduced prices and incentives to put production pressure on the land. A land reserve was established, which took many of the more erodible acres out of production for varying periods of time. Soil deterioration slowed encouragingly.

In 1973, after a threatened petroleum shortage, the nation's land resources again came under severe pressure to produce. A growing world food shortage, followed by favorable commodity prices, provided a strong incentive for export. The United State's grain exports alone doubled in four years. Soon nearly all of the reserve cropland was again brought under cultivation. More than 1 million acres (0.4 million ha) were taken from fence rows and land protection programs. Many were the more steeply sloping and erosion-prone soils, and soil losses grew alarmingly. An estimated 6 billion t (5.4 billion mt) of soil per year were lost by water erosion in 1980 from nonfederal lands alone, 79% of the total excluding Alaska (Holeman, 1980). Thirty-eight percent of the losses were from cropland, 26% from rangeland, 8% from forest, 7% from pasture, and the remaining 21% from streambanks, gullies, roads and urban construction sites. Seventy-nine percent of the soil lost was by sheet and rill erosion, removed from the land surface in thin layers and deposited in lower portions of fields, fence rows, and flood plains. About 20-30% reached the waterways and was carried downstream. In a 40-yr study in the central United States up to an 8.7 bushel/acre (546 kg/ha) reduction in corn yield was calculated to follow the loss of 1 in. (2.5 cm) of topsoil (Williams et al., 1981). If erosion is allowed to continue at the 1977 rate on the 43% of the most erodible land in the Corn Belt, it is estimated that corn and soybean yields will be reduced by from 15 to 30% by the year 2030 (USDA, 1980).

SOIL CONSERVATION

Building terraces and farming contoured strips are effective and add artistic designs to the landscape, but they are difficult and costly to establish and are cumbersome to manage on some fields. Furthermore, they must be maintained from season to season, and plows and other tillage implements remain integral parts of such systems. Exposed, clean cultivated surfaces planted to row crops are unprotected much of the year. The most vulnerable areas on farms, terrace outlets, steeper slopes and large windswept fields can be protected with perennial grass sods or windbreaks; however, as the urgency for greater production becomes more intense, some of these erosion-prone areas too will be needed for harvested crops.

Soil and water conservation requires a workable, continuing program, not a single effort, as an integral part of crop management. Furthermore, merely saving the soil from further erosion is not sufficient to sustain a growing population. The soil also deteriorates in other ways. Excessive aeration leads to loss of organic matter, and compaction affects water runoff, soil drainage, rooting habits, and plant growth. Damaged and depleted soils need to be rebuilt, improved, and used more efficiently if they are to support a permanent agriculture. The new tillage systems discussed in this report are designed for specific sites and situations and promise to contribute much toward rebuilding and retaining soil productivity.

SUBSTITUTES FOR TILLAGE

Serious challenges to excessive use of the plow began slowly. After all, for thousands of years the plow has been the very symbol of agriculture. Cook (1922) in West Virginia successfully reseeded hilly pastures with surface disk cultivation, but the rough surface was inadequate for good seed placement and more seed was needed. Graber in 1927 introduced what he called "pasture renovation" to introduce clover to overgrazed, grub-injured, hilly pastures in Wisconsin. "Whatever the method of seedbed preparation may be," he wrote, "the objective is to cut and fertilize the surface sufficiently to provide a favorable medium for germination, emergence and subsequent growth of seedlings while deterring erosion . . .".

Reasons for tillage slowly unfolded as investigators observed closely the growth and development of young seedlings and the environments that affected them. Dramatic successes were obtained in early renovations where thin, overgrazed, and grub-injured bluegrass (*Poa pratensis* L.) sods were easily cut with small, light farm disks. Dead vegetation remaining on the surface presented no competition to young seedlings, while less water runoff reduced erosion to a small fraction of that from plowed surfaces. More fertile pastures with thicker grass sods also demanded attention. Several times more disking was required to subdue these thicker sods sufficiently to permit new seedlings to survive. Heavy grazing or burning prior to disking also helped. However, from four to six passes over the field, on each of several occasions during hot summer weather, were generally required to subdue the more fertile, vigorous perennial sods adequately to permit new seedlings to become established. Larger and heavier disks, deeper reaching chisel cultivators, and heavily weighted cutaway disks required more powerful tractors, more fuel, and more time than were economic (Sprague et al., 1947). On land that was plowable the plow was more effective and economical than other implements. Most important, inadequate kill of well-established, unwanted vegetation on the now fertilized soils permitted rapid regrowth of the weedy species. The return of unwanted grasses and broadleaf weeds to their former dominance resulted in poor growth and survival of the newly planted forage seedlings. The now costly renovation needed to be repeated too soon to be practical (Ahlgren et al. 1944).

CHEMICAL GROWTH CONTROL

The precursor to substitutes for tillage in crop management had its beginnings with the emergence of a new science during the 1920s. Plant physiologists G. Haberlandt (1922) and H. Soding (1926) in Germany and F. W. Went (1926) in Holland identified the existence of naturally occurring compounds in plants that functioned as internal regulators of plant growth. This constituted growth control with organic chemicals. Inorganic acids and some of their metallic salts for many years had been used as toxic soil sterilant weed killers. Now physiologists shifted their attention to systemic growth regulators, as phytoactive organic chemicals, very small quantities of which, when translocated to or absorbed by certain leaves, stems, or roots, produced marked changes in growth. Small dosages were lethal when applied to certain species yet had little or no effect on others. This provided a unique opportunity for regulatable selectivity.

The release in 1944 of 2,4-D,* as a translocatable selective herbicide with certain broadleaf species, introduced a new era of weed control to agriculture (Hamner and Tukey, 1944; Crafts and Harvey, 1949). Applied to a more

*Common and chemical names of herbicides are listed in Appendix A.

tolerant grass crop, such as corn, at extremely low rates, without cultivation, it virtually eliminated competition from certain troublesome broadleaf weeds. Soon other chemicals were discovered that were selectively toxic to a wide variety of species, both crops and weeds. First they were used to lessen competition following emergence of cereal crops, thus eliminating the need for cultivation. Similarly they replaced mechanical tillage for summer fallow in orchards. Next, chemicals were applied to the soil surface soon after planting (preemergence), which killed weeds as crops emerged (Anderson and Wolf, 1947). During 1948–1949, herbicides were tested that killed all herbaceous plants and left dead plant residue well anchored to the soil. Forage crop seeds placed to provide seed–soil contact germinated and grew unimpeded by competition from other plants.

SEEDBED PREPARATION

At this point an alternative to the use of the plow and disk for seedbed preparation was demonstrated, and the principles of no-tillage systems became apparent (Sprague, 1952). An experiment was later reported of its use with corn (Barrons and Fitzgerald, 1952). Chemical energy was substituted for tractor power to accomplish the first priority purpose of plowing, to eliminate weed competition. The exact nature of competition was not yet clear, but release from it produced dramatic responses in crop growth (Sprague et al., 1962a). Plants now could be provided with a more nearly optimum environment for growth and development from the moment of germination. Changes in planting equipment were needed to place the seed properly in the mulched seedbed with as little soil disturbance as possible. Better herbicides, lethal to a broad spectrum of weeds, were needed that would disappear from the soil quickly so new crops could be planted without delay. Soil aggregate structure was retained and improved, water runoff on steep slopes was reduced as absorption was enhanced, and the surface of the soil was protected from wind and rain (Sprague et al., 1962b).

Early research was additive in its revelation of the value of less tillage in crop management. Water, microclimate, and fertilizer managements became as important to establishment and survival as the initial objectives of erosion control and weed competition. Experiments using a minimum of surface tillage led to no tillage in seedbed preparation for many crops, made possible by new herbicides and new planters. Thus seeds were placed and plant competition was eliminated at the critical early stages of growth. No-tillage seedbed preparation was motivated by a belief that there must be a better way to establish a crop than by plowing. Much research was needed to establish workable systems with all crops during the decades that followed.

Important questions needed answers. What had been the fundamental purpose of plowing through the ages? Were there valid reasons for repeated mechanical disturbance of the surface soil layer that aided crop growth? Are deep plowing and clean cultivation per se beneficial in other ways than weed control? Does crop refuse need to be buried and what will be the effect on the soil, the crop, and pests if it is killed but retained on the surface? Does seedling development benefit from the soil being loosened and aerated mechanically? Is it important that fertilizer materials be mixed throughout the soil profile mechanically or can surface applications be made effective? What management practices are most conducive to recharging a root zone with soil organic matter, minerals, and moisture? Are there better methods for controlling insects and diseases than interrupting their growth cycles with tillage? How effective has plowing been as a means of eliminating competition from troublesome weeds? Would alternative methods of preparing a seedbed be equally effective or better than tillage to establish an optimum environment for growth and survival of a young seedling? In essence, why do we plow and what are the alternatives? These were the questions raised by researchers seeking better systems of cultural management.

It is no surprise that forage agronomists were among the first to search for an alternative to plowing and repeated surface tillage. The conventional system of seedbed preparation consisted of plowing, fertilizing, disking several times, firming, and smoothing prior to planting. The lands assigned to perennial pastures on farms often have productive soil, yet are the least suited for cultivation, most of them being steeply sloping, inaccessible, hard to work with tillage equipment, and the most erosion-prone. Many are shallow, stony, or excessively wet and very difficult or impossible to plow. For all these reasons it was wise to maintain the most difficult soils in perennial sod, thereby reducing the need for seedbed preparation. Surface-tillage techniques demonstrated fairly good control of soil erosion when plant materials covered the soil surface. Reducing competition permitted the seedlings to grow. Numerous studies later revealed additional benefits to be gained from a dead mulch anchored to the surface. These will be discussed in more detail in later chapters.

WHY PLOW?

Why did man develop and continue to use the plow? First and foremost, plowing and subsequent cultivation were the only means available to establish and maintain a monoculture, for example, to keep out unwanted vegetation. Nature's first principle of ecology is that two plants growing in the same place at the same time compete for space and raw materials. Except for some tropical situations, both will grow better when uninhibited by a companion.

Second, the moldboard plow was designed primarily to bury plant materials that had collected during and since the previous planting. Effectively done it creates a neat and orderly appearance in the field, which is artistically pleasing. Denying weedy plants sunlight also sometimes kills them. Furthermore, it allows shallow cultivation and planting equipment to perform more uniformly by removing obstacles.

Third, deep plowing was intended to mix lower ho
with surface layers. This permitted deep placement o
als and aided in distributing them throughout the pr
mixing improved rooting and availability of miner
Fourth, and possibly of less importance, plowin
the surface to encourage water absorption, aid pre
encourage emergence.

Finally, plowing was intended to implement g
weed, insect, and disease cycles. Also, turning a furrow fo
crisp spring morning awakened the spirit following a long dull winter. Em
cally, however, tilling the soil long has been tedious hard work.

Whatever the reasons have been, tillage has served these functions with varying degrees of success for many centuries. Also, under some conditions of soil and climate, whether or not tillage has been carelessly undertaken, soil deterioration has been excessive, most weedy species have been inadequately controlled, moisture management has been less than desired and time and energy requirements have been inefficiently high. An alternate means of providing an optimum environment for crop growth long has been needed to improve production, conserve and recharge our soil and water resources and add greater permanency to agricultures throughout the world. The new tillage systems discussed in this volume offer opportunities to attain these objectives.

A NEW AGRONOMY

If such a major change as replacement of the moldboard plow in the culture of our major crops is to take place, most of the prior research findings impinging on the many aspects of cultural management need reevaluation. Microclimates are greatly altered, which affects temperatures, evaporation rates, and soil moisture. Fertilization techniques and mineral nutrient availability all need to be reassessed. Soil aeration, rooting habits, and organic matter accumulation are affected. New soil structures and profile characteristics will develop. Patterns of seed germination, seedling development, and diseases will be affected. Weed seed, insect, and predator life cycles may be changed. Planting depths, dates, and rates need to be restudied. New opportunities unfold for improved crop rotations as seedbed preparation times shorten and new cultivars appear. Planting equipment needs to be redesigned to perform on different surfaces and different soil sites at higher speeds. New materials, labor, and energy budgets need to be explored and evaluated.

These and other reassessments have been under way for over three decades, sufficient for in-depth evaluation. Scientists from many disciplines have contributed to major changes in cultural management. Different crops, soils and climates throughout the world are represented. Seven major annual crops were described in a report on minimum tillage by the U.S. Department of Agriculture (1975) as well adapted to these new tillage systems. These seven—corn,

, soybeans, wheat, rye, oats, and barley—represent most of the an- arvested acreage except hay and about two-thirds of the total crop uction in the United States. The report estimated that about 2% of the opland (2.23 million ha of annual and winter-annual crops in the United States) was under surface- or no-tillage cultivation in 1974 and anticipated that 45% (62 milion ha) would be involved by the year 2000. In addition, much of the remainder of the cropland, used for perennial forages, likely will be represented. Some commercial vegetables, such as tomatoes, also have been planted with herbicides in place of tillage for seedbed preparation following a cover crop. A separate vegetable culture, employing opaque plastic film in place of herbicides, has been used. Variations are being developed for the home gardener and small farmer in both the tropics and middle latitudes. Forest plantings and turf and fiber crops have used herbicides as a substitute for tillage during establishment. Herbicides have also been applied beneath trees as a maintenance practice in orchards for improved moisture and pest control.

TILLAGE SYSTEMS TERMINOLOGY

There is at present no adequate national census of the crop acreage represented by surface-tillage and no-tillage management systems. Estimates by state soil conservationists (Lessiter, 1982) from 1974 to 1981 and CTIC (see Chapter 14) are useful, but are considered conservative in some respects and incomplete in others. It is difficult to obtain reliable estimates, since terms often are inconsistent or imprecisely represent the crop or farm situations that exist. Prevailing systems for seedbed preparation alone differ greatly among climatic zones, geographic areas, soils, and species present. For these and other reasons much is wanting in the clarity and consistency of the prevailing terminology related to tillage. The multiplicity of terms in use represents a variety of solutions to special problems and situations characteristic of diverse agricultures. Many terms are colloquialisms that have come into use as abbreviated descriptions or even slogans intended to identify local cultural practices.

To promote continuity in the following discussions three terms are defined here to identify the kind, amount, and sequence of soil disturbance during seedbed preparation. These are presented as benchmark descriptions with the understanding that the individual chapters will expand their descriptions of soil manipulation and cropping systems as necessary.

Plow-Tillage. Initial use of a moldboard or disk plow to invert the upper soil profile sufficiently to bury or incorporate materials on the surface. This is normally followed by cutting and smoothing operations prior to planting. In short, it involves plowing prior to planting and a soil surface free of trash as the crop grows.

Surface-Tillage. Breaking, tearing, cutting, or otherwise loosening of the surface layers of soil with variously shaped disks or sweep or chisel cultivators prior to seed placement. Normally some plant residues are buried to shallow depths, while the remainder shield the soil surface as the crop becomes established.

No-Tillage. The use of herbicides or other methods to kill all live plants on or near the surface of the soil, followed by as little disturbance of the soil as possible to provide good seed placement. A mulch is well anchored to the surface. The soil is undisturbed prior to seed placement.

In the plow-tillage system all or most of the live plant parts are buried, providing a uniform trash-free soil surface. In surface- and no-tillage systems, various amounts of plant materials will be retained on or within the surface soil. The latter two systems can be collectively referred to as *conservation tillage* or *reduced tillage*. Plant materials may consist of sown cover crops, crop residues, or weed species. Any of the three tillage systems may use herbicides later at any stage of growth throughout the life of the crop.

SEEDLING ESTABLISHMENT

Tillage systems are generally considered to begin at the time of seedbed preparation, whether or not additional treatments may be required. Similarly, stand establishment is regarded as the single most important stage of growth in the life cycle of a crop. Considering seedling physiology and the risks of seedling failure, stand establishment is also considered to be the most vulnerable stage of development. Where establishment costs are high (for seed, fertilizer, pest control, labor, and land use), good growth and survival are essential. The primary function of seedbed preparation is to create an environment for rapid crop growth that approaches optimum.

The vigor and survival of the young seedling influence the development of the crop throughout its entire life. A germinating seed contains a limited pool of reserve foods for support during the period of establishment. Survival and subsequent development depend solely on early initiation of the seedling's capacity to provide its own source of photosynthates before the reserve sink is exhausted. Whether the crop is annual or perennial in growth habit, rapid development of a vigorous seedling and optimum stand density are the first prerequisites to profitable yield. For this reason tillage systems are considered first for their effectiveness in seedling establishment.

OBJECTIVES

Many research reports dealing with tillage systems are distributed through a wide range of journals, bulletins, and proceedings. The principles involved in the many disciplines represented need to be coordinated and brought into perspective. This treatise is intended to evaluate the physical, biological, engineering, and economic research that has been conducted and the principles involved in the systems now being tested where no-plow cultivation has replaced tillage-intensive culture. It is further intended to describe the major findings and their usefulness, the limitations of different systems in various cropping sequences and the effects that the new management systems have on soil and water resource renewal. It will consider the effects of no-tillage in different climates and on soils found in various geographical regions. Herbicides and their functional use in different systems will be brought into focus. Weed-control strategies will be discussed and special problems identified.

Surface mulches and related predator, insect, and disease problems will be described, along with strategies to keep them under control in the new environments. Physical and chemical changes above and in the soil as a result of adjusted tillage patterns will be discussed for their influence on erosion, water absorption, fertilizer movement, and mineral availability. Areas of soil improvement will be considered not only for their influence on current production, but also for their impact on agricultures of the future. Equipment and engineering requirements to produce the specific environments that will favor the crop will be described. Seedbed preparation to enhance seedling survival by adjusting microclimates will receive special attention. Time schedules and crop rotation sequences provide new opportunities.

New labor and energy budgets are discussed as measures of efficiency and cost effectiveness. Tillage adjustments alter such inputs as labor, fuel, pesticides, fertilizers, and equipment and such outputs as yield and quality. Management depends on operator skills adjusting to soils, climates, costs, and markets. All these combined determine short-term profitability and long-term productivity.

Cultural management, oriented around plow and cultivator operations through long usage, settled into reasonably stable systems of crop production. These discussions are intended to represent the base on which a new agriculture is unfolding through the substitution of herbicides for tillage in seedbed preparation resulting in more favorable total environments for crop growth with less tillage.

LITERATURE CITED

Ahlgren, H. L., M. L. Wall, R. J. Muckenhirn, and F. V. Burcalow. 1944. Effectiveness of renovation in increasing yields of permanent pastures in southern Wisconsin. *J. Am. Soc. Agron.* **36**:121–131.

Anderson, J. C. and D. E. Wolf. 1947. Pre-emergence control of weeds in corn with 2,4-D. *J. Am. Soc. Agron.* **39**:341–342.

Barney, G. O. 1980. *Global 2000 Report to the President.* Vol. 1 Summary. Superintendent of Documents, Washington, DC.

Barrons, K. C. and C. D. Fitzgerald, 1952. An experiment with chemical seedbed preparation. *Down Earth* **8**:2–3.

Bennett, H. H. 1939. *Soil Conservation.* McGraw-Hill, New York.

Brink, W. 1951. *Big Hugh*. Macmillan, New York.

Carter, V. G. and T. Dale. 1974. *Topsoil and Civilization*. Revised edition. University of Oklahoma Press, Norman.

Cook, I. S. 1922. *West Virginia Pastures*. W. Va. Agric. Exp. Stn. Bull. 177.

Cotton, J. S. 1910. Improvement of pastures in eastern New York and New England states. USDA Circ. 49.

Crafts, A. S. and W. A. Harvey. 1949. Weed Control. *In Advances in Agronomy*. Vol. 1. Academic Press, New York, pp. 289–320.

Dudal, R. 1976. Inventory of the major soils of the world with special reference to mineral stress hazards. In Wright, M. J. (ed.). *Plant Adaptation to Mineral Stress in Problem Soils*. Cornell University Agricultural Experimental Station, Ithaca, NY.

FAO United Nations Production Yearbook. 1978. FAO Statistical Series No. 22. United Nations, Rome.

Fussell, G. E. 1965. *Farming Technique from Prehistoric to Modern Times*. Pergamon, Oxford.

Graber, L. F. 1928. Evidence and observations on establishing sweet clovers in permanent bluegrass pastures. *J. Am. Soc. Agron.* **20**:1197–1205.

Haberlandt, G. 1922. Cell division hormones and their relation to wound repair, fertilization, parthenogenesis and adventive embryogeny. *Biol. Zeut.* **42**:145–172 (Exp. Stn. Record **49**:219).

Hamner, C. L. and H. B. Tukey. 1944. The herbicidal action of 2,4 dichlorophenoxyacetic acid and 2,4,5 trichlorophenoxyacetic acid on bindweed. *Science* **100**:154–155.

Holeman, J. BN. 1980. Erosion rates in the United States. Press Release December 10, 1980. USDA, SCS, Washington, D. C.

Jardin, W. M. 1913. Management of soils to prevent blowing. *J. Am.Soc. Agron.* **5**:213–218.

Jope, E. M. Agricultural implements. 1956. Chapter 3 in Singer, C. E. J. Holmyard, A. R. Hall, and T. Williams (eds.). *A History of Technology, Vol. II, The Mediterranean Civilizations and Middle Ages, 700 BC to AD 1500*. Oxford Press (Clarendon), England, pp. 81–93.

Klemm, Friedrick. 1964. A history of western technology. Translated by Dorothea Waley Singer. MIT Press. Cambridge, MA.

Lessiter, F. (ed.). 1982. An eleven year tillage comparison 1972–82. *No-Till Farmer* **10**(3):5.

Moorhouse, L. A. 1909. Some soil problems in Oklahoma. *Proc. Am. Soc. Agron.* **I**:234–238.

Silcox, F. A. 1936. *The Western Range*. Senate Document No. 199. 74th Congress, 2nd Session. Superintendent of Documents, Washington, DC.

Soding, H. 1926. Uberdeneinfluss der Junger Infloreszenz and das Wachstum ihres Schaftes. *Jahrb. Wiss. Bot.* **65**:611–635.

Sprague, M. A. 1952. The substitution of chemicals for tillage in pasture renovation. *Agron. J.* **44**:405–409.

Sprague, M. A., R. D. Ilnicki, R. W. Chase, and A. H. Kates. 1962a. Growth of forage seedlings in competition with partially killed grass sods. *Crop Sci.* **2**:52–55.

Sprague, M. A., R. J. Aldrich, R. D. Ilnicki, A. H. Kates, T. O. Evrard, and R. W. Chase. 1962b. Pasture improvement and seedbed preparation with herbicides. N. J. Agric. Exp. Stn. Bull. 803.

Sprague, V. G., R. R. Robinson, and A. W. Clyde. 1947. Pasture renovation: I. Seedbed preparation, seedling establishment, and subsequent yields. *J. Am. Soc. Agron.* **39**:12–25.

Tull, Jethro. 1733. *Horse Hoeing Husbandry*. London.

U. S. Department of Agriculture, Office of Planning and Evaluation. 1975. Minimum tillage: A preliminary technology assessment. Part II of report for Committee on Agriculture and Forestry, United States Senate. Publ. No. 57-398, Superintendent of Documents, Washington, DC.

U.S. Department of Agriculture, Soil Conservation Service. 1980. America's soil and water: Conditions and trends. Superintendent of Documents, Washington, DC.

Thimann, K. V. 1933. Studies of the growth hormone of plants. *Proc. Natl. Acad. Sci.* **18**:692–701.

Voigt, W. Jr. 1976. *Public Grazing Lands*. Rutgers University Press, New Brunswick, NJ.

Went, F. W. 1926. On growth accelerating substances in the coleoptile of *Avena sativa* L. *Proc. K. Akad. Wetensch. Amsterdam Verslag Gwone Vergader. Afd. Natuurk* **35**:723–732 (Biol. Abst. **7**:1293).

Williams, J. R., et al. 1981. Soil erosion effects on productivity: A research perspective. *J. Soil Water Conserv.* **36**:82–90.

2

SOIL AND MOISTURE MANAGEMENT WITH REDUCED TILLAGE

D. R. GRIFFITH
Research and Extension Agronomist
Department of Agronomy
Purdue University
West Lafayette, Indiana

J. V. MANNERING
Professor of Agronomy
Purdue University
West Lafayette, Indiana

J. E. BOX
Director
Southern Piedmont Conservation Research Center
USDA-ARS
Watkinsville, Georgia

INTRODUCTION

The importance of the soil as a medium for root growth and a source for nutrients and water has long been recognized by researchers and modern

---

The views expressed here are those of the authors and may not reflect the policy of the U.S. Department of Agriculture or the Purdue University Agricultural Experiment Station.

farmers. Well-established physical, chemical, and biological factors contribute to the development and productivity of soils. Changing from traditional tillage methods can alter these factors and, as a result, alter productivity. Information in this chapter is focused on soil physical properties and their relation to tillage.

Both the amount and depth of preplant tillage affect many soil physical properties. However, crop response to these changes in soil properties depends on length of growing season (north to south), amount of rainfall (east to west), and native soil properties. Thus, effects of changes in soil properties due to tillage must be interpreted differently for regions, localities within a state, and often for soils on a farm, when they differ appreciably in drainage, texture, depth, and topographic characteristics.

Soil properties that may be altered with changes in tillage include organic matter, erodibility, moisture, temperature, density, and aggregation. Tillage systems vary in the amount of soil pulverization they induce, but placement of the previous crop residue (surface, incorporated, or partially incorporated) often has a greater influence on these soil properties than degree of pulverization. Tillage systems available to farmers today represent a range from subsoiling to a 35–40-cm depth, to complete loosening of a 25-cm surface layer of soil, to simply opening a 1-cm slot for each row. Residue incorporation may range from complete to none.

TILLAGE SYSTEMS

Most studies of soil physical properties refer to the soil condition after a series of operations with tillage equipment—the *system* used for land preparation and planting. While specific examples of tillage equipment are discussed in Chapter 3, a classification of some commonly used tillage systems as to degree of mechanical manipulation of soil (Table 2.1) will help to explain their effect on soil physical properties.

The systems listed are commonly used, but farmers often combine operations from two or more of these systems to meet specific needs. For instance, rotary tillers or harrows can be used in addition to disks and field cultivators to prepare a seedbed after the primary tillage operation. Ridges or "beds" are formed on plowed ground for cotton seeding in the south. "No-till drills" provide narrow strip tillage, but the strips are so close together that much more tillage results than with "no-till" planting in 75–100-cm rows. These altered or combination tillage systems are not represented in Table 2.1.

RESIDUE PLACEMENT AND SOIL ORGANIC MATTER

Residue placement preceding and during the growing season, especially the amount remaining on the soil surface, affects accumulation of soil organic matter, soil erodibility, soil temperature, and soil moisture. There may also be

TABLE 2.1. Tillage Systems and Soil Manipulation

System	Typical Depth of Tillage (cm)	Amount of Pulverization (rank)[a]	Amount of Inversion (rank)[a]
Primary full width tillage (15-cm depth or more)			
Moldboard plow, disk and/or field-cultivate twice	15–25	1	1
Chisel plow (10-cm twisted points), disk twice	20–25	2	2
Chisel plow (5-cm straight points), disk twice	20–25	4	4
Primary tillage disk, shallow disk twice	15–20	3	3
Shallow full width tillage (< 15-cm depth)			
Disk and/or field cultivate twice	10–15	6	5
Rotary tillage once or twice	5–10	5	6
Stubble-mulch for wheat (wide V-sweep plus rod weeder)	5–10	9	9
Wide strip tillage (10–38 cm)			
Till-plant in ridge	2–8	8	7
Strip rotary tillage	5–10	7	8
Narrow strip tillage (< 10 cm)			
Subsoil, plant	30–35	10	10
No-tillage, plant	5–10	11	11

[a]1 = greatest.

an affect on crop growth, maturity, and yield. Soil organic matter resulting from residue decomposition affects soil aggregation and stability. In order to evaluate the effect of a particular tillage system on these soil properties, an accurate estimate of surface residue is necessary. Weight of residue per unit area has traditionally been used to evaluate soil cover. However, recent studies show that percentage of the soil surface covered is better correlated with the effect on soil properties, amount of soil water, and erosion control.

Soil Surface Cover

In Indiana studies (Mannering et al., 1975), percentage of cover was measured immediately after planting for six tillage systems. The data (Table 2.2) are from continuous corn (*Zea mays* L.) plots and are an average for four experiments. Note that there is not only a great range in surface cover among tillage systems, but among experiments in the same tillage system.

Residue cover with all of the no-plow systems may vary with amount and uniformity of distribution from the previous crop. In addition, surface residue

TABLE 2.2. Residue Cover Immediately after Planting, Corn After Corn, Indiana[a]

Tillage System	Residue Cover (%)[b]
Moldboard plow, disk twice	1.0 (0.6–1.8)
Till-plant	8.4 (3.5–13.5)
Disk twice (standard tandem)	12.9 (10.0–17.5)
Chisel (2-in. straight points), disk twice	19.0 (6.0–29.0)
Strip rotary tillage (8-in. strip)	62.0 (38.0–89.0)
No-tillage	76.0 (59.0–92.0)

[a] Average of four experiments. Measured by projecting pictures on a grid and counting intersections with cover.
[b] Ranges are given in parentheses.
SOURCE: Mannering et al. (1975).

with till-planting varies with the depth at which sweeps or disks operate in the ridge. Surface residue with disking or chiseling varies with size and shape of disk blades or chisel points, and with amount of secondary tillage for seedbed preparation. Tandem disks 40 cm in diameter and straight chisel points 5 cm wide were used in this study. Till-plant residue data reflect ridge scalping of 5–7 cm, resulting in less surface cover than for more shallow till-planting.

Prior crop species may have a great effect on surface residue with the no-plow tillage systems, with soybeans usually leaving less residue than corn. This was documented on four soils having relatively low yield potential in northeastern Indiana (Mannering et al., 1976). Percentage of cover following soybeans compared to cover following corn was reduced from 4% to 2% for fall plowing, from 37% to 8% for fall chiseling, and from 64% to 21% for no-tillage. Measurements were made in the spring before seedbed preparation and planting.

An Iowa study (Erbach, 1982) on a highly productive soil shows that *undisturbed* soybean (*Glycine max* L.) residue can nearly equal corn residue in surface cover (Table 2.3). Disturbance with any kind of tillage, however, greatly reduced surface cover from soybeans compared to that from corn. Note that till-plant left significantly more residue here than in the Indiana study (40% cover after corn) and that only slot plant (similar to no-tillage) left greater than 20% cover after planting in soybean residue.

Surface residue cover after small grain, soybeans, and corn was compared for different tillage systems in southern Minnesota (Sloneker and Moldenhauer, 1977). Cover following oats (*Avena sativa* L.) taken for grain was similar to that following corn, but, again, only no-tillage provided greater than 20% cover after soybeans.

Crops such as cotton (*Gossipium* spp.) and dry beans leave surface residue after harvest similar to that from soybeans. Of course, any annual crops taken for silage, such as corn or peas (*Pisum* spp.), leave little surface cover (less than 20%) unless regrowth is possible. Planting row crops without tillage into killed

TABLE 2.3. Residue Cover after Soybeans and Corn, Central Iowa[a]

	Residue Cover (%)			
	After Soybeans		After Corn	
Tillage system	Before Spring Tillage	After Planting	Before Spring Tillage	After Planting
Fall plow, disk once, harrow once	4	3	11	6
Fall chisel,[b] field cultivate once	23	12	45	30
Spring offset disk once, standard disk once	71	12	78	31
Till-plant	76	19	75	40
Slot-planted on ridges	73	37	78	53

[a] Average of 4 years. Residue estimated by meter stick method.
[b] Chisel had 7.5-cm-wide twisted points.
SOURCE: Erbach (1982).

sod, whether established perennials or annual cover crops, usually leaves 90–100% surface cover shortly after planting.

Using the previous data and other studies as guides, percentage of surface cover after planting for several tillage system versus previous crop combinations are given in Table 2.4 for "typical" situations. No-plow systems in specific situations could easily vary ±30%, however, based on management of the crop and tillage system.

Soil Organic Matter

The amount and distribution of soil organic matter resulting from decay of residues will be altered after several years in a reduced-tillage system. Two Indiana studies (Cruz, 1982, Fernandez, 1976), on soils differing widely in natural organic matter content, show that organic matter increased significantly near the soil surface (0–10 cm) with reduced tillage, and was maintained or increased slightly below the 8–10-cm level with no-tillage (Table 2.5). While residues were not placed below the surface by tillage, apparently organic matter was maintained due to additions from the decaying plant roots and lower soil temperature in the no-tillage system, which reduced organic matter loss from oxidation. Even though residues are incorporated to the depth of tillage with moldboard plowing, higher soil temperatures lead to increased oxidation and a lower organic matter level.

In Ohio, organic matter was measured in the top 1.25 cm of soil after 18 years of continuous corn without tillage and with plowing (Dick, 1983). Increases for no-tillage over plowing were 2.5 times on a dark silty clay loam and 2.2 times on a light-colored silt loam. Below 7.5 cm, organic matter was about

TABLE 2.4. Residue Cover After Planting for Typical Tillage Versus Previous Crop Combinations

	Residue Cover (%) for Previous Crop of:			
Tillage System	Corn	Soybeans	Small Grain	Sod
Moldboard plow, disk, field cultivate	5	2	5	10
Chisel (10-cm twisted points)				
+ disk twice	15	2	10	20
+ disk once, field cultivate once	20	5	15	25
Chisel (5-cm straight points)				
+ disk twice	20	5	15	25
+disk once, field cultivate once	30	10	25	30
Primary tillage disk (deeper than 15 cm)				
+ disk twice (standard tandem)	10	5	15	20
+ disk once, field cultivate once	20	10	20	—
Shallow disking (less than 15 cm)				
once (standard tandem)	40	20	45	50
twice (standard tandem)	20	10	25	30
Field cultivate once	—	30	—	—
Till-plant in ridge	30	20	—	—
In-row subsoil, plant	70	50	80	85
No-tillage, plant	80	60	90	95

TABLE 2.5. Amount and Distribution of Soil Organic Matter with Plow, Chisel and No-tillage Systems in Indiana[a]

	Tracy Sandy Loam		Chalmers Silty Clay Loam	
Tillage systems	Depth (cm)	OM (%)	Depth (cm)	OM (%)
Moldboard plow	0–10	1.51	0–7.5	4.1
	10–20	1.53	7.5–15	4.1
	20–30	0.82	15–22.5	3.7
Chisel (2-in. twisted points)	0–10	2.48	0–7.5	4.6
	10–20	1.68	7.5–15	4.1
	20–30	0.82	15–22.5	3.6
No-tillage	0–10	1.94	0–7.5	4.8
	10–20	1.67	7.5–15	4.2
	20–30	0.94	15–22.5	3.8

[a] After growing continuous corn for 7 years.

SOURCE: Cruz (1982) and Fernandez (1976).

equal for the two systems on the light soil, but was reduced for no-tillage on the dark soil, possibly due to restricted rooting.

In the Southeastern United States, maintaining or increasing soil organic matter is especially important, since many soils are naturally low in organic matter and high temperature leads to its rapid breakdown.

Research in Kentucky (Blevins et al., 1977) shows that organic matter level in the top 5 cm of soil under an original bluegrass (*Poa Pratensis* L.), sod was reduced only slightly after 5 years of continuous corn when no-tillage planting, cover crops, and high fertility levels were used (Table 2.6). Moldboard plowing reduced organic matter by 46% in the 0–5-cm layer.

A 10-year continuous corn study in the Southern Piedmont region of South Carolina (Beale et al., 1955) compared the effect of plowing and disking with and without cover crops on soil organic matter. Both a vetch-rye (*Vicia-Secale* spp.), mixture and crimson clover, (*Trifolium incarnatum* L.), used as cover crops increased organic matter under both tillage systems after 10 years, but greatest increases were with shallow or "mulch" tillage.

Ohio researchers (Triplett et al., 1968) found that increased surface residue, and the resulting increase in organic matter, had a positive effect on corn yield on certain soils but not on others under no-tillage planting. Light-colored silt loam soils, which tend to crust easily, benefited most from surface residue, but 50–60% cover after planting was needed for no-tillage yields to be equal to or better than yields with conventional tillage. Corn yields on dark-colored silty clay loams were not improved by adding surface cover. These soils normally do not form strong crusts. They shrink while drying, forming cracks in the soil surface, providing points of entry for water infiltration.

The increase in surface residue, and the resulting long-term increase in organic matter near the soil surface, has mostly positive effects on soil and water management, especially on soils naturally low in organic matter, but may necessitate increased rates of certain preemergence herbicides. Periodic plowing, whether planned or necessary due to unexpected pest problems, will reverse both the positive and negative effects of organic matter buildup near the surface.

TABLE 2.6. Effect of Tillage and Nitrogen Rate on Soil Organic Matter for Continuous Corn in Kentucky

Soil Depth (cm)	Original Bluegrass Sod	Organic Matter (%)							
		Nitrogen rate (kg/ha)							
		0		84		168		336	
		NT[a]	CT[a]	NT	CT	NT	CT	NT	CT
0–5	5.18	3.68	2.37	3.96	2.60	4.11	2.78	4.53	2.79
5–15	2.47	2.22	2.23	2.28	2.53	2.15	2.60	2.46	2.52
15–30	1.36	1.36	1.17	1.37	1.57	1.24	1.47	1.45	1.37

[a]NT = no-tillage, CT = conventional tillage (moldboard plow).
SOURCE: Blevins (1977).

SEEDBED PREPARATION AND SURFACE CLODDINESS

Preparing a "proper" seedbed has often been a perplexing problem for farmers since the moldboard plow and other full-width tillage tools came into common use. The farmers's objective has been to prepare a seedbed fine enough for good soil–seed contact, yet coarse enough not to allow crusting after heavy rains.

The amount of tillage to accomplish this depends on soil texture, soil moisture, organic matter, and intensity of rainfall after planting; thus "proper" seedbed preparation may vary between soil types and between years for the same soil. Amount of tillage necessary has been a subjective decision for most farmers—and the decision has often been wrong.

There have been few research attempts to define the best seedbed in terms of seed germination and emergence, and soil crusting. An Indiana study (Mannering et al., 1975) concluded that a seedbed containing 20–30% aggregates smaller than 2 mm in diameter is well suited to corn germination. Of the four tillage systems studied (plow, chisel, till-plant, and no-tillage), till-plant most consistently provided this aggregate range and the best germination on five soil types ranging from sandy loam to silty clay loam. When aggregates of this size fell below 15%, germination was lowered. Greater than 30% small aggregates (< 2 mm) caused more soil crusting and reduced seedling emergence.

After plowing and chiseling, farmers have traditionally made three to five passes with disks, field cultivators, harrows, rollers, etc., to prepare a seedbed for corn or soybeans. Up to eight passes were used to prepare a seeded for cotton and small seeded vegetable crops. Beginning in the mid-1960s farmers began reducing seedbed preparation operations to gain more timely planting and cut costs. In the 1970s, new combination tillage tools provided more aggressive tillage in one pass, and planter modifications allowed good germination in less uniform seedbeds. As a result, one to three passes to prepare a corn or soybean seedbed is now common.

Although the traditional full-width fine seedbed provides some things farmers want (kills growing weeds, allows herbicide incorporation, allows uniform seed placement), the list of negatives is much longer. They include: subject to severe erosion; compaction often results when soil is worked too wet; soil surrounding seeds may dry excessively; a crust may form over seeds if intense rains occur before plant emergence; and costs in terms of time, labor, and fuel may be excessive.

The newer one-spring-pass strip tillage systems such as till-plant and no-tillage have advantages over traditional seedbed preparation in addition to reduced time, labor, and fuel use. Seeds are almost always placed in moist soil, since there is no preworking to dry the soil. Therefore, seeding depth can and often should be more shallow than in tilled seedbeds. For corn, however, seed should be covered by at least 1 cm of soil to ensure maintenance of moisture at the plant crown where the permanent root system develops. Crusting is seldom a problem with these systems owing to increased crop residues and organic matter near the soil surface. As with conventional tillage, however, best soil-to-seed contact occurs when moisture in the top 5 cm of soil reaches the ideal "working" range.

TILLAGE AND SOIL MOISTURE

Tillage practices may influence soil moisture throughout the growing season. Reduced-tillage systems decrease evaporation losses, if residue remains on the surface, and increase rainfall infiltration owing to surface roughness and/or residue. Water runoff is slowed by both roughness and residue, allowing more time for infiltration, and surface residues tend to prevent soil crusting, thus increasing infiltration. The net effect from surface roughness and/or residue is usually less variation in soil water during summer months and more plant-available water.

Evaporation, Transpiration, and Infiltration

Evaporation is a primary source of water loss during the first half of the growing season before the crop canopy closes. The effect of surface residue in reducing losses from evaporation is well documented. A greenhouse study in Texas (Unger and Parker, 1968) showed that cumulative evaporation after 16 weeks was 57% less when wheat residue was on the surface rather than mixed with soil.

Both evaporation and transpiration were estimated in no-tillage and conventionally tilled corn in Kentucky (Phillips, 1980). Soil water evaporation was less from no-tillage corn in every month (May through September) during the 4-year study. Average annual evaporation was reduced by 15 cm in no-tillage plots (Table 2.7). Therefore, more water was available for transpiration in the

TABLE 2.7. Estimated Soil Water Evaporation and Transpiration[a] by Corn from No-Tillage and Conventionally Plow Tilled Plots on Maury Silt Loam Soil, 1970–1973, Kentucky

		No-Tillage		Conventional	
Month	Rainfall (mm)	Transpiration (mm)	Evaporation (mm)	Transpiration (mm)	Evaporation (mm)
May	179	00	21	00	63
June	97	76	10	64	68
July	101	124	03	95	21
August	41	92	02	72	14
September	91	15	05	11	25
Total	509	307	41	242	191

[a] Based on 900 mm of rooting depth.

SOURCE: Phillips (1980).

no-tillage plots, especially in July and August, often resulting in higher corn yields. In 1970, a very dry year, no-tillage versus plow-tillage yields were 6.75 vs 6.14 mt/ha and in 1971, a year of adequate rainfall, no-tillage versus plow-tillage yields were 11.59 vs 10.84 mt/ha.

Simulated rainfall studies in Ohio (Triplett et al., 1968) (Table 2.8) showed that infiltration was greatly increased with surface residue. Studies were on a Wooster silt loam soil with 1−2% slope, and residue was from the previous year's corn crop. Infiltration was greater for no-tillage with surface residue than for plowed bare soil, and it increased as residue increased. This occurred even though bulk density was higher in the no-tillage plots. The researchers suggest that better structural stability in the no-tillage, heavier residue plots aided infiltration.

The Ohio study showed that freshly tilled soil has a very high water intake rate, but this declined rapidly due to surface sealing from rainfall. The mulch-covered surface maintained a relatively steady intake rate throughout the season. A long-term study at Coshocton, Ohio found much greater infiltration under natural rainfall with surface mulch and no-tillage planting than with conventional plow-tillage planting.

Many researchers have reported more large pores in mulched, untilled soil than in tilled soil. Earthworm channels that remain open to the soil surface (not sealed by implement wheel traffic) no doubt play an important role in increasing infiltration. How much of the increase actually results from large pores is not well documented.

Water runoff from both simulated and natural rainfall has been measured in many tillage-erosion studies. No-tillage with surface residue usually reduces runoff, compared to plowed soil, but the degree of reduction varies greatly due to rainfall intensity and amount, soil type, amount and kind of residue, slope steepness, and row direction. A combination of surface residue, surface roughness, and contour rows is likely to provide the greatest reduction in runoff. However, as little as 0.5 mt/ha of residue per acre (without roughness) has been shown to significantly reduce runoff (Mannering, 1979).

Actual soil moisture differences due to tillage treatment have been measured in many experiments. In the Ohio trial on Wooster silt loam, mean

TABLE 2.8. Effect of Tillage and Residue on Infiltration with Simulated Rainfall, Ohio

Treatment	Total Infiltration After 1 hr (mm)	
	Initial Run	Wet Run[a]
Plowed, bare	18.0	10.4
No-tillage, bare	12.2	6.4
No-tillage, 40% surface residue cover	23.4	13.5
No-tillage, 80% surface residue cover	43.9	34.8

[a]Twenty-four hours after initial run.

SOURCE: Triplett et al. (1968).

available moisture from June 15 to September 15 in the 0–46-cm zone was 4.27 cm for plowed bare soil, 4.37 cm for no-tillage with residues removed, 5.34 cm for no-tillage with 40% cover, and 5.88 cm for no-tillage with 80% cover. More vigorous growth in the no-tillage corn with a residue cover led to greater transpiration, thus limiting even greater differences in soil moisture.

Extensive soil moisture studies in Kentucky (Blevins et al., 1971) have compared no-tillage corn planted into a killed perennial sod with corn in plowed bare soil. Volume percent moisture was greatest throughout the season under the killed sod. Differences were greatest near the soil surface, but were noted to a depth of 45 cm on one soil and to a depth of 90 cm on another soil with higher conductivity. Soil moisture with no-tillage planting into a killed annual cover crop was slightly less than under the perennial sod, but much greater than in a plowed seedbed. This water reserve carried the crop through periods of short-term drought without detrimental moisture stress developing in the corn.

Other studies in the Southeastern United States have shown increased soil moisture due to surface mulch. In Virginia (Moody et al., 1963) wheat straw spread on top of plowed soil was compared with the same amount of straw plowed down. The 3-year study showed seasonal soil moisture to average 2.13 cm higher and corn yield to increase from 4.13 to 7.01 mt/ha with mulch on the surface.

In another 3-year Virginia trial (Jones et al., 1969), conventional and no-tillage were compared with and without surface residue. Mulches used were killed grass sod with no-tillage and straw with conventional plow tillage. Surface mulch influenced corn grain yield more than tillage in each of the three years (Table 2.9). Runoff measured during two years of the study represented 27% of precipitation on unmulched plots and only 4.5% on mulched plots.

A Minnesota study (Gantzer and Blake, 1978) found volumetric water content averaged 10% greater for no-tillage versus fall plowing from May through September in a continuous corn trial.

In northern Indiana (Mannering et al., 1975) soil moisture was measured under no-tillage and moldboard plow tillage during July and August, the peak use period for corn (Fig. 2.1). On a well-drained sandy loam, soil moisture in

TABLE 2.9. Effect of Tillage and Surface Residue on Corn Yield, 3-year Average, Virginia

Tillage—Residue	Grain Yield (mt/ha)
Conventional[c]	6.29[b]
Conventional + mulch	8.00[a]
No-tillage + killed sod	7.81[a]
No-tillage, sod removed	5.66[b]

[a,b]Values followed by the same letter are not significantly different (0.01).

[c]Turn-plow plus two diskings.

SOURCE: Jones et al. (1969).

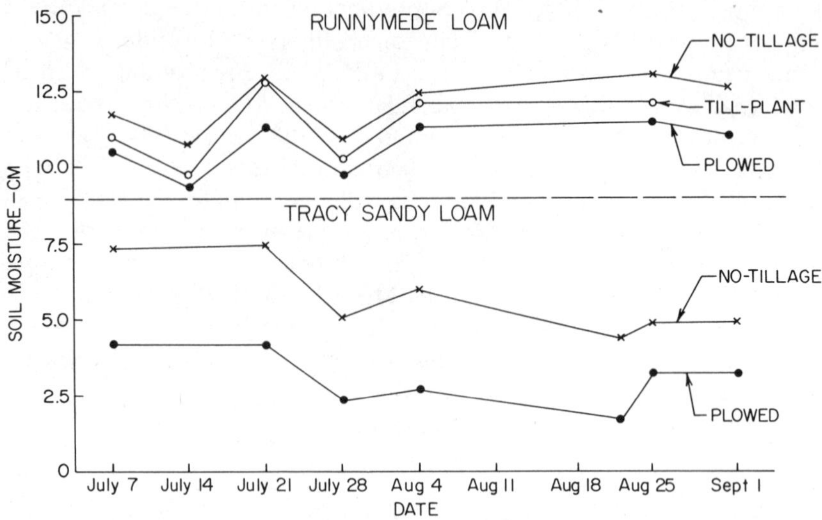

FIGURE 2.1. Effect of tillage on moisture in the upper 45 cm of soil, Indiana (Mannering et al., 1975).

TABLE 2.10. Effect of Tillage on Surface Residue and Water Stored, Western Nebraska

Fallow System	Residue Cover (%)	Water Stored (%)
No-tillage	78	130
Stubble mulch	38	123
Bare fallow	5	100

SOURCE: Peterson and Fenster (1982).

no-tillage plots averaged more than 2 cm greater during this period, to a depth of 45 cm. On a poorly drained soil, differences were less pronounced but were consistent throughout the period. Till-planting, intermediate between plowing and no-tillage in amount of residue cover, was consistently intermediate in soil moisture also.

In the Great Plains (Peterson and Fenster, 1982), maintaining wheat residues on the surface during the fallow season increased soil moisture available for the succeeding crop by 30% (Table 2.10). Less evaporation, more snow trapped in the fields, and less water use by weeds all contributed to increased moisture.

Cover Crops and Soil Moisture

Growing a cover crop of wheat, rye, or legume for over-winter protection from soil erosion usually increases soil moisture by adding to the surface mulch after

TABLE 2.11. Effect of Cover Crop Management Techniques on Gravimetric Soil Water Content 17 Days after Planting and on Corn Grain Yield, 1981, Florence, South Carolina

Cover Crop Management	Soil Water (%) by Depth 0–15 cm	15–30 cm	30–45 cm	45–60 cm	Corn Yield (mt/ha)
Clean tillage, no cover	8.9	9.9	19.4	21.4	7.39
Disk 2 weeks preplant	5.7	6.6	15.7	19.5	6.59
No-tillage, herbicide at plant	5.4	5.6	13.7	18.5	5.91

SOURCE: Campbell et al. (1982).

the cover is chemically killed. Where crops are removed for silage on low-organic-matter soils, cover crops appear to be necessary for no-tillage yields to be competitive with conventional tillage yields in the following year. The added cover increases soil moisture and improves structural stability.

Time of killing the cover crop, prior to no-tillage planting, can be critical in maintaining desired moisture levels, however.

Fragipan soils in southern Ohio, Indiana, and Illinois often have perched water tables in the spring, leading to wet surface soils and delayed planting. In this area, farmers who no-tillage plant into cover crops usually let them grow until planting time to get maximum transpiration and soil drying. In the high rainfall areas of Georgia, Alabama, and Mississippi, killing the cover crop ahead of planting usually is not necessary.

However, on the coastal plain soils of the Southeastern United States, which have a low water holding capacity and shallow plant rooting depth, allowing a cover crop to grow until planting corn or soybeans will deplete soil water and may cause yield reduction of the summer crop due to drier soil conditions. In a South Carolina study (Campbell et al., 1982) corn was planted on April 4th with different cover crop management practices; then soil water was measured 17 days after planting (Table 2.11). Both soil water and grain yield were significantly reduced when the cover crop was allowed to grow until the corn was planted. Another study at the same location indicated that soybeans were less affected than corn by cover crop management and soil water level.

In-Row Subsoiling and Soil Moisture

Coastal plain soils in the Southeastern United States often have root-restricting layers 15–25-cm below the surface that reduce plant-available water. Many farmers in this area use a subsoiler under the row, with no-tillage between rows, to allow plant roots to penetrate compacted zones. A large number of experiments have tested the effect from in-row subsoiling with variable results.

Effect of subsoiling on cotton root growth and yield was studied at eight widely separated areas in Alabama (Taylor and Bruce, 1968). Subsoiling increased cotton yield by 40% at two locations, did not affect yield at four locations, and lowered yield at two locations. Suman and Peele (1974) found that subsoiling without irrigation increased soybean yields 4 out of 6 years in South Carolina.

In Georgia, no-tillage corn grain yields were compared with and without subsoiling on Orangeburg fine sandy loam soil at nitrogen rates of 200–400 kg/ha (Langdale et al., 1981). Yields were significantly higher with subsoiling. Conventional plow tillage, no-tillage, and no-tillage with in-row subsoiling were compared in an adjacent 2-year Georgia study (Box and Langdale, 1984). No-tillage corn treatments were planted in killed wheat residue. Subsoiling increased yields over both no-tillage alone and conventional tillage, and for both irrigated and nonirrigated treatments.

Importance of Water Saved

Soil moisture saved through reduced tillage systems assumes great importance in regions of low rainfall and high evapotranspiration, on soils low in water-holding capacity, and in years with below normal rainfall. In the Corn Belt, excessive soil moisture in the spring months often has a negative effect on crop growth since it slows soil warming and delays planting. However, on soils where drought stress often occurs during summer months, having more available water during crop pollination and seed filling usually offsets the early season negative influence, leading to better yields.

In the Southeastern United States moisture saved should have an even more positive influence on production although average precipitation is higher than in more northern latitudes. Season-long evaporation losses are much greater at southern latitudes, and soils with low-water-holding capacity are much more prevalent.

The probability of drought has been calculated for 70 locations in North Carolina (van Bavel and Verlinden, 1956) and Georgia (van Bavel and Carreker, 1957). Based on 25 years of weather records, the percentage of drought days varies greatly from region to region and from month to month within each of these states. For example, in the mountains of North Carolina, drought is expected only in June and for only 15% of the days in June. However, in the Piedmont area of North Carolina, drought can be expected for 15% of the days in May, 30% in June, and 20% during the remainder of the season. Such rainfall variability may help explain the varied response to no-tillage and subsoil planting in the Southeastern United States.

In low-rainfall regions such as the Great Plains, moisture saved should have a positive influence on production of any crop grown on any soil type and in any year. For a high percentage of farmland in the United States, moisture saved should be a primary reason for farmers to consider reduced-tillage systems.

TILLAGE SYSTEM AND SOIL TEMPERATURE

Soils that have a partial or complete residue cover will warm more slowly in spring than soils with residues removed or incorporated by tillage. Since maximum root growth for most crops occurs between 25 and 30°C, temperature variations below or above this range can cause reduced root growth. The amount and length of temperature reduction in spring, for a particular reduced-tillage system, compared to plowed soil, depends on latitude, soil drainage characteristics, slope and slope exposure, and kind and amount of previous crop residue.

Cornstalk residues were shown to reduce soil temperature by 0.39°C at the 10-cm depth for each 2.25 mt/ha added (up to 9 mt/ha) during May and June in the northern Corn Belt (Amemiya, 1977). A reduction of 1.1°C greatly reduced early season corn growth in this region. Tillage, soil drainage, and latitude influenced temperature and corn growth in an Indiana study (Griffith et al., 1977) during the first 8 weeks after planting in three continuous corn tillage trials (Table 2.12). The studies were on well-drained and poorly drained soils in northern Indiana and on a well-drained soil in southern Indiana, a difference in latitude of about 320 km.

Daily maximum soil temperatures were lower in the reduced-tillage systems in all trials, with no-tillage planting always the lowest. Till-planting on a ridge had temperatures closest to those with plowing. Corn growth was well correlated with soil temperature at the northern location, with no-tillage planting providing slowest growth. However, at the southern location, the combination of a longer growing season and a well-drained soil produced best corn growth with no-tillage planting, even though early soil temperature was reduced.

Figure 2.2 shows the effects of soybean residue and corn residue on soil temperature after planting at Lafayette, IN (Cruz, 1982). Average maximum temperature during the first 6 weeks was almost 2°C warmer following soybeans. Again corn growth was positively correlated with soil temperature.

TABLE 2.12. Average Daily Maximum Soil Temperature and Height of Corn 8 Weeks After Planting at Two Latitudes in Indiana[a]

	Northern Indiana				Southern Indiana	
	Tracy Sandy Loam		Runnymede Loam		Bedford Silt Loam	
Tillage System	Temperature (°C)	Height (cm)	Temperature (°C)	Height (cm)	Temperature (°C)	Height (cm)
Spring plow, disk twice	22.4	112	21.7	109	26.1	208
Fall plow, disk twice	22.6	114	22.0	117	—	—
Chisel, field cultivate	20.1	99	19.6	107	24.2	193
Till-plant	21.1	107	20.8	102	25.1	206
No-tillage	18.8	97	18.2	86	23.4	221

[a] Average of third and fourth years in same tillage system, continuous corn.

SOURCE: Griffith et al. (1977).

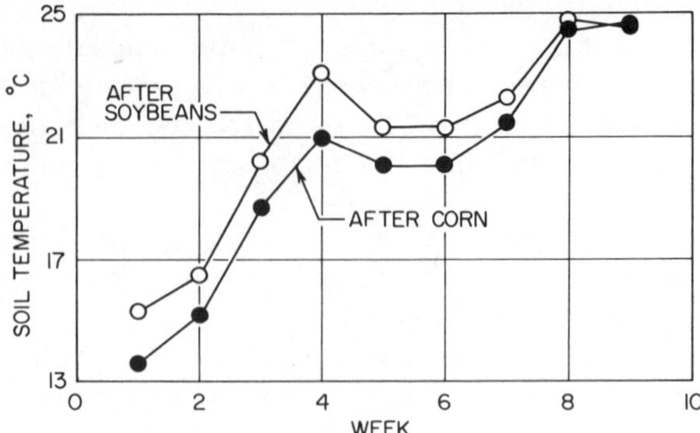

FIGURE 2.2. Weekly means for daily maximum soil temperatures in no-tillage corn (Cruz, 1982).

Daily minimum soil temperature varies little among tillage systems or types of surface residue (Burrows and Larson, 1962); thus, the diurnal variation is less with no-tillage planting than with "clean" or plow tillage. Maximum temperature differences between no-tillage and plowed soils are likely to be greatest on dark soils, since a plowed dark surface absorbs more radiant energy than a plowed light surface, and the absorption of radiant energy with no-tillage is controlled by residue color rather than soil color.

The importance of these temperature differences on plant growth and yield may vary greatly with latitude in the Eastern United States. Germination, growth, and yield are often depressed by low soil temperatures in Minnesota. At normal planting dates and with similar residue cover, low spring temperatures may have little effect in Kentucky, and may have a positive effect in Georgia, where high soil temperatures in June often inhibit root growth and cause excessive evaporation losses.

The adverse effect of surface residue on soil temperature and plant growth in the northern Corn Belt can be mostly offset by till-planting in an elevated ridge and scraping residues off the row area. As a result, this form of reduced tillage is more popular than no-tillage planting above about 42°N latitude.

TILLAGE SYSTEM AND SOIL DENSITY

Soil density affects water movement and root growth, and sometimes affects yield potential of crops. Since tillage systems vary in depth of tillage and amount of implement traffic, they also affect soil density in the rooting zone during the growing season.

Table 2.13 gives density measurements (0–20 cm) for four tillage systems on five soils that had been in continuous corn for 5 years in Indiana (Griffith et al., 1977). Two to three weeks after planting, measurements were made between rows that did not have wheel traffic at planting. Density was greatest with no-tillage, intermediate with till-planting, and least with plowing and chiseling.

TABLE 2.13. Effect of Tillage System on Bulk Density (0–20 cm) 2 to 3 Weeks after Planting[a]

Tillage System	Tracy Sandy Loam (g/cm^3)	Runnymede Loam (g/cm^3)	Bedford Silt Loam (g/cm^3)	Blount Silt Loam (g/cm^3)	Pewamo Silty Clay Loam (g/cm^3)
Spring plow, disk twice	1.20	1.13	1.13	1.21	1.06
Fall chisel, field cultivate	1.25	1.11	1.06	1.13	1.01
Till-plant	1.35	1.33	1.16	1.28	1.15
No-tillage	1.56	1.43	1.33	1.39	1.17

[a]Density determined by Troxler subsurface density gauge.

SOURCE: Griffith et al. (1977).

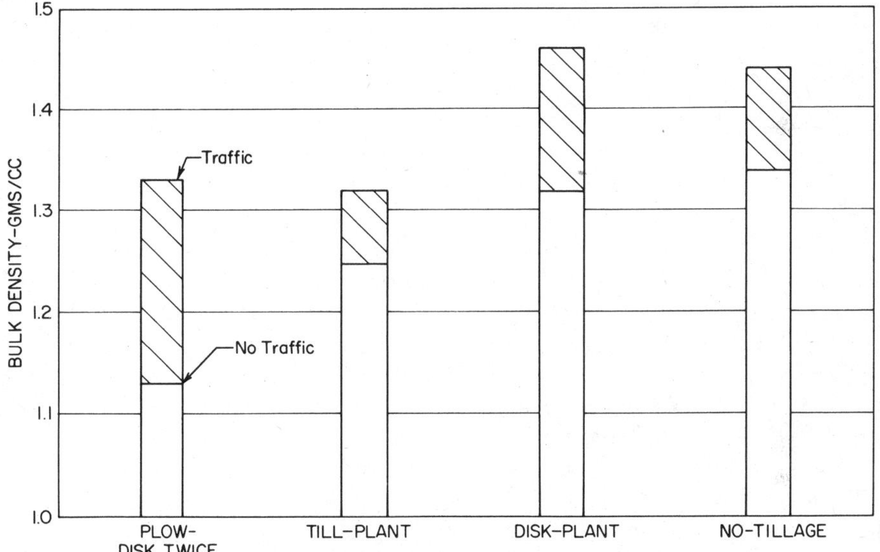

FIGURE 2.3. Effect of tillage systems and implement traffic on soil density, top 20 cm, after planting (Mannering et al., 1975).

Density between rows *with* wheel traffic in these same studies (Fig. 2.3) increased over nontracked soil with all tillage systems, but the greatest increase was with the plow disk-twice system. Density in the plow system with traffic was about equal to density in the no-tillage system without traffic.

Measurements after harvest on four of these same soils (Table 2.14) show no-tillage soil density to be only slightly higher than that with plowing or chiseling after several years in the same tillage system. Thus, most of the difference in density due to tillage was an annual effect and gradually decreased from planting through harvest.

Crop rotation tends to reduce soil density, no matter what the tillage system (Table 2.15). Density was measured shortly after planting in the fifth year of a

TABLE 2.14. Effect of Tillage Systems on Bulk Density (0−20 cm) after Harvest[a]

Tillage System	Tracy Sandy Loam (g/cm^3)	Runnymede Loam (g/cm^3)	Bedford Silt Loam (g/cm^3)	Pewamo Silty Clay Loam (g/cm^3)
Spring plow, disk twice	1.63	1.48	1.35	1.29
Fall chisel, field cultivate	1.59	1.50	—	1.31
No-tillage	1.77	1.55	1.39	1.34

[a] Density determined by Troxler subsurface density gauge.

SOURCE: Mannering, 1974.

TABLE 2.15. Effect of Tillage and Rotation on Soil Bulk Density (0−20 cm) at 4 Weeks after Planting on Chalmers Silty Clay Loam, Lafayette, IN.[a]

Tillage System	Continuous Corn (g/cm^3)	Corn/Soybeans (g/cm^3)
Fall plow, disk, field cultivate	1.26	1.17
Fall chisel, disk, field cultivate	1.15	1.00
No-tillage	1.33	1.18

[a] Density determined by Troxler subsurface density gauge.

SOURCE: Cruz (1982).

study on a well-structured soil with 4% organic matter near Lafayette, IN (Cruz, 1982). Density in the top 20 cm of soil was reduced in rotation for plow, chisel, and no-tillage systems, compared to continuous corn. For no-tillage planting, this probably contributes to improved corn growth in rotation compared to continuous corn.

A more dense soil usually means less pore space for air and water movement. A Minnesota study (Gantzer and Blake, 1978) showed that both total pore space and size of pores were affected in comparing no-tillage versus fall plow systems on a clay loam soil (Table 2.16).

Air-filled porosity at low water tensions at the 7.5−15-cm depth in no-tillage plots was less than one-half that with plowing on May 20. This may contribute to reduced crop growth for no-tillage planting under wet soil conditions. Differences were greatly reduced by September 20, but there was still significantly less pore space in no-tillage plots. There were only minor differences below plow-layer depth.

Number of large pores or biochannels resulting from earthworm activity and plant roots were counted in these same soil cores (Table 2.17). No-tillage plots had the greater number of pores larger than 1 mm in diameter in all comparisons. Greatest differences were on May 20 at the 7.5−15-cm depth.

Interpreting the importance of soil density as it relates to tillage is not easily accomplished. The density level (g/cm^3) at which plant rooting is significantly reduced depends on soil texture, structure, and drainage. Rooting could be significantly inhibited at a density of 1.4−1.5 on a well-structured silty clay

TABLE 2.16. Air-Filled Porosities of Undisturbed Cores (7.6 × 7.6 cm) of Le Sueur Clay Loam, MN[a]

	May 20		September 20	
Soil Water Potential (mb.)	No-Till (cm^3/cm^3)	Fall Plow (cm^3/cm^3)	No-Till (cm^3/cm^3)	Fall Plow (cm^3/cm^3)
		7.5–15-cm depth		
−20	0.051	0.118	0.094	0.122
−50	0.072	0.176	0.123	0.172
Oven-dry at 105°C	0.470	0.570	0.502	0.554
		30–37.5-cm depth		
−20	0.061	0.051	0.056	0.060
−50	0.096	0.086	0.088	0.100
Oven-dry at 105°C	0.470	0.470	0.481	0.487

[a]Fifth year for tillage treatments in continuous corn.

SOURCE: Gantzer and Blake (1978).

TABLE 2.17. Geometric Mean Numbers of Biochannels of Differnet Sizes for No-Tillage and Plow-Tillage

	May 20		September 20	
Channel Diameter (mm)	No-Tillage ($channels/m^2$)	Fall Plow ($channels/m^2$)	No-Tillage ($channels/m^2$)	Fall Plow ($channels/m^2$)
		7.5–15-cm depth		
>3	112	58	69	14
2–3	366	136	105	45
1–2	799	234	217	150
Total	1317	420	635	216
		30–37.5-cm depth		
>3	133	87	128	27
2–3	492	390	252	166
1–2	968	846	660	481
Total	1699	1443	1247	767

SOURCE: Gantzer and Blake (1978).

loam, while rooting may be adequate up to 1.7 on a low-organic-matter sandy loam.

The amount of roots necessary for maximum crop yield is also not well defined. This might vary with genetic yield potential and rooting morphology of particular hybrids and varieties, and with water available in the root zone (also dependent on tillage system).

The negative effect from high soil density on rooting and water movement is partially offset by the increase in large pores (earthworm channels and previous root channels) in untilled soil. For these pores to remain open and effective, however, a high percentage of the soil should remain free from implement wheel traffic, especially in the row area.

Negative effects from high soil density would be most important on soils that are naturally low in permeability and on soils with natural or man-made "pans."

TILLAGE SYSTEM AND SOIL AGGREGATION

Soil aggregation is the cementing or binding together of soil particles into a secondary unit or granule. A high level of aggregation is considered to be an indication of good soil structure and a positive influence on plant growth. The stability of aggregates in water is used as an index of (1) the resistance of soils to dispersion, (2) a soils' susceptibility to compaction, (3) the degree of soil aeration, (4) soil drainage, (5) water intake rate, (6) susceptibility to soil erosion, and (7) plant emergence. Thus, the degree of aggregation is an excellent indication of the physical condition of a soil.

Tillage, or the lack of it, has a major effect on soil aggregation. An Indiana study (Mannering et al., 1975) measured water-stable aggregates in four tillage systems after 5 years of continuous corn (Table 2.18). Soil aggregates for chiseling and till-planting increased by about one-third in the top 5 cm, compared to moldboard plowing, while aggregates were more than doubled for no-tillage planting. More surprising, perhaps, was an increase in aggregation in the 5–15-cm zone in the no-plow systems, even though little or no residue was placed there. Increases in this zone were 20% for chiseling and till-planting and 50% wth no-tillage planting.

TABLE 2.18. Effect of Tillage System on Soil Aggregation[a]

	Aggregation Index[b]	
Tillage System	0–5 cm	5–15 cm
Moldboard plow	0.347	0.468
Chisel (2-in. straight points)	0.456	0.561
Till-plant	0.467	0.560
No-tillage, plant	0.768	0.699
No-tillage, plant (first year)[c]	0.350	0.604

[a] After 5 years of continuous corn; average of four soil types; samples taken just before planting.

[b] Samples passed through 8 mm sieve while moist, air dried, and stability measured by wet sieve method.

[c] Not tilled for 3 years, plowed fourth year, planted without tillage in fifth year.

SOURCE: Mannering et al. (1975).

Improved aggregation not only helps in water management, through better infiltration, but it should help to offset the negative effect on rooting of high soil density in untilled soil. This positive influence takes several years to develop after implementing reduced tillage, however. Where moldboard plowing is practiced every second or third year, benefits in terms of soil physical properties will be greatly reduced. This was illustrated in the Indiana study by comparing aggregation in plots that were continuously not tilled for 5 years versus plots that were not tilled for 3 years, plowed the fourth year, then not tilled the fifth year. Plowing for 1 year reduced aggregation in the top 5 cm of soil to the same level as continuous plowing.

The benefits of improved aggregation, although a positive influence on all soils, would be of greatest importance on low-organic-matter soils (less than 3%). These soils are often erodible, have low infiltration rates, and form crusts easily if high in silt. Improved soil structure and aggregation, gained through reduced tillage, is a valuable asset on these soils; an asset that is lost if they are plowed periodically.

TILLAGE SYSTEM AND PLANT ROOTING

Soil density, porosity, aggregation, infiltration, and water distribution in the soil profile resulting from a particular tillage system affect amount, size, and pattern of crop roots. In addition, cultivation for weed control—a common practice in full-width tillage systems—reduces rooting near the soil surface by pruning roots and drying the soil.

Long-term Indiana studies show corn roots are often reduced in total, are larger in diameter, and are concentrated near the soil surface in a no-tillage system. Figure 2.4 shows corn roots following moldboard plowing and no-tillage planting after 3 years of continuous corn (Navas, 1969). The no-tillage

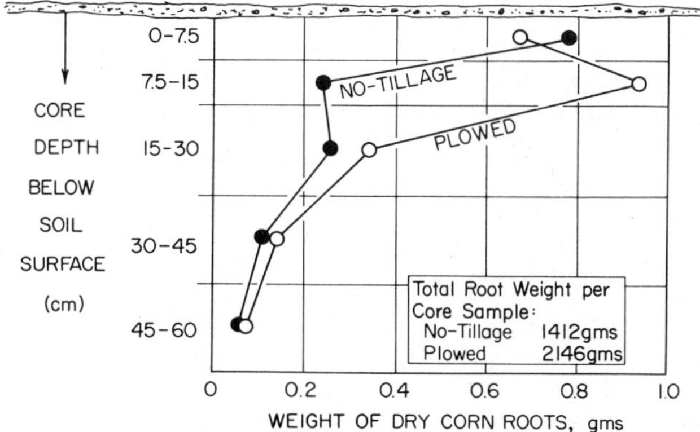

FIGURE 2.4. Effect of plowing and no-tillage planting on amount and distribution of corn roots (Navas, 1969).

corn produced 34% less roots to a 60 cm depth than corn in a plowed soil, but a much higher percentage of these roots were near the soil surface. The data were an average of measurements in rows and centered between rows on four soil types.

On a well-structured 4% organic matter soil near Lafayette, IN, total corn root weight was not reduced in the seventh year of no-tillage planting, but, again, a much higher percentage of the roots were found in the top 7.5 cm of soil (Cruz, 1982). In this same study, roots of till-planted corn were about equal to those with plowing and chiseling in the top 7.5 cm, but were significantly greater than with plowing, chiseling, or no-tillage planting in the 7.5–15-cm zone. Measurements included both in-row and between-row samples.

Root size may also be changed due to tillage. Finer roots are considered more desirable since they contact more soil surface and provide greater opportunity for nutrient uptake. Another Indiana study (Barber, 1971) (Table 2.19) showed no-tillage corn roots to be larger in diameter in the top 10 cm of soil than roots with plowing or chiseling, but about equal to roots from other tillage systems below that level.

Rooting characteristics for crops other than corn may not be affected by reduced tillage in a similar way. Limited soybean data show less effect from tillage system on total root weight and rooting patterns than on corn roots. Soybeans have a tap root rather than the fibrous root system of corn and are more likely than corn to produce lateral rooting near the surface after cultivation in response to late summer rains. These characteristics would not be greatly affected by tillage system.

Another factor, in addition to soil physical properties, may have a significant effect on corn root growth. Decomposition products from corn residues appear to inhibit root growth in a succeeding corn crop (Yakle and Cruse, 1983). This allelopathic effect has also been documented in wheat after wheat. The relation of this growth effect to soil properties is not clear, but residue placement in relation to seed placement could have a significant effect on early root growth.

TABLE 2.19. Effect of Tillage on Corn Root Length/gram of Dry Roots[a]

Depth (cm)	Plow (m/g of roots)[b]	Chisel (m/g of roots)[b]	No-Tillage (m/g of roots)[b]
0–5	38.7	52.7	22.6
5–10	39.2	32.7	21.6
10–15	27.6	24.0	25.0
15–30	25.3	21.6	21.6
30–45	21.9	25.9	29.9
45–60	27.6	29.8	27.3
Average	30.1	31.1	24.7

[a]Ninth year of continuous corn, Raub silt loam soil (2.6% organic matter).
[b]Core samples taken 12.5, 25, and 50 cm from the rows.
SOURCE: Barber (1971).

While crop root growth is an important aspect of production, its effect on yield will depend on the availability of moisture and nutrients in the root zone. Often, crop yield has not been well correlated with root weight or depth in tillage studies (Griffith, et al., 1977). Specifically, shallow rooting as in no-tillage systems, has not been a deterrent to high yield where surface residue maintains soil moisture near the surface throughout the growing season.

TILLAGE SYSTEM AND SOIL EROSION

Soil erosion—the removal of surface soil by water and wind to expose subsoil that is usually less productive—has plagued civilizations for centuries. For most countries of the world soil erosion is still a serious problem. In the United States, 30% of the total cropland is losing soil to erosion faster than it is being reformed through natural processes. The tolerable soil loss limits (the rate above which crop productivity will decrease over time) has been estimated to range from 5 mt/ha on very shallow soils with unfavorable subsoils to 11 mt/ha on deep soils with favorable subsoils. State average annual soil losses from cropland reported in the USDA Resource Conservation Act survey of 1982 ranged from 1 to more than 30 mt/ha. Greatest losses often occur on the most fragile land.

The major force in soil erosion from water is a process of detaching soil particles through raindrop impact, then moving them downslope by overland water flow. In an area with about 100 cm of rainfall annually, each square kilometer is bombarded by a least a quadrillion raindrops annually. Their impact energy is roughly equivalent to 4.5 mt of TNT.

Splash erosion is the first step in soil aggregate breakdown that results in soil crust formation, thus reducing the soil's ability to absorb water, and increasing runoff. Sheet erosion, the removal of a fairly uniform layer of soil through splash and runoff, leads to rill erosion (numerous small channels) and to deeper gullies as the erosion process worsens.

There are several forms of soil transport by wind. In *saltation*, direct wind pressure moves fine sand (0.1–0.5 mm in diameter) by a series of short bounces along the soil surface. This form accounts for 50–75% of soil movement by wind. *Surface creep* describes the rolling of soil particles greater than 0.5 mm in diameter along the soil surface. *Suspension* in the air of soil particles less than 0.1 mm in diameter moves soil long distances, but accounts for less than 10% of soil loss by wind.

Several tillage system characteristics that affect soil physical properties, residue incorporation, and surface roughness and density, also influence soil erosion by water or wind.

Surface residue is effective in reducing erosion because it shields the soil particles from detachment by raindrops or wind. It also minimizes surface sealing, allowing more water to soak into the soil, and reduces velocity of runoff, thus its ability to transport sediment.

Surface roughness is effective in controlling water erosion because it increases water-storage capacity in the tilled layer and reduces velocity of runoff and rate of surface sealing. Roughness reduces wind velocity at ground level, thus reducing soil detachment.

The *more dense* soil surface resulting from no-tillage planting has more resistance to detachment by raindrop impact than soils that have been loosened by tillage.

While erosion from both wind and water are serious in specific situations throughout the United States and elsewhere, water erosion is far more damaging in regions of high rainfall. In the Great Plains, with annual rainfall between 25 and 50 cm, wind erosion is more severe.

Many farmers feel that traditional methods of erosion control such as strip cropping, terraces, grassed water ways, and sod-based rotations are not compatible with the large farms, large equipment, use of herbicides, and continuous grain cropping of modern agriculture. Controlling erosion through reduced-tillage practices appears to be more practical for today's farmers and has the potential for increased profit for many of them.

No one conservation tillage system is clearly superior in reducing soil erosion under all possible cropping, soil, and weather conditions. Variations in slope, soil type, row direction, previous and present crop grown, rainfall intensity, and frequency will alter the extent of soil-conserving benefits that might accrue from a particular system.

Research has been done, however, on the relative effectiveness of the various tillage systems under some of the conditions listed above. The results of this research not only point up the value of conservation tillage as a whole, but also provides a basis for selecting a particular system to match a producers particular farming situation.

Water Erosion

Studies conducted in both Indiana (Mannering, 1979) and Illinois (Siemens and Oschwald, 1976) using simulated rainfall confirm the erosion control effectiveness of conservation-tillage systems on row-crop land, especially when tilled and planted across a slope.

The Indiana experiment (Table 2.20) showed that the spring chisel, till-plant, and no-tillage systems reduced soil loss from an initial "test" storm by 94%, 60%, and 85%, respectively, compared with conventional spring plowing, when tilled and planted across slope. The spring chisel system had the most pronounced influence in reducing runoff (18% of plowing). Till-plant and no-tillage also reduced runoff amounts, but by a much smaller degree.

The erosion-control effectiveness of these three systems largely resulted from lowered runoff amount and velocity and reduced sediment concentration. The effectiveness of both chisel and till-plant was greatly enhanced by the across-slope row direction. Surface roughness in till-plant and surface residue cover in no-tillage were responsible for the reduced soil loss.

TABLE 2.20. Runoff and Soil Loss on 9% Slope Continuous Corn Land Tilled and Planted Across Slope (Bedford, IN)[a]

	First 6.25 cm Storm		Second 6.25 cm Storm	
Tillage System	Runoff (% of rain)	Soil Loss (mt/ha)	Runoff (% of rain)	Soil Loss (mt/ha)
Spring plow/disk/plant	71	23.3	88	27.1
Spring chisel/field cultivate/plant	18	0.7	60	2.5
Till-plant	59	7.4	78	8.5
No-tillage	42	3.6	62	3.8

[a]Bedford silt loam with slope length 22.3 m. Test storms applied within 4 weeks after planting.

SOURCE: Mannering (1979).

TABLE 2.21. Runoff and Soil Loss from 5% Slope Continuous Corn Land Tilled and Planted Up-and-Down Slope (Lafayette, IN)[a]

	Crop Stage 1		Crop Stage 2	
Tillage System	Runoff (% of rain)	Soil Loss (mt/ha)	Runoff (% of rain)	Soil Loss (mt/ha)
Spring plow/disk/plant	56	46.0	70	29.4
Spring chisel/field cultivate/plant	36	24.4	57	20.4
No-tillage	46	8.1	52	3.8
Light disk tillage	43	15.7	60	9.9

[a]Sidell silt loam with slope length 21.3 m. Crop stage 1 tests made about 2 weeks after planting, stage 2 tests about 6 weeks after planting. A 6.25 cm storm was applied at each stage.

SOURCE: Mannering (1979).

The Illinois study used two storms of 6.25 cm rainfall each. Disk-chisel and no-tillage systems used across-slope reduced soil loss 89% and 91%, respectively, compared to fall moldboard plowing (Siemens and Oschwald, 1976). Effectiveness of the disk-chisel (a gang of disks precede the chisel) in almost completely eliminating runoff and soil loss during the first storm was due to a combination of surface roughness, row direction, and surface residue. No-tillage was effective because residue cover greatly reduced sediment concentration of runoff.

In another Indiana test (Mannering, 1979) where tillage and rows ran up-and-down slope (Table 2.21), a no-tillage system proved much more effective than spring chisel tillage in reducing soil loss. Soil loss from a simulated rainstorm 2 weeks after planting was 47% and 82% less, respectively, with chisel tillage and no-tillage than with conventional spring plowing. Obviously, chisel tillage up-and-down a slope will not reduce runoff amount and velocity to the extent it would if across a slope. No-tillage, because its erosion control depends largely on surface cover, is not affected as much by row direction.

Soil erosion is much more severe following soybeans than following corn, regardless of the tillage system used. Two studies, from Indiana (Mannering, 1979) (Table 2.22) and Illinois (Siemens and Oschwald, 1976) (Table 2.23), document this effect.

Although reduced tillage (especially fall disking and no-tillage) did decrease erosion significantly on both types of cropland, the amount of soil loss was substantially more from the soybean field than from the corn field. Most of the difference can be explained by the amount of residue cover—that is, the greater the cover, the less the erosion.

The influence of row direction on effectiveness of a reduced tillage system was again very apparent. Soil losses in Indiana, with tillage up-and-down slope, were consistently higher than those from the same systems in Illinois, where rows ran across slope.

TABLE 2.22. Surface Cover and Soil Loss from Various Tillage Systems on 4% Slope Land Tilled Up-and-Down Slope Following Corn and Soybeans (Harlan, IN)[a]

	Surface Cover		Soil Loss	
Tillage System	After Corn (%)	After Beans (%)	After Corn (mt/ha)	After Beans (mt/ha)
Fall moldboard plow	7	1	22.0	41.0
Fall chisel tillage	25	12	15.0	30.3
Light fall disk tillage	70	17	2.5	12.6
No-tillage	69	26	2.5	13.5

[a]Morley clay loam with slope length 10.7 m. Tests made after overwinter weathering but prior to spring tillage. Two storms were applied at 6.25 cm of rainfall each.

SOURCE: Mannering (1979).

TABLE 2.23. Surface Cover and Soil Loss from Various Tillage Systems on 5% Slope Land Tilled Across Slope Following Corn and Soybeans (Urbana, IL)[a]

	Surface Cover		Soil Loss	
Tillage System	After Corn (%)	After Beans (%)	After Corn (mt/ha)	After Beans (mt/ha)
Fall moldboard plow	4	2	12.8	25.6
Fall disk-chisel tillage	50	11	1.3	7.4
No-tillage	85	59	1.1	3.8

[a]Catlin silt loam with slope length 10.7 m. Tests made after overwinter weathering but prior to any spring tillage; 12.5 cm of simulated rainfall were applied in two storms.

SOURCE: Siemens and Oschwald (1976).

Both simulated and natural rainfall studies at Watkinsville, GA, show the effectiveness of surface residue in controlling erosion on soils of the Southeastern United States (Langdale et al., 1978).

Using the in-row subsoil system to plant soybeans in rye stubble reduced soil loss to insignificant levels compared to conventional clean tillage in simulated rainfall studies (Table 2.24). Loss rates ranged from 36 to 50 mt/ha with plowing and from 0.05 to 0.72 mt/ha with in-row subsoiling. Runoff rates were also greatly reduced for in-row subsoiling when simulated rain was applied to dry soil.

In another study at the same location, soybeans were no-tillage planted into four levels of rye residue ranging from 2.46 to 4.70 mt/ha (Table 2.25). Simulated rainfall showed excellent erosion control at all levels of residue and at three crop growth stages during the growing season.

The effect of reduced tillage in reducing runoff and soil loss is often more dramatic with natural rainfall than in simulated studies. Hydrologic data were monitored on three watersheds in Georgia, two with parallel terraces and a grassed waterway, the other without these conservation practices (Langdale et al., 1978). One of the "conservation" watersheds was no-tillage planted and the other conventionally planted each year. The "nonconservation" watershed was no-tillage planted and conventionally planted in alternate years. Rainfall, runoff, and sediment data are presented in Tables 2.26 and 2.27.

TABLE 2.24. Effect of Tillage, Soil Cover, Antecedent Moisture, and Slope Length on Runoff and Soil Loss from Simulated Rainfall, GA[a]

	In-Row Subsoil and No-Tillage Planting					
	Rye Stubble		Stubble plus Soybean Three-quarter Canopy		Plow-Tillage and Fallow	
Slope length (m)	Low Moisture	High[b] Moisture	Low Moisture	High Moisture	Low Moisture	High Moisture
			Runoff (%)			
10.7	4.2	36.7	19.2	62.0	71.5	87.5
21.4	12.9	67.4	43.8	80.8	85.0	91.8
			Soil loss (mt/ha)			
10.7	0.05	0.40	0.09	0.22	36.3	39.2
21.4	0.29	0.72	0.11	0.18	50.1	39.4

[a]Soil type was Cecil silty clay loam with 6% slope.
[b]Second run was within 24 hr of initial rainfall.
SOURCE: Langdale et al., (1978).

TABLE 2.25. Effect of Rye Residue Level and Time of Application on Runoff and Soil Loss with Simulated Rainfall in No-tillage Planted Soybeans

Time of rainfall application[a]	Rye Residue (mt/ha) 2.46	3.02	3.58	4.70
		Runoff (%)		
At soybean planting[b]	57.1	57.6	57.6	54.2
Half-canopy soybeans	57.1	58.6	59.7	51.4
Full-canopy soybeans	58.2	52.2	56.6	58.1
		Soil Loss (mt/ha)		
At soybean planting	0.11	0.09	0.07	0.07
Half-canopy soybeans	0.11	0.09	0.09	0.09
Full-canopy soybeans	0.05	0.07	0.05	0.09

[a]All simulated rainfall with low antecedent moisture.
[b]No-tillage planted behind a fluted coulter.
SOURCE: Langdale et al. (1978).

TABLE 2.26. Hydrologic Parameters Associated with Conventional and No-tillage Planting on "Conservation" Watershed, GA[a,b]

Season	Rainfall (cm)	Runoff Events (number)	Runoff (cm)	Sediment (mt/ha)
		Tilled Watershed		
May–July	33.6	4	4.2	2.06
August–October	21.7	2	1.3	0.25
November–January	30.3	4	2.8	0.11
February–April	37.7	5	6.4	0.39
Mean Annual	123.3	15	14.7	2.81
		No-Tillage Watershed		
May–July	30.4	1	0.5	0.02
August–October	30.1	3	0.8	0.01
November–January	34.0	5	3.3	0.11
February–April	24.0	1	1.7	0.05
Mean Annual	118.5	10	6.3	0.19

[a]These watersheds had parallel terraces and grassed waterways.
[b]Cecil sandy loam soil.
SOURCE: Langdale et al. (1978).

TABLE 2.27. Hydrologic Parameters Associated with Plow-tillage and No-tillage Planting on a "NonConservation" Watershed, GA[a]

Season	Rainfall (cm)	Runoff Events (number)	Runoff (cm)	Sediment (mt/ha)
		Years with Tillage		
May–July	38.1	10	13.3	22.22
August–October	20.2	4	2.3	0.86
November–January	37.0	6	2.0	0.51
February–April	32.3	6	4.5	2.67
Mean Annual	127.6	26	22.1	26.26
		Years when Planted without Tillage		
May–July	32.5	3	1.5	0.02
August–October	20.5	2	0.1	0.01
November–January	27.3	3	0.5	0.01
February–April	43.5	8	6.6	0.10
Mean Annual	123.8	16	8.7	0.14

[a]Cecil sandy loam soil.

SOURCE: Langdale et al. (1978).

Annual sediment loss was a disaster with conventional tillage in the "nonconservation" watershed—26.26 mt/ha. Terraces and grassed waterways reduced sediment to 2.81 mt/ha. No-tillage planting reduced soil loss even further to 0.19 and 0.14 mt/ha for the "conservation" and "nonconservation" watersheds. In these watersheds, additional conservation practices were not needed in years when no-tillage planting was used.

Wind Erosion

Soil erosion from wind occurs in any region where broad areas exist without windbreaks, especially on mucks or sandy soils. In the Great Plains, however, wind erosion is far more serious than water erosion on soils not covered by residues or growing crops.

"Stubble-mulch" farming in the Great Plains region has proven to be very effective in reducing wind erosion as well as in moisture conservation. During the year of fallow between wheat crops, weeds are controlled by shallow tillage with V sweeps and a rod weeder, leaving much of the previous crop wheat residue on the soil surface.

An 8-year study (Fenster, 1970) at Alliance, Nebraska, compared wind erosion during fallow from stubble-mulch, one-way disk fallow and bare fallow using a moldboard plow. Average annual soil losses from wind erosion were 1.93, 3.14, and 6.50 metric tons/ha, respectively.

Wind tunnel tests in the Nebraska panhandle (Chepil and Woodruff, 1963) showed that fallow soil tilled with a moldboard plow had an average soil loss of 24.0 mt/ha compared with 1.8 mt/ha for stubble-mulch fallow. Fallow using no-tillage, with weed control from chemicals only, was even more effective in wind-erosion control. Where little or no residue is left on the soil surface, periodic tillage to roughen the soil surface provides some temporary control of wind erosion, but this is much less effective than maintaining residue cover.

Most Erosive Periods

Soil erosion does not occur uniformly throughout the year, but is much more extensive in certain months. In the east-central Corn Belt, 37% of total soil loss occurs during April, May, and June (Wischmeier and Smith, 1978). Loss during this period was even greater, 45%, in central Iowa. Total rainfall is greatest during this period, and rains tend to be of greater intensity. Tilled soils are most susceptible during this period because secondary tillage has left a fine seedbed and crop canopy closure has not yet occurred. Surface roughness from fall tillage has been reduced due to overwinter weathering.

Erosion in the Eastern and Southeastern United States is greatest during the spring–summer months, although significant erosion may also occur during winter, since soils are frozen for only brief periods. Soil loss was closely associated with number and intensity of thunderstorms in a Georgia study (Barnett and Hendrickson, 1960) (Table 2.28). July was the most erosive month with June, May, and August following. Over a 20-year period an average of 11 thunderstorms each year caused 86% of the erosion and accounted for 56% of the runoff even though these storms contained only 25% of the rainfall.

On the Great Plains, high wind velocities of short duration may occur during any month, but most prolonged windy periods occur in March, April, and May (Fenster and McCalla, 1977). Thus, erosion from tilled soil is most serious during these months.

In order for conservation tillage to be effective in solving erosion problems in any region it must provide surface cover or roughness during these most erosive periods. Fall chiseling, which provides roughness and some residue over winter, is a popular reduced-tillage system in the Corn Belt. However, if secondary tillage during March or April levels the soil and covers most of the residue, there is little conservation value from this system during the late spring and early summer.

For any region of the United States and whatever the crop grown, it seems obvious that best protection from water or wind erosion results from maintaining the proper amount of surface residue cover from harvest until canopy closure of the succeeding crop.

TABLE 2.28. Average Monthly Distribution of Soil Loss and Number of Thunderstorms, Watkinsville, GA

	January	February	March	April	May	June	July	August	September	October	November	December
Soil Loss (mt/ha)[a]:	3.352	2.466	2.690	3.587	4.483	6.501	11.697	4.483	1.569	.448	.897	.897
Thunderstorms (number):	0.4	0.3	0.5	0.8	1.4	1.6	2.4	2.1	0.8	0.4	0.3	0.2

[a]On Cecil sandy loam soil.

SOURCE: Barnett and Hendrickson (1960).

TILLAGE SYSTEM AND THE TOTAL SOIL ENVIRONMENT

Ultimately, the soil environment created by any particular tillage system will be reflected in plant growth and crop production. Field trials comparing different tillage systems have been conducted in most states in the United States and many countries around the world. Most of the major crops grown in the world have been included. While this research is new to some areas, many experiment stations and some farmers have had experience with reduced-tillage systems for 15–20 years or longer.

Crop response to tillage has often varied among these experiments. Systems leaving a large amount of surface residue have produced yields that were sometimes higher and sometimes lower than yields with clean tillage. In many areas the major causes for this differential response are now clearly identified.

In the Corn Belt a favorable yield response to surface residue systems has been associated with well-drained soils, low-organic-matter soils, crop rotation rather than continuous cropping, and more southern latitude which produces a longer growing season. In any particular situation, of course, all of these factors interact to influence the response to reduced tillage. In the Southeastern United States, soils which are subject to frequent drouth stress appear to be most responsive to surface-residue tillage systems.

Matching Tillage to Soil Type

Several scientists have rated adaptability of certain tillage-planting systems to groups or classes of soils. In Ohio, (Triplett et al., 1973) soil series were placed in five tillage groups according to soil properties and their influence on no-tillage planting for corn. About 1,335,510 ha were considered adapted to no-tillage, with another 1,011,750 ha adapted if artificial drainage was present.

The 346 soil series in Indiana, have been placed in 23 groups based on drainage characteristics, organic matter, texture, and slope (Galloway et al., 1977). Nine tillage-planting systems, ranging from plowing to no-tillage, were rated as to their adaptability, for *corn after corn*, to each of the 23 soil groups. Both crop yield and soil erosion were considered in making the ratings. No-tillage planting was rated "one" (on a scale of one best, five worst) on 38% of Indiana's "row crop" acreage. Till-planting and other reduced-tillage systems were much more widely adapted. Suggestions were given for altering the ratings when corn was to be grown in rotation rather than in monoculture. The rating system is widely used by advisory personnel when working with individual farmers who are considering a major shift in tillage, but who have little experience with the new system.

Since most research in the Corn Belt shows similar response in relating tillage systems to soil type, the Indiana classification system has been adapted to a broader area (Galloway and Griffith, 1978). Benchmark soils from both the eastern and western Corn Belt were named for the 23 tillage-management soil groups. A similar tillage system classification is presented here in Table 2.29.

TABLE 2.29. Adaptability of Eight Tillage-Planting Systems for Corn to Major Tillage-Management Soil Groups of the Corn Belt[a]

Tillage-Management Soil Groups by Natural Drainage, Texture, and Permeability	Organic Matter (%)	Dominant Slopes (%)	Adaptability Rating for Systems[b]							
			Moldboard		Chisel		Disk			
			Fall	Spring	Fall	Spring	Fall and Spring	Spring Only	Till-Plant	No-Tillage
I. Poorly drained, level and depressed, dark-colored uplands, terraces, and lakebeds										
A. Mucks (over 12 in. deep)	>30	0–1	5	1	4	2	4	2	5(4)	5(4)
B. Clay loam on slowly permeable clay	>4	0–1	1	4	1	4	2	2	1[c]	4(2)
C. Clay loam and loam on moderately permeable clay loam	>4	0–1	1	3	1	3	2	2	1[c]	3(2)
D. Sandy loam on loam over sand	>4	0–1	4	2	3	2	3	2	1[c]	3[e](2)
E. Loamy sand over sand with band of loam	>4	0–1	4	2	3	2	3	2	1[c]	3[e](2)
II. Poorly and somewhat poorly drained nearly level, grayish uplands										
A. Silt loam on slowly permeable fragipan or claypan	1.5–3	0–2	5	2	5	2	3(4)	3(2)	2[c]	3(1)
III. Somewhat poorly drained, nearly level, grayish to dark colored upland, terrace, and lakebeds										
A. Silt loam on slowly permeable clay	2–4	1–3	2	3	1	3	2	2	1[c]	3(2)
B. Silt loam or loam on moderately permeable clay loam	2–4	1–3	2	3	1	3	2	2	1[c]	2(1)
C. Sandy loam on moderately permeable clay loam	2–3	1–3	3	2	2	2	3	2	1[c]	2[e](1)
D. Loamy sand over sand with bands of loam	2–3	0–2	5	2	3	2	3	2	1[c]	1[e]

TABLE 2.29. *(Continued)*

Tillage-Management Soil Groups by Natural Drainage, Texture, and Permeability	Organic Matter (%)	Dominant Slopes (%)	Adaptability Rating for Systems[b]							
			Moldboard		Chisel		Disk			
			Fall	Spring	Fall	Spring	Fall and Spring	Spring Only	Till-Plant	No-Tillage
IV. Well and moderately well drained, sloping, brownish uplands and terraces										
A. Silt loam on clay loam on very slowly permeable fragipan	2–3	0–6	5	3(4)	2(3)	2(3)	3(4)	2(3)	1^d	1^e
		6–12	5	3(4)	3(4)	2(3)	3(4)	2(3)	$1^d(2)$	1^e
B. Silt loam on slowly permeable clay	2–4	0–6	5	2(3)	2(3)	2(3)	3(4)	2(3)	1^d	1^e
		6–12	5	3(4)	3(4)	3(4)	3(4)	2(3)	2^d	1^e
C. Silt loam on moderately permeable clay loam or clay	2–4	0–6	5	2(3)	2(3)	2(3)	3(4)	2(3)	1^d	1^e
		6–12	5	3(4)	3(4)	2(3)	3(4)	2(3)	$1^d(2)$	1^e
D. Sandy loam on moderately permeable clay loam	2–3	0–6	5	$2^e(3)$	3(4)	$2^e(3)$	3(4)	$2^e(3)$	1^d	1^e
		6–12	5	$3^e(4)$	3(5)	$3^e(4)$	3(5)	$2^e(4)$	2^d	1^e
V. Well drained smooth to undulating, brownish, loamy terraces and outwash										
A. Silt loam and loam on moderately permeable clay loam	2–4	0–6	4(5)	2(3)	2(3)	2(3)	2(3)	2(3)	1	1
		6–12	5	3(4)	3(4)	2(3)	3(4)	2(3)	$1^d(2)$	1^e
B. Sandy loam on permeable loam	2–3	0–6	4(5)	$2^e(3)$	2(3)	$2^e(3)$	2(3)	$2^e(3)$	1	1
		6–12	5	$3^e(4)$	3(4)	$2^e(4)$	3(4)	$2^e(3)$	2^d	1^e
C. Loamy sand and sandy loam on thin loamy subsoils	2–3	0–6	5	$3^e(4)$	3(4)	$2^e(3)$	3(4)	$2^e(3)$	2(3)	1
		6–12	5	$3^e(4)$	3(5)	$2^e(4)$	3(5)	$2^e(3)$	$2^d(3)$	1^e

VI. Excessively drained, nearly level to duney brownish sands on thin sandy loam subsoil	<3	0–12	5	3^e(4)	3(4)	2^e(3)	3(4)	2^e(3)	2(3)	1^e
VII. Poorly dained dark and clayey bottomland	>4	0–1	5	3	4	2	5	2	2^c	4(3)
VIII. Poorly drained, light gray, silty bottomland	<2	0–1	5	3	4	3	5	2	2^c	4(3)
IX. Somewhat poorly drained grayish, generally loamy bottomland	2–3	0–1	5	2	4	2	5	2	2^c	3(2)
X. Well and moderately well-drained brownish, loamy bottomland	2–3	1–3	5	1	3	1	3	2	2^c	1

[a] Corn after soybean ratings in parentheses.

[b] 1. Highly adapted by all standards. 2. Well adapted, limitations at low frequency or over small part of area. 3. Moderately well adapted, limitations more frequent, or affect greater area. 4. Marginally adapted, more limitations or affect larger area than for 3. 5. Unadapted, due to high frequency of limitations or their occurrence over an entire area.

[c] Rating assumes rows follow direction of natural drainage.

[d] Assumes rows cross-slope or on contour to lessen runoff, channeling by erosion and washout of seed.

[e] Consider use of winter cover crop to improve tilth, nutrient retention and erosion control.

SOURCE: Galloway and Griffith (1978).

Tillage system ratings are based on both yield potential and erosion potential for corn after corn. Where ratings change for corn after soybeans, they are given in parentheses. In general, corn after soybean ratings improve for reduced-tillage systems on poorly drained soils due to reduced surface residue and earlier warmup. But, the corn after soybean ratings are lower for full-width tillage systems on sloping soils due to greater erosion hazards. While research and experience with reduced tillage for soybeans is not adequate to support similar ratings, limited data suggest that soybeans after corn ratings would be similar to those for corn after corn.

Note that no-tillage planting has the highest rating on most of the sloping or drouthy soils while moldboard plowing or chiseling often has the highest rating on dark, poorly drained soils. Till-planting has more "one" or "two" ratings than other systems across all soils, provided row direction is across slope on erodible soils and with slope on poorly drained, untiled soils.

While this classification and rating system was developed for Corn Belt soils, there is considerable evidence that similar methods for rating tillage systems could be used in humid regions around the world, including the Southeastern United States and more tropical climates (Lal, Chapter 10).

Where detailed soil surveys are available, benchmark soils for each tillage-management soils group could be provided by area soils specialists. Without localized soil surveys, detailed descriptions of the soils groups will allow knowledgeable soils specialists to properly classify individual fields or landscapes.

Using such rating systems in choosing a tillage system ensures good soil and water management for a particular soil. However, soils may differ drastically within a farm operation, and often within the same field. Thus, many farmers need to be equipped for more than one tillage system (easily achieved with modern equipment) or choose a "compromise" system. Often, the compromise, where rolling and flat soils occur about equally in the same field, involves intermediate tillage systems using chisels or disks.

While important, optimizing soil and water management is only one of several factors that will influence a farmer's choice of tillage system. Equipment on hand, pest problems (weed, insect, disease, and rodent), and operator skills may all influence profitability and ultimate adoption of a new tillage system.

SUMMARY

Reduced tillage, especially those systems that leave part or all of the previous crop residue on the soil surface, alters many soil physical properties. Some of these changes have a positive effect on plant growth; others have a negative effect.

☐ Surface placement or shallow incorporation of residues leads to a buildup of organic matter near the soil surface. This generally has a positive effect on soil physical properties. Amount of surface residue varies greatly, however, with crop, tillage method, yield level, and harvest method.

- ☐ Strip tillage and no-tillage planting present radically different methods of seedbed preparation, but, with proper management, they provide germination which is equal to that with conventional plow methods.
- ☐ Tillage systems leaving 50% or more of the soil surface residue covered after planting generally increase soil moisture throughout the season due to increased infiltration and decreased evaporation. In areas with low annual rainfall and on soils with low-water-holding capacity, the added water should increase yield potential. On poorly drained soils in northern latitudes the extra water may delay planting and reduce yield potential.
- ☐ Systems with surface residue have lower soil temperature early in the growing season. This delays growth in the northern United States, but could be a benefit in the southern United States and more tropical climates.
- ☐ Strip or no-tillage systems leave a more dense soil through much of the season, sometimes leading to reduced root mass and concentration of roots near the soil surface. Better soil aggregation and more undisturbed large soil pores after several years in the reduced-tillage systems tend to offset the negative effect of increased density on water movement and root growth.
- ☐ Tillage systems that leave surface residue and/or roughness will reduce soil erosion by water or wind. No-tillage planting into residues that cover 80% or more of the soil surface will reduce erosion to below tolerable limits in most situations.

Interpreting the effect of changes in soil physical properties due to reduced tillage on yield potential is not a simple process. The effect of these changes in soil properties is often related to soil texture, drainage, crop grown, previous crop, and latitude. However, research to date suggests that the positive aspects of strip tillage and no-tillage planting outweigh the negative aspects on a high percentage of cropland including virtually all land with serious erosion problems.

ACKNOWLEDGMENTS

The authors wish to thank Drs. W. C. Moldenhauer, E. J. Kladivko, G. B. Triplett, and M. A. Sprague for their review and suggestions for improving this paper.

LITERATURE CITED

Amemiya, M. 1977. Conservation tillage in the Western Corn Belt, *J. Soil and Water Cons.* **32**:29–36.

Barber, S. A. 1971. Effect of tillage practice on corn root distribution and morphology, *Agron. J.* **63**:724–726.

Barnett, A. P. and B. H. Hendrickson. 1960. Erosion on Piedmont Soils. *Soil Conservation, XXVI*, No. 2:31–34 USDA. U.S. Government Printing Office, Washington, DC.

Beale, O. W., G. B. Nutt, and T. C. Peele. 1955. The effects of mulch tillage on runoff, erosion, soil properties, and crop yields, *Soil Sci. Soc. Am. Proc.* **19**:244–247.

Blevins, R. L., G. W. Thomas, and P. L. Cornelius. 1977. Influence of no-tillage and nitrogen fertilization on certain soil properties after 5 years of continuous corn, *Agron. J.* **69**: 383–386.

Blevins, R. L., Doyle Cook, S. H. Phillips, and R. E. Phillips. 1971. Influence of no-tillage on soil moisture, *Agron J.* **63**:593–596.

Box, J. E. Jr., and G. W. Langdale. 1984. The effects of in-row subsoil tillage and soil water on corn yields in the Southeastern Coastal Plain of the United States, *Soil and Tillage Research*, **4**:67–78.

Burrows, W. C. and W. E. Larson. 1962. Effect of amount of mulch on soil temperature and early growth of corn, *Agron. J.* **54**:19–23.

Campbell, R. B., R. E. Sojka, and D. L. Karlen. 1982. Residue Management, Cropping Systems, and an Overview of No-till and Conservation Tillage Research in the Coastal Plains, *Proceedings of the 5th Annual Southeastern No-Till Systems Conference*, South Carolina, Agri. Exp. Sta., Florence, SC.

Chepil, W. S. and N. P. Woodruff. 1963. The physics of wind erosion and its control, *Adv. in Agron.* **15**:211–302.

Cruz, J. C. 1982. *Effect of Crop Rotation and Tillage Systems on Some Soil Physical Properties, Root Distribution and Crop Production*, Ph.D. Thesis, Purdue University, W. Lafayette, IN.

Dick, W. A. 1983. Organic Carbon, Nitrogen and Phosphorus Concentrations and pH in Soil Profiles as Affected by Tillage Intensity. *Soil Sci. Soc. Am. J.* **47**:102–107.

Erbach, D. C. 1982. Tillage for continuous corn and corn-soybean rotation, *ASAE Trans.* **25**: 906–911.

Fenster, C. R. 1970. Conservation tillage in the Northern Plains, *J. Soil and Water Cons.* **32**:37–42.

Fenster, C. R. and T. M. McCalla. 1970. *Tillage Practices in Western Nebraska with a Wheat-Fallow-Rotation*, Bul. 597, Nebr. Agr. Exp. Sta., Lincoln, NE.

Fernandez, B. 1976. *The Effect of Tillage Systems on Soil Physical Properties*, Ph.D. Thesis, Purdue University, W. Lafayette, IN.

Galloway, H. M. and D. R. Griffith. 1978. Tillage—Which is best for each soil type? (Part 2), *Crops and Soils Magazine* **30**:10–14.

Galloway, H. M., D. R. Griffith, and J. V. Mannering. 1977. *Adaptability of Various Tillage-planting Systems to Indiana Soils*, Coop. Ext. Serv. Pub. AY-210, Purdue Univ., W. Lafayette, IN.

Gantzer, C. J., and G. R. Blake. 1978. Physical characteristics of Le Sueur clay loam soil following no-till and conventional tillage, *Agron. J.* **70**:853–857.

Griffith, D. R., J. V. Mannering, and W. C. Moldenhauer. 1977. Conservation tillage in the Eastern Corn Belt, *J. Soil and Water Cons.* **32**:20–28.

Jones, J. N., J. E. Moody, and J. H. Lillard, 1969. Effects of tillage, no tillage, and mulch on soil water and plant growth, *Agron. J.* **61**:719–721.

Langdale, G. W., A. Barnett, and J. E. Box. 1978. Conservation Tillage Systems and their Control of Water Erosion in the Southern Piedmont, Proceedings of the First Annual Southeastern No-Till Systems Conference, Georgia Exp. Sta., Watkinsville, GA, special pub. 5, pp. 20–29.

Langdale, G. W., J. E. Box, Jr., C. O. Plank, and W. G. Fleming. 1981. Nitrogen requirement Associated with improved conservation tillage for corn production, *Communications in Soil Sci. and Plant Anal.* **12**:1133–1149.

Mannering, J. V. 1974. Effect of Tillage System on Surface Cover, Soil Physical Properties and Corn Roots, unpublished manuscript, Purdue University, W. Lafayette, IN.

Mannering, J. V. 1979. *Conservation Tillage to Maintain Soil Productivity and Improve Water Quality*, Coop. Ext. Serv. pub. AY-222, Purdue Univ., W. Lafayette, IN.

Mannering, J. V., D. R. Griffith, C. B. Johnson, and R. Z. Wheaton. 1976, *Conservation Tillage—Effects on Crop Production and Sediment Yield*, Am. Soc. Agr. Eng. Paper No. 76-2551.

Mannering, J. V., D. R. Griffith and C. B. Richey. 1975. *Tillage for Moisture Conservation*, Am. Soc. Agr. Eng. Paper No. 75-2523.

Moody, J. E., J. N. Jones, Jr., and J. H. Lillard. 1963. Influence of straw mulch on soil moisture, soil temperature and the growth of corn, *Soil Sci. Soc. Am. Proc.* **27**:700–703.

Navas, J. 1969. *The Effect of Several Tillage Systems on Some Soil Physical Properties and on Corn Growth*, M.S. Thesis, Purdue University, W. Lafayette, IN.

Peterson, G. A. and C. R. Fenster. 1982. No-till in the Great Plains, *Crops and Soils Magazine* **43**:7–9.

Phillips, R. E. 1980. *Soil Moisture*, In R. E. Phillips, G. W. Thomas, and R. L. Blevins (eds.). *No Tillage Research: Research Reports and Reviews*, University of Kentucky, College of Agr. and Agr. Exp. Sta., Lexington, pp. 23–42.

Siemens, J. C., and W. R. Oschwald. 1976. *Corn-Soybean Tillage Systems: Erosion Control, Effects on Crop Production, Costs*, Am. Soc. Agr. Eng. Paper No. 76-2552.

Sloneker, L. L. and W. C. Moldenhauer. 1977. Measuring the amounts of crop residue remaining after tillage, *J. Soil Water Cons.*, **32**:231–236.

Suman, R. F. and T. C. Peele. 1974. *Limiting Agronomic Factors in Soybean Production*, South Carolina Agr. Exp. Sta. Bul. 1051.

Taylor, H. M. and R. R. Bruce. 1968. Effects of soil strength on root growth and crop yield in the Southern United States, *Int. Soil Sci. Soc. Trans., 9th (Adelaide, Australia)* **I**:803–811.

Triplett, G. B., Jr., D. M. VanDoren, Jr., and S. W. Bone. 1973. *An Evaluation of Ohio Soils in Relation to No-tillage Corn Production*, Ohio Agri. Res. Dev. Cent., Wooster, OH, Res. Bul. No. 1068.

Triplett, G. B., Jr., D. M. VanDoren, Jr., and B. L. Schmidt. 1968. Effect of corn stover mulch on no-tillage corn yield and water infiltration, *Agron. J.* **60**:236–239.

Unger, P. W. and J. J. Parker, Jr. 1968. Residue placement effects on decomposition, evaporation, and soil moisture distribution, *Agron. J.* **60**:469–472.

Van Bavel, C. H. M. and J. R. Carreker. 1957. *Agricultural Drought in Georgia*. Georgia Agr. Exp. Sta. Tech. Bull. No. N.S. 15.

Van Bavel, C. H. M. and F. J. Verlinden. 1956. *Agricultural Drought in North Carolina*, North Carolina Agr. Exp. Sta. Tech. Bull No. 122.

Wischmeier, W. and B. D. Smith. 1978. *Predicting Rainfall Erosion Losses—a Guide to Conservation Planning*, USDA Agr. Handbook No. 537, pp. 45–49.

Yakle, G. Y. and R. M. Cruse. 1983. *Effects of Extracts of Fresh and Decomposing Corn Residue on Corn Seedling Development*, Agronomy Abstracts, Am. Soc. Agron., Madison, WI.

3

TILLAGE AND PLANTING EQUIPMENT FOR REDUCED TILLAGE

R. I. THROCKMORTON
Consultant, Planting and Tillage Equipment
357 Harris Ave.
Clarendon Hills, Illinois

INTRODUCTION

Much emphasis has been placed on surface tillage to lower farming costs, to provide timely field operations, and to improve soil structure, while directing attention to soil and water conservation.

Current surface-tillage farming systems, such as those using chisel plowing or disking as primary tillage, are effective to retain a surface mulch for controlling wind and water erosion through winter and early spring. Secondary tillage in spring greatly reduces surface residue leaving the soil open to wind or water erosion at a time when strong winds and heavy rainfalls are prevalent. The situation has been the same in wheat-producing areas, where, to seed successfully the new crop, the protective mulch is often greatly reduced before seeding.

WHERE WE HAVE BEEN

Human's search for adequate food is continuous as the world population grows. Hand tools and power implements long have been an integral part of this effort as a means to improve efficiency.

Early humans learned that food supplies were more available when they brought production near their domain rather than to forage in search of it. They soon discovered the need to place seeds under the soil surface using a stick to dig a hole—hence, the first agricultural tool came into being. Progressively, they removed pests from their crops and brought them water when rain was inadequate. When they could no longer manually clear the enlarged garden plot, animals and heavier tree forks were used to "plow" the soil. Mechanical soil working became the primary operation for planting and control of weed growth.

Up to the middle 1950s, maximum tillage produced the best yields for most farmers. Soil was opened in fall to absorb moisture over winter and a bare dark surface warmed it rapidly in spring. Weeds not controlled in the previous crop were buried. The moldboard plow provided the primary tillage and accomplished most, if not all, of the following:

- Buried surface trash and crop residues, which harbored insects and diseases.
- Buried weeds and weed seeds.
- Incorporated broadcast lime and fertilizers.
- Loosened the soil and reduced clod size.
- Increased water infiltration rate, soil aeration, soil drying rate, and soil temperature.

However, tillage also:

- Encouraged crusting.
- Enhanced wind and water erosion.
- Brought rocks and weed seeds to the surface.
- Produced a plow sole in some soils.
- Spread problems weeds vegetatively.

During the same period, secondary tillage and cultivating operations increased to as many as five diskings before planting and five or more cultivations during crop growth. Disking controlled early weed growth, provided uniform planter opener penetration, and helped control planter seed depth placement. A smooth fine seedbed was produced with firmness near the planting depth to bring moisture to the germinating seed. The surface was dry to deter germination of shallow weed seeds. Frequent cultivation killed weeds and left a rough surface to hold light rain where it fell. Such operations produced good yields as long as crop requirements were provided and sufficient farm labor was available.

Awareness of the "yield loss date for planting" pressed the farmer to be more timely in field operations. To reduce the time required to get crops planted, new systems of reduced tillage appeared in the late 1950s and early 1960s, including wheel-trackplant and plowplant. These systems reduced the time required to get the seed planted in tilled soil by combining the primary and secondary tillage operations, seedbed preparation, and the planting operations. Although acceptance of these systems did not expand to a considerable acreage, farmers did observe that extensive tillage operations were not required to produce good yields.

Pesticides to deal with weeds, insects, and diseases expanded into common usage. These were often surface applied and incorporated during final preplant secondary tillage. With weed pressure reduced by herbicides, farmers found they could eliminate secondary disking operations as well as later cultivations while providing a good environment for the growing crop.

Reduced tillage expanded during the 1960s and 1970s allowing farmers to increase farm acreage, become less dependent on hired farm labor, perform operations more timely, and obtain higher yields. Scientific farming emerged rapidly due to research and extension in the Land Grant College system and industry to make products available.

WHERE WE ARE TODAY

Today, farmers may find the total costs of producing a crop exceed the income obtainable from its sale; a reduction of production cost is imperative. In addition to machinery, fuel and petroleum-based input costs are expensive. Soil scientists are finding that compaction of soils from equipment traffic to be a problem in many areas, and has the potential to reduce yield and increase costs to remove traffic pans. In light of these problems, surface-tillage systems are evolving. The farmer has made much progress in reducing tillage operations to near the minimum necessary for crop production. Surveys show that the farmers's motivating forces for surface tillage are: (1) to reduce time per unit area, (2) to save fuel, (3) to reduce labor and machinery costs, and (4) to control erosion. A closer look at today's agriculture indicates that farmers are concentrating on the elimination of primary and secondary tillage through surface-tillage and no-tillage practices. A number of farmers have changed from energy-intensive moldboard plowing to chisel or disk tillage operations for residue destruction, effective herbicide incorporation, and easier and more accurate planting. Cultivation remains a common practice in some situations.

Much of today's surface-tillage equipment is similar to past conventional equipment yet is stronger to till a wider width. The principal difference is in how the equipment is used selectively to produce a specific field condition for accurate planting, seeding, and pest control.

In this chapter, information is presented on how different kinds of tillage and planting equipment are used in surface-tillage systems. For continuity and simplicity, equipment is divided into the following categories:

1. Moldboard plow.
2. Full width tillage (not Moldboard Plow):
   a. deep tillage (plow depth)
   b. shallow primary tillage
   c. shallow secondary tillage
3. Strip tillage and planting.
4. No-tillage.

MOLDBOARD PLOWING

Although surface tillage excludes the moldboard plow, it does and will continue to have an important role in agriculture for many years to come. Modern moldboard plows have extra clearance, which allows the farmer to partially or completely bury the large residues of bumper crops. This can be done without extra field trips previously needed with stalk choppers or disk harrows to cut residues and crop crowns.

Fall plowing relieves the pressures of spring tillage for timely planting (Fig. 3.1). By operating the plow to leave some of the stubble uncovered between each furrow slice, both wind and water erosion are reduced until spring secondary tillage is performed. Spring seedbed preparation consists of using a disk harrow, field cultivator, and/or seedbed conditioner for firming, final preparation, and herbicide incorporation.

The fuel requirement of this surface tillage is approximately 30% less than combined stalk shredding and plowing. The elimination of one trip over the field saves time when time is critical. Good tilth of fall-plowed soil reduces spring soil working.

The potential benefits of fall moldboard plowing for spring planting are:

- Timely operations spread labor demand.
- Good soil tilth.
- Reduced surface residue to interfere with herbicide incorporation and planting.
- Less disease and insect inoculum.
- Fertilizer incorporation.
- Weed seeds are buried.
- Good moisture penetration.

FIGURE 3.1. Fall plowing—leaving the ground rough with trash (Deere and Co.).

- Quickly warmed soil in spring.
- Redistribution of fall applied herbicides.

The potential problems are:

- High erosion potential.
- Buried surface residues are unavailable to save soil moisture and reduce midseason soil temperatures.
- May be energy intensive.
- May not be suitable for certain soils.

FULL WIDTH TILLAGE (NOT MOLDBOARD PLOWED)

Deep Tillage (Plow Depth)

Full-width tillage tools used in a surface-tillage system work the entire soil surface but do not turn it under.

Chisel Plow

The most popular replacement for the moldboard plow, on sloping lands, is the chisel plow. The chisel plow works best in drying soils and fits well in a fall primary tillage system since soil lifts and shatters best under dry conditions. Chisel points are the most common tool in Corn Belt conditions; however, in a wheat fallow system, cultivation for weed control may utilize sweeps.

Fall chisel plowing (Fig. 3.2) in heavy row crop residue generally requires a disking or stalk shredding operation to prevent plugging between shanks placed on 30 cm centers. Usually 50–70% of crop residue remains on the surface after chisel plowing, helping to control erosion from wind or water. Spring rains will unearth another 5–10% of the original residue. Erosion protection continues until residue reduction with secondary tillage begins. Because of greater amounts of surface residue, more extensive secondary tillage may be needed for successful herbicide incorporation and planting operations. Coulters can be used on the planter preceding the opener; however, stalks are difficult to cut against loosened soil and can easily foul the planter and, later, the cultivator. Secondary tillage often reduces surface residues to 5–10% of the original amount.

The fuel requirement of the chisel plow system (chisel plus stalk reduction) is approximately 5–10% less than a moldboard plow system.

FIGURE 3.2. Chisel plowing small grain stubble (Krause Plow Corp.).

The potential benefits of a chisel plow over a moldboard plow are:

- Surface residues retained for erosion control.
- Potential savings in fuel.
- Time and energy savings (wider tillage with same tractor).
- Fertilizer incorporation.
- Roughness increases water infiltration.

The potential problems are:

- Equipment plugs in heavy stalk residues if not shredded or disked.
- More intense secondary tillage may be needed to reduce clods and residues as aids to pesticide incorporation, planting, or seeding.
- Less flexibility, not as well suited in spring as fall.
- Soil stays wetter and cooler in spring, possibly delaying secondary tillage, incorporation, and planting.

Combination Chisel Plow

As was mentioned previously, the chisel plow may require a disking or shredding operation prior to use where heavy plant residues exist. A combination chisel plow mounts coulters ahead of the chisel shanks. Coulters are either individually mounted ahead of the chisel shank or are arbor bolt mounted on approximately 20 cm centers. Stalks are more easily cut on unworked soil and the 20 cm residues easily flow between chisel shanks spaced 38–46 cm apart. Stalk reduction and chisel plowing are accomplished in one operation with savings in both time and fuel.

The combination chisel plow (Fig. 3.3) often uses 8–10-cm-wide moldboards which turn the soil and partially bury surface residues. This mixing action both fixes in place and incorporates leaves and other light residues. Previously broadcast fertilizers are also worked into the soil mix. This mixing action reduces the surface residues to 40–50%; however, in spring an additional 5–10% of the surface residue will become exposed as rains wash soil from shallow buried stalks.

The fuel requirements for the combination chisel plow are less than the two operations of stalk reduction and chisel plowing; however, with the addition of the 10 cm turning moldboards, draft increases markedly and both fuel use and time to till a field increases. These vary widely across different designs and soil types.

The potential benefits of the combination chisel plow system are:

- Can cut heavy residues without previous residue reduction.
- Tends to hold and mix small residue particles and fertilizers into the soil better than a conventional chisel plow.

FIGURE 3.3. Combination chisel plow in heavy residues (Portable Elevator Div., Dynamics Corp of America).

- Leaves adequate residues on the surface to hold rainfall and reduce erosion.

The potential problems are:

- Less surface residue throughout the winter than with conventional chisel plow.
- Higher draft requirement than a chisel plow especially when equipped with 10 cm moldboards.
- Requires extensive secondary tillage for pesticide incorporation and planting.
- Soils tend to remain cooler and wetter in spring, delaying secondary tillage and planting.

Plowing Disk Tillage

Both heavy-duty plowing tandem disk harrows and heavy-duty offset disks are used in reduced-tillage systems. Both are characterized by large diameter disk blades (66 cm plus) and wide disk blade spacing (25 cm plus). The large diameter is needed to till soil to plow depth and cut surface residues. The wide disk spacing provides clearance for soil and cut residues to easily pass through the equipment without plugging especially in heavy residues and moist soils.

The plowing disks (Fig. 3.4) should not be confused with conventional tandem disk harrows used for incorporation and/or final seedbed preparation. The latter disks, with 19–20 cm blade spacing, will easily become plugged in heavy residues especially under moist soil conditions.

Fall and spring primary tillage with plowing disks can be successful, although excessive soil moisture can be a problem in spring operations. Frequently, the plowing disk is used when residue destruction and mixing throughout the tilled depth are desired. Residue destruction is usually sufficient to prevent interference with incorporation, planting, seeding, or cultivation.

Plowing disk tillage is appropriate when harvest and fall tillage cannot be completed. The plowing disks bury residues, weed seeds, fertilizer, and previous crop grains lost in harvest.

A conventional disk will reduce surface residues 50% or more on each pass; however, much like the moldboard plow, the plowing disk will leave only a small fraction of the residue on the surface (Fig. 3.5). The soil is loose and

FIGURE 3.4. Plowing disk in heavy residues (International Harvester Co.).

FIGURE 3.5. Offset plowing disk (Miller Mfg. Co.).

superabsorbent. Planting can be done directly behind plowing disks, although an incorporation pass, which also firms the seedbed, is generally used.

The potential benefits of the plowing disk systems are:

- Cuts and mixes heavy surface residues and fertilizer throughout the tilled profile in one operation.
- Buries weed seeds and shattered or unharvested crop grains.
- Reduces tillage time from that required by the moldboard plow.
- Provides conditions for easy incorporation and planting.

The potential problems are:

- Large reduction or elimination of surface residues for erosion control.
- Somewhat more sensitive to soil moisture conditions.
- Disks tend to overwork soil.
- Possible delay in subsequent operations when used as primary tillage in spring.

Subsoil Bedder

Bedding is a much different form of tillage and is used in both humid and arid regions. In humid regions, the soil is bedded so excess soil moisture will move from the elevated seed bed. In arid regions, the furrow between the raised beds provides gravity irrigation paths.

A subsoil bedder combines subsoiling and bedding operations. The advantage of this system is the location of the subsoil slot relative to the crop row (either between beds or directly under the row to be planted). Increasing attention to subsoiling to plow pans or compacted soils, especially in irrigated areas, led to the development of this combination tool (Fig. 3.6). The subsoil bedder eliminates one pass over the field.

Crop roots easily find the slot for deeper penetration providing increased drouth resistance. Research indicates that the effectiveness of subsoiling can be lost in some soils if a heavy tractor tire passes closer than 25 cm to the slot. A tire wider than 50 cm in a 100-cm-row spacing can materially reduce the effectiveness of the subsoiling operation.

The potential benefits of the subsoil bedder system are:

- Combines two field passes into one.
- Locates the subsoil slot in relation to the crop row for subsequent planting and/or irrigating.
- Buries residues for easier bed working, incorporating, planting, and gravity irrigation.

The potential problems are:

- Buries residue, which makes it unavailable for erosion control or other benefits.
- Establishes fixed rows during primary tillage.

V-Sweep Plow

Wheat production in the Great Plains area relies on stubble mulch systems to control weeds, increase soil moisture storage, and reduce wind and water erosion. The moldboard plow may be used infrequently to plow under severe weed infestations and some farmers continue to use clean plow-tillage methods exclusively.

V-sweep plows (Fig. 3.7) have 1.22–1.83-m V-blades and retain more standing stubble on the surface than chisel plows. V-plows are operated 8–13 cm deep to break up compacted soils and cut weed roots to conserve moisture. V-plow usage leaves the soil open sufficiently to absorb rainfall and snowmelt while the standing stubble reduces evaporation of soil moisture, traps snow, moderates soil temperature, and reduces blowing.

The potential benefits of the V-sweep plow system are:

- A minimum number of shanks moving through the soil and residue-covered surface.

FIGURE 3.6. Combination subsoil bedder (E.L. Caldwell & Sons, Inc.).

FIGURE 3.7. V-sweep plow with mulch treader (Richardson Mfg. Co., Inc.).

- Large portion of standing surface stubble maintained.
- Provides for control of most weeds.
- Opens soil for moisture infiltration and storage.
- Control of wind and water erosion.
- Standing stubble protects seedlings through the fall and winter.

The potential problems are:

- Requires large tractors, since this is a time-sensitive operation.
- Small grass weeds may not be controlled in one pass, requiring a herbicide application or additional tillage.
- Most grain drills in use will not seed through the heavy residues preserved by V-plow tillage.

Shallow Primary Tillage

Another form of reduced tillage eliminates the deep tillage operations and concentrates on shallow tillage on the seedbed. Some shallow-depth tillage equipment includes large volume tanks for a combination tillage/herbicide operation.

Tandem Disk Harrow

Standard tandem disk harrows are used to reduce previous crop residues and work the soil to only about 10 cm. The first disking may be performed after harvest in the fall immediately following a broadcast fertilizer application.

Residue is generally retained until spring. Frequent diskings kill growing weeds and germinating weed and crop seedlings. Disking and field cultivator operations incorporate herbicides to complete seedbed preparation.

A disk tillage system minimizes the time problems as disking is a fast wide-pass operation; however, wet soil can halt disking. Frequently, a field cultivator is substituted for a disk for weed control or as part of the incorporation process. Potential benefits of the disk tillage system are:

- Fall disking is quickly accomplished; it is a much faster, less precise operation than deep tillage.
- Most residue is preserved through the winter and early spring.
- Large tractors are not required.
- Mechanical weed control residues reliance on herbicides.

Potential problems are:

- Surface residue is greatly reduced in spring increasing potential for erosion.
- Disks are easily plugged in heavy residues and on moist soils, which delays planting.
- Frequent disking breaks down soil particles resulting in induced crusting.
- Deep tillage operations may be required every two to four years to open pans and soil compaction.

Field Cultivator

Field cultivator tillage (Fig. 3.8) is used most frequently in combination with a disk tillage system. Both tools are for shallow operation and the field cultivator usually is not overloaded at the shallow depth. The field cultivator leaves more residues on the surface for erosion control than a disk and is an excellent tool to control weeds by undercutting and leaving dry soil on the surface where germinating weed seeds die. The undercutting, with less inversion of soil, preserves soil moisture.

The field cultivator, like the disk harrow, can be a high-speed wide-pass operation. When used in damp soils, the sweeps tend to leave crusted soil clods producing a rough surface that requires additional clod reduction such as disking.

The potential benefits of the field cultivator tillage system are:

- Preserves plant residues during final seedbed preparation.
- Good retention of soil moisture.
- High-speed operation.
- Has some incorporation capabilities.

FIGURE 3.8. Seedbed preparation with field cultivator (Wil-Rich, Inc., Lear Siegler Inc.).

- Does not require a large tractor.
- Dries soil above operation depth to inhibit weed seed germination.

The potential problems are:

- Shanks rake and gather heavy residues.
- Not adapted to moist soils.
- Weed seeds are not buried.
- Requires additional operation for incorporation of herbicides.

Shallow Secondary Tillage

Seedbed Conditioner

The seedbed conditioner (Fig. 3.9) was originally used to level beds and provide shallow tillage on the bed top for planting cotton in the Delta area. This tool has undergone many changes while being adapted to other crops as a secondary tillage tool.

The modern seedbed conditioner typically has two or three front rows of field cultivator shanks and sweeps followed by a rotary cutting mixing blade. Ranks of tine teeth or peg teeth follow the rotary blade to provide further

FIGURE 3.9. Seedbed conditioner (Noble Division, Lear Siegler Inc.)

stirring, clod reduction, and smoothing. In addition, mulchers or rotary baskets can be rear mounted. Some designs have chisel teeth in front and disk gangs in the middle, replacing the rotary blade.

The seedbed conditioner is an excellent tool to reduce clods both on the surface and in the seedbed zone. It can incorporate herbicides, but may require two passes for full effectiveness. Use of this tool should eliminate one or more field passes with other secondary tillage equipment.

The potential benefits of the seedbed conditioner system are:

- Reduces clods in one pass in a rough primary tillage system.
- Has herbicide incorporation capabilities.
- Produces an even, consistent seedbed for better planter operation.
- Less soil moisture loss compared to multidiskings.

The potential problem is:

- Tends to foul when operated in surface or shallowly incorporated residues.

Roller Harrow

The advent of reduced tillage, including the elimination of some tillage operations, has generally produced rougher, more cloddy soils prior to final preparation of the seedbed. The roller harrow (Fig. 3.10) effectively reduces both dry surface clods and clods just below the surface in the seedbed zone to produce a fine soil particle and firm seedbed. This is especially important in planting solid seeded soybeans with grain drills.

The front gang of rollers pulverizes the surface clods. The center-mounted spring teeth work just past the seedbed zone to raise clods for the rear gang to crush.

The potential benefits of the roller harrow tillage system are:

- Prepares a fine soil particle firm seedbed in one pass for easy accurate planter or drill operation.
- Loose soil on top, plus a firmed seedbed, provides moisture at seeding depth for rapid even germination.

The potential problems are:

- Shallow plant residues incorporated by the primary tillage operation tend to plug the spring teeth.
- Is most beneficial in clean secondary tillage systems.

Small Grain Combination Tillage/Seeders

When preemergence herbicides have not been used, operation is usually preceded by shallow tillage to kill growing weeds and to dry the surface to

FIGURE 3.10. Roller harrow preparing a fine seedbed (Farmhand, Inc.).

retard weed seed germination. The tillage/seeder performs both shallow tillage and seeding in one operation. The tillage section can be used separately as a conventional field cultivator.

The tillage portion of the combination tillage/seeder (Fig. 3.11) is generally similar to field cultivator shanks with an assortment of ground tooling for different types of seeding. Some designs have heavier shanks similar to chisel plows. The tillage ranks have extra clearance for residue flow. This is accomplished by using four (or more) ranks with fore and aft spacing.

Tillage/seeders are equally effective for seeding through residues, for no-tillage seeding, and for clean ground seeding. The wide fore and aft spacing of ranks provides more trash clearance than a standard field cultivator; however, some reduction of heavy residues may be desirable.

The design of the combination tillage/seeder permits a variety of row spacings. These machines are air seeders, which provide increased row width flexibility by blocking seed delivery to rows not being seeded. Cultivation of the entire surface continues as the shanks are not removed.

The seed is blown from metering heads to drop behind the shanks, which are mounted on rigid bars of the frame. Tandem walking-beam axles tend to level the frame and follow the ground contour. Floating wing sections are gauged similarly. Accuracy of seed depth placement is not as consistent as with drills having independently mounted seed openers, which better conform to uneven soil surfaces.

FIGURE 3.11. Combination tillage/seeder (Wil-Rich, Inc., Lear Siegler Inc.).

Fertilizer can generally be applied, either band or broadcast, as required. Herbicides can be surface broadcast during the same tillage/seeding operation. Some designs can band fertilizer up to 20 cm deep; however, this entails a separate operation from seeding.

No-tillage operations require a stiff shank, similar to chisel shanks, which will maintain tillage depth in hard soils. If soils or portions of a field are hard or stony, the shanks will automatically move back and up resulting in shallow seed placement.

The potential merits of the tillage/seeder system are:

- Performs both a tillage and seeding operation in one pass.
- Central large hopper for grain and fertilizer provides convenience and time saving.
- Can apply fertilizer band or broadcast while seeding.
- Can apply broadcast herbicides while seeding.
- Open shank pattern allows crop residues to flow easily.
- Less machine cost, better utilization, tillage portion can be used for separate secondary tillage.

The potential problems are:

- More complex when tillage and seeding in one operation.
- Operation in heavy plant residues will likely require separate residue reduction operation(s).
- Accurate seed depth and coverage.

Planters and Drills for Surface Tillage

Planters

As farmers changed from clean tillage to preserving surface residues, the planter needed modifications and additional attachments, primarily to allow the planter to operate through surface or semiincorporated residues.

A coulter is used on the planter in a system of reduced tillage with residues on and near the surface. The 5-cm-wide fluted or rippled coulter cuts through clods and residue providing a narrow tilled strip for the planter opener. The coulters are generally mounted on a tool bar frame member preceding the planter openers. Coulter depth control in a full width tilled soil is not critical because planter designs have sufficient weight on the planting unit for the opener-mounted gauge wheels to firm the seed bed, controlling seed depth placement. The firming action of the gauge wheels also forms a column of firm soil under the seed, which hastens germination by improving moisture migration to the seed zone. Firming wheels press seeds into the firm soil for better seed–soil contact.

Modern planters with heavy (approximately 180 kg) downpressure coulter capacity are effective in no-tillage planting into most soils. The firm soil provides support so the residues can be cut easily and the planter opener can place the seed in clean moist soil. Even minor tillage ahead of the coulter will impede cutting of the surface residues. When this occurs, the surface residues may be pushed into the seed slot, placing some seed in contact with residues instead of soil, with the poor seed placement resulting in uneven emergence. In no-tillage double cropping of soybeans into wheat stubble, residue pushed into the seed slot may interfere with coverage and hold the seed on top after the planter passes, leaving seeds exposed.

Planter settings in a true no-tillage operation generally require a more shallow seed placement, since soils are cooler and moisture for germination is generally present near the surface.

Reduced-tillage systems generally rely on both herbicides and cultivation for weed control. The type of cultivation depends on the amount and kind of surface residues and the firmness of the soil. Cultivation of some soils can increase yields and is easily accomplished if surface residues are sparse (corn following soybeans). Cultivation of heavy residues require cultivators with sweep (or disk) patterns on the cultivator, which will not rake and accumulate the residue. Disk cultivating blades easily clear trash but are not widely used. Alternative sweep patterns may use wider sweeps with fewer shanks to avoid gathering residues.

Drills

Much effort has been directed by researchers, product engineers, and farmers to drill openers and drill components that can accommodate large amounts of loose surface residue. Attachments such as row straw cleaners, a flat disk gang, and elements turning on the opener disks have been used. None of these have received wide acceptance by farmers. Farmers continue the common practice of reducing surface cover by planting time so the drill will not plug. Some drills have greater fore and aft clearances for better trash flow; however, the problem of plugging in heavy residues remains.

STRIP TILLAGE PLANTING

Strip-tillage systems may or may not use a stalk reduction or shallow-tillage operation. The amount and nature of previous crop residues and soil conditions determine the need for preplant operations. The two major types of strip tillage are flat planting and ridge planting—some strip-tillage designs have both flat and ridge plant capabilities.

Rotary Strip Tillage

In a flat planting operation, the strip rotary tillage units are centered ahead of each row planter with no tillage performed between the strips at planting time.

When heavy residues are present, a stalk-reduction operation aids the rotary tillage units to perform satisfactorily. A previous shallow-tillage operation may be required in hard soils to allow the rotary tillage unit to operate to the desired depth.

The early tillage pass(es) may incorporate herbicides and broadcast fertilizers. Since soils are covered with previous crop residue until spring, they are usually cool and wet. Wet soils, combined with soft residues, contribute to plugging the rotary tiller, therefore, planting, and even emergence, can be delayed. Once the strip is tilled and residues are buried, the bare soil will absorb heat rapidly, and plant growth can catch up with conventionally planted crops in a few weeks.

At midseason, strip-tilled soils generally are cooler, owing to the residue cover, and preserve more moisture than a clean-tilled surface.

On sloping erosion-prone land strip tillage should be performed on the contour. When planted up and down the slope, excess surface water will quickly erode the tilled strips removing seed and soil.

The potential benefits of the strip tillage system are:

- Primary tillage operation eliminated.
- Most tillage is confined to the seedbed area.
- Improved water infiltration and storage through winter and early spring.
- Erosion control when planted on the contour.
- Surface residues hold spring and summer rainfall.
- Cooler and more moist soils at midseason.

The potential problems are:

- Planting may be delayed in spring on wet soils.
- Generally a slower planting operation than other systems.
- Cold damp soils may slow crop seed germination.
- Rotary tillage units can fail when encountering buried objects.
- Early tillage operations to reduce residue may allow some erosion to occur.

Sweep Strip Tillage Planting

This system is adaptable to either ridge or flat planting. The system requires an initial residue-reduction operation when following a heavy residue crop. In flat planting, the broadcast fertilizer is applied in late fall or early spring. Anhydrous ammonia can be knifed in after chopping previous crop stalks. The planting arrangement usually places the new row over the previous one. A sweep with trash rods clears a path and the planter follows in one operation. Disk row cleaners may be substituted for the sweep and trash bars. No tillage is performed on the surface under the sweep, which leaves a smooth firm surface to assist in planter performance and also leaves a clod-free strip for good herbicide application. The exposed smooth moist path quickly warms and the crop seeds germinate rapidly. Weed seeds also germinate rapidly but are controlled by the herbicide. The residue pushed between rows during planting helps hold rainfall and control erosion. Contour planting on slopes is recommended.

One cultivation using disks to move soil away form the row and then back controls small weeds in the row. A sweep is operated in the row middle to control larger weeds. One cultivation is usually sufficient, but more can be performed if needed.

In a flat planting sweep strip-tillage operation, the previous year's cultivation does not build a high ridge; therefore, when planting over the old row, the new seedbed is depressed placing seeds in cool moist soil. Good germination can be attained in areas where the soil has warmed quickly in spring and does not contain excess moisture. This is why flat sweep planting was first adopted in western Iowa and Nebraska. In high-rainfall areas, water ponding over the depressed seedbed can reduce crop emergence.

The sweep strip-tillage planting system can also be used in a ridge planting system (Fig. 3.12). Planting on a ridge produces rapid crop germination, probably earlier access to the field to plant, and easier harvest.

A ridge system can be started in a flat planted crop by using a heavy-duty disk cultivator rather than sweeps. The disk cultivator throws soil to the crop

FIGURE 3.12. Sweep till plant on old ridge (Fleischer Mfg. Inc.).

row building a ridge about 20 cm high. Through the balance of the growing season and over winter, the soil firms into a solid ridge.

Following harvest, the stalks are chopped in either the fall or spring. Residues accumulate between ridges to hold moisture and to slow surface water movement. Fertilizer can be broadcast applied providing the spreader wheel treads match row spacings. Anhydrous ammonia is knifed in between ridges.

At planting time, the sweep with trash bars cuts 5–10 cm off the top of the ridge throwing old crowns and a light soil covering to the row middles. This leaves a bare soil strip that is drained of excess moisture and warms earlier in the spring. This seedbed also quickly germinates weed seeds which are killed by a banded preemergence herbicide. Cultivation with a heavy-duty disk cultivator at layby rebuilds the ridges for the following crop.

The potential benefits of a sweep strip tillage plant system are:

- Adapted to flat or ridge planting.
- Ridges and surface residues hold erosion to a minimum.
- Primary tillage not required.
- Lack of extensive soil turning keeps virile buried weed seeds dormant.
- Surface weeds in the strip-tilled area easily controlled by a band herbicide application.
- Elevated seedbed warms and drains excess moisture earlier in season.
- Fertilizer more available earlier in season owing to higher temperature.
- Strip on firmed ridge provides for precision planting.
- Soil from row middles annually returned to the ridge.
- Mechanical cultivation controls weeds when herbicides do not.
- Reduced fuel and time required to grow the crop.
- Labor inputs more evenly distributed throughout the year.
- Usually earlier entry to field for planting.
- On sloping land, contour ridges act as small terraces to deter erosion.
- Good yields on both poorly and well-drained soils.
- Corn harvester header snouts can be run in lower row middles for easier pick up of lodged crops.

The potential problems are:

- Ridges required a permanent traffic pattern for field operations, which may be inconvenient.
- Crop rotation requires destruction of ridges (an extra field operation).
- Systems not adapted to rows under 75 cm.
- Current equipment is tractor mounted (aid to staying on ridge), whereas most farmers prefer pull-type equipment.
- Incorporated herbicides cannot be used.

- Surface residues may harbor insects and crop diseases and attack the growing crop even though not in direct contact.

Disk Strip-Tillage Planting

The disk strip-tillage planter can be used to flat plant or plant on a ridge. A residue-reduction operation may be needed in heavy residues.

In a surface-tillage system (no primary tillage) the row cleaning disks cut through surface residues and move the residues and clods to the row middles where they are mixed with soil. The operation is the same whether flat planting or ridge planting. In a ridge planting operation, the ridges are rebuilt during the layby cultivation with heavy-duty cultivators (Fig. 3.13). A fluted coulter, followed by the planting unit, is used to prepare a narrow tilled seedbed below the disk row cleaner depth.

The planting strip is in bare, moist soil, which warms quickly (flat planting) or has already warmed (ridge planting). Weeds and crop seeds quickly react to the warm moist soil and germinate. The surface-applied preemergence herbicide is used to kill weeds in the row.

The potential benefits and problems of the disk strip till planting system are the same as for the sweep strip till planter. The sweep strip till system may leave a smoother strip for the planter and preemergence herbicide application.

NO-TILLAGE PLANTING

No-tillage planting systems are characterized by no overall-tillage or stalk-reduction operations prior to planting. A narrow tilled strip, 5 cm wide maximum, is the only tillage operation performed. In some designs, a flat coulter is the only tool preceding the planter opener.

Row-Crop No-Tillage Planting

No-tillage is ideally suited to sloping lands where extensive tillage results in very high erosion rates. No-tillage has made productive fields of land that formerly could produce only pasture crops or very low yields of arable crops. This system is adaptable to well-drained soils that warm quickly in spring or crops that are midseason planted such as double-crop soybeans or fall-planted winter small grains.

No-tillage farming is also adapted to rocky fields such as found in the northeast where outcropping rocks make plow tillage impossible or far too costly owing to rapid equipment destruction. The cutting coulter and planter double-disk openers will ride over rocks until normal soil conditions are again encountered and planting continues.

It is not well adapted to the northern part of the cornbelt for spring-planted

FIGURE 3.13. Disk till plant on old ridge (Hiniker Co.).

crops owing to cool moist soil conditions that delay planting (up to two weeks or more) and result in poor germination and uneven stands.

Soil erosion is greatly reduced compared to other farming systems. Water runoff is clear of soil sediment but may carry dissolved fertilizer, insecticide, or herbicide components. Both high- and low-water-runoff rates may be incurred with no-tillage. This varies with the form of surface residues, soil-moisture-storage level, soil type, and duration of continuous no-tillage.

Under no-tillage conditions soil settles and becomes more firm requiring up to 180 kg pressure to obtain planter coulter penetration. In no-tillage planting systems, a variety of coulters may be used depending on soil conditions and the amount of tillage needed in the narrow strip for the planter opener. Penetration and cutting action of the coulter selected is vital so the planter opener can move uniformly through the soil and place the seed at the proper depth. The flat coulter should operate only slightly below the planter opener depth and no deeper as seed will fall to coulter depth in plastic soils. A wide (5-cm) fluted coulter can pick up moist soil and deposit it outside the narrow tillage path. This action increases the speed of operation for some machine designs. Adequate furrow closing is important to obtain good soil seed contact for quick, even emergence. This may require crushing the sides of the tilled strip together and firming.

No-tillage planting (Fig. 3.14) requires precise management and timing. Weed and insect infestations must be dealt with quickly to prevent excessive yield loss through a buildup in successive crops. The control of weed, insect, and animal damage is important to crop performance (see Chapters 11 and 12).

The potential benefits of the no-tillage system with row crops are:

- Increased yield on adapted soils.
- Vastly reduced soil erosion.

FIGURE 3.14. No-till planter (Allis-Chalmers Corp.).

- No primary tillage required.
- No root pruning cultivation.
- Reduced time and labor inputs.
- Reduced soil moisture loss from evaporation.
- Reduced fuel requirements.
- Large tractors not needed.
- Easier to maintain timeliness of planting, fertilization, and pest control.
- Planter can be used in all tillage systems.
- More intense use of land now considered marginal because of erosion hazard.
- Firm residue-covered soil supports traffic during wet seasons.

The potential problems are:

- Requires high level of management.
- Not well adapted to poorly drained soils or early spring plantings in cool, moist climates.
- Largely dependent on chemical control of weeds, insect, and animal damage.
- All weeds may not be controlled easily.
- Loss of fertilizer, herbicide, or insecticide components in runoff.
- Nitrogen losses may require higher applications.
- Build up of surface acidity requires altered liming practices.
- Inversion tillage and mixing of surface materials may be necessary periodically.
- High rates of water carrier needed for some pesticide applications.
- Surface residues may harbor insects and diseases to attack the growing crop.

Double-Crop No-Tillage Planting

The practice of double-cropping soybeans into small grain stubble has rapidly increased with the availability of no-tillage planters that have made planting more accurate, timely, and less expensive.

Previously, farmers would disk the residues so conventional planters could penetrate the soil to place and cover the seed properly. This required considerable time delay, and disking reduced soil moisture. The delay often missed vital rainfalls needed for germination and seedling growth, and quicker no-tillage planting captured the maximum growing season to increase yield.

No-tillage double cropping can use the same row-crop planting equipment used in no-tillage full-season crops. The equipment is generally spaced for 76 cm rows; 38-cm row widths are generally the minimum available. Farmers also have the option of no-tillage grain drills, for much narrower rows, which normally produce higher yields.

The no-tillage drill (Fig. 3.15) is a new drill concept brought about by the necessity to provide high down pressure on each opener coulter. Conventional drills seldom have the weight or opener down-pressure springs capable of that needed in no-tillage operations. No-tillage drills can be used for full season or in double-cropping operations where no prior stalk reduction is necessary.

The ground units are designed with a cutting coulter mounted just ahead of the disk seed opener. This controls the depth of the cutting coulter relative to the seed opener. The ground-engaging units are arranged so a hydraulic cylinder applies the pressure necessary for adequate penetration. Opener protection is provided by a spring or trip mechanism to prevent damage from buried obstructions. Depth gauging is through an adjustable rear firming wheel.

The potential benefits of no-tillage double cropping are:

- Conserves soil moisture.
- Tillage or stalk-reduction operations not needed.
- Can use conventional no-tillage planters or no-tillage drills.

FIGURE 3.15. No-till grain drill (Lilliston Corp.).

- Saves time and energy compared to tilled doubled cropping.
- Makes maximum use of cropping season by planting immediately following previous crop harvest.
- Herbicides available provide good weed control.
- Soybean cultivation is eliminated in 18 cm rows resulting in easier combine harvest of lower pods.

The potential problems are:

- Reliance on timely herbicide application(s).
- Seeds may be planted on top of residues forced into the planter opening by the coulter, reducing germination, especially on loose soils.
- Specialized no-tillage planting equipment required.
- Some weeds may not be adequately controlled by herbicides.

Full-Season No-Tillage Drill Planting

Full-season soybean cropping with the no-tillage drill requires herbicide actions outlined for no-tillage row crops. Post emergence herbicides can be applied by broadcast ground sprayers in early growth stages as young soybeans recover from the wheel traffic. Later herbicide applications must be made by air or by equipment using traffic paths left in the crop (see Chapter 11).

The elimination of tillage ahead of the drilling operation preserves soil moisture and provides extra time for planting. This is helpful when wet cool soils delay the planting operation. Surface residues reduce erosion losses and an 18–20-cm row will form a closed canopy in approximately 5 weeks to further reduce weed germination and soil-moisture evaporation.

No-tillage drilling of wheat is just recently finding acceptance. As farmers moved from clean tillage to stubble mulch tillage, the art of maintaining a residue surface cover has developed through the use of chisel plows, V-sweep plows, chisel rod weeders, and rod weeders. Farmers have come to preserve residues so well that disking is frequently required to reduce the stubble so conventional drills can seed without plugging.

A no-tillage or ecofallow system retains the standing stubble. Herbicides or one undercutting with a V-sweep plow controls weeds. When compaction pans are present, the V-plow must be used to break up the pan to aid root growth.

The tillage and/or herbicide is applied immediately following harvest of the previous crop. Weed control is effective through the fall, winter, and early spring when a second herbicide application is generally required. Special no-tillage drills then seed directly through the standing stubble.

The potential benefits of no-tillage or ecofallow wheat systems are:

- Stubble holds snow where it falls.

- Stubble prevents wind erosion.
- Stubble maintains cooler soil in spring, warmer soil in winter.
- Stubble prevents wind/soil contact reducing evaporation.
- Increases soil water storage by as much as 5 cm.
- Reduces tillage to one operation and, frequently, none.
- Saves time and fuel.
- Operations can be more timely.
- Better seedling environment during emergence and over winter.

The potential problems are:

- Wide (12 m) no-tillage wheat drills are not commercially available.
- Herbicide applications require very precise rate control.
- Weather variations can result in too little or too much herbicide carryover.
- More fertilizer required for the first 2–3 years.
- Herbicides may not control all weeds.

Row-Crop No-Tillage Subsoil Planters

The soils in lower coastal plains crop producing areas of southeastern United States respond to subsoil operations in crop production systems. Subsoiling extends the depth of rooting and provides access to moisture during periods of drought.

In a broadcast operation subsoiling frequently results in subsoil slots that are not located close enough to the plant to provide the deeper rooting desired. Furthermore, compaction under tractor wheels close the subsoil slots. The subsoil no-tillage planter combines the subsoil operation with drill planting to place seed over the slot in one operation.

Planting can be preceded by a disking or stalk-reduction operation if residues are heavy. The coulters ahead of the subsoil shank cut through the residues and a preemergence herbicide is applied. The system is credited with increasing yields under both normal rainfall and irrigation. A conventional farming system for these soils would require subsoiling, primary, and several secondary tillage operations prior to planting.

The potential benefits of a no-tillage subsoil system are:

- Increased crop yields.
- Other tillage operations greatly reduced or eliminated.
- Savings in time and fuel.
- Majority of tractor traffic precedes subsoil planting preserving subsoil slots.
- Crops more tolerant of drought periods.

The potential problems are:

- Weed control relies mostly on herbicides.
- System is most effective only in specific soil types or subsoil conditions.
- Larger than usual tractor may be required owing to high draft of subsoiler.

Forage Seeders

Forage crops are commonly relegated to extensive acreages, rough or stony fields, steep slopes, and other land classes unsuitable for row-crop production. Equipment functions with difficulty on such land. Because of the nature of the landscape, no-tillage methods are often the only means available for seedling establishment. A major objective in forage improvement is the replacement of the existing sward with more productive species. Fertilizer application is an important part of forage improvement and can be broadcast or drilled during seeding or in a separate operation. Details on herbicide selection and timing are discussed in Chapter 11.

Seed of most perennial forage species are small with an optimum seeding depth of 5–10 mm. Precision depth control is difficult, at best, yet important, in rough terrain. Optimum row spacings range from 15 to 25 cm. As with row-crop seeding, the opener must cut through surface cover, place seed in moist soil, and cover it. Mowing and removal of topgrowth or close grazing before seeding makes sod penetration by equipment easier.

Equipment to perform these tasks has been developed to cope with the soil and seed situations.

1. Drills with disk openers preceded by a coulter, designed for planting wheat and narrow-row soybeans, often can be used for forage crops on the better land when equipped with legume or grass seed hoppers. When planting into a well-developed sward, minimal amounts of soil are moved from the seed trench. Gauge-firming wheels following the opener provide seed coverage.

2. Another design uses a knife or shoe opener preceded by a coulter. In some older machines, the tip of the shoe was curved forward to aid in soil penetration but this increased the hazard of damage from underground obstructions. Some openers were mounted on a rigid bar, decreasing precision depth control on uneven surfaces. Accordingly, seeding units were of narrow widths to attain acceptable seed placement. Shoe openers are capable of applying both fertilizer and seed in one operation.

3. Powered, narrow rotary cutters slice through surface cover to open the soil; seeds are dropped into the slit and loose soil provides some coverage. The rotary cutter penetrates the soil without the weight required to force disk openers into the soil. However, they have high maintenance requirements under some soil conditions.

4. A knife opener is mounted beside a disk in one drill for forage seeding. Pairs of openers are individually mounted and penetration into hard soil is facilitated by adjustments to transfer weight of cast iron presswheels to the seed opener. Thus, overall machine weight can be less than normal for disk machines. Individual mounting of pairs of openers provides more precise depth control than is possible with tool bar mountings.

5. Another recently invented concept has potential for improving germination of small forage seeds. The openers are individually mounted to better follow the ground contour and are held in the soil by two pitched blades, one on either side of the opener. The blades provide suck to hold the openers in the soil as well as undercut a narrow path in the established sward. The undercutting reduces competition and provides loosened soil to cover the seed. A rear gauge wheel controls depth and firms soil over the seed.

In recent years, livestock returns have not been adequate to support large investments in forage planters. As seed costs increase and livestock become more profitable, no-tillage forage seeders will become more numerous, precise, and effective.

REDUCED TILLAGE AND PLANTING EQUIPMENT FOR THE FUTURE

One important factor in long-term economical crop production is preservation of the productive top soil being lost by erosion. Reduced tillage can help reduce erosion losses, and promises to help prolong production on a long-term basis. Such conservation tillage, to be effective, must retain the crop residue on the surface from the initial tillage to the closing of the crop canopy between the row.

A further step in reduced tillage farming is providing winter cover. Future successful farming systems will be dependent on equipment that will preserve this cover yet be able to move through it without clogging. A major objective in all reduced-tillage systems is performing necessary operations without disturbing the surface cover.

New concepts are possible when equipment is designed to operate under predictable conditions. One example is the Aitchinson no-tillage drill opener concept from New Zealand. Rather than depend on heavy weighting for coulter penetration into the soil, this drill uses a narrow pitched blade on each side of the opener. The pitch (low in front) holds the opener in the soil. Zero drill-opener loading for coulter penetration becomes very important in controlling cost for new design concepts of no-tillage drills used in the U.S. Corn Belt (6 m wide, 18 cm rows) and Winter Wheat Belt (12 m wide, 30 cm rows) are needed.

The tillage planting systems discussed in this chapter are, of necessity, greatly simplified. As research progresses systems will become faster, easier, more precise, appropriately applied, and, it is hoped, have more widespread application. As an understanding of what is required by the crop to encourage germination and seedling establishment becomes available, engineers will be able to design new concepts to accomplish the objectives.

The information in this chapter is based on the author's experience in equipment planning covering almost three decades. Of necessity, the tillage and planting systems have been simplified and no attempt has been made to include the multitude of variations in today's farming systems.

4

MINERAL NUTRITION AND FERTILIZER PLACEMENT

GRANT W. THOMAS
Professor of Agronomy
Department of Agronomy
University of Kentucky
Lexington, Kentucky

INTRODUCTION

When one changes from conventional plow systems to no-tillage, much more is involved than simply a cessation of tillage. Soil structure is once again allowed to develop naturally with plant root penetration and the leaf fall of crops and cover plants; soil water evaporation is reduced by the shielding cover of plant material; and the soil temperature regime shifts to the cooler side. In addition, because the soil is little disturbed, amendments such as fertilizers and lime remain longer near the surface.

The overall effects of this change on plant nutrition are not known, but after some 20 years of experience, a number of points seem to repeat themselves. Experiments on phophorus, potassium, and lime are plentiful, and the results are only moderately conflicting. Experiments on micronutrients are few, indeed, and many of the ideas concerning their behavior under no-tillage are estimates.

This chapter will cover the behavior of plant nutrients, with the exception of nitrogen, under no-tillage. Nitrogen will be covered in Chapter 5. Emphasis will be placed on the behavior of phosphorus and potassium and, to a lesser extent, on calcium and magnesium; sulfur and the micronutrients will be discussed only in general terms.

SOIL SITUATION UNDER NO-TILLAGE

Structure

Soil structure originally was developed under the native plants of a given region, whether they were trees, grass, or brush. The development of soil structure implies that roots repeatedly follow old root channels, that water moves through the soil also following these channels, and that clay suspended in the water follows the same channels and is gradually deposited on the pore walls. Under such an environment, overland flow is a rare occurrence and erosion, if there is an adequate cover, is very low.

The invention of the plow had a fatal effect on this structural pore system. Plowing largely disrupts the pores in the plow layer and tends to plug the macropores after years of repeated plowing. The more accurate the job of plowing, the worse the effect on soil water movement. The practical effect of this is to retard rapid drainage into the subsoil, to leave the plow layer saturated, and, with enough rain and enough intensity, to lift the plow layer (on a perched water table) and float it away. See Chapter 2.

In addition to this commonly observed effect, there are others: (1) the plow layer of a tilled soil will be wetter than the respective depth of a nontilled soil during the rain and immediately afterward; (2) evaporation will be high from a plowed soil at first, because the first-stage evaporation will be sustained longer from a wetter soil; (3) the subsoil will not receive as much water as with a structured and nontilled soil; (4) as a result of all these effects, the water content in the plow layer will vary more widely than the water content at a similar depth of unplowed soil; and (5) this variability has effects on plant nutrients, in particular, diffusion and leaching will be affected.

Water Content

In addition to the short-term water effects listed above, due largely to structural differences, there are water-content differences almost entirely due to surface cover. The nature of this cover may vary considerably, but the most important factor is the proportion of the soil surface covered. Depth of the plant reside has relatively little effect on evaporation. The effect of plant residue on evaporation is complex and related to the microclimate arising from (1) reflection of incoming sunlight, which effectively reduces the net incoming radiation and temperature that regulates relative humidity and thereby evaporation and (2) the diffusion path of water vapor through the residue. Unpublished work in Kentucky many years ago clearly demonstrated this effect on early morning soil water content, which increased 2% or more on a gravimetric basis from afternoon to the next morning.

The water content of the upper part of the soil (0–75 cm) is changed radically by the cover of plant residues. This effect is shown in Fig. 4.1 for an entire growing season on a Maury soil, Lexington, KY. It is apparent that the largest differences between no-tillage and conventionally tilled soil occur early in the season, when evaporation is relatively much more important than transpiration. However, this difference persists throughout the season, so that one can say that the soil profile under no-tillage always remains wetter than the upper part of a soil under conventional plow-tillage (notwithstanding the short-term differences already described in the section on structure).

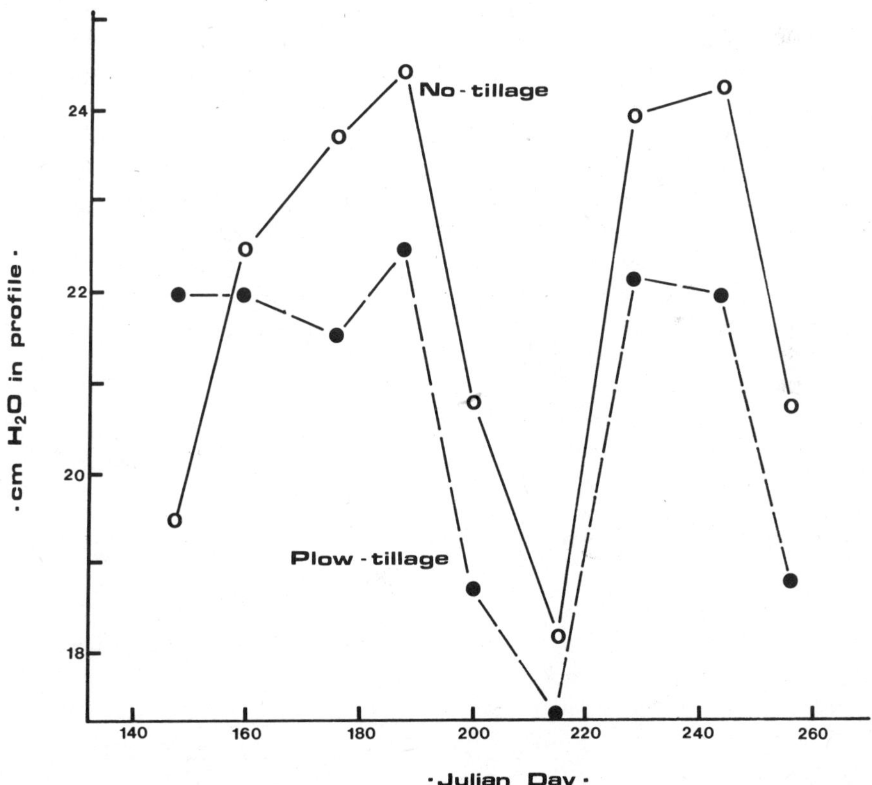

FIGURE 4.1. Total soil water in the 0–75-cm profile of Maury silt loam soil planted to corn with plow-tillage and no-tillage in 1977 (Thomas and Phillips, 1981).

No-tillage has direct effects on uptake of plant nutrients. First, there is a generally higher rate of transpiration during the entire growing season (NaNagara et al., 1976). Nutrients (calcium, for example) at high concentration in the soil solution, will tend to have a higher uptake under no-tillage than under plow-tillage because more water enters the plant, carrying the nutrients with it by mass flow (Barber, 1962). Secondly, the diffusion of nutrients such as phosphorus that are sparingly soluble in the soil solution is enhanced greatly by additional soil water. Mahtab et al. (1971), for example, showed a near linear increase in the diffusion coefficient of phosphorus with increasing soil water.

In addition to the favorable effects on nutrient uptake listed above, the effects of structure and soil water evaporation on the more-soluble nutrients tends to be negative under conditions of high rainfall. Because the structural properties, earlier alluded to, tend to allow deeper movement of nutrients during rainfall, they already are deeper in the soil profile. The reduced evaporation rate thereby offers less chance that they will be moved upward again. Over a period of years this should tend to reduce calcium and magnesium (but not phosphorus or potassium) contents of the soil. That this is, indeed, the case is shown clearly in Table 4.1, which is taken from 10 year results by Blevins et al. (1983a). Note that higher nitrogen rates, which both produce hydrogen ions and increase calcium and magnesium solubility, intensify the loss of these nutrients.

Taken together, then, soil structure and plant residue (mulch) on the soil surface have both positive and negative effects on plant nutrient uptake by plants. These effects will be examined in more detail in referring to specific nutrients.

TABLE 4.1. Exchangeable Calcium and Magnesium in Maury Silt Loam After 10 Years of Conventional Plow-Tillage and No-Tillage Corn

		Ca		Mg	
Depth (cm)	N Rate (kg/ha)	NT[a] (meq/100 g)	PT[b] (meq/100 g)	NT (meq/100 g)	PT (meq/100 g)
0–5	0	5.48	7.84	0.33	0.70
	168	2.61	6.13	0.47	0.59
5–15	0	5.96	7.48	0.62	0.72
	168	4.74	6.19	0.55	0.55
15–30	0	6.61	7.49	0.58	0.71
	168	6.10	6.91	0.54	0.63

[a] No-tillage.
[b] Plow-tillage.
SOURCE: Blevins et al. (1983a).

Temperature

Plant residue that remains at the soil surface has two properties which tend to change the soil temperature compared with a bare soil. First, the color of the residue is usually, although not always, lighter than the soil itself. When this is true, incoming radiation tends to be reflected more than it would be from the surface of the soil itself. Second, the plant residue acts as an insulator, since it is largely filled with still air, so heat produced by solation does not reach the soil.

The net effect is that daytime soil temperatures tend to be lower under no-tillage than plow-tillage. In spring, it is as if the season were retarded in that soil temperatures reach their optimum for planting from 7 to 14 days later under no-tillage than plow-tillage. Nor is this the only reason for lag in soil warming under no-tillage. It has already been noted that the average soil water content is higher under no-tillage. The heat required to warm a wet soil is much greater than that required to warm a dry one and this, depending on relative water contents of the tilled and untilled soils, can be the most important factor (Marelli et al., 1981). Therefore, the interacting effects of mulch and water combine to provide lower daytime soil temperatures, and this effect is especially notable during the early part of the season following winter precipitation and when there is very little plant canopy. These effects can limit plant nutrient uptake when soil temperatures are marginal, say, for phosphorus. On the other hand, in more southerly regions, or on southfacing slopes, the effects may be beneficial, since soil temperatures under a bare soil are often too high for maximum plant nutrient uptake.

The effect of no-tillage on nighttime soil temperature is just the opposite of that in daytime. The mulch serves as an insulator; reradiation to the nighttime sky is greatly reduced. Therefore, minimum early morning soil temperatures are *higher* under no-tillage than plow-tillage (Table 4.2).

TABLE 4.2. Maximum and Minimum Soil Temperatures at 5 cm Depth Under No-Tillage and Plow-Tilled Corn in Argentina

	Plow-Tillage			No-Tillage		
Date	Max (°C)	Min (°C)	Mean (°C)	Max (°C)	Min (°C)	Mean (°C)
Oct. 31	27.8	11.3	19.6	23.8	12.8	18.3
Nov. 1	27.5	12.8	20.2	23.3	14.3	18.8
2	30.5	13.6	22.2	25.5	15.0	20.2
3	31.3	16.3	23.8	25.6	17.5	21.6
4	28.3	26.8	22.6	24.8	17.1	21.0
5	28.3	18.8	23.6	25.4	19.0	22.2
6	29.8	18.3	24.0	26.1	18.5	22.3

SOURCE: Marelli et al. (1981).

Because of the moderating effect of a mulch on soil temperature, the *roots* of plants under no-tillage are much less subject to temperature extremes than are the roots of plants under plow-tillage. This is not true of plant shoots, however, which are correspondingly much warmer during the day (owing to reflected solar radiation) and much cooler during the night (owing to reradiation). On young tender plants these daytime–nighttime shifts are sometimes critical. Frost damage to leaves of young corn plants under no-tillage is always worse and the tendency to burn leaf tips is always more. As far as plant nutrient uptake itself is concerned, the roots are subjected to less extreme variation in average temperature under no-tillage than plow-tillage.

Soil Organic Matter

Because the soil is disturbed so little under no-tillage and because erosion is reduced, the level of soil organic matter tends to rise or to stabilize rather than fall as is the case with continuous plow-tillage. Experiments in Kentucky show dramatic differences in both organic carbon and organic nitrogen after 10 years (Table 4.3). Difference in organic matter content not only affects the difficult to measure properties of soil tilth, structure, feel, and appearance but also has direct effect on plant nutrition because of the nutrients carried in the organic matter itself. Chief among these are nitrogen, sulfur, phosphorus, and zinc. The presence of these nutrients in organic matter is not an unmixed blessing as can be seen in the case of nitrogen (Chapter 5), for there is a tendency to improve nutrient availability with its use provided other conditions are right for their release. Certainly, over a long period of time, the changes in soil organic matter content result in a more fertile soil with no-tillage than with conventional tillage involving a plow.

TABLE 4.3. Organic Carbon and Nitrogen in No-Tillage and Plow-Tilled Corn Soils After 10 Years

	Organic N (0–5 cm)		Organic C (0–5 cm)	
N Rate (kg/ha)	NT[a] (%)	PT[b] (%)	NT (%)	PT (%)
0	0.23	0.14	2.22	1.30
84	0.25	0.15	3.05	1.45
168	0.24	0.15	2.89	1.44
336	0.30	0.15	3.03	1.51
Original sod	—	—	3.11	

[a] No-tillage.
[b] Plow-tillage.
SOURCE: Blevins et al. (1983a).

The very conditions that lead to the accumulation of organic matter in the soil also lead to reduced availability of the nutrients contained in the organic matter. Much of these effect depends on the weather, there generally being a tendency toward less availability under low temperatures than high. In theory, at least, the slow release of nutrients from a large and growing pool should eventually equal the rapid release of nutrients from a small and declining pool. The problem with this reasoning is the almost total lack of reliable data about nutrient contents of organic matter, release rates, and effective pool size. It suggests an area of "fertile" research in comparing the effectiveness of the two systems.

More understandably, the positive effects of plowing up an old sod on the following crop are frequently observed, and these effects are largely caused by decomposition of organic matter, not its accumulation. The question becomes how to take advantage of the increased organic matter level observed rather than to presume that an increase in organic matter, *per se*, is an assurance of improved crop yield. It may or may not be, depending on other considerations that will be discussed with the appropriate nutrients.

Acidity

In the temperate humid zones (Blevins et al., 1977), but apparently not in the tropics (Lal, 1981), there is a marked tendency toward the acidification of the upper part of the soil with no-tillage. This is caused by (1) the increased leaching experienced under no-tillage and (2) the fact that the soil is not turned and mixed. This tendency is accelerated by the addition of nitrogen fertilizer (Table 4.1). Thus, one would expect more acidification under corn than wheat and more under wheat than soybeans, to which nitrogen is rarely applied.

The apparent reason this does not occur in the tropics is that there are longer dry periods during which salts of basic cations are brought back to the surface by plant roots. When the plant residue is left on the surface, there is a tendency to concentrate bases there. In contrast, in humid temperate zones, there is a general tendency to flush out salts during the winter and spring so a gradual acidification occurs.

Aside from the direct effects of calcium and magnesium loss from the soil surface, acidification has secondary effects on the availability of other nutrients, on the quantities of exchangeable aluminum and manganese present and on the efficiency of triazine herbicides. The effects on calcium and magnesium loss are well documented, as are the increase in aluminum and manganese (Blevins et al., 1978). The ineffectiveness of triazines at low pH also has been studied enough to know that weed control is poor below a pH of about 5.5 (Kells et al., 1980). What is unknown is the extent of the effect of increased acidity on availability and uptake of phosphorus, potassium, and micronutrients. Because this has been studied so little we resort to general principles of nutrient behavior under varying pH. Thus, this change in soils under no-tillage must be discussed in light of work done not under no-tillage but under plow-tillage.

It should be stated that the acidification of soil under no-tillage is not, in principle, different from that which occurs with plow-tillage. However, since it occurs principally at the soil surface, it can be treated at the soil surface by topical applications of lime, without resorting to tillage. While it may be argued that lime is less effective when applied in this fashion (which is true), it also is worth noting that if the problem is solved, it is probably better to handle it this way than to till. However, with lime as with other nutrients there exists a considerable diversity of opinion between adherents of not tilling and those who advocate mixing nutrients with the soil.

PHOSPHORUS

Native Phosphorus

The behavior of native phosphate in a soil under no-tillage is governed by the changes in the soil that already have been detailed. Of these, the most important is the usual increase in soil water. The effect of soil water on diffusion can be estimated from the following equation:

$$D_s = D_w \gamma^2 \theta$$

where D_s = diffusion coefficient of phosphate in soil
D_w = diffusion coefficient of phosphate in water
γ^2 = tortuosity factor of soil paths
θ = gravimetric water content of the soil

From this equation, it can be seen that the effective diffusion coefficient in the soil is directly proportional to the soil water content provided γ^2 is constant. Olsen and Watanabe (1963) and Baldovinos and Thomas (1967) found uptake of phosphorus at a given level in the soil solution proportional to $\sqrt{D_s b}$, where b is the slope of a curve relating labile phosphate to phosphate in the soil solution (b = labile P/soil solution P). Since D_s is related directly to soil water, θ, this infers that phosphorus uptake by plants is enhanced by higher water content. This is, of course, the case. Phosphorus stress is always worse under dry soil conditions, so much so that phosphorus deficiency is often called "dry weather" disease. Therefore, one way of improving phosphorus availability is to increase the soil water content, which is done under no-tillage. Shenk (1979) showed yield of corn in Costa Rica to be higher under no-tillage at all phosphorus levels. Other work shows similar trends. It is important to note that even without phosphorus applied, no-tillage gives superior yields, providing sufficient quantities of other nutrients are present. Essentially, this effect can be attributed to the higher water content of the soil water under a no-tillage system.

Another factor that is very important in determining phosphorus uptake is soil temperature. As has been noted, it is nearly always lower in spring with no-tillage. The effect is frequently noted on corn (*Zea mays*, L.) a C4 grass, which has been moved to colder regions because of its high yield potential. However, because of its relatively long growing season, it must be planted earlier than optimum in the northern parts of the United States to avoid damage from fall frosts. If the planting date is such that soil which is conventionally tilled has a temperature of 18°C while that of a nontilled soil has a temperature of 15°C and the temperature at which corn can absorb phosphorus is 16°C (which is normal), the corn on the conventionally tilled soil will grow well, while that on the no-tilled soil will be delayed until the temperature increases. Spring is a critical time for corn, since many predators, whether they be insects, slugs, or small animals, delight in eating it to the ground. In Kentucky planting is delayed, but where the growing season is short, an application of phosphorus to the seed before planting is sometimes helpful.

Without question, one of the limitations of no-tillage is its inability to "let the crop grow" under conditions of low soil temperature. The reasoning is often justified on the basis of poor phosphate uptake under cold conditions. It is a plant related factor, not wholly due to soil conditions.

Rates of Phosphate Fertilizers

Although there has been little work done on rates of phosphorus comparing no-tillage and plow-tillage, that which has been done suggests that to obtain the same yield, less phosphate needs to be applied under no-tillage. The major reason for this appears to be the higher water content of the upper part of the soil under no-tillage. This improves the diffusion rate of phosphorus toward the plant root and, hence, phosphorus uptake. Since this principle operates with native phosphorus, one would expect added phosphorus would be affected in the same way.

Another factor that favors phosphate uptake by plants is the degree of "fixation" of the added phosphorus but the soil itself. Phosphorus fixation is the result of adsorption and/or precipitation of phosphorus by the metal cations calcium, aluminum, and iron. In some soils, one of these metal cations dominates in phosphorus fixation; in others, two or all three may be important. In any case, the solubility of the compounds formed is the key to phosphorus concentration in the soil solution. Looking at the kinds of compounds likely to be formed, one may estimate the effects of soil changes under no-tillage on the solubility of phosphorus.

In calcareous soils (those with an excess of calcium carbonate) and to a lesser extent in soils with a pH of 6.5 or above, the major cation involved in reducing phosphorus solubility is calcium. Compounds likely to be formed are dicalcium phosphate, octacalcium phosphate, and hydroxy apatite. The solubility of phosphorus decreases with each compound named. Lindsay and Moreno (1960) have shown a direct relationship between pH and the solubility of each of these compounds. As pH decreases one unit, the solubility of phosphorus *increases* 10 times. Thus, anything that would tend to reduce pH in a high pH soil would tend to increase phosphorus solubility. With no-tillage one can expect a lower pH because of the tendency toward loss of bases in the soil surface, already alluded to, and because of the increase in CO_2 content of the soil air due to increased water content and (presumably) to a higher organic matter content. Therefore, specifically in calcareous soils, there are physical grounds to expect increased phosphorus solubility.

In soils with pH 6.5 or below, one can take little comfort from the behavior of phosphorus with aluminum or iron. The same solubility principles prevail, only with a decrease of one pH unit the solubility of phosphorus *decreases* 10 times. As with high pH soils, no-tillage tends to reduce pH. However, the solubility of phosphorus may or may not follow the elegant graph constructed by Lindsay and Moreno. This graph was constructed assuming that the solubility of aluminum is governed by the compound $Al(OH)_3$, gibbsite; it is doubtful that this is the case. Moreover the solubility of aluminum in soils can be changed markedly by the presence of organic matter in the soil (Thomas, 1975).

Hargrove and Thomas (1981) showed that the solubility of aluminum in organic matter–soil, or organic matter–clay mixtures, was greatly reduced, roughly proportionately to the organic matter content. They also showed that phosphorus "fixed" by aluminum on organic matter is not nearly as insoluble as in phosphorus "fixed" by clay or hydrous oxides.

If organic matter is an important factor in the solubility of phosphorus, no-tillage could be expected to enhance its availability because of the higher organic matter contents found at the soil surface under no-tillage. On field plots at Lexington, Kentucky we found, after 10 years of no-tillage, the organic matter content of the surface centimeter of soil to be 12%, compared with about 3% on plow-tilled soil. This suggests, but does not prove, that organic matter could play a large role in phosphorus availability under no-tillage.

The foregoing discussion leads to the following plausible (but as yet unproven) reasons for higher phosphorus availability in soils under no-tillage:

1. There is more water under no-tillage, so that diffusion of phosphorus to plant roots is higher.
2. The pH of soil is generally reduced under no-tillage, which improves phosphorus availability in soils with pH above 6.5, but reduces it in soils below about 6.5.
3. The presence of a higher level of organic matter in soils under no-tillage suppresses aluminum solubility and tends to enhance phosphorus availability.

The result of these conclusions is that phosphorus fertilizer rates required should be less in all soils with no-tillage. A further reason for this conclusion will be discussed in the section that follows.

Placement of Phosphorus

When work on no-tillage began, one of the earliest questions asked was "How will we apply phosphorus?" It must be remembered that in the early 1960s, the prevailing practice was to band phosphorus fertilizer "2 in. to the side and 2 in. below" the seed, even though, given the equipment available, this was seldom ever accomplished. The question of phosphorus placement in no-tillage was unanswered. We selected a site on Brush Mountain above the appropriately named town of Poverty, VA. The soil must have given the town its name because it registered approximately zero in soil test phosphorus. The state of the native vegetation proved that the soil tests were not misleading. On one set of plots we incorporated fertilizer phosphorus with ^{32}P tracer and on the other set we simply applied it to the soil surface, a soil surface that had a cover of red clover, *Trifolium pratense*, L., killed with a massive dose of atrazine. Corn yields were pretty terrible, but the P uptake data are of considerable interest (Table 4.4). Note that the surface-applied phosphorus gave higher uptake for the first 60 days as well as a higher proportion of phosphorus coming from the applied fertilizer (at 30 days, more than three times as much). Surface-applied phosphorus fertilizer is readily available to plants under no-tillage. There was no comparison of banding phosphorus fertilizer with surface broadcasting.

Belcher and Ragland (1972) compared surface-broadcast and surface-broadcast plus banded application of phosphorus on no-tillage corn. Their results showed no statistical difference but a slightly higher yield with only surface broadcast at low rates. Triplett and VanDoren (1969) showed results indicating better uptake of phosphorus from surface-applied phosphorus at an early (8–10 leaf) growth stage of corn. To the author's knowledge, there is no evidence that surface-broadcast phosphorus is not equal to or better than other methods of phosphorus fertilization. However, there may be situations where deep placement of phosphate is advantageous.

TABLE 4.4. Percentage of P from Fertilizer and Percentage of P in Corn Plants

Days after Planting	% P from Fertilizer		% P in Plant	
	No-Tillage	Plow-Tillage	No-Tillage	Plow-Tillage
30	54	16	0.07	0.04
46	43	32	0.18	0.18
60	25	21	0.16	0.13
67	36	37	0.15	0.15

SOURCE: Singh et al. (1966).

Phosphorus Profile in the Soil

When phosphorus is applied yearly to a soil that is not disturbed, the phosphorus profile will show very high levels at the soil surface and decline rapidly with depth. Because this is so strongly at variance with the phosphorus profile in plowed soils, where phosphorus is relatively uniform in the plow zone and then drops to a low level below, many people have suggested that plowing to mix the phosphorus with a larger soil zone is desirable, if not necessary. This is a difficult problem to address because of the multiplicity of climates and soils in which no-tillage is practiced. It certainly is conceivable that a soil subjected to a sustained drought could supply less phosphorus to plants if the phosphorus were concentrated at the soil surface. On the other hand, we have experienced many years of extreme drought while studying no-tillage with no indication of phosphorus problems due to surface application. It seems rather more likely that the fervor for mixing phosphate with soil has more to do with the fervor for mixing soil. Old habits die hard.

One thing to be said in favor of surface application of phosphorus (high levels of phosphorus at the surface) is that organic matter level and water profiles quite often are similar. This means that the phosphorus is located in a zone of high organic matter, which tends to enhance phosphorus uptake by increasing phosphate solubility in the soil solution. If, in addition, there is more water in the upper part of the soil, phosphate uptake will also be enhanced by the increased diffusion rate of phosphate to the plant root.

In practical terms, deep placement of phosphate to ensure availability during drought probably does not pay. Not only will a drought severe enough to impede phosphate uptake limit yield, but availability of plowed-down phosphate will likely be so much less than surface-applied phosphate that the so-called insurance value will simply disappear with time. Certainly, experience with perennial crops such as alfalfa, *Medicago sativa* L., or grasses does not lead to the conclusion that deep placement of phosphorus is necessary. Top dressing has served for a long time as the major means of adding phosphorus to such crops, indicating that roots grow where the nutrients are, assuming water is sufficient for survival.

Roots of row crops explore the soil where nutrients are located. When phosphorus and water are located near the surface of the soil, root density of corn, for example, will be very high there (Phillips, et al., 1981). On the other hand, this does *not* mean that all the roots will be located there and that in a sudden drought the plants will be destroyed. On the contrary, the overall water withdrawal by both corn and soybean *Glycine max* L., roots from soils where they have been grown under both systems is remarkably similar. In fact, in most cases, except for the surface soil, where the lower water content of plow-tilled soil appears to be due to higher evaporation, the water withdrawal profiles of no-tillage and plow-tilled crops are essentially identical.

POTASSIUM

General Behavior

Potassium, a cation with an affinity for soil which ranges from medium to very high, usually is held tightly by soils when present in small amounts. Therefore, its leachability is low except on sandy soils which lack exchange capacity. In a 2-year lysimeter study under conventional and no-tillage corn, Tyler and Thomas (1977) found that hardly any potassium was lost with either system. In the same soil, Blevins et al., (1983b) found after 10 years that potassium level was the same in the tilled and no-tilled Maury soil in Kentucky. Distribution with depth was markedly different, however (Table 4.5). As with the phosphorus, potassium is high at the surface when the soil is not disturbed.

Looking at potassium from the standpoint of its availability in the soil, there is a tendency for its affinity for the soil to decrease exponentially with increasing amounts. If one wants to compare affinities of calcium and potassium, for example, the exchange reaction between these two cations can be written as

$$\text{Ca} - \text{soil} + \text{K} \longrightarrow \tfrac{1}{2}\text{Ca}^{2+} + \text{K} - \text{soil}$$

Arranging this in the form of the Gapon equation gives

$$\frac{(\text{K})\,[\text{Ca}^{2+}]^{1/2}}{(\text{Ca})\,[\text{K}^{+}]} = \text{k}_G$$

where k_G is the Gapon exchange coefficient. The larger the value of k_G, the more tightly potassium is held by the soil. On four Texas soils (Table 4.6) the value of k_G drops from 20 to 30 to 4 to 6 as the percent potassium ([K/CEC] $\times$ 100) increases to 8% or more. This implies that there are some sites that specifically bind potassium. When these sites are filled, potassium is held about as tightly as calcium.

TABLE 4.5. Exchangeable Potassium in Maury Silt Loam After 10 Years of Plowed and No-Tillage Corn

Depth (cm)	No-Tillage (meq/100 g)	Plow-Tillage (meq/100 g)
0–15	0.68	0.39
5–15	0.34	0.43
15–30	0.21	0.29

SOURCE: Blevins et al. (1983b).

TABLE 4.6. Values of k_G (Gapon Constant) for Different (K_{exc}/CEC) Values for Some Texas Soil Clays

	(K) (CEC)		
Soil name	0.02 $kg \cdot L/(mol)^{1/2}$	0.04 $kg \cdot L/(mol)^{1/2}$	0.08 $kg \cdot L/(mol)^{1/2}$
Houston Black	20	12	8
Montell	20	12	8
Miller	25	12	8
Beaumont	6	5	4

SOURCE: Knibbe and Thomas (1972).

TABLE 4.7. Effect of Tillage on Percentage of P and K in Corn Plants

Nutrient	Plow-Tillage (%) 8–10 leaf stage	No-Tillage (%) 8–10 leaf stage	Plow-Tillage (%) After tasseling	No-Tillage (%) After tasseling
P	0.29	0.35	0.27	0.27
K	3.92	4.20	2.00	2.06

SOURCE: Triplett and Van Doren (1969).

This type of behavior implies that the banding or localization of potassium should make it more highly available to plants. Higher concentrations of potassium in less soil should lead to a situation where at least some of the potassium is held relatively loosely by soil. There is some evidence to support this view; work by Rasnake and Thomas (1976), for example, shows that the value of k_G (from the equation above) is inversely related to potassium uptake in a group of Kentucky soils. Exchangeable K itself was less well related to potassium uptake. The quantity–intensity relationships of Beckett (1964) for potassium are simply another way of expressing the same idea. In his plots, the slope of the line relating "change" in exchangeable potassium with potassium activity ratio ($[K^+]/[Ca + Mg]$) in solution is related directly to the value of k_G *and* cation exchange capacity. The *change* in the slope with increased potassium applied is related to the change in k_G.

In no-tillage with the potassium applied primarily to the soil surface, there is formed near the surface a band of potassium at a higher level than in the bulk of the soil. If the ideas presented above are valid, higher potassium uptake should occur. Skimpy evidence suggests that this is true. Work by Triplett and VanDoren (1969) (Table 4.7) showed greater uptake of both potassium and phosphorus with no-tillage than plow tillage. However, there is less than adequate support for this idea since few data exist. In addition, with no-tillage we are not dealing with a simple system. Many things change, including distribution of potassium, water content, and temperatures.

Potassium Diffusion

As with phosphorus, potassium diffusion to plant roots is important for normal plant uptake. Without doubt, the major factors influencing diffusion rate are the level of potassium in the soil solution and the ability of the soil to maintain that level over an extended period of time. A means of predicting these factors is with the Gapon equation or the Beckett *Q/I* graphs of soils already discussed. A further important factor, as with phosphorus, is the amount of soil water present. These principles have already been discussed thoroughly in the section on phosphorus. To review, since potassium diffusion can only occur in water, the more water present (or the more continuous and broad the path), the more potassium can be expected to diffuse. As with phosphorus, it is often noted that potassium deficiencies are more striking with severe drought.

To summarize, potassium diffusion rate depends on the level of potassium in the soil solution, the dependability of its renewal, and the amount of water available. The first two factors are governed by soil potassium content and the exchange reaction coefficient, while the third is a matter of soil water present. The factors changed by no-tillage are soil potassium level and soil water content. Since both are changed in the direction of increasing diffusion, one would suppose that potassium uptake would always increase under no-tillage. This is not always the case.

Potassium Uptake in Cold, Wet Soils

In the upper Midwest of the United States especially, there has been concern for adequate potassium uptake under no-tillage. This concern arises from cold soil temperatures often encountered at corn planting time. The effect is also thought to be related to higher bulk density and/or excessive soil water. This seems to be largely a plant-related uptake problem and can be overcome to some extent by raising the level of potassium in the soil (Moncrief and Schulte, 1982).

Table 4.8 shows uptake of both potassium and phosphorus with three tillage systems in Wisconsin. It is apparent that, unlike phosphorus, potassium was absorbed less readily under no-tillage conditions. The poorer uptake of potassium responds to row placement of potassium, which is relatively easy to do with no-tillage and also to higher levels of potassium in the soil. Therefore, whichever approach is cheaper would be suitable for areas where cold, wet soils cause problems with potassium uptake.

In cold areas, poor uptake of potassium under no-tillage is a recurring problem but not always the rule. Further south, this problem is rare indeed. In Kentucky, this behavior has been noted only once, when corn planted on a soil medium to high in potassium failed to grow and showed potassium deficiency. In a part of the field that had been fertilized heavily with potassium for a previous tobacco crop, the corn grew normally and was high in potassium. On

TABLE 4.8. Uptake of P and K by Corn as Affected by Tillage, Fayette Silt Loam, Wisconsin

	Uptake[a] (mg/plant)	
Tillage	Phosphorus	Potassium
No-tillage	1.80	12.3
Chisel	1.35	13.8
Moldboard plow	1.22	15.9

[a]At soil test level of "high."

SOURCE: Moncrief and Schulte (1982).

returning some 2 weeks later, corn was growing normally in all parts of the field with almost no difference in plant size. This shows how important weather and attendant soil conditions can be in potassium uptake.

Despite the extensive investigations of Moncrief and Schulte (1982), it is impossible to say at present the exact conditions that control potassium uptake. As of now we can only say that slowed uptake can be expected when temperatures are cool and soils are excessively wet and/or compacted. Conditions such as these are more apt to occur with no-tillage, than plow-tillage during the spring.

Summary

Potassium, which is held relatively tightly by soils, does not tend to be lost more readily under no-tillage than under plow-tillage. In fact, its distribution in the soil when no soil mixing is done, tends to be higher in the surface few centimeters than when plowed. However, under some conditions, notably cold, wet, soils, early in the spring, there is sometimes a problem with potassium uptake by crops. This can be overcome with higher rates and/or banding of potassium. It appears to be a plant-related problem occurring most frequently in excessively wet soils.

In general, potassium is a nutrient little affected by tillage. Compared to other nutrients, potassium can be managed under no-tillage much as it is with plow-tillage. An exception may be in soils that are both wet and cold.

CALCIUM

Leaching of Calcium

Calcium is the cation most abundant is most agricultural soils. As such, it is susceptible to leaching with anions such as nitrate, chloride, and sulfate as well as a normal loss as bicarbonate. Therefore, when high rates of nitrogen are applied, there is always more loss of calcium than before. Most nitrogen is applied as ammonia, urea, or ammonium nitrate. All of these fertilizers form hydrogen ions when oxidized by bacteria to nitrate. The biological reactions can be shown as follows:

$$NH_4^+ + 1\tfrac{1}{2}O_2 \longrightarrow O_2^- + H_2O + 2H^+$$

$$NO_2^- + \tfrac{1}{2}O_2 \longrightarrow NO_3^-$$

Calcium nitrate will simply remain in the soil solution unless an excess of water moves it down and out of the soil. In humid regions of the United States, all soils tend to lose their nitrate (and calcium) during winter and spring, times of high rainfall and relatively low evapotranspiration. With no-tillage, and a mulch present, this period of potential leaching loss extends longer into the spring and early summer. Tyler and Thomas (1977) and Thomas et al. (1973), for example, found substantial leaching losses of nitrate with no-tillage during the month of June. Plow-tillage, with the same precipitation, lost almost no nitrate. Tyler and Thomas (1977) also showed that calcium was the cation accompanying nitrate, for the most part. Losses of other cations were small.

Basically there are two causes of calcium leaching loss during spring and early summer. The first, probably most important, is that under no-tillage, with a cover of plant residues killed by herbicides, there is much less evaporation of soil water. Spring-planted crops are still small so transpiration is not a large factor. Thus, when rain occurs, there is more tendency for leaching from a wet soil than a dry soil. The normal winter–spring period of leaching is extended into early summer owing to the surface mulch of dead plant residue.

The second cause of greater calcium loss through leaching is related to the natural soil structure that is found under no-tillage. As a number of publications have shown (Thomas et al., 1973; McMahon and Thomas, 1974; Thomas and Phillips, 1979) in soils much of the incoming water moves through the

TABLE 4.9. NO_3 Loss from No-Tillage and Plow Tilled Soils Using Lysimeters

	NO_3-N leached	
Month	No-Tillage	Plow-Tillage
January	9	11
February	1	1
March	13	10
April	7	4
May	8	3
June	18	6
July	0	0
Total	56	35

SOURCE: Tyler and Thomas (1977).

cracks, structural faces, and holes in the soil. Naturally, when a soil is undisturbed, as with no-tillage, this tendency is increased markedly. The result is that calcium nitrate located at the surface of the soil can be carried out of the soil with much less water. Table 4.9 shows the effect with lysimeters for a plowed and no-tillage corn soil (Tyler and Thomas, 1977).

Together these two effects cause the loss of more calcium from a soil under no-tillage than from the same soil under plow-tillage. Table 4.1 (Blevins et al., 1983a) shows the loss of calcium from a 30-cm profile over a 10-year period in Kentucky. Losses from the surface layers with no-tillage averaged more than twice the losses from plow-tillage. Unless the calcium is replaced by salts or liming this, of course, causes an increase in soil acidity.

Soil Acidity

Whenever basic metallic cations are exchanged from the soil by hydrogen ions, there results an increase in soil acidity. This change can be measured by a simple pH taken in a soil water paste. Table 4.10 shows the pH of soils with various nitrogen treatments after 10 years of no-tillage and plow-tillage. It is evident that the drop in pH is much more pronounced with no-tillage, especially at the soil surface. The extremely low pH at the soil surface is a consequence of not mixing the soil by tillage. However, at greater depths the pH is still lower under no-tillage suggesting (Table 4.1) that the calcium, once it left the soil surface, left the entire profile.

Other measurements of soil acidity are exchangeable aluminum and exchangeable manganese. These values (Blevins et al., 1983b) are shown after 5 years of no-tillage and plow-tillage in Table 4.11. It is evident that exchangeable aluminum and manganese are closely related to pH (Table 4.10) and that all three measurements indicate greater soil acidity with no-tillage than with conventional tillage.

Generally speaking, the most serious problem caused by acidity in soils is the toxicity of aluminum. Manganese is a secondary and at times equally serious problem depending on the soil content of manganese. Both of these metal

TABLE 4.10. Soil pH of 0–5 and 5–15-cm Layers After 10 Years of Corn as Affected by Nitrogen and Tillage

Yearly N rate (kg/ha)	No-Tillage		Plow-Tillage	
	0–5 cm	5–15 cm	0–5 cm	5–15 cm
0	5.75	6.05	6.45	6.45
84	5.20	5.90	6.40	6.35
168	4.82	5.63	5.85	5.83
336	4.45	4.88	5.58	5.43

SOURCE: Blevins et al. (1983b).

TABLE 4.11. Exchangeable Al and Mn After 10 Years of Corn as Affected by Tillage and Nitrogen Rate

Nitrogen rate (kg/ha)	Al (meq/100 g)		Mn (meq/100 g)	
	NT[a]	PT[a]	NT	PT
0	0.06	0.03	0.02	0.01
168	1.22	0.04	0.10	0.02
336	2.50	0.14	0.19	0.03

[a]NT and PT = no-tillage and plow-tillage, respectively.

SOURCE: Blevins et al. (1983b).

TABLE 4.12. pH of Tilsit Silt Loam Soil After 5 years of Variable Lime Rates Under No-Tillage and Plow-Tillage

Soil depth (cm)	pH					
	No-Tillage			Plow-Tillage		
	0	3.4	10.1	0	3.4	10.1
0–5	4.6	5.5	6.4	5.2	5.5	5.7
5–10	5.5	5.7	6.4	5.4	6.0	6.5
10–20	5.5	5.6	5.9	5.6	5.8	6.6
20–30	5.0	5.1	5.2	5.2	5.3	5.5

[a]Surface applied, not mixed.

SOURCE: Blevins et al. (1978).

cations increase exponentially in availability below a pH of about 5.5. Therefore, pH is an excellent indicator of likely problems once it drops below 5.5.

Another effect of soil acidity which can be serious with no-tillage corn is the inactivation of triazine herbicides with low pH (Kells et al., 1980). In our long-term experiments, in fact, this was the first evidence of low pH, lower yield and poor control of weeds.

Liming No-Tillage Soils

As has been noted previously (Table 4.10) the most acid part of soil under no-tillage is the surface few centimeters. Thus, it appears possible to apply lime to the soil surface without mixing it in, as one might do on a pasture or an alfalfa meadow. Results of lime application on a western Kentucky site are shown in Table 4.12. It can be seen that the lime effect is moderately shallow (<15 cm), but it corrected the serious problem of low pH at the soil surface. Hargrove et al. (1982) found a more acid subsoil in their Georgia work on which surface application of lime did not improve this condition. It is probable that with acid subsoils some deeper mixing of lime should be done occasionally, especially if the subsoil pH is declining.

In soils where pH changes are greatest near the surface, the application of lime without mixing will be sufficient to correct the problem. Probably most important is to make fairly frequent soil tests to maintain proper soil acidity. With this surveillance, soil pH will not be a problem with no-tillage. There will be, in humid areas, a slightly higher use of lime under no-tillage conditions than with plow-tillage. This tendency will be greater as the rate of nitrogen increases. Even at very high nitrogen rates it is doubtful that liming alone would be an economic factor important enough to discourage the use of no-tillage.

SECONDARY AND MICROELEMENTS

Magnesium

The behavior of magnesium in soils is similar to that of calcium except that, in most soils, it is present in much less quantity. While it has about the same affinity for soil and therefore should be lost by leaching roughly proportionately, data never seem to show this. For example, after 10 years of continuous corn the percentage of calcium in no-tillage compared with plow-tillage with 168 kg N/ha was 42%. With the same treatment, no-tillage magnesium was 80% that of plow-tillage. What accounts for the difference?

Clay minerals contain a variable but significant amount of magnesium in their structure. As a soil becomes acid, the clay begins to dissolve by the following reactions:

$$H_3 - \text{clay}\ [Al(OH)_3] \longrightarrow Al{-}\text{Clay} + 3H_2O$$

or

$$H_2 - \text{clay}\ [Mg(OH)_2] \longrightarrow Mg{-}\text{Clay} + 2H_2O$$

The first reaction produces exchangeable aluminum, the major problem of acid soils. The second reaction gives exchangeable magnesium, which is why exchangeable magnesium levels do not fall as fast as calcium levels in most soils. An exception is sandy soil where there is no source of mineral magnesium. In certain soils, the quantity of exchangeable magnesium increases as the soils acidify. This indicates the clays are high in mineral magnesium, which is dissolved and becomes exchangeable.

Because of the continuous source of magnesium, no-tillage poses little threat to exchangeable magnesium levels except in very sandy soils. Hence, it would be very unlikely to expect a response from magnesium-containing lime or magnesium salts under no-tillage unless the soil were sandy. And here, a response might also be expected with conventional tillage.

Sulfur

There are three major sources of sulfur in soils (a fourth, if irrigation is practiced). They are organic matter, soil minerals, and air pollution. Of these three, only soil organic matter is affected by changing from plowing to no-tillage. As pointed out earlier, under no-tillage, soils tend to lose less or to gain organic matter. Since sulfur in organic matter is only made available to plants upon decomposition releasing SO_4^{2-}, there is a possibility of greater sulfur deficiency under no-tillage than plow-tillage. This is true especially when the air is clean and soils are low in sulfur-bearing minerals. Such conditions are found in the Northwest and Southeast United States. Therefore, the chances for response to sulfur additions should be greater in those areas for no-tillage than for plow-tillage. There have been reports from Washington and Minnesota of sulfur deficiency on wheat and corn, respectively, under no-tillage.

Treating soils for sulfur deficiency is conveniently done by adding gypsum at a moderate rate. Gypsum is inexpensive, if not free, in most parts of the country, since it is a by-product of the fertilizer industry. A typical rate would be 200 kg/ha providing 36 kg of S per hectare. Unless unusual leaching occurs, this is enough for any major crop. Although calcium is added as well, soil acidity will not be affected.

Zinc

Of the microelements, zinc and iron are those most often deficient in some areas. Practically nothing is known of iron problems with no-tillage, possibly because so little no-tillage is practiced where iron deficiency is a problem. With zinc there is only scattered evidence also. However, knowing better the principles of zinc deficiency it is possible to predict very generally the probability of its occurrence under no-tillage.

Zinc deficiency usually occurs in soils with a pH of 6.5 or greater and where organic matter is relatively low such as on the thin soils of Yucatan (pH 7.8–8). A familiar pattern is to find zinc deficiency where a soil has been scraped to level it or where it has suffered erosion. Because no-tillage tends to lower the pH of the surface soil and because it tends to preserve soil organic matter, it would be expected that zinc deficiency would be a lesser problem in soils under no-tillage.

While there is little evidence that this statement is true, it is interesting to record observations in central Kentucky. The limestone-derived soils often have a high incidence of zinc deficiency with corn. Over the past 15 years I have observed only one no-tillage field where zinc deficiency was evident. In that field corn was cut for silage, no cover crop was planted, and fairly serious erosion occurred. It seems safe to say, even from this small sample, that zinc deficiency will be no more prevalent and perhaps less prevalent with no-tillage compared to plow-tillage. Further observations and studies are warranted to sharpen these statements.

Molybdenum

Molybdenum is the one microelement that increases in availability as soil pH rises. For full availability, the pH must be slightly above 6 (Thomas and Hargrove, 1984). Because it is difficult to keep surface soils at this level under no-tillage, it might be expected that no-tillage crops would have more problems with molybdenum deficiency than conventionally grown crops, particularly with legumes and tobacco, crops sensitive to molybdenum deficiency. There is no evidence to date that this is a problem. Again, using principles, one would expect more problems with molybdenum with no-tillage than with conventional tillage.

Methods to overcome molybdenum deficiency vary from liming, to small additions of sodium molybdate to fertilizer, foliar spraying, mixing molybdenum with the inoculant for legumes, the last of which usually kills the bacteria. In any event, molybdenum deficiency can be countered with reasonably low expenditures.

SUMMARY

In this chapter an attempt has been made to discuss soil fertility issues (other than nitrogen) affected by no-tillage. It can be said with some confidence that phosphorus availability is enhanced by the use of no-tillage. This apparently is due to the "banding" effect of surface application and to the greater diffusion rate engendered by increased soil water.

With potassium, the picture is not as clear. Surface application and higher water content should generally make potassium more available under no-tillage. But, it appears that under some wet and cold conditions potassium uptake under no-tillage is less. This condition apparently can be overcome either by unusually high potassium application rates or by banding potassium near the seed. Also, potassium deficiency is by no means universal, even in cold, wet soils.

Soil acidity becomes greater faster under no-tillage, especially at the soil surface. It can be corrected with surface-applied lime. The management of low pH values requires frequent soil testing and application of lime before damage to yield occurs.

Of the secondary and microelements, magnesium is little affected except, perhaps, in very sandy soils; sulfur is likely to be less available from the soil organic matter. Zinc tends to be more available due to higher organic matter and lower pH, while molybdenum (due to the lower pH) might be less available.

All in all, soil fertility under no-tillage is affected strongly by more water in the soil and a high level of organic matter which decomposes more slowly. Add to this induced acidification and lower temperatures encountered in spring and one can well deduce the type of soil environment that will result. Data are scarce on many aspects of mineral nutrition, and more research is required to confirm some current deductions.

LITERATURE CITED

Barber, S.A. 1962. A diffusion and mass-flow concept of soil nutrient availability. *Soil Sci.* **93**:39–49.

Baldovinos, F. and G. W. Thomas. 1967. The effect of soil clay content on phosphorus uptake. *Soil Sci. Soc. Am. Proc.* **31**:680–682.

Beckett, P. H. T. 1964. Studies on soil potassium II. The immediate *Q/I* relations of labile potassium in the soil. *J. Soil Sci.* **15**:9–23.

Belcher, C. R. and J. L. Ragland. 1972. Phosphorus absorption by sod-planted corn (*Zea mays* L.) from surface applied phosphorus. *Agron. J.* **64**:754–756.

Blevins, R. L., L. W. Murdock, and P. L. Cornelius. 1977. Influence of no-tillage and nitrogen fertilization on certain soil properties after five years of continuous corn. *Agron. J.* **69**: 383–386.

Blevins, R. L., L. W. Murdock, and G. W. Thomas. 1978. Effect of lime application on no-tillage and conventionally tilled corn. *Agron. J.* **70**:322–326.

Blevins, R. L., M. S. Smith, G. W. Thomas, and W. W. Frye. 1983a. Influence of conservation tillage on soil properties. *J. Soil Water Cons.* **38**:301–305.

Blevins, R. L., G. W. Thomas, M. S. Smith, and W. W. Frye, 1983b. Changes in soil properties after 10 years of no-tillage and conventionally tilled corn. *Soil Tillage Res.* **3**:135–146.

Hargrove, W. L., and G. W. Thomas. 1981. Effect of organic matter on exchangeable aluminum and plant growth in acid soils. Chapter 8 *Chemistry in the Soil Environment.* American Society Agronomy, Madison, Wisconsin.

Hargrove, W. L., J. T. Reid, J. T. Touchton, and R. N. Gallagher. 1982. Influence of tillage practices on the fertility status of an acid soil double-cropped to wheat and soybeans. *Agron. J.* **74**:684–687.

Kells, J. J., C. E. Rieck, R. L. Blevins, and W. M. Muir. 1980. Atrazine dissipation as affected by surface pH and tillage. *Weed Sci.* **28**:101–104.

Knibbe, W. G. J. and G. W. Thomas. 1972. Potassium-calcium exchange coefficients in clay fractions of some vertisols. *Soil Sci. Soc. Am. Proc.* **36**:568–572.

Lal, R. 1981. No-tillage farming in the tropics. In *No-tillage Research: Research Reports and Reviews.* University of Kentucky, pp. 103–151.

Lindsay, W. L., and E. C. Moreno. 1960. Phosphate phase equilibria in soils. *Soil Sci. Soc. Am. Proc.* **24**:177–182.

Mahtab, S. K., C. L. Godfrey, A. R. Swoboda, and G. W. Thomas. 1981. Phosphorus diffusion in soils: I. The effect of applied P, clay content and water content. *Soil Sci. Soc. Am Proc.* **35**:393–397.

Marelli, H., B. M. deMir, and A. Lattanzi. 1981. La temperatura del suelo y su relacion con los sistemas de labranza. Informe Especial 14. EERA. Marcos Juarez, Argentina.

McMahon, M. A., and G. W. Thomas. 1974. Chloride and tritiated water flow in disturbed and undisturbed soil cores. *Soil Sci. Soc. Am. Proc.* **38**:727–732.

Moncrief, J. F., and E. E. Schulte. 1982. The effect of tillage and fertilizer source and placement on nutrient availability to corn. Iowa's 34th Fertilizer and Ag. Chem. Dealers Conf., Iowa State Univ., Ames, Iowa.

NaNagara, T., R. E. Phillips, and J. E. Leggett. 1976. Diffusion and mass flow of nitrate nitrogen into corn roots grown under field conditions. *Agron. J.* **68**:67–72.

Olsen, S. R. and F. W. Watanabe. 1963. Diffusion of phosphorus as related to soil texture and plant uptake. *Soil Sci. Soc. Am. Proc.* **27**:648–653.

Phillips, R. E., G. W. Thomas, and R. L. Blevins. 1981. In *No-Tillage Research: Research Reports and Reviews*. University of Kentucky.

Rasnake, M. and G. W. Thomas. 1976. Potassium status of some alluvial soils in Kentucky. *Soil Sci. Soc. Am. Proc.* **40**:883–886.

Shenk, M. S. 1979. Unpublished data. CATIE Turrialba, Costa Rica.

Singh, T.A., G. W. Thomas, W. W. Moschler, and D. C. Martens, 1966. Phosphorus uptake by corn (*Zea mays* L.) under no-tillage conventional practices. *Agron J.* **58**:147–148.

Thomas, G. W. 1975. The relationship between organic matter content and exchangeable aluminum in acid soil. *Soil Sci. Soc. Am. Proc.* **39**:591.

Thomas, G. W. and W. L. Hargrove. 1984. The chemistry of soil acidity. *In* Soil Acidity and Liming. 2nd ed. American Society of Agronomy, pp. 3–56.

Thomas, G. W. and R. E. Phillips. 1979. The effect of macro-pores on movement on solutes in soils. *J. Environ. Qual.* **8**:149–152.

Thomas, G. W. and R. E. Phillips. 1981. Modeling soil water contents and their effects on stream flow in Kentucky. Research Report No. 128. Water Resources Research Institute, University of Kentucky.

Thomas, G. W., R. L. Blevins, R. E. Phillips, and M. A. McMahon. 1973. Effect of a killed sod mulch on nitrate movement and corn yield. *Agron. J.* **65**:736–739.

Triplett, G. B., Jr. and D. M. VanDoren, Jr. 1969. Nitrogen, phosphorus and potassium fertilization of non-tilled maize. *Agron. J.* **61**:637–639.

Tyler, D. D. and G. W. Thomas. 1977. Lysimeter measurements of nitrate and chloride losses from soil under conventional and no-tillage corn. *J. Environ. Qual.* **6**:63–66.

5

NITROGEN UTILIZATION WITH NO-TILLAGE

R. H. FOX
Professor of Soil Science
Department of Agronomy
Pennsylvania State University
University Park, Pennsylvania

V. A. BANDEL
Professor of Soil Fertility
Department of Agronomy
University of Maryland
College Park, Maryland

NITROGEN TRANSFORMATIONS

The conditions associated with no-tillage that affect the availability of N to plants are a modified soil environment that influences microbial activity and inorganic N movement, and reduced opportunities for incorporation of N fertilizers in potentially volatile forms. The major inputs and outputs of the soil inorganic N pool (Fig. 5.1) provide a framework to discuss the processes affected by the elimination of tillage and their influence on N availability to plants. The following section titles correspond to the processes identified in Fig. 5.1. Only no-tillage and conventional plow-tillage will be discussed in this chapter since they are the two extremes of the degree of tillage encountered. Minimum tillage covers a wide range of intermediate tillages, and N transformations and reactions in soils will be intermediate between those of no-tillage and plowed soils.

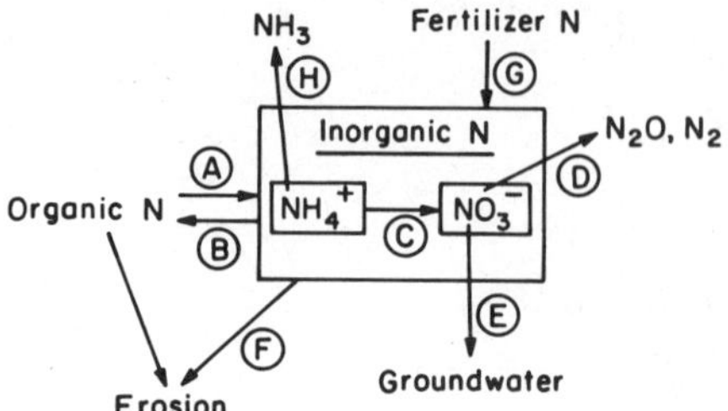

FIGURE 5.1. Inputs and outputs for the soil inorgnaic N pool that may be affected by tillage: A = mineralization, B = immobilization, C = nitrification, D = denitrification, E = leaching, F = erosion, G = fertilizer additions, H = volatilization.

Mineralization

Mineralization is the microbial transformation of organic N to inorganic N. As a result, both substrate and environmental factors control the mineralization rate in soils. Substrate properties influencing this rate are the total amount of soil organic N, the C/N ratio of the organic matter, and the nature of the organic matter (e.g., freshly added plant material versus aged humus) (Alexander, 1977; Jenkinson and Rayner 1977; Van Veen et al., 1981). Important environmental parameters are soil temperature, soil moisture status, and soil pH.

Most investigators have found that, compared to conventionally tilled soils, the organic N concentration in the surface 15 cm of nontilled soils is significantly greater (Blevins et al., 1977; Dick, 1983; Doran, 1980; Juo and Lal, 1979; Kupers and Ellen, 1970; Lal, 1976; Linn and Doran, 1984). Corn was the test crop in the majority of these experiments, but small grains were included in several reports. In contrast, Powlson and Jenkinson (1981) found that when changes in bulk density were accounted for, there were no significant differences in the organic N content (on a weight/weight basis) of the plow layer in three of four long-term (5–10 years) no-tillage versus plow-tillage experiments with small grains in the United Kingdom. Carter and Rennie (1982) also found that tillage did not affect the total C or N contents in several tillage experiments with wheat in the Canadian prairies.

There is general agreement, however, that no-tillage results in a redistribution of the organic matter within the plow layer. In the absence of tillage a relatively high concentration of organic matter accumulates in the surface few centimeters of soil, while concentration then decreases sharply with depth. The organic matter content of the plow layer is much more uniformly distributed with depth in conventionally tilled soils. Consequently, in comparison, no-tillage soils have considerably higher organic matter contents in the surface few centimeters of soil and lower contents in the remainder of the plow layer. Examples of increased organic matter contents and the redistribution of the organic matter with depth as a result of no-tillage in several long-term experiments are shown in Table 5.1 and Fig. 5.2. In the 18 and 19 year experiments

TABLE 5.1. Effect of Tillage on Average Soil Bulk Density and Organic C and N Contents from Six (Four Continuous Corn, Two Wheat) Long-Term (4–12-Year) Experiments

Soil Parameter	Ratio NT/PT[b] at Three Soil Depths[a]		
	0–7.5 cm	7.5–15 cm	15–30 cm
Bulk density	1.04**	1.05***	1.00
Organic C	1.41***	0.99	0.94*
Organic N	1.29***	1.01	0.96

[a]Levels of significance: * = $p < 0.05$, ** = $p < 0.01$, *** = $p < 0.001$.
[b]NT = no-tillage, PT = plow tillage.
SOURCE: Linn and Doran (1984).

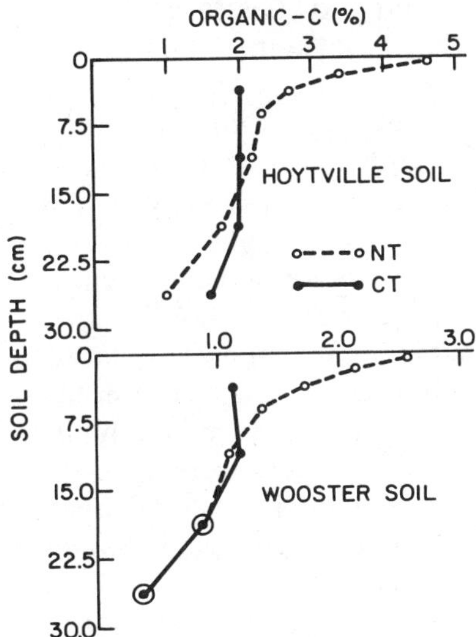

FIGURE 5.2. Organic C concentrations in soil profiles as affected by 18 or 19 years of conventional or no tillage (Dick, 1983).

with continuous corn shown in Fig. 5.2, the organic matter content of the surface 22.5 cm of soil was 16–19% higher in no-tillage soils than in plowed soils (Dick, 1983).

Converting to no-tillage will probably not greatly affect the nature or the C/N ratio of soil organic matter. Measurements of more labile pools of organic C and N, such as biomass and potentially mineralizable N, in studies comparing no-tillage and plow-tillage have shown that the size of these labile pools is highly correlated with the total soil organic C and N contents. Converting to no-tillage had approximately the same effect on them that it did on the total soil C and N contents (Bennett et al., 1975; Carter and Rennie, 1982; Doran, 1980).

The effect that this redistribution of, and often increased, organic matter content has on the availability of N to crops in untilled soils has not been fully established. The most common observation is that during the first few years after initiating no-tillage there is less N available to plants from soil organic matter in untilled soils than in plowed soils (Blevins et al., 1983; Davies and Cannell, 1975; Dowdell and Crees, 1980; Eylands and Gallaher, 1983; Phillips et al., 1980; Stanford et al., 1979; Vaidyanathan and Davies, 1980). The difference has been most notable in soils receiving little or no fertilizer N. There is no general agreement on the cause, or causes, for this reduction in N availability in untilled soils. Since in most long-term tillage experiments the soil organic N content in untilled soils became greater than in plowed soils after several years, a logical cause of the lower N availability would be a slower mineralization rate. Other possibilities, however, are higher immobilization rates or more denitrification or leaching of NO_3-N in untilled soils.

Comparisons of the microbial populations and activities in untilled and plowed soils in long-term tillage experiments have shown that aerobic microbial populations are correlated with total organic C and N levels in the soil (Doran, 1980; Linn and Doran, 1984). In other words, they found significantly more aerobes in the surface 7.5 cm and significantly fewer in the 7.5–15-cm layer of untilled compared to tilled soils (Table 5.2). Doran and Power (1983) concluded that the main cause of reduced availability of N in untilled soils is that the less oxidative environment leads to less nitrification and more denitrification. They also believed there was reduced mineralization in nontilled soils. Dowdell and Cannell (1975) also concluded that mineralization was reduced in undisturbed soils. They reached this conclusion after finding that there was more than twice as much NO_3^- in the soil profile under conventionally tilled winter wheat than with no-tillage wheat when soil conditions were not conducive to denitrification or NO_3^- leaching. Stinner et al. (1983) also found more inorganic N in the soil profile and greater crop N uptake from plowed soil for the first two years in a tillage experiment on a 12-year fallow field. In contrast, Rice (1983) found no significant difference in the NO_3^- content of the surface 30 cm of soil over a 2-year period in the long-term no-tillage versus plow-tillage continuous corn experiment in Kentucky. Other recent investigations have also failed to find any consistent difference in mineralization rates in laboratory incubations of intact cores (Rice et al., 1983) or soil samples (Deans et al., 1983) from tilled and nontilled fields.

Predicting how changed soil environmental conditions in untilled fields will affect net mineralization rate is difficult because the major no-tillage effects of increased water content (Blevins et al., 1971; Blevins et al., 1983; Doran, 1980; Hill and Blevins, 1973; Jones et al., 1969; Lal, 1974; Linn and Doran, 1984; Rice and Smith, 1982), and decreased temperature (Blevins et al., 1983; Lal, 1974; Linn and Doran, 1984; Willis, 1977) have contrasting effects on mineralization rate. The optimum water content for mineralization is approximately field capacity (Alexander, 1977). The normally drier, plowed soils would therefore have a lower mineralization rate than moister untilled soils, everything else being equal. Linn and Doran (1984) reported that the average water

TABLE 5.2. Average Soil Microbial Populations and Water Content as a Function of Tillage and Soil Depth in Several Tillage Experiments Throughout the United States

	Doran (1980)		Linn and Doran[a] (1984)	
Soil Parameter	Soil Depth, 0–7.5 cm	Soil Depth, 7.5–15 cm	Soil Depth, 0–7.5 cm	Soil Depth, 7.5–15 cm
	NT/PT Ratios[b]			
Total aerobes	1.35	0.71	1.19	0.72
Facultative anaerobes	1.57	1.23	1.36	0.98
Denitrifiers	7.31	1.77	—	—
Water content	1.47	0.98	1.24	1.05

[a]Microbial ratios are average of 1980 and 1982 surveys, water content, 1981 survey.
[b]NT = no-tillage; PT = plow-tillage.
SOURCE: Doran (1980); Linn and Doran (1984).

content of untilled soils was more conducive to both aerobic and anaerobic activity than in the drier, tilled soils in the six experiments they evaluated from around the United States.

The optimum temperature for mineralization is between 40 and 60°C (Alexander, 1977), so cooler untilled soils in spring would be expected to have a lower mineralization rate if temperature were the rate-limiting factor. Reduced mineralization rate due to lower soil temperature under mulch cover is thought to be one of the principal causes of lower rates of organic matter loss in untilled soils in the tropics (Juo and Lal, 1979; Lal, 1974). Lower O_2 concentration or aeration (Doran, 1980; Linn and Doran, 1984; Mielke et al., 1982) and acid soil surface (Bandel, 1979; Blevins et al., 1977; Dick, 1983; Fox and Hoffman, 1981) of untilled soils may also have an effect on the mineralization rate of organic N.

Doran (J. W. Doran, personal communication, 1984) believes that relative mineralization rates in soil planted to corn under the two tillage systems may behave as illustrated in Fig. 5.3. There is a peak of mineralization activity in tilled soil immediately after plowing when the soil has been disturbed and aerated. However, during the latter part of the growing season there is greater mineralization in untilled soil due to greater moisture availability.

One way to determine the net effect of environmental parameters on N mineralization in untilled versus plowed soils would be to carefully monitor them and use the observations in a proven simulation model of soil N transformations such as those discussed by Tanji (1982) and Molina et al. (1983). It would then be possible to observe how varying these parameters would affect predicted mineralization rates and N availability.

In summary, empirical evidence indicates that mineralization may be slightly decreased or not affected when tillage is eliminated. It is not possible to generalize on how the altered environment under no-tillage will affect mineralization rate because a number of the environmental changes have contrasting effects. In order to predict how converting to no-tillage will alter mineralization

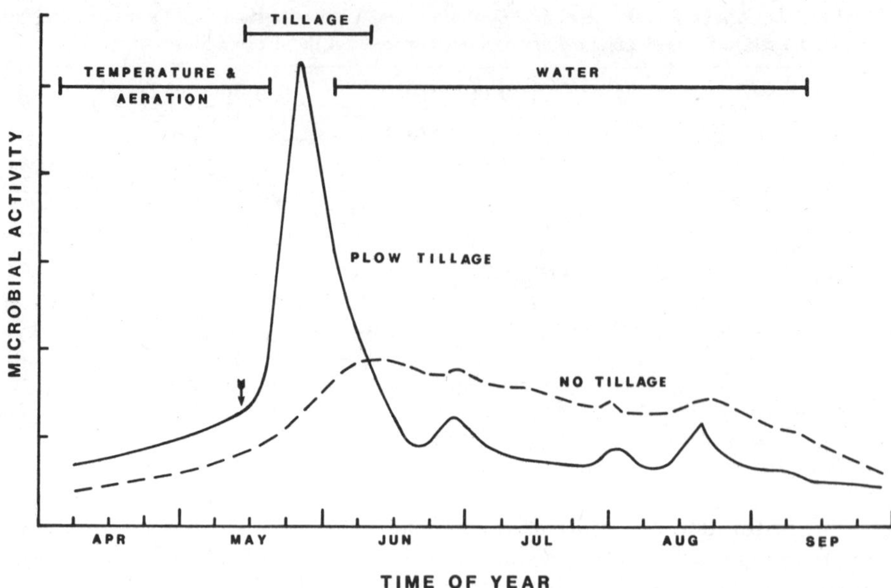

FIGURE 5.3. Hypothetical relationship between relative microbial activity and time of year in a plowed and no-till soil; factors controlling activity are shown on top of graph, arrow indicates time of plowing (J. W. Doran, personal communication).

rate in a given soil, it will be necessary to measure the induced changes in all environmental parameters affecting mineralization throughout the growing season and use this information in a suitable simulation model. Results will probably show that the net effect will vary under different soil and environmental conditions. For example, one would expect the effect no-tillage has on mineralization in a cool, moist climate and a clay-textured, poorly drained soil would differ from that observed in a warm arid climate and a coarse-textured, well-drained soil.

Mineralization of Residual Legume Nitrogen

Triplett et al. (1979) and Fox and Piekielek (1983) found tillage had no apparent affect on the residual N available to corn following alfalfa sods. However, because of the importance of this source of N in many areas and the possible reduction of mineralization rates in some untilled soils, further research should be conducted.

Immobilization

Immobilization is essentially the opposite process of mineralization in that it involves the microbial conversion of inorganic N into organic N. One of the principal causes of higher soil organic N levels and reduced N availability in untilled soils may be that immobilization is more rapid in nontilled soils (Rice and Smith, 1984; Smith et al., 1983). These investigators found that the short-term immobilization rate of surface applied fertilizer ^{15}N was much greater in nontilled than plowed soil under continuous corn. Total inorganic N concentrations were not reported in these experiments, however. As Jansson (1958) and Jansson and Persson (1982) have shown, because of the complexity of the mineralization–immobilization–turnover processes occurring in a soil, it is not possible to accurately measure mineralization or immobilization rates in a soil by only measuring the distribution of ^{15}N among soil pools after an elapsed time period. It is still possible, however, that immobilization rates can be greater in untilled soil. Doran (1980) and Linn and Doran (1984) demonstrated that larger microbial populations were present in the surface of their no-tillage soils (Table 5.2), and Rice and Smith (1984) and Smith et al. (1983) reported that differences between tillage systems in rate of immobilization of ^{15}N were sufficiently large that there may well have been real differences between immobilization rates. Greater microbial activity and rate of N turnover in the surface few centimeters of untilled soils would be expected because of the greater organic matter and biomass contents in this layer. However, even if there is greater microbial activity near the surface of untilled soil, whether or not tilling the soil causes a net increase in immobilization depends on the C/N ratio of the organic matter in this surface layer. Since the C/N ratio of surface litter in no-tillage grain crops is generally high, there is potential for greater immobilization.

Additional evidence supporting increased immobilization with no-tillage may be found in the work by Mengel et al. (1982), who measured larger corn yields and greater N recovery by no-tillage corn when fertilizer N was injected several centimeters below the soil surface. These investigators believed that the increased efficiency of injected N fertilizer was due to reduced immobilization resulting from fertilizer being placed below the high-organic matter-containing surface layer in untilled soils.

In contrast to these observations of apparent increased immobilization of N in no-tillage corn, Dowdell and Crees (1980) found very little difference in uptake of fertilizer ^{15}N by wheat from conventionally or nontilled soil. They concluded that tillage did not greatly affect immobilization in their UK soil.

Since optimum growing conditions are essentially the same for the microbial population immobilizing N as it is for the population mineralizing it, the preceding discussion on how the changes in the soil environment under no-tillage affect mineralization rate would apply equally well to immobilization rates.

In summary, there is evidence indicating greater immobilization of N in no-tillage corn, but available data are insufficient to conclude with certainty that greater immobilization does indeed occur and, if so, to what extent. On the other hand, there is also evidence that tillage had no effect on immobilization with some crops, soils, and climates. As with mineralization, obtaining precise measurements of soil organic matter and environmental parameters under different tillage systems and using them in a suitable simulation model could perhaps produce better estimates of the differences in net immobilization that would be expected.

Nitrification

Nitrification is the microbial conversion of NH_4^+ to NO_3^-. Unlike mineralization and immobilization, which are carried out by a wide variety of microorganisms, nitrification in natural ecosystems is effected only by nitrifying autotrophs of the family *Nitrobacteriaceae*. Nitrification is usually more rapid than mineralization and is therefore substrate-limited. Ammonium can accumulate to some degree in cold soils because of the greater sensitivity of nitrifiers to low temperature. However, the normal condition is that, except for a few weeks after an NH_4^+-containing or -producing amendment has been added to a soil, there is very little NH_4^+ present when a well-drained agricultural soil is warm and moist enough for biological activity. Another difference from mineralization and immobilization is that nitrifiers require O_2, which means that nitrification will be limited or stopped in poorly drained soils that have become anaerobic.

No evidence could be found in the literature indicating that the nitrification rate in untilled soil was consistently greater or smaller than that occurring in tilled soil. Several investigators have found lower NO_3^- contents in the soil profile under no-tillage (Dowdell and Cannell, 1975; Randall, 1980; Stinner et al., 1983; Thomas et al., 1973), but this has generally been attributed to lower mineralization rates, more NO_3^- leaching, or increased denitrification in the untilled soil. Caskey (1983) and Rice and Smith (1983) have reported that under controlled environmental conditions there were no consistent differences in nitrification rates in soil from continuous corn tillage experiments in Minnesota and Kentucky, respectively. Rice and Smith (1983) found that the NO_3^- to NH_4^+ ratios were generally lower in the no-tillage soils but that the total inorganic N contents under the two tillage systems were approximately equal. They found that the factor most often limiting nitrification was reduced moisture availability and, because of the generally greater moisture content in the untilled soil, nitrification rate in the field was often higher in the nontilled soils.

As with the microbial processes of mineralization and immobilization, it is difficult to predict the net effect of the altered environment of untilled soil on the nitrification process. The greater moisture content and fraction of water-filled pore space in untilled soils may limit the supply of O_2 needed by nitrifying bacteria in poorly drained and poorly aerated soils and thus reduce nitrification. However, in well-drained soils where moisture is sometimes limiting, the greater moisture associated with no-tillage will increase nitrification as was observed by Rice and Smith (1983). Optimum temperature for nitrification is usually in the range of 30–35°C (Alexander, 1977), so the effect on nitrification of lower temperatures in untilled soil will depend on whether the lower soil temperatures are closer or farther from this optimum than in tilled soils. Nitrifying bacteria are more sensitive to low soil pH than are mineralizing–immobilizing microbes, and the lower pH near the surface of no-tillage soils may reduce nitrification rates somewhat in this layer. However, there is no evidence that this occurs.

In summary, research to date indicates that the nitrification rate in most untilled soil is probably not markedly different from the rate observed in conventionally tilled soil in the same environment. Thus, any changes in crop response to N fertilizer or modification in soil organic matter depletion rates as a result of no-tillage can likely be better explained by alterations in one of the other N transformation processes under no-tillage conditions.

Denitrification

Denitrification as it most commonly occurs in soils is the microbial reduction of NO_3^- or NO_2^- to N_2O or N_2. Denitrification is carried out by several genera of aerobic bacteria that can use NO_3^- or NO_2^- as terminal electron acceptors in the absence of O_2 (Alexander, 1977; Firestone, 1982). These bacteria are classified as facultative anaerobes. Both energy and electron donors for soil denitrifying populations are normally supplied by soil organic matter. Thus, soil conditions conducive to denitrification are low O_2 concentration, presence of bacterial populations capable of denitrification, a source of energy (electron donors) such as organic C, the presence of NO_3^- or NO_2^- ions, and environmental conditions suitable for biological activity. These conditions most often occur in agricultural soils when the soil becomes saturated with water, the microbial population consumes the O_2 available in the pore space, and the system turns anaerobic. Denitrification can occur in soils at water contents below saturation. This may result from microsites within the soil, such as the interior of aggregates, being anaerobic and capable of supporting denitrification, even though the soil as a whole is not O_2 deficient.

The research group in Nebraska (Doran, 1980; Doran and Power, 1983; Linn and Doran, 1984; Mielke et al., 1982), which has extensively studied the environment and microbiology of soils in a number of long-term tillage experiments throughout the United States, believes that the greater water content, higher proportion of water-filled pore space, and significantly larger populations of facultative anaerobes and denitrifiers in untilled soils (Table 5.2) may lead to more denitrification. Other investigators have also found either a greater denitrification potential (Caskey, 1983; Rice and Smith, 1982) or observed lower NO_3^- contents in the soil profile with no-tillage treatments in poorly drained soils and attributed the loss to denitrification (Cannell et al., 1980; McMahon and Thomas, 1976; Randall, 1980).

Evidence thus indicates that in untilled soils, the greater moisture content, higher bulk density, greater proportion of water-filled pore space, and larger denitrifying population all contribute to a potential for greater N losses due to denitrification. These results also suggest that the probability of actual, as opposed to potential, denitrification loss is much higher in more poorly drained, clay-textured soils in a humid environment than it is for well-drained soils in an environment where excess soil moisture is seldom encountered. The lower soil temperature and pH in untilled soils probably have minor effects on denitrification rates compared to the moisture and aeration effects.

Leaching

Leaching is the downward movement of soluble compounds present in water percolating through the soil profile. Nitrate ions are very soluble in water, and their negative charge is not attracted to the negatively charged cation exchange surfaces in soils. Consequently, nitrate is particularly susceptible to leaching, and in nonsaline soils generally comprises the dominant anion in leachate waters (Black, 1968). Since most inorganic N present in soils is in the NO_3^- form, the potential exists for substantial losses of inorganic N from the soil profile when excess water percolates beyond the root zone.

The research group at Kentucky (McMahon and Thomas, 1976; Phillips et al., 1980; Thomas et al., 1973; Tyler and Thomas, 1977) has found more NO_3^- leaching in the continuous corn no-tillage plots than in the plowed plots and that this greater leaching often occurred in June. An example of their results is shown in Fig. 5.4. They believe that the greater loss is due to (1) increased water movement because of the higher moisture content of untilled soil during the early part of the growing season (reduced evaporation due to surface mulch and greater infiltration due to the higher moisture content) and (2) the larger number of continuous pores that develop with no-tillage and lead to rapid, channelized water movement (and NO_3^-) downward in the soil profile. Randall (G. W. Randall, University of Minnesota, personal communication) has also observed slightly greater losses of NO_3^- into tile drains with no-tillage treatments in a clay loam soil.

The lower soil temperature and lower pH in untilled soils would be unlikely to greatly affect NO_3^- leaching except as they influence the formation of NO_3^-. Lower soil temperature would contribute to lower evaporative loss in no-tillage, although it probably is much less important than the mulch layer in reducing evaporative moisture loss from untilled soils.

In summary, although there have been few studies comparing NO_3^- leaching losses in tilled versus untilled soils, all results indicate a potential for more loss with no-tillage. The greater water content of untilled soils favors greater leaching loss, but the often higher NO_3^- content in tilled soil may lead to greater NO_3^- leaching losses in these soils. Thus, if there is sufficient excess moisture to have leaching under both tillage systems, there may be more NO_3^- leaching loss in tilled soil. To predict whether there will be more or less leaching loss with no-tillage than plow-tillage on a given soil, however, it will be necessary, as with the previously discussed N transformations, to measure or estimate with a model how the altered environment of no-tillage affects the supply of NO_3^- available for leaching. For example, does the well-established

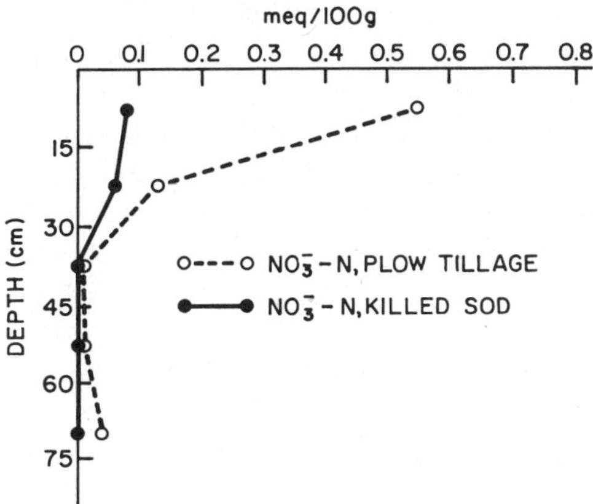

FIGURE 5.4. Nitrate N distribution in the profile of a Lowell silty clay loam 76 days after a June 1 application of 336 kg/ha of N as NH_4NO_3 to corn growing with conventional tillage or in a no-tilled, killed sod mulch (McMahon and Thomas, 1976).

greater moisture content of untilled soil lead to greater NO_3^- leaching or to greater denitrification?

Erosion

One of the greatest merits of no-tillage agriculture is its potential to reduce erosion losses (Chapter 2). With reduced erosion there will also be reduced losses of soil N. Juo and Lal (1979) believe that one of the principal contributors to less soil organic matter loss in their no-tillage treatments in West Africa was reduced erosion. Angle et al. (1984) found that losses of total and inorganic N from no-tillage corn were only 12% and 11% of that lost from tilled corn in a 3-year study of erosion and runoff losses on a silt loam soil in Maryland. McDowell and McGregor (1980) reported that total N lost by erosion from a loessial soil planted to tilled soybeans in Mississippi was 10 times greater than from untilled soil growing soybeans. Hardin (1978) reported a fivefold reduction in topsoil loss with no-tillage compared with tilled corn in Missouri. Comparable reductions in loss of soil organic matter would have undoubtedly also occurred in this experiment.

It can thus be safely concluded that there are significantly lower erosion losses of N from untilled soils. The magnitude of the reduction will depend on the amount of erosion that occurs under plow-tillage. The more loss, the greater the reduction that can be expected by converting to no-tillage.

NITROGEN FERTILIZER USE EFFICIENCY

Except for NH_3 volatilization from unincorporated urea or NH_4^--containing N fertilizers applied to the surface of nontilled soils, the effect of N fertilizer efficiency of converting to no-tillage agriculture will be determined by the interaction of many of the factors affecting soil N transformations and movement. For example, immobilization, denitrification, and leaching all affect added inorganic N in the same manner that they affect native inorganic N. Ammonia volatilization from unincorporated fertilizers in no-tillage agriculture, however, is a special problem discussed in a later section.

Results from experiments comparing the N fertilizer response or efficiency (yield or N uptake per unit of N fertilizer applied) between no-tillage and plow-tillage agricultures can be divided into five categories (Fig. 5.5). In the first category yield and/or N-uptake response to applied fertilizer N were the same in both tillage systems. Nitrogen fertilizer efficiency would therefore be similar in both systems. In the second, yields and/or N uptakes were lower with no-tillage at zero or low N fertilizer rates but the same amount of N fertilizer was required to attain a common maximum yield in both tillage systems. This would result in lower N fertilizer efficiency at low N rates with no-tillage but similar N efficiencies with both tillages at optimum N rates. In the third category, yields and N uptakes were lower with no-tillage at low N rates, but similar maximum yields were obtained with both tillages at higher N rates, although it required more N to attain this maximum with no-tillage. In this

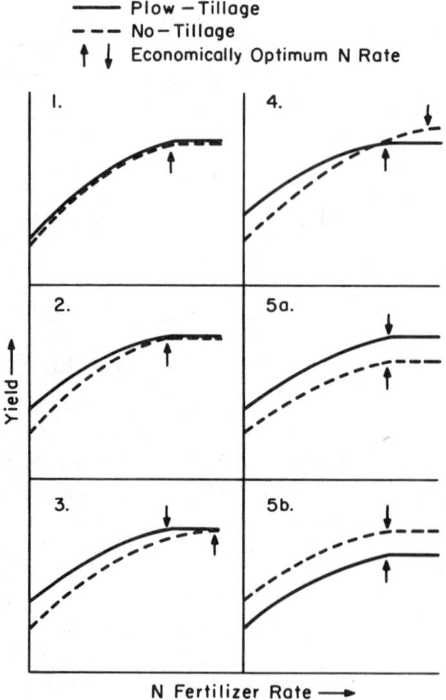

FIGURE 5.5. Five categories of comparative N fertilizer response between no-till and conventional-till. Numbers by individual graphs refer to categories described in the text.

category, N efficiency would be lower with no-tillage at all fertilizer rates. In the fourth, yields and N uptakes were lower with no-tillage at low N rates but at higher N rates both maximum yields and N fertilizer required were greater for no-tillage. The N fertilizer efficiencies for results that fell in this category would be less in no-tillage at lower N rates and either lower or similar for both tillage systems at economically optimum N rates. In the fifth situation, yields and/or N uptakes and N efficiencies were lower with one of the tillage systems at most N rates. This occurs when one of the tillage systems produced higher yields at all N rates. The difference between this category and category 3 are that the same maximum yields can be obtained with both tillage systems in category 3, but more N fertilizer is required for the untilled soil. In category 5 some factor other than N availability limits yield in one of the tillage systems and therefore similar maximum yields cannot be attained even at very high N fertilizer rates.

There are several examples of research reports where the N fertilizer response was the same under both tillage systems. Baeumer (1970) reported no apparent difference in N response in early no-tillage experiments in Germany. Triplett and Van Doren (1969) found tillage had no effect on the N concentration of ear leaves after tasseling in a 6-year N-rate experiment with continuous corn on an imperfectly-drained silt loam soil in Ohio. Reeves and Ellington (1974) reported that grain yield response to fertilizer N added to wheat was the same for both tillage systems in a 3-year experiment in Victoria, Australia. They also noted, however, that plant N uptake in the early part of the growing season was less in the no-tillage system. Ellis and Howse (1980) described very similar results in England. In their experiment, tillage had no effect on yield response to N fertilizer by small grains on three soils, but N concentration of plants grown on untilled soils was less during the early growth stages. In a 4-year experiment with continuous corn in West Virginia, Legg et al. (1979) found there was no difference in fertilizer response between the two tillage systems in 2 years when rainfall was adequate but that no-tillage yields were substantially higher in 2 years when rainfall was subnormal. Stanford et al. (1979), working with several rates of ^{15}N-depleted fertilizer for 4 years on two soils in Maryland, reported relatively little difference in corn yield or N uptake response between the two systems on the soil with the greater total N content.

The most commonly reported effect of tillage on N fertilizer response or efficiency has been lower N uptake or yield for no-tillage at low N rates, but equal yields at higher N rates. These experiments would be placed in category 2 in the classification scheme presented. An example from a long-term N response experiment with continuous corn in Kentucky is shown in Fig. 5.6 (Blevins et al., 1983). Others reporting similar results with continuous corn were Bandel (1979) and Stanford et al. (1979) in Maryland for the lower total N containing soil. Kitur et al. (1984) and Thomas et al. (1973) reported similar results in Ken'ucky. Lower wheat yields and/or N uptakes with no-tillage at low N rates but similar maximum yields and optimum fertilizer N rates with both tillage systems were noted by Cannell et al. (1980), in a wet year, and Dowdell and Crees (1980), both in the UK.

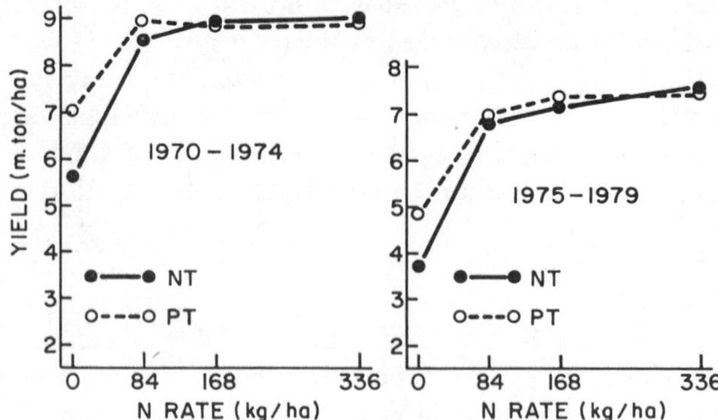

FIGURE 5.6. Corn grain yield with conventional and no-tillage at different N fertilizer rates for two time periods of a continuous corn tillage experiment. The second 5-year-period averages are for the limed plots (Blevins et al., 1983).

In summaries of no-tillage experiments in Europe, Davies and Cannell (1975) and Bakermans and DeWit (1970) found that, on the average, approximately 10 and 20–40 kg/ha more fertilizer N were required for similar optimum yields of small grains in the UK and the Netherlands, respectively. These results would be placed in category 3 in the classification scheme presented.

The second most commonly reported N fertilizer response with no-tillage compared to plow-tillage is lower yields and N efficiency at low fertilizer N rates and higher maximum yields at higher N rates with no-tillage. More N fertilizer is required for these higher no-tillage maximum yields, however. These experiments belong in category 4. Examples of this type of response in Kentucky and Maryland are illustrated in Figs. 5.7 and 5.8. A summary of tests over 46 location-years comparing the effect of tillage on N fertilizer response by corn in Maryland shows that the yields were higher with plow-tillage 70% of the time at the 0 kg/ha N rate, but at the 180 kg/ha rate the no-tillage corn produced greater yields than tilled corn 70% of the time (Table 5.3). Hardin (1978) and Kang and Yanusa (1977) also reported this type of interaction between tillage and N response. The same response was also found in the UK with wheat by Vaidyanathan and Davies (1980) and in a dry year by Cannell et al. (1980).

One explanation for this type response is that at low N rates, N availability is the factor most limiting yields. In the no-tillage system more N is lost by denitrification and leaching and there may be more immobilization and less mineralization of N. This results in lower yields in untilled soils than in tilled soils at the same N rate. At higher N rates the generally increased moisture availability in no-tillage increases crop yield potential and more fertilizer N is thus required for this higher yield potential (Phillips et al., 1980). Because of higher yields and N requirements with no-tillage corn, N fertilizer recommendations are 28 and approximately 40 kg/ha N higher than for conventionally tilled corn in Kentucky (Phillips et al., 1980) and Maryland (Bandel et al.,

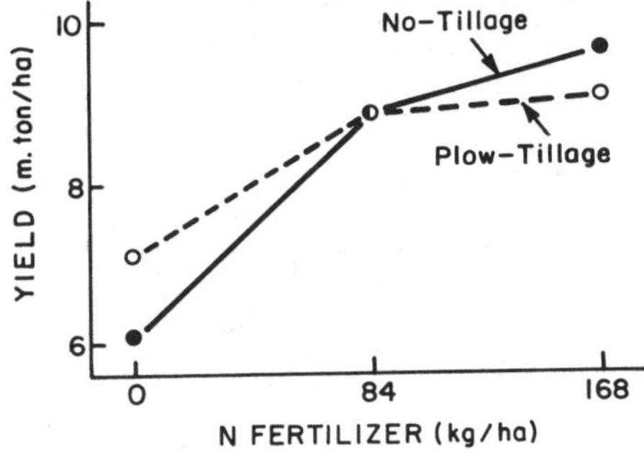

FIGURE 5.7. Influence of tillage on the average response of corn to nitrogen fertilizer in eight locations in Kentucky (Phillips et al., 1980).

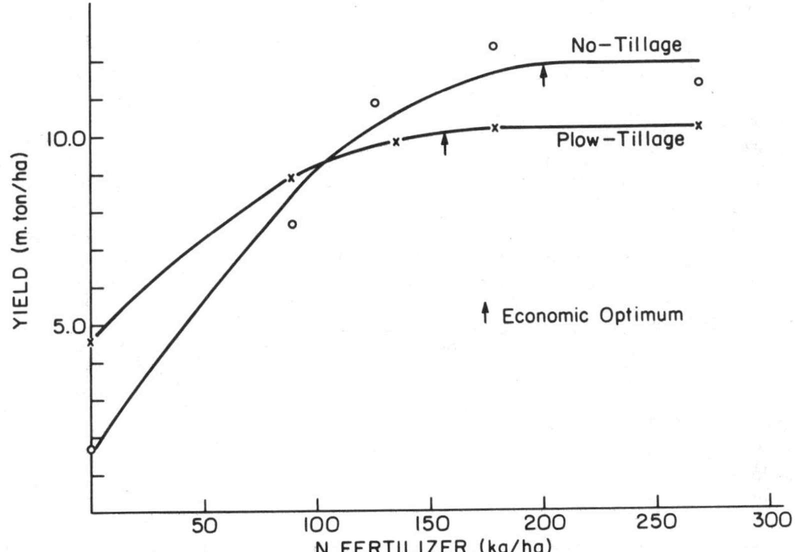

FIGURE 5.8. Corn grain yield as a function of N fertilizer rate and tillage, Poplar Hill Research Farm, MD, 1981 (Bandel, unpublished data).

1984), respectively. Whether the greater N needed for optimum yield in these cases is only due to the higher yield potential with no-tillage, or whether it is also needed to compensate for the greater N loss and reduced net mineralization, or both, is unresolved.

Using a quadratic-linear plateau response function, a fertilizer N:grain cost-price ratio of 2.78 (U.S. $/kg for both units; equivalent to a N cost of U.S. $0.30/lb and a corn grain price of $3.02/bu), and ignoring differences in application or harvest costs, the economic optimum N rates for the experiment in Maryland (Fig. 5.8) were 157 and 202 kg/ha for maximum economic yields of

TABLE 5.3. Influence of Tillage on Probability of Obtaining Maximum Corn Yield at Variable Nitrogen Rates at Five Maryland Locations, 1973–1983

Nitrogen Rate (kg/ha)	No-Tillage[a]		Plow-Tillage[b]		Total Tests
	Number of Tests	%	Number of Tests	%	
0	14	30	32	70	46
45	9	36	16	64	25
90	23	50	23	50	46
135	30	65	16	35	46
180	32	70	14	30	46
270	15	71	6	29	21
Mean	—	53	—	47	—

[a]Number of tests and percentage of times in which no-tillage corn out-yielded plow-tillage corn.
[b]Number of tests and percentage of times in which plow-tillage corn out-yielded no-tillage corn.
SOURCE: Bandel (unpublished).

10.1 and 11.9 mt/ha for plow-tilled and untilled corn, respectively. The grain produced per kg of N applied at the economically optimum rate would be 64 kg and 58 kg for the plow- and no-tillage corn, respectively, indicating a slight superiority of efficiency with plow-tillage at the economically optimum rate. Even though the N use efficiency is slightly less with no-tillage in these field experiments, the economic return from N fertilizer would be considerably higher. In this particular example, ignoring differences in all other costs, the economic gain from fertilizer N addition (grain value minus fertilizer cost at the economic optimum N rate) would be U.S. $183 per hectare greater with untilled than tilled corn.

The last type response reported in no-tillage by N rate experiments is where yields and/or N uptakes were lower with one or the other tillages at all N rates, Category 5. Fredrickson et al. (1982) found greater fertilizer ^{15}N uptake by no-tillage wheat in comparison to plow-tillage wheat at both N rates in their study in Washington. Moschler et al. (1972), Mochler and Martens (1975), and Bennett et al. (1975) found that no-tillage corn yields (and, consequently, N fertilizer efficiency) were higher at all N rates. In these latter experiments, greater yields and N efficiency with no-tillage were attributed to greater moisture availability in the untilled soils. These results are not surprising because when soil moisture deficiency limits yields, the likelihood of denitrification and leaching are minimal. In addition, increased soil moisture in untilled soil would also increase the mineralization rate of organic N. An example of greater N efficiency with conventional plow-tillage at all fertilizer N rates was reported by Smith and Howard (1980) who found that yields and N uptake by barley were greater with tilled soils at all four N fertilizer levels. It is likely that greater N fertilizer efficiency with plow-tillage as compared to no-tillage would mainly occur in poorly drained soils in humid environments where water deficiency generally does not limit yield, and where increased water content and reduced aeration in untilled soil leads to more denitrification, leaching, and less net mineralization at all N rates.

In summary, the following reasons are proposed for the widely divergent effects of eliminating tillage on N fertilizer yield responses and efficiencies in different climatic and soil environments. In warmer, drier environments where soil moisture deficiency usually limits yield, greater water availability with no-tillage would increase yield potential, and rarely create the wet, O_2-deficient conditions that result in N loss by denitrification in no-tillage. Consequently, under these conditions, the efficiency of N fertilizer will be equal to or greater with no-tillage than with plow-tillage at most N rates. With opposite conditions of a cool, moist climate and a heavy, poorly drained soil, the environment under no-tillage would be much more conducive to N loss by leaching and denitrification and to reduced mineralization of soil organic N. With these conditions, N fertilizer efficiency, and perhaps yield, would be expected to be lower with no-tillage at all N rates.

Soil environments between these extremes are typical of much of temperate agriculture; the soil is often wet and cool at the beginning of the growing season and moisture is marginally adequate during the latter part of the season. For these conditions, the following effects of no-tillage on availability and efficiency of fertilizer N are hypothesized. During the cool, wet, early portion of the growing season there would be more leaching and denitrification potential for both fertilizer and soil NO_3^-; less mineralization of organic N; and possibly more immobilization of fertilizer N in the untilled soil. This would lead to reduced availability of N to the crop during the early growing season, as observed by Reeves and Ellington (1974). Later, when soil moisture is lower and perhaps limiting to both crop growth and microbial activity in the tilled soil, the higher moisture availability in the untilled soil would contribute to increased mineralization of both native organic N plus that immobilized from fertilizer applied earlier in the year. Greater growth and increased response to the fertilizer N (as discussed above in describing categories 2, 3, and 4 type responses and illustrated in Fig. 5.6, 5.7, and 5.8) could be expected. The net effect of this would be reduced yield with no-tillage at low or zero N rates and equal or greater yield with no-tillage with high N rates. Whether yield and/or fertilizer N requirements would be greater with no-tillage at near optimum N rates depends on the balance between the larger losses and immobilization of N that occur at the beginning of the season and the greater plant growth and N availability from mineralization that take place later in the season in untilled soil.

Whether or not yields are improved, the net effect of no-tillage on crop production appears to be a reduction in native soil organic N loss due to less erosion and possibly lower net mineralization of organic matter. Therefore, when yield, optimum fertilizer N rate, and crop N uptake from untilled soils are similar to those with plow-tillage (N response categories 1 and 2, Fig. 5.5), the overall retention of N within the soil rooting depth is greater with no-tillage.

Thus, it can be concluded that for the long-term N balance, overall N efficiency in untilled soils is probably higher for most well-drained soils in temperate regions of the world.

As the difference in soil organic N level between the two tillage systems becomes large enough with time, one would expect that N availability in untilled soil receiving no N fertilizer would be as great as that in comparably unfertilized tilled soils. The mineralization rate may be slower in untilled soil but the organic N pool should at some time become large enough that the same amount of N is mineralized as in the faster mineralizing tilled soil. Rice et al. (1983) reported that after 12 years of no-tillage, relative corn yield in zero-N untilled soil was finally as high as in tilled soil and suggested that the increased size of the organic N pool in the untilled soil may have been the cause.

In conclusion, for nonvolatilizing sources, N fertilizer use efficiency in no-tillage agriculture may be higher, lower, or the same as in tilled soils. Nitrogen efficiency is likely to be higher with no-tillage in climates where soil moisture is limiting and lower in cool, moist climates with poorly drained soils where large denitrification losses may occur. With climates and soils between these extremes, N efficiency may possibly be lower for several years after converting to no-tillage. After a period of time, increased soil organic N content in the untilled soils should result in comparable N availability in tilled and untilled soils and, consequently, similar N fertilizer efficiency would be attained in the two systems.

NITROGEN FERTILIZER MANAGEMENT

Rates

Assuming unlimited capital, the most profitable application rate of N fertilizer is the economic optimum rate, which by definition, occurs when the cost of the last increment of fertilizer applied equals the value of the last additional yield resulting from that increment of fertilizer. Any comparison of economic optimum N rates for nontilled versus tilled cropping systems will depend on how no-tillage affects N fertilizer response. An example of this type analysis of data from an experiment in Maryland was discussed above (Fig. 5.8). In this experiment, corn yields were lower at low N rates, higher at high N rates, and more fertilizer N was required for greater maximum yield with no-tillage (category 4 response). More fertilizer N was needed for the economic optimum but the N use efficiencies were approximately the same with both tillages, and profit per unit of crop area was greater with no-tillage. It should be emphasized, however, that these results only apply to this specific N fertilizer–tillage interaction category, and that for the other four interaction categories discussed, the effect of eliminating tillage on the economic optimum N rates and yields would depend on how no-tillage affected the N response curves. For example, in category 1, the economic optimum rate would be the same for both tillage systems. Economic optimum N rates would also be similar for category 2, but more fertilizer N would be required for the economic optimum with no-tillage in category 3 (see Fig. 5.5).

Time of Application

Applying N fertilizer just prior to the time of greatest plant N uptake results in its most efficient utilization (Olson et al., 1964a; Olson and Kurtz, 1982). For corn, this means applying the greatest portion of N fertilizer as a sidedress application 3–5 weeks after emergence when the corn is 30–45 cm tall. This practice reduces the time that the fertilizer N is subjected to leaching, denitrification, volatilization, and runoff losses. With no-tillage, the probability of loss by many of these pathways is increased because of wetter soil and the fact that fertilizer N may not be incorporated. There is also the chance that more of the fertilizer N may be immobilized during the time it remains on or near the microbially active surface layer of the untilled soil. An example of the greater efficiency from sidedressing N with no-tillage as compared to tilled corn is shown in Table 5.4. This timing by tillage interaction was significant in three of the 17 experiments conducted in Maryland between 1978 and 1982. Phillips et al. (1980) also reported that N applied 30 days after planting produced yields in the no-tillage treatments as great or greater than similarly treated tilled plots. In some cases where the N supplying capability of the soil is low it may be necessary to split the application and apply some N at planting to provide sufficient N for unrestricted seedling growth until the time the majority of N is applied as a side dressing. This is especially true for such crops as winter small grains where a long time exists between planting and maximal growth and N uptake the following spring.

TABLE 5.4. Average Grain Yield of No-Tillage (NT) and Plow-Tillage (PT)[a] Corn as Influenced by Fertilizer N Application Time[b]

	Time of Application			
	At planting (mt/ha)		Sidedress (mt/ha)	
Site, Year	NT	PT	NT	PT
Poplar Hill, 1978	5.86	6.81	7.17	7.07
Plant Res. Farm, 1979	6.55	6.64	7.12	6.68
Poplar Hill, 1982	8.49	9.72	9.45	9.28
Mean	6.97	7.72	7.91	7.68

[a]Tillage by time of application interactions were significant in each experiment.
[b]Average of five N application rates 0, 90, 134, 179, and 269 kg/ha.
SOURCE: Bandel (unpublished).

Sources

Since fertilizers are frequently not incorporated with no-tillage, the potential exists for much greater NH_3 volatilization losses from NH_4^+- containing or producing fertilizers than from plow-tillage where fertilizers are generally incorporated (Nelson, 1982). If the pH of the soil surface is 7.2 or less, the loss from sources that do not contain urea is minimal (Martin and Chapman, 1951; Mills et al., 1974). Soils with a pH greater than 7.2 generally contain free $CaCO_3$ and are classified as calcareous. Ammonia volatilization losses from calcareous soils are greater from unincorporated urea than from any unincorporated NH_4^+-containing sources (Nelson, 1982). Furthermore, volatilization losses from surface-applied NH_4^+ N sources are greatest from calcareous soils when the accompanying anion forms an insoluble precipitate with Ca such as $(NH_4)_2SO_4$ and $(NH_4)_2H_2PO_4$ (Fenn and Kissell, 1973). Fenn and Kissel (1973) believe this is caused by the pH of a calcareous soil being controlled by the equilibrium between Ca^{2+} and CO_3^{2-} ions and their relatively insoluble reaction product, $CaCO_3$. When Ca^{2+} ions form an insoluble precipitate with the anion in the NH_4^+ fertilizer source, it allows a higher CO_3^{2-} concentration in the soil solution. The CO_3^{2-} ions react with H^+ ions in soil solution forming HCO_3^- and H_2CO_3, resulting in a higher soil pH. This leads to a greater proportion of the NH_4^+ plus NH_3 being NH_3 and thus more NH_3 volatilization.

Urea $[CO(NH_2)_2]$ applied to a soil is normally enzymatically hydrolyzed to $(NH_4)_2CO_3$ within several days if the soil temperature and moisture are adequate for crop growth. The $(NH_4)_2CO_3$ dissociates and the CO_3^{2-} induces a pH increase to above 9.0 near the urea granule. Using the equilibrium constant for the reaction $NH_4^+ + OH^- \longleftrightarrow NH_3 + H_2O$, it can be calculated that the proportion of the total ammoniacal N that is NH_3 at pH 9.0 is 36%, whereas pH 7.0 it is only 0.36%. Therefore, NH_3 volatilization of surface-applied urea can occur from most agricultural soils regardless of their pH. Soil and weather factors that cause an increase in volatilization loss from surface applied urea are (1) high soil pH; (2) low cation exchange capacity; (3) low pH buffering capacity in the soil; (4) sufficient soil moisture to allow urea hydrolysis; (5) warm soil temperatures; (6) soil moisture loss by evaporation; (7) high urea application rates; (8) a high urease activity in the soil; and (9) some air movement. This subject has recently been reviewed by Nelson (1982) and Terman (1979).

An increasingly popular N source is urea–ammonium nitrate solution (UAN) containing approximately half urea and half NH_4NO_3 in a water base. The urea in the solution undergoes the same hydrolysis reaction when applied to the soil, and, consequently, NH_3 volatilization losses can also occur from unincorporated UAN. Olson et al. (1964b) demonstrated that volatilization losses from UAN were approximately one-third as great as from urea on a bare soil, but that when broadcast on crop residue the UAN loss was as great as from urea. Therefore, with residue-covered, untilled soils, NH_3 volatilization losses

TABLE 5.5. Influence of Nitrogen Rate and Source on No-Tillage Corn Grain Yields at Three Locations in Maryland, 1976–1979

	N Rate (kg/ha)					
N Source[a]	0	45	90	135	180	Mean
			mt/ha			
Ammonium Nitrate	5.37[b]	6.72	8.20	8.88	9.28	7.69
Urea	5.37	6.32	7.51	7.97	8.52	7.14
UAN	5.37	6.12	7.58	8.43	8.91	7.28
Mean	5.37	6.39	7.76	8.43	8.90	7.37

[a] $LSD_{0.05}$: N source = 0.20 mt/ha, N rate = 0.20 mt/ha, N source × N rate = 0.40 mt/ha.
[b] values in mt/ha.

SOURCE: Bandel (unpublished).

from broadcast UAN could be expected to be potentially as large as from urea. Fox and Hoffman (1981) reported similar apparent volatilization losses of up to 30% from urea and UAN broadcast on soil and residue in an experiment with no-tillage corn in Pennsylvania. In Maryland, Bandel (unpublished data) observed that the apparent loss from broadcast UAN averaged about one-half that from urea in numerous field experiments from 1976–1979 (Table 5.5).

Both urea and UAN are being manufactured and used in increasingly greater quantities in the United States and elsewhere. In 1982, 11.5% of the direct applied N fertilizers on U.S. farms was from urea and 25.1% from UAN (Hargett and Berry, 1983). In Maryland and Pennsylvania where no-tillage is utilized on over one-half and one-third of the land, respectively, 68% and 84% of the direct applied N is either as urea or UAN. Moreover, it is anticipated that urea sources will become more available due to industrial and economic advantages. It will therefore be necessary to devise specific best management practices for these N sources on untilled land. Methods currently proposed to minimize loss from these sources with no-tillage are discussed below.

Anhydrous NH_3 can be injected into untilled soil and produces yields and N uptake comparable to other injected sources (Mengel et al., 1982). Anhydrous NH_3 has not generally been used for no-tillage corn because corn residue interferes with the knives required for injection. It is often difficult to seal the knife slot in firm, residue covered soil. However, coulters preceding the knives, sealing wings, and packing wheels to ensure closure are overcoming these problems (Fee, 1983).

Specific Management Methods

Methods of improving N fertilizer use efficiency in no-tillage generally involve delayed application until immediately prior to the period of heaviest plant demand. Examples include localized placement, timing in relation to climatic conditions, and fertilizer additives. The first method of delayed application applies to all N sources as discussed above. The remaining approaches are especially important with urea-containing sources and specifically include: (1) injection; (2) dribble or surface band applications of UAN; (3) ensuring there are at least 10 mm of precipitation within 48 hr after application; and (4) adding amendments to the urea or UAN.

Injecting urea-containing fertilizers several centimeters under the soil surface greatly reduces NH_3 volatilization losses (Nelson, 1982). Although most studies have been in the laboratory or greenhouse, several recent field investigations have confirmed that urea injection into no-tillage corn essentially eliminates apparent volatilization loss (Mengel et al., 1982; Bandel et al., 1984; Fox, R. H., unpublished data). Maryland research showed that injecting urea increased average corn yields by more than 1.6 mt/ha over broadcast urea (Table 5.6). Injecting UAN increased yields by approximately 1 mt/ha, but injecting nonvolatilizing NH_4NO_3 did not significantly increase yield. Pennsylvania results also show improved corn yield from urea injection with no yield difference between surface and injected NH_4NO_3. The increased efficiency of injected urea-containing sources in these studies was most likely due to a reduction in NH_3 volatilization rather than a decrease in immobilization of injected N as was suggested by Mengel et al. (1982). However, there are problems with injection of urea-containing fertilizers on untilled fields such as availability of suitable equipment and the increased time and energy required to inject compared with broadcast application.

Banding or "dribbling" UAN on the soil surface markedly reduces volatilization compared to a broadcast spray (Touchton and Hargrove, 1982; Bandel, V. A., unpublished data; Fox, R. H., unpublished data). Maryland results (Table 5.7) show that dribbled UAN produced yields 0.8–2.3 mt/ha with an average of 1.4 mt/ha greater than broadcast UAN. The yields with dribbled UAN were comparable to those obtained with injected UAN and broadcast NH_4NO_3. A 1983 Pennsylvania study comparing several rates of UAN as either spray or dribble applied to no-tillage corn (Fig. 5.9) found that the

TABLE 5.6. Four-Year Summary of the Influence of N Source and N Placement on No-Tillage Corn Yields, Poplar Hill Research Farm, Maryland, 1979–1982.

	N Source			
Nitrogen Placement[a]	Ammonium Nitrate	Urea	UAN	Mean
		mt/ha		
Broadcast	9.66	7.98	8.96	8.87
Injection	9.74	9.60	9.93	9.76
Mean	9.70	8.78	9.44	9.31

[a]N placement means significant at 5% level.

$LSD_{0.05}$: N source = 0.26 mt/ha, N source by N placement = 0.37 mt/ha.

SOURCE: Bandel (unpublished).

TABLE 5.7. Influence of N Source and Placement on No-Tillage Corn Grain Yield at Several Locations in Maryland, 1982

N Treatment	Wyet[a]	Poplar Hill[a]	Poplar Hill[a]	Poplar Hill[b]
	Grain Yield (mt/ha)			
Check	2.09	1.95	2.66	2.66
Ammonium nitrate	7.06	9.74	8.91	10.27
UAN broadcast	6.22	7.53	8.56	9.98
UAN dribbled	7.53	9.83	9.35	11.05
UAN injected	7.80	10.50	9.81	11.20
$LSD_{0.05}$	0.87	0.82	0.90	0.90

[a]N rate = 135 kg/ha.
[b]N rate = 180 kg/ha.
SOURCE: Bandel (unpublished).

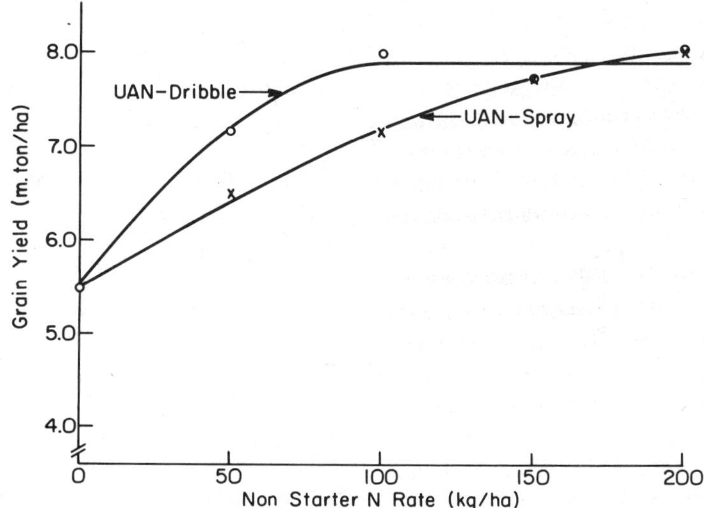

FIGURE 5.9. Influence of method of application of urea-ammonium nitrate solution (UAN) on corn grain yield, Agronomy Research Farm, Pennsylvania, 1983 (Fox, unpublished data).

economic optimum N rate for sprayed UAN was 234 kg/ha but was only 110 kg/ha for dribble applied (based on a quadratic−linear plateau response model and a fertilizer-N-cost to corn-grain-price ratio of 6.72, both units in U.S. $/kg). In other unpublished studies in Pennsylvania, on a silt loam soil where relatively large apparent NH_3 volatilization losses were observed from surface-applied urea, no differences were found in the apparent N utilization of dribbled versus injected UAN at several rates and two times of application. Touchton and Hargrove (1982), working on a sandy Georgia soil, found that dribbled UAN resulted in greater no-tillage corn yields and ear-leaf N concentrations than spray-applied UAN. Injecting the UAN produced significantly greater yields than surface, dribbled UAN at the lower two N rates in one of the two years of their study. Thus, although it appears that injecting UAN may produce slightly greater N fertilizer use efficiency than dribbled UAN, it remains to be determined whether the increased energy and machinery costs necessary for injection are worth this slightly greater efficiency.

Forster and Lippold (1975) found that NH_3 volatilization loss ceased after 10 mm of rain fell on surface-applied urea at several sites in East Germany. Urea and UAN are both very soluble, and rain will leach urea into the soil where any NH_3 produced will be absorbed by soil constituents before it becomes lost to the atmosphere. In other words, rain is serving as a means of injecting urea-containing N sources. An examination of the data from Pennsylvania, Maryland, and Germany (Fox and Hoffman, 1981; Bandel et al., 1980; Forster and Lippold, 1975) showed that if at least 10 mm of rain falls within 48 hr after surface application of urea-containing fertilizers, there is no apparent NH_3 volatilization loss (Fox and Hoffman, 1981). They also speculated that the longer the rain-free period following application, the greater the likelihood and magnitude of loss. Catchpoole et al. (1983) and Harper et al. (1983) also observed that NH_3 losses from urea applied to a pasture in Australia were much smaller when 5 mm or more of rain fell soon after application.

Much current research is devoted to finding an economical amendment or coating for urea-containing fertilizers that will reduce or eliminate NH_3 volatilization losses from unincorporated urea (Terman, 1979). The four general categories of amendments that have been used to reduce volatilization losses in research are (1) urease inhibitors; (2) pH reducing amendments; (3) surface coatings; and (4) amendments that cause the formation of Ca and Mg carbonates.

A recent comparison of urease inhibitors (Martens and Bremner, 1984) reported that the most effective of those tested was phenylphosphorodiamidate (PPD), a compound patented by East German researchers. Martens and Bremner (1984) reported additions as low as 1 μg/g of soil retarded urea hydrolysis in five soils by an average of 47%. These authors cite several studies showing that a retardation of urea hydrolysis increased N recovery in greenhouse and field studies. Thus, if it becomes commercially feasible to use this additive, a hope exists for reducing volatilization losses from surface-applied urea-containing fertilizers without changing application methods.

Terman (1979) summarized research on various acid-producing amendments to urea fertilizers. In general, any compound that lowers the pH surrounding the urea particle also reduces NH_3 volatilization from unincorporated urea. To our knowledge there currently are no commercial urea sources specifically treated with acid-forming compounds to reduce volatilization losses. Gasser and Penney (1967) demonstrated that urea-phosphate made from a 1:1 mole ratio mixture of urea and phosphoric acid and urea–urea–

phosphate (3 urea:1 phosphoric acid) worked well as fertilizer sources for grass pastures and barley. The National Fertilizer Development Center of the Tennessee Valley Authority (NFDC/TVA) has produced experimental 3:1 urea–urea–phosphate compounds (UUP) that have reduced NH_3 volatilization loss from noncalcareous soils in the laboratory and field (Fox, unpublished data; Khasawneh, NFDC/TVA, personal communication). Field research designed to estimate NH_3 volatilization losses from UUP under a range of soil and weather conditions is continuing to determine if it can be economically marketed as an alternative urea-containing source for no-tillage crop production that will result in less volatilization loss and greater use efficiency. One problem with this source is that its analysis is approximately 37–12–0 ($N-P_2O_5-K_2O$) and, as a P as well as N source, would not be suitable for those wanting to apply only N.

Matocha (1976) found that NH_3 volatilization losses from sulfur-coated urea (SCU) were only a fraction of those from surface-applied urea in a laboratory experiment. The problem with SCU is that it is too expensive to use for field crops, although it is currently being used commercially as a slow release N source for turf.

Fenn et al. (1981) demonstrated that adding soluble Ca or Mg salts to urea reduced NH_3 volatilization losses. The Ca^{2+} and Mg^{2+} ions react with the $CO_3{}^{2-}$ released from urea hydrolysis forming insoluble carbonates, thus preventing the soil solution pH from rising to above 9.0 as is the case with unamended urea particles. The lower pH reduces the fraction of $NH_4{}^+$ plus NH_3 that is NH_3 and therefore lowers volatilization loss. Urea treated with Ca- or $Mg(NO_3)_2$ is not generally available. Fenn et al. (1982) recently reported that including a K^+ or $NH_4{}^+$ salt with urea also lowers NH_3 volatilization. The theory is that the K^+ or $NH_4{}^+$ will displace some Ca^{2+} or Mg^{2+} from exchange sites which will then react with the $CO_3{}^2$ as above. Rappaport and Axley (1984) reported that adding a solution of KCl to urea on a 1:1 weight basis resulted in lower NH_3 losses in a laboratory experiment, and greater yields of hay and corn per unit of N than urea in the field. Research in Pennsylvania (Fox, unpublished data) with an experimental dry urea–KCl mixture from NFDC/TVA found in a laboratory experiment that, per unit of N applied, there was slightly less NH_3 volatilization from this source than from urea. However, in a field experiment there was very little difference in apparent efficiency between the experimental urea–KCl and surface-applied urea.

In summary, it appears that to ensure maximum crop yield per unit of surface-applied N with nontilled, noncalcareous soils, a nonvolatilizing N source should be applied just prior to the time of maximum plant N uptake. If a nonvolatilizing source is not available or too expensive and a urea-containing source must be used, efficiency can be improved by delaying application until the beginning of the period of high plant N demand. Other strategies to increase use efficiency of urea-containing fertilizer sources with no-tillage are to inject the fertilizer or, for UAN, to apply it in a surface band (dribbled).

NITROGEN FERTILIZER EFFECTS ON SOIL pH

Hydrogen ions are released during the nitrification of NH_4^+ to NO_3^- in soil. Therefore, the application of NH_4^+-containing or -producing fertilizers to soils increases the acidity of soils. Because fertilizer applied to no-tillage soil is not incorporated, one would expect acidity produced from the added NH_4^+ to be concentrated in the surface few centimeters of soil. This predicted acidification of the soil surface has been observed in most long-term no-tillage experiments where NH_4^+- containing or -producing fertilizers were used (Bandel, 1979; Blevins et al., 1977; Blevins et al., 1983; Dick, 1983; Fox and Hoffman, 1981). Fox and Hoffman (1981) found that applying 202 kg/ha of N as $(NH_4)_2SO_4$ for 5 years to a silt loam soil in continuous no-tillage corn culture lowered the pH of the surface 2.5 cm of soil from 6.9 to 4.7. Bandel (1979) and Blevins et al. (1983) have observed soil pH levels below 5 in the surface few centimeters of soil at higher N rates even with sources less acidifying than $(NH_4)_2SO_4$. An important problem produced by these acid soil surfaces in nontilled soils is the inactivation of herbicides, especially the triazines (Lowder and Weber, 1982; Chapter 11). Also, at soil pH values less than 5.5 soluble and exchangeable Al levels increase and Al toxicity may adversely affect crops (Blevins et al., 1983). This acidification of the soil surface in no-tillage can be overcome by liming (Blevins et al., 1983). The rapid build up of acidity that is possible with high rates of residually acidic N fertilizer makes it advisable, however, to determine pH of the surface 2–5 cm of soil every year or two. The total amount of lime needed in no-tillage management may be more than with plow-tillage, since in both the long-term continuous corn tillage studies in Ohio (Dick, 1983) and Kentucky (Blevins, et al., 1983), there was apparently greater total soil acidification in the no-tillage plots than in the plow-tillage plots. After 18 or 19 years in the Ohio study, soil pH in nontilled soils was significantly lower, to 30 cm at one site and 22.5 cm at the other. Both tillage treatments received the amount of lime recommended to maintain the plowed soil at or above pH 6.0 in the plow layer. After 10 years in the Kentucky study, pH in unlimed treatments was lower in the surface 15 cm of the no-tillage soil than in the plowed soil and the pH in the 15–30 cm layer was approximately the same with both tillages.

SUMMARY

Processes that control the content and availability of soil N to crops are mineralization, immobilization, nitrification, denitrification, leaching, volatilization, and erosion. Conversion from plow-tillage to no-tillage results in a cooler, wetter, less aerobic soil environment that affects many of these processes. The most common observation has been that the availability of soil N was less in nontilled than tilled soils for the first several years after converting to no-tillage. Most researchers have also found that soil organic matter levels are higher in nontilled soils than tilled soils growing the same crops. All studies have shown that the distribution of soil organic matter and microbiological activity becomes much more stratified within the soil profile after several years of no-tillage. With no-tillage management, soil organic matter concentration and microbial activity are relatively high in the surface few centimeters of soil and decrease rapidly with depth.

The causes of lowered N availability in untilled soils for the first several years after tillage has stopped have not been resolved. The cooler, wetter soil resulting from no-tillage could lead to greater leaching and denitrification and less mineralization. More immobilization may also occur in untilled soils because of the accumulation of surface residue with a high C/N ratio. However, when soil moisture is limiting, the greater moisture content of untilled soils should result in greater mineralization in these soils. It is suggested that the only way to know how the changed soil environment under no-tillage affects N availability under a specific set of soil and meteorological conditions is to precisely measure the appropriate soil and meteorological parameters and use them in a proven simulation model of soil N transformations.

Nitrogen fertilizer is generally not incorporated in nontilled soils, so the possibility exists for NH_3 volatilization losses from NH_4^+- containing or -producing fertilizers. With nonvolatilizing N sources, the N fertilizer efficiency (yield and N uptake per unit of N applied) in untilled soils may be greater, less, or the same as in tilled soils. Nitrogen fertilizer efficiency is likely to be greater with no-tillage in climates where soil moisture is limiting and less in cool, moist climates with poorly drained soils where large denitrification losses may occur. With climates and soils between these extremes, N efficiency, especially at lower rates, may be less for several years after converting to no-tillage.

When NH_4^+ containing sources are applied to the surface of soils with a pH greater than 7.2 or urea-containing sources are not incorporated with any soil, the potential exists for large NH_3 volatilization losses. With these conditions, N fertilizer efficiency is likely to be much less with no-tillage than with conventional tillage where the N fertilizer is incorporated. Strategies to reduce this potential for NH_3 volatilization loss with no-tillage include injecting the fertilizer, delaying application of the major portion of the N fertilizer until just prior to the time of greatest plant N uptake (e.g., postplant sidedress applications to corn when it is 30–45 cm high), and dribble instead of spray applications of urea-containing N solutions. A rainfall of at least 10 mm within 48 hr after fertilizer application will also significantly lower volatilization losses. Research on developing amendments that can be added to reduce volatilization from these N sources is continuing.

LITERATURE CITED

Alexander, M. 1977. *Introduction to Soil Microbiology*, 2nd ed. Wiley, New York.

Angle, J. S., G. McClung, M. S. McIntosh, P. M. Thomas, and D. C. Wolf. 1984. Nutrient losses in runoff from conventional and no-tilled corn watersheds. *J. Environ. Qual.* **13**:431–435.

Baeumer, K. 1970. First experiences with direct drilling in Germany. *Neth. J. Agric. Sci.* **18**: 283–292.

Bakermans, W. A. P. and C. T. DeWit. 1970. Crop husbandry on naturally compacted soils. *Neth. J. Agric. Sci.* **18**:225–246.

Bandel, V. A. 1979. Nitrogen fertilization of no-tillage corn. 1979 Abstracts, N. E. Branch ASA:15–20.

Bandel, V. A., F. R. Mulford, and H. J. Bauer. 1984. Influence of fertilizer source and placement on no-tillage corn. *Fert. Issues* **1**:38–43.

Bennett, O. L., G. Stanford, E. L. Mathias, and P. L. Lundberg. 1975. Nitrogen conservation under corn planted in quackgrass sod. *J. Environ. Qual.* **4**:107–110.

Black, C. A. 1968. *Soil Plant Relationships*, 2nd ed. Wiley, New York.

Blevins, R. L., D. Cook, S. H. Phillips, and R. E. Phillips. 1971. Influence of no-tillage on soil moisture. *Agron. J.* **63**:593–596.

Blevins, R. L., G. W. Thomas, and P. L. Cornelius. 1977. Influence of no-tillage and nitrogen fertilization on certain soil properties after 5 years of continuous corn. *Agron. J.* **69**:383–386.

Blevins, R. L., G. W. Thomas, M. S. Smith, W. W. Frye, and P. L. Cornelius. 1983. Changes in soil properties after 10 years continuous non-tilled and conventionally tilled corn. *Soil Tillage Res.* **3**:135–146.

Cannell, R. Q., F. B. Ellis, D. G. Christian, J. P. Graham, and J. T. Douglas. 1980. The growth and yield of winter cereals after direct drilling, shallow cultivation, and ploughing on noncalcareous clay soils. *J. Agric. Sci. Camb.* **94**:345–359.

Carter, M. R. and D. A. Rennie. 1982. Changes in soil quality under zero tillage farming systems: distribution of microbial biomass and mineralizable C and N potentials. *Can. J. Soil Sci.* **62**:587–597.

Caskey, W. H. 1983. Temporal effects and tillage practice on nitrification and denitrification. *Agron. Abst.* **1983**:154.

Catchpoole, V. R., D. J. Oxenham, and L. A. Harper. 1983. Transformation and recovery of urea applied to a grass pasture in south-eastern Queensland. *Aust. J. Expt. Agric. Anim. Husb.* **23**:80–86.

Davies, D. B. and R. Q. Cannell. 1975. Review of experiments on reduced cultivation and direct drilling in the U.K., 1957–1974. *Outlook Agric.* **8**:216–220.

Deans, J. R., C. E. Clapp, J. A. E. Molina, and N. J. Durben. 1983. Effects of nitrogen, tillage, and residue management on net mineralized nitrogen. *Agron. Abstr.* **1983**:155.

Dick, W. A. 1983. Organic carbon, nitrogen, and phosphorus concentrations and pH in soil profiles as affected by tillage intensity. *Soil Sci. Soc. Am. J.* **47**:102–107.

Doran, J. W. 1980. Soil microbial and biochemical changes associated with reduced tillage. *Soil Sci. Soc. Am. J.* **44**:765–771.

Doran, J. W. and J. F. Power. 1983. The effects of tillage on the nitrogen cycle in corn and wheat production. *In* R. R. Lawrence et al. (eds.). *Nutrient Cycling in Agricultural Ecosystems*, Univ. Georgia Agric. Expt. Sta. Special Pub. 23, Athens, Ga., Sept. 1980.

Dowdell, R. J. and R. Crees. 1980. The uptake of ^{15}N-labeled fertilizer by winter wheat and its immobilization in a clay soil after direct drilling or ploughing. *J. Sci. Food Agric.* **31**:992–996.

Dowdell, R. J. and R. Q. Cannell. 1975. Effect of plowing and direct drilling on soil nitrate content. *J. Soil Sci.* **26**:53–61.

Ellis, F. B. and K. R. Howse. 1980. Effects of cultivation on the distribution of nutrients in the soil and the uptake of nitrogen by spring barley and winter wheat on three soil types. *Soil Tillage Res.* **1**:35–46.

Eylands, V. J. and R. N. Gallaher. 1983. Grain sorghum response to winter crops, tillage, and N management. *Agron. Abstr.* **1983**:167.

Fee, R. 1983. Nitrogen, new rules for conservation tillage. 1983. *Successful Farming*, March, pp. 15–21.

Fenn, L. B. and D. E. Kissel. 1973. Ammonia volatilization from surface applications of ammonium compounds on calcareous soils: I. General theory. *Soil Sci. Soc. Am. Proc.* **37**:855–859.

Fenn, L. B., J. E. Matocha, and E. Wu. 1982. Substitution of ammonium and potassium for added calcium in reduction of ammonia loss from surface-applied urea. *Soil Sci. Soc. Am. J.* **46**:771–776.

Fenn, L. B., R. M. Taylor, and J. E. Matocha. 1981. Ammonia losses from a surface-applied nitrogen fertilizer as controlled by soluble calcium and magnesium; general theory. *Soil Sci. Soc. Am. J.* **45**:777–781.

Firestone, M. K. 1982. Biological denitrification. *In* F. J. Stevenson (ed.). Nitrogen in agricultural soils. *Agronomy* **22**:289–326. Am. Soc. Agronomy, Madison, Wis.

Forster, I. and H. Lippold. 1975. Ammonia losses from urea fertilizers. II. Determining ammonia losses under field conditions as affected by the weather (in German). *Archiv. für Acker. und Pflanzenban und Bodenkunde* **19**:631–639.

Fox, R. H. and L. D. Hoffman. 1981. The effect of N fertilizer source on grain yield, N uptake, soil pH, and lime requirement for no-till corn. *Agron. J.* **73**:891–895.

Fox, R. H. and W. P. Piekielek. 1983. Response of corn to nitrogen fertilizer and the prediction of soil nitrogen availability with chemical tests in Pennsylvania. Penna. State Univ. Agric. Expt. Stat. Bull. 843.

Fredrickson, J. K., F. E. Koehler, and H. H. Cheng. 1982. Availability of ^{15}N-labeled nitrogen in fertilizer and in wheat straw to wheat in tilled and no-till soil. *Soil Sci. Soc. Am. J.* **46**: 1218–1222.

Gasser, J. K. R. and A. Penny. 1967. The value of urea nitrate and urea phosphate as nitrogen fertilizers for grass and barley. *J. Agric. Sci.* **69**:139–148.

Hardin, G. B. 1978. Conserving nitrogen for no-till corn fertilizers. *Agric. Res.* **27**(2):6–7.

Hargett, N. L. and J. T. Berry. 1983. 1982 fertilizer summary data. Natl. Fert. Dev. Ctr., T.V.A., Muscle Shoals, Ala.

Hargrove, W. L., R. A. Rauniker, and B. R. Bock. 1983. Ammonia volatilization from urea in no-tillage production systems. *Agron. Abst.*, **1983**:170–171.

Harper, L. A., V. R. Catchpoole, R. Davis, and K. L. Weir. 1983. Ammonia volatilization: soil, plant, and microclimate effects on diurnal and seasonal fluctuations. *Agron. J.* **75**:212–218.

Hill, J. O. and R. L. Blevins. 1973. Quantitative soil moisture use in corn grown under conventional and no-tillage methods. *Agron. J.* **65**:945–949.

Jansson, S. L. 1958. Tracer studies on nitrogen transformations in soil with special attention to mineralization-immobilization relationships. *Kungl. Lantbrukshögskolans Annales.* **24**: 101–361.

Jansson, S. L. and J. Persson. 1982. Mineralization and immobilization of soil nitrogen. *In* F. J. Stevenson (ed.). Nitrogen in Agricultural Soils. *Agronomy* **22**:229–252. Amer. Soc. Agron., Madison, Wis.

Jenkinson, D. S. and J. H. Rayner. 1977. The turnover of soil organic matter in some of the Rothamsted classical experiments. *Soil Sci.* **123**:298–305.

Jones, J. N., Jr., J. E. Moody, and J. H. Lillard. 1969. Effects of tillage, no-tillage, and mulch on soil water and plant growth. *Agron. J.* **61**:719–721.

Juo, A. S. R. and R. Lal. 1979. Nutrient profile in a tropical Alfisol under conventional and no-till systems. *Soil Sci.* **127**:168–173.

Kang, B. T. and M. Yunusa. 1977. Effects of tillage and phosphorus fertilization on maize in the humid tropics. *Agron. J.* **69**:291–294.

Kitur, B. K., M. S. Smith, R. L. Blevins, and W. W. Frye. 1984. Fate of ^{15}N-depleted ammonium nitrate applied to no-tillage and conventional tillage corn. *Agron. J.* **76**:240–242.

Kupers, L. J. P. and J. Ellen. 1970. Experience with minimum tillage and nitrogen fertilization. *Neth. J. Agric. Sci.* **18**:270–276.

Lal, R. 1976. No-tillage effects on soil properties under different crops in Western Nigeria. *Soil Sci. Soc. Am. J.* **10**:762–768.

Lal, R. 1974. No-tillage effects on soil properties and maize (*Zea mays* L.) production in Western Nigeria. *Plant Soil* **40**:321–331.

Legg, J. O., G. Stanford, and O. L. Bennett. 1979. Utilization of labeled-N fertilizer by silage corn under conventional and no-till culture. *Agron. Abstr.* **1979**:175.

Linn, D. M. and J. W. Doran. 1984. Aerobic and anaerobic microbial activity in no-till and plowed soils. *Soil Sci. Soc. Am. J.* **45**:794–799.

Lowder, S. W. and J. B. Weber. 1982. Atrazine efficacy and longevity as affected by tillage, liming, and fertilizer type. *Weed Sci.* **30**:273–280.

Martens, D. A. and J. M. Bremner. 1984. Effectiveness of phosphoroamides for retardation of urea hydrolysis in soils. *Soil Sci. Soc. Am. J.* **48**:302–305.

Martin, J. P. and H. D. Chapman. 1951. Volatilization of ammonia from surface fertilized soils. *Soil Sci.* **71**:25–34.

Matocha, J. E. 1976. Ammonia volatilization and nitrogen utilization from sulfur-coated ureas and conventional nitrogen fertilizers. *Soil Sci. Soc. Am. J.* **40**:597–601.

McDowell, L. L. and K. C. McGregor. 1980. Nitrogen and phosphorus losses in runoff from no-till soybeans. *Am. Soc. Agric. Eng. Trans.* **23**:643–648.

McMahon, M. A. and G. W. Thomas. 1976. Anion leaching in two Kentucky soils under conventional tillage and killed sod mulch. *Agron. J.* **68**:437–442.

Mengel, D. B., D. W. Nelson, and D. M. Huber. 1982. Placement of nitrogen fertilizers for no-till and conventional till corn. *Agron. J.* **74**:515–518.

Mielke, L. N., J. W. Doran, and K. A. Richards. 1982. Physical environment of plowed and nontilled surface soils. *Agron. Abstr.* **1982**:253.

Mills, H. A., A. V. Barker, and D. N. Maynard. 1974. Ammonia volatilization from soils. *Agron. J.* **66**:355–358.

Molina, J.A.E., C. E. Clapp, M. J. Shaffer, C. W. Chichester, and W. E. Larson. 1983. NCSOIL, A model of nitrogen and carbon transformation in soil; Description, calibration and behavior. Soil Sci. Soc. Am. J. 47:85–91.

Moschler, W. W., G. M. Shear, D. C. Martens, G. O. Jones, and R. R. Wilmonte. 1972. Comparative yield and fertilizer efficiency of no-tillage and conventionally tilled corn. *Agron. J.* **64**:229–231.

Moschler, W. W. and D. C. Martens. 1975. Nitrogen, phosphorus and potassium requirements in no-tillage and conventionally tilled corn. *Soil Sci. Soc. Am. Proc.* **39**:886–891.

Nelson, D. W. 1982. Gaseous losses of nitrogen other than through denitrification. *In* F. J. Stevenson (ed.). Nitrogen in agricultural soils. *Agronomy* **22**:327–364.

Olson, R. A., A. F. Dreier, C. Thompson, K. Frank, and P. H. Grabowski. 1964a. Using fertilizer nitrogen effectively on grain crops. Nebr. Agric. Exp. Stn. Bull. 479.

Olson, R. A., K. D. Frank, and A. F. Dreier. 1964b. Controlling losses of fertilizer nitrogen from soils. 8th Intern. Cong. Soil Sci., Romania, IV:1023–1031.

Olson, R. A. and L. T. Kurtz. 1982. Crop nitrogen requirements, utilization, and fertilization. *In* F. J. Stevenson (ed.). Nitrogen in agricultural soils. *Agronomy* **22**:567–604. Am. Soc. Agron., Madison, Wis.

Phillips, R. E., R. L. Blevins, G. W. Thomas, W. W. Frye, and S. H. Phillips. 1980. No-tillage agriculture. *Science* **208**:1108–1113.

Powlson, D. S. and D. S. Jenkinson. 1981. A comparison of the organic matter, biomass, adenosine triphosphate and mineralizable nitrogen contents of ploughed and direct drilled soils. *J. Agric. Sci.* **97**:713–721.

Randall, G. W. 1980. Fertilization practices for conventional tillage. Presented to Iowa Fertilizer and Agricultural Chemical Dealers Conference.

Rappaport, B. D. and J. H. Axley. 1984. Potassium chloride for improved urea fertilizer efficiency. *Soil Sci. Soc. Am. J.* **48**:399–401.

Reeves, T. G. and A. Ellington. 1974. Direct drilling experiments with wheat. *Aust. J. Exp. Agric. Animal Husb.* **14**:237–240.

Rice, C. W. 1983. Microbial nitrogen transformations in no-till soils. Ph.D. Dissertation, University of Ky.

Rice, C. W. and M. Scott Smith. 1982. Denitrification in no-till soils. *Soil Sci. Soc. Am. J.* **46**:1168–1173.

Rice, C. W. and M. S. Smith. 1983. Nitrification of fertilizer and mineralized ammonium in no-till and plowed soil. *Soil Sci. Soc. Am. J.* **47**:1125–1129.

Rice, C. W. and M. S. Smith. 1984. Short-term immobilization of fertilizer N at the surface of no-till soils. *Soil Sci. Soc. Am. J.* **48**:295–297.

Rice, C. W., M. S. Smith, and J. H. Grove. 1983. Mineralization of N in no-till and plowed soils. *Agron. Abstr.* **1983**:179.

Smith, K. A. and R. S. Howard. 1980. Field studies of nitrogen uptake using ^{15}N-tracer methods. *J. Sci. Food Agric.* **31**:839–840.

Smith, M. S., C. W. Rice, B. K. Kitur, J. H. Grove, and R. l. Blevins. 1983. Immobilization of fertilizer N in no-till soils: Effects of sub-surface N placement on crop recovery. *Agron. Abstr.* **1983**:180.

Stanford, G., V. A. Bandel, J. J. Meisinger, and J. O. Legg. 1979. N behavior under no-till and conventional corn culture. II. Grain and forage yields in relation to amounts of N applied and total N uptake. *Agron. Abstr.* **1979**:183.

Stinner, B. R., G. D. Hoyt, and R. L. Todd. 1983. Changes in soil chemical properties following a 12 year fallow: A 2-year comparison of conventional tillage and no-till agroecosystems. *Soil Tillage Res.* **3**:277–290.

Tanji, K. K. 1982. Modelling of the soil nitrogen cycle. *In* F. J. Stevenson. Nitrogen in agricultural soils. *Agronomy* **22**:721–722. Am. Soc. Agron., Madison, Wis.

Terman, G. L. 1979. Volatilization losses of nitrogen as ammonia from surface-applied fertilizers, organic amendments, and crop residues. *Adv. Agron.* **31**:189–224.

Thomas, G. W., R. L. Blevins, R. E. Phillips, and M. A. McMahan. 1973. Effect of killed sod mulch on nitrate movement and corn yield. *Agron. J.* **65**:736–739.

Triplett, G. B., Jr., F. Haghiri, and D. M. Van Doren, Jr. 1979. Plowing effect on corn yield response to N following alfalfa. *Agron. J.* **71**:801–803.

Triplett, G. B., Jr. and D. M. Van Doren, Jr. 1969. Nitrogen, phosphorus, and potassium fertilization of non-tilled maize. *Agron. J.* **61**:637–639.

Touchton, J. T. and W. C. Hargrove. 1982. Nitrogen sources and methods of application for no-tillage corn production. *Agron. J.* **74**:823–826.

Tyler, D. D. and G. W. Thomas. 1977. Lysimeter measurements of nitrate and chloride losses from soil under conventional and no-tillage corn. *J. Environ. Qual.* **6**:63–66.

Vaidyanathan, L. V. and D. B. Davies. 1980. Response of winter wheat to fertilizer nitrogen in undisturbed and cultivated soil. *J. Sci. Food. Agric.* **31**:414–415.

Van Veen, J. A., W. B. McGill, H. W. Hunt, M. J. Frissel, and C. V. Cole. 1981. Simulation models of the terrestrial nitrogen cycle. *In* F. E. Clark and T. Rosswell (eds.). *Terrestrial Nitrogen Cycle*. Ecol. Bull. (Stockholm) 33.

Willis, W. O. 1977. Soil temperature and tillage. *In* Research progress and needs, conservation tillage. Agric. Res. Serv. ARS-NC57:19–22., North Central Region, USDA.

6

CROP MANAGEMENT PRACTICES FOR SURFACE-TILLAGE SYSTEMS

G. B. TRIPLETT
Professor of Agronomy
Department of Agronomy
Mississippi State University
Mississippi State, Mississippi

Crop management may be defined as man's favorable enhancement of crop growth. Cultural management consists of man's influence to direct, nurture, and control the growth of a crop through its life cycle to favorably encourage production; choosing the crop, tilling the fields, and planting each crop in its proper place and season at a specified density. It may also include application of mineral nutrients, supplemental water, and measures to protect the crop from various pests. The sum and sequence of these activities comprise the crop management system.

Components of a crop management system may have a well-proven scientific basis or may have been developed empirically or be part of the folklore, such as planting corn crops according to phases of the moon or when the leaves of the bur oak are the size of a squirrel's ear. Management varies with the crop species, the climatic region, and the degree of mechanization so that methods in one region appear far different from those in another. Nevertheless, a common thread connects methods followed with corn (*Zea mays* L.) in Iowa, Australian wheat (*Triticum aestivum* L.), and a Central American subsistence farmer. Each in turn identifies factors that limit crop production and corrects those within constraints imposed by economic considerations. Some needs are satisfied by inputs of nutrients or pesticides, while others involve timing of operations and cropping sequences.

This chapter seeks to identify and examine factors known to influence crop productivity through changes in tillage practices and the impact of tillage changes on development of new management systems. Some have major focus in other chapters and accordingly are treated briefly here.

REQUIREMENTS FOR GROWTH

The basic requirements for plant growth and development include an adequate supply of moisture, oxygen, sunlight, carbon dioxide, and mineral nutrients. Optimum temperatures are specific for each crop. Insofar as possible, it must be protected from parasitic pests. The moisture requirement of the crop can be managed by rate of planting and supplemental irrigation. Management also can be directed to increase infiltration of rainfalls and to reduce evaporation losses of soil moisture. Both approaches are used to complement each other and a similar approach is used to meet crop nutrient needs.

Cultural management effects on the crop and environment are interrelated in that one factor induces change in others. Thus, specific management systems must be evaluated for suitability of each crop in type locations. Tillage modifications can have a major influence on the crop environment, and crop production must be evaluated to assess these effects. Emphasis in this chapter is placed on grain crops and growing conditions in humid, temperate regions.

PLANT BREEDING FOR TILLAGE SYSTEMS

Identification, development, and dissemination of superior crop cultivars have been major factors in more than doubling per-hectare yield of many crops during the past few decades. Cardwell (1982) attributed more than 50% of the increase in Minnesota corn yield to improved germplasm. However, improvement in crop management practices also were developed during this period, including increased fertilizer use, improved weed control, earlier planting, higher populations, and improved drainage—all of which are necessary for improved corn hybrids to exhibit their superior yield. Kronstad et al. (1978) state that "historically, breakthroughs in plant breeding have been accompanied by research efforts designed to identify cultural practices for maximum productivity."

Crop breeding for cultivar improvement has many goals including high yield, stable production, disease and insect resistance, early maturity, stem strength, plant height, rooting habits, winter hardiness, and drought resistance. In addition, quality factors (namely, color, taste, protein and oil content, milling and baking properties) are part of the selection criteria irrespective of the tillage system used (Kronstad et al., 1978). Multiple goals extend the selection process but greatly improve the value of new, superior cultivars.

Nearly all crop cultivars available today were developed in a tilled-seedbed environment. The selection process developed superior performance under that environment, but did not address performance under other tillage practices. Cultivar evaluations under reduced tillage systems have been made with a few crops by planting available cultivars in tilled and untilled soil and comparing their performance. Similar yield rankings between the two has led to the conclusion that cultivars superior for tilled seedbeds also perform well in untilled soil. Some of the hazards peculiar to untilled seedbeds have been recognized, and cultivars with strong seedling vigor or superior tolerance to cold seedbeds have been identified and recommended. It is fortunate that existing germplasm performs well under both tilled and untilled conditions, but the root of the question remains. Cultivar development took place under intensive tillage and selection in no-tillage environments was not part of the process. Progress in identifying superior germplasm for reduced or no-tillage systems will come with tillage intensity a factor in the selection process.

Kronstad et al. (1978) outline basic steps for an improvement program. First, growth factors associated with reduced tillage must be identified. Second, it must be determined if there exists within a species sufficient genetic variation to breed effectively for tolerance or resistance to those conditions. Third, the breeder must develop selection criteria and techniques to identify superior progeny in clonal material and segregating populations. Fourth, once identified, the desired progeny must retain other agronomic and quality attributes of genetic integrity that the cultivars will find useful.

One area of opportunity for improved performance may be adaptation to poorly drained soils. Currently proposed recommendations (Chapter 2) specify that poorly drained soils be tilled for optimum performance. Crop rotation and/or tillage rotation help mitigate this limitation, and cultivar improvement may reduce the tillage requirement without loss of yield on poorly drained soils. Soybean (*Glycine max* L.) yield has been affected less than corn in untilled environments, but the additional moisture conserved by a mulch cover may encourage excessive vegetative growth and cause lodging of some cultivars. Crop improvement for an untilled environment could be important with soybeans and corn, and possibly other crops.

Some differences do exist between corn hybrids grown in tilled and untilled environments (Table 6.1). Trials were conducted in Ohio on a well-drained silt loam soil (considered first choice for no-tillage crop production, Chapter 2). No-tillage and tilled areas were adjacent, unirrigated, and alternated each year. Each test contained between 75 and 150 hybrid entires. Both tillage systems were planted on the same day in early May and harvested at the same time. Grain yields were taken and harvest moisture and days to mid-silk were recorded as a maturity index.

Grain yields varied between years reflecting growing conditions that prevailed. Stands were generally adequate and not a major influence on crop

TABLE 6.1. Grain Yield of Three Corn Hybrids in Two Tillage Systems

	Year							
	1980		1981		1982		1983	
Hybrid	Plow	No-Tillage	Plow	No-Tillage	Plow	No-Tillage	Plow	No-Tillage
				Grain Yield (mt/ha)				
Funk G4323	10.28	8.53	8.15	—	11.04	10.16	5.20	5.14
French 210	9.97	9.47	9.09	9.40	9.28	11.72	5.45	6.02
DeKalb XL55A	11.10	9.53	7.84	8.65	10.70	10.34	5.77	6.27
$LSD_{0.05}$[a]	0.85	0.85	0.78	0.89	0.94	0.88	1.12	0.94
				Stalk Lodging (%)				
Funk G4323	15.4	1.2	6.6	—	21.5	12.9	26.5	23.6
French 210	12.2	5.8	11.0	2.5	58.4	14.3	20.7	19.0
DeKalb XL55A	5.5	10.3	3.5	2.6	22.0	20.7	35.8	37.8
$LSD_{0.05}$[a]	8.6	4.4	7.9	4.1	18.1	11.2	15.0	12.0

[a]Difference required for significance at the 5% level of probability.

SOURCE: Jordan (1980, 1981, 1982, 1983)

productivity during any season. Relative yield with no-tillage and moldboard plowing also varied between growing season. Yield of the hybrid "French 210" during 1982 was significantly lower than that of the other two hybrids on the plowed site and significantly higher on the untilled site. This interaction shows that differences do exist in corn germplasm response to tilled and untilled environments. Germplasm screening for tillage systems should be conducted over several seasons. Days to mid-silk and grain moisture at harvest were similar for the three hybrids. Thus, any reduction in early growth rate because of cold soil or other conditions associated with no-tillage did not affect crop maturity. Similar results were reported by Eckert (1984).

The hybrid performance trial was not designed to answer the question of why one hybrid responded more favorably to a particular environment than another. Information presented on stalk lodging, however, may offer an insight into crop response. As corn plants senesce, stalk rot organisms, such as *Diplodia*, invade and weaken the stems, leading to increased breakage. Maintaining an environment favorable to longer stalk life reduces lodging. In 1982, the French's hybrid had 58% lodging in tilled soil and 14% in untilled soil, indicating more healthy plants in untilled soil.

PLANTING

Crop establishment is requisite to satisfactory production in all management systems, irrespective of tillage. The viable seed has specific requirements for germination (water, oxygen, and proper temperature) that must be met before new plants can emerge. Planting equipment developed for a plowed seedbed often functions poorly in undisturbed soil. Modifications of equipment and techniques have been necessary for seed placement in untilled soil.

The seed represents the resting stage of the plant, containing a food source, a root initial, a growing point, and structures to effect emergence. The dormant seed contains moisture at levels below those required for cell division and may have other mechanisms of inhibition. Dormancy mechanisms may include mechanical inhibitions to entry of water and/or oxygen (hard seed) or biological growth regulators, which inhibit germination. While seed of some crops or weeds have special requirements to break dormancy for germination and growth, all seeds have the basic requirements of moisture, oxygen, a live seed, and a suitable temperature.

Time of Planting

Optimum time of planting for field crops coincides with the advent of a warm rainy season of the duration and temperature to permit the crop to mature. The growing season for many crops in temperature regions represents the interval between the last freeze in spring and the first in autumn. In more tropical areas seasons are determined mostly by precipitation. Ideally, crops occupy these periods as fully as possible, extended with irrigation. Early planting shifts crop growth into periods of maximum available sunlight and favorable temperatures. Delayed planting frequently moves reproductive phases into time periods when weather (temperatures, moisture conditions, and day length) are less than ideal for pollination, fruit development, and ripening. The penalty for delayed planting is a shorter season, lower yield, and higher grain moisture contents at harvest. The optimum planting time for most annual field crops usually spans but a few weeks.

A soybean study in Mississippi (Hodges et al., 1983) reported that soybeans planted during May and early June resulted in maximum yields of several varieties, row spacings, and locations. Each day planting was delayed beyond 10 June reduced yield potential by 47 kg/ha. Other crops follow similar patterns. The optimum corn planting date in the midwestern United States is early May, determined partly by soil temperature, with severe yield penalties for delays past May 20–25. Planting date represents an important factor in cultural management. Winter wheat planting in autumn is delayed to avoid Hessian fly damage. The role of tillage in management of this pest has not been well defined.

The influence of tillage on the time of establishment varies. Tillage to finish a seedbed requires less time during the busy planting season in areas where primary tillage is accomplished in fall or winter before spring planting. Early tillage in other areas is difficult because of wet soils and may create an unacceptable erosion hazard. This shifts primary tillage closer to planting time, which may thus be delayed due to the additional workload.

The time problem becomes more acute as crop area increases relative to machine capacity and available labor. The planting operation into either tilled or untilled soil is rapid, requiring far less time and power than seedbed preparation, since soil conditions suitable for tillage are also suitable for planting. Fewer tillage operations expedite planting of large acreages, particularly where fall or winter tillage is impractical.

Seed Placement

Planter operation must satisfy seed placement requirements for quick germination and emergence. The planter must cut through surface residues, place the seed in contact with soil at the proper depth, and cover the seed. Planting depth represents a compromise between a more dependable moisture supply and an increased hazard of soil crusting to restrict emergence as depth increases. Seed should be covered for protection against predation by wildlife. Tillage increases soil drying rate, and untilled, weed-free soil commonly contains more moisture than does soil that has been tilled. Because of this, the ideal planting depth may not be as great for untilled as for tilled soil.

Modern planter design has helped to overcome most problems of planting into untilled soil. Coulters are adjusted to function at the desired seeding depth. Gauge wheels positioned near the seed openers maintain precise seeding depths and reduce the amount of soil moved away from the row. Angled presswheels available on some planters push the sides of the seed trench together, covering the seed.

Chisel plowing and disking, or field cultivation, in intermediate tillage systems, loosen soil, bury residue, and permit better functioning of conventional planters. Till-planting designates a system that uses a special planter. Cultivation in the till-plant system builds ridges in the crop row during the growing season. Crop residue is pushed aside in the next season, and planting is in the old row. This operation removes many weed seeds and provides a moist soil zone for the seed. The bare soil in the row warms more rapidly, which increases the rate of seedling emergence.

Crop emergence represents the ultimate test of equipment design and the operator's skill. In early development of reduced-tillage systems emergence of row crops planted in untilled soils was routinely 10–20% lower than in tilled soil, and the seeding rate was increased to compensate. Currently available equipment still may provide 5–10% lower emergence in untilled soil.

PLANT POPULATIONS AND SPACING

Ideal plant populations vary with growth requirements of plant species and cultivars within species. A few large fruit or nut trees per hectare may be adequate, whereas small grains may require several million plants. Some crops readily compensate for low populations in some environments through vegetative development; others are more rigid. Optimum populations of annual crops may range from 50,000 to 80,000 plants/ha with corn; 300,000 to 400,000 plants/ha with soybeans; and 1.5 to 2 million plants/ha with wheat. Wheat adjusts to wide ranges in plant density through extensive tillering. Well-spaced plants at 10% of an optimum population may yield 80% of those seeded at recommended rates.

Plants at low populations explore their surroundings more completely for nutrients, light, and water. Reduced competition permits individual plants to grow larger and yield more than at higher plant densities. However, if population density is below the genetic capacity of the individual to expand vegetatively, yield per unit area is reduced. At low populations, the crop canopy may not develop completely, and weed suppression is lessened.

Plants compete with each other for space and raw materials. As populations increase, individual plant productivity declines, while area productivity increases. Individual plants of most species increase in height and decrease in stem diameter with further crowding until individual plants become barren, weakened, etiolated, and lodge, and the crop is lost.

Plants compete above and below ground level. Root-zone competition is not fully understood, but growth is limited by close proximity of neighbors. Soil exploration and uptake of nutrients and water per plant are limited. Accordingly, optimum populations are smaller in dry climates than in more humid areas. As the canopy develops, lower leaves become increasingly shaded, and light levels drop below the compensation point of photosynthesis.

Spacing and arrangement of plants are important to good management. Equidistant arrangements normally produce maximum yield thereby permitting maximum interception of sunlight and forming a complete canopy more quickly than when planted closely in rows. This, in turn, aids in suppression of weed growth. There are, however, valid reasons for planting in rows with plants spaced closely, leaving interrow spaces unoccupied.

Most available planting equipment uses openers with finite spacings. If postemergence cultivation is used for weed control, the rows must be spaced for safe passage of the cultivator. Harvesting equipment is designed with fixed spacings, which determines row widths for crops such as corn, although most other crops do not have this limitation. Without cultivation as a reason for planting in rows, producers have a wider latitude in selecting planting patterns. Planting is a precise operation to ensure good stand establishment in a suitable pattern. Its requirements are not altered greatly by tillage.

SOIL MOISTURE

Water as a raw material is essential for photosynthesis, turgor support, and mineral, O_2, and CO_2 transport to growing cells. Photosynthesis and growth slow considerably in plants under water stress. The amount and seasonal distribution of moisture is a major factor determining which, where, and how crops are grown. Excesses and deficiencies frequently limit crop yield. Moisture management is determined in large part by tillage intensity and weed control.

Plants are prolific users of water, and moisture movement through plants after canopy closure approaches the evaporation rate from a free water surface of 7–8 mm/day/cm^2. Roots explore the soil and extract water, only a small percentage of which is used in metabolism. Most is lost by transpiration and evaporation. Depth of plant rooting varies from one to several meters, depending on species, moisture, and soil characteristics. Most mineral nutrients are located near the soil surface, the primary reason for deeper rooting is the greater availability of water at deeper soil profiles.

Water Availability

Water available for crop growth is a function of water present in the soil at the time of planting plus water received from rainfall, irrigation, and other sources during the crop season. From this, subtract water used by crop and weed growth, evaporation from the soil surface, and runoff to obtain the crop moisture budget. Management goals are maximum infiltration of rainfall while minimizing losses by evaporation and use by weeds. Ideally, water should be uniformly available to the crop plants until maturity. Tillage practices can influence several of these factors.

Use Efficiency

Mulch cover on soil slows the rate of runoff and retains cracks and large pores, worm holes, and root channels as convenient pathways for water infiltration. Coupled with a lower evaporation rate, water is held near the soil surface for longer periods during and after rainfalls, which results in a net increase in total water storage (Chapter 2). This is particularly true of soils that crust when subjected to raindrop impact.

Crops in mulch-covered untilled soil experience less drought stress than in plowed soil during the growing season, and productivity is improved. Yields from tilled and untilled sites are similar during years with ample rainfall.

TEMPERATURE

Cardinal temperatures for growth, development, and survival of plants varies with species and stages of maturity. Temperature at the upper and lower tolerance levels are critical. Plant, air, and soil temperatures are related but vary independently due to solation, conduction, radiation, evaporation, and insulation. Tillage has little effect on air temperature above the crop but does affect the microclimate near the surface and the soil temperature. Mulch cover and greater soil moisture content of untilled soil serve to both insulate the soil surface and increase energy input required to raise soil temperature. Untilled soil warms less than tilled soil during the day and cools more slowly at night. This effect can be either detrimental or beneficial to seedling emergence and crop growth. In spring in more northern latitudes, slower warming of soil delays crop emergence. In Ohio the delay by corn has been 1–2 days, but in Minnesota the emergence delay may be longer and require later planting to ensure rapid seedling emergence. As the growing season progresses, temperatures in both tilled and untilled soils rise to levels no longer limiting to plant growth. In tropical areas or during midsummer in temperate regions, surface temperatures of bare, tilled soil may reach 50°C, adequate to damage crop seedlings. Here the insulating effect of mulch and lower soil temperature help protect plants from injury.

TILLAGE AND PEST CONTROL

Protection of the crop from weeds, insects, and disease pests is important to crop management. Tillage destroys unwanted vegetation, disrupts the habitats of beneficial and destructive soil insects, and buries crop residue with innocula of disease organisms.

Eliminating tillage reduces the producer's management of disease and insects through disruption of their natural habitats, but other methods can be used to minimize losses (Chapter 12 and 13). Breeders select crop cultivars with resistance to specific pests, and, since many pests are host specific, rotating crops reduces pest populations. The requirement for application of fungicides and insecticides in crop production with reduced tillage is little greater than for plowed systems in most crops.

Drought stress and weed competition increase the severity of damage from diseases and insects. More moisture available to crops in untilled, mulch-covered soil provides healthier, more vigorous plants better able to resist and outgrow injury from diseases and insects present. The net effect of reducing tillage has been increased numbers of some pests, no change in others, and decreased severity with others. Producers must recognize pest problems as they develop and adjust their management programs to deal with them.

Near complete weed control in crop management systems is essential, which accounts for the development of no-tillage or surface-tillage systems only with those crops and in geographic locations where alternative means of weed control function adequately. Uncontrolled weed growth reduces crop yield to near zero as they compete with the crop for space, light, moisture, and nutrients. Weeds also may serve as alternate hosts for disease and insect pests.

OTHER SOIL ASPECTS OF SURFACE TILLAGE

Crop Nutrition

Good seed contains a store of energy (plant food) and enough minerals to meet a plant's early development needs. After a few weeks, roots explore an increasing volume of soil and extract mineral nutrients and water to support further growth. Supplementary mineral nutrients constitute a major input in crop production, and inadequate nutrition is often a limitation to crop development. The nutritional requirements of young plants are low owing to their small size and limited growth. The demand for minerals increases as the plant expands and subsides with maturity. Most nutrients are absorbed from the soil solution near the root tip; therefore, roots must explore areas where nutrients are located and in an environment suitable for root growth.

Tillage has profound direct and indirect effects on soil nutrients, and their distribution and availability. Surface applications of some nutrients remain at or near the soil surface and are available only to roots growing near the soil–mulch interface.

Considerable research has been focused on methods of feeding plants growing in tilled and untilled soils. While some adjustments of fertilizer application methods and rates may be necessary when shifting from tilled to untilled production, crop nutritional needs can be readily satisfied with both tillage systems. (See Chapter 4 and 5.)

Traffic Support

Untilled, mulch-covered soil supports traffic far better during rainy periods than does tilled soil. This permits more timely planting and timely harvest during wet seasons. Both tilled and untilled soils can become too soft to support traffic, but untilled soil firms more rapidly after wetting. Mulch cover reduces problems with mud adhering to equipment during operations. Soil that is not deformed or compressed interferes little with root growth. Deep wheel marks in the soil, left by harvest traffic, require tillage and smoothing before planting the following season.

Plant Support

Plants lodge as stems bend or break (stem lodging) and as roots pull from the soil (root lodging). Both occur with corn, most frequently during heavy rains accompanied by wind, and are related to plants weakened by disease or insects. Crops growing in untilled soil resist root lodging, and the incidence of this problem is reduced several fold when compared with adjacent tilled areas. Stem lodging is less affected by tillage methods than disease.

Obstructions

Obstructions in the soil such as roots, stumps, or stones make moldboard plowing and other deep tillage practices very difficult on some sites and practically impossible on others. Planters with disk openers roll over shallow obstructions. Buried stones present no hazard to harvesting equipment. Sites not considered for tilled crop production because of obstructions are suitable for no-tillage or surface tillage.

A modest percentage of forested acreage is cleared each year and used for production of annual crops. After the trees are cut, stumps are removed, the soil is tilled, and roots are raked and removed to permit final tillage. These operations are expensive and must be done carefully where conventional tillage systems are anticipated. No-tillage requires only removal of above-ground obstructions before cropping.

Soil Compaction

Soil compaction is a major problem in certain coastal plain soils in southeastern United States. Traffic of farm implements and foraging livestock compress the soils and increase density. Slicing action of equipment breaks the structure that restricts root growth and water infiltration. Fields with soils where this occurs normally are prepared for crop production by deep tilling, subsoiling, and restricting traffic as much as practical. No-tillage also is useful in these situations.

A special planter has been developed for reduced and no-tillage planting on soils with compaction problems. The seed opener includes a deep tine that penetrates and loosens the compacted zone beneath the seed row. Plant roots easily follow this opening into the compacted zone, and the loosened area improves rainfall infiltration. Some use of gypsum also is being made to improve drainage through clay layers overlying deeper well-drained layers.

TILLAGE AND CROP ROTATION

Planting crops in a designated sequence represents a time-hallowed crop management practice that has found new parameters with no-tillage. Crop rotations of forage legumes followed by grain crops represented a practical means of maintaining crop productivity before inexpensive fertilizer nitrogen became widely available. In such a system, the forage legume occupied the field for 1, 2, or more years. The meadow was destroyed and nitrogen was released to supply the annual grain crop that followed. After one, two, or three annual crops, meadow was reestablished to continue the cycle. Nitrogen nutrition of the grain crop was an important contribution of the legume in rotations. Corn yield following 1 year of alfalfa (*Medicago sativa* L.) meadow was as much as triple that of continuous corn grown without added nitrogen (Triplett, 1962). With current management levels, and corn grain yields of 9 mt/ha not uncommon, forage legumes (alfalfa) have supplied most of the nitrogen needs of the grain crop (Triplett et al., 1979). Tillage is not required to obtain release of legume nitrogen.

Results from a crop rotation study for plowed and no-tillage corn on two soils are shown in Table 6.2. Nitrogen was applied to all rotations at rates considered not limiting for any system. With plowing on both soils, yield trend with crop rotation was positive and ranged from 5 to 13%. The corn–soybean yields were intermediate between the corn–oats–hay rotation and continuous

TABLE 6.2. Average Corn Grain Yield on Two Soils, in Three Rotations and Two Tillage Systems in Ohio (Average of Years 7–12)

		Tillage (mt/ha)		
Soil	Rotation	No-Tillage	Plow	Significance
Wooster	Continuous corn	9.4	8.4	**
silt loam	Corn–soybeans	9.5	8.7	*
	Corn–oats–hay[b]	10.4	9.7	**
Hoytville	Continuous corn	6.8	8.0	*
silty clay loam	Corn–soybeans	7.9	8.3	NS
	Corn–oats–hay	8.2	8.4	NS

[a]Probability level that differences in means are significant: * >5%; ** >1%; NS: not significant at <20%.

[b]Oats (*Avena sativa* L.).

SOURCE: Adapted from van Doren et al. (1976)

corn. With nitrogen not a factor, the yield response of rotation may reflect pest control.

Soils in this study differed in their response to tillage. The Wooster soil is well drained and considered first choice for no-tillage. (See soil classification scheme, Table 2.29.) Yields with no-tillage were greater than with plow-tillage with all cropping sequences on Wooster soil. Hoytville soil is poorly drained and not a first-choice soil for no-tillage. On this site, continuous corn yields without tillage were significantly lower than when plowed. Yields of no-tillage corn following soybeans or meadow were equal to those with tillage. No-tillage yields can be equal to yield after fall plowing if the corn follows some other crop, an important management factor for this soil (Van Doren et al., 1976).

In the example shown (Table 6.2), corn appeared every other year when alternated with soybeans or every third year in the meadow cycle. If cover crops followed annual crops of corn, would there be a rotational benefit seen in the longer cycle of corn–soybeans? Such an effect, if it existed, could be especially important in maintaining no-tillage corn yields on poorly drained soil.

There are few published studies dealing directly with cover crops and corn in monoculture on poorly drained soils. In studies with cover crops planted as intercrops on better-drained soil, yields of tilled corn receiving adequate nitrogen applications were equal with and without cover crops but were lower than following meadow. In studies on a poorly drained soil, yields of tilled continuous corn were similar with or without cover crops (Triplett, 1962). Thus, the same crop appearing no more often than alternate years may be necessary to fully realize rotational effects.

Despite their obvious benefits, crop rotations call for extra planning. A sequence that includes a less profitable crop, and gains a mere 5–10% yield improvement with a rotation may encourage a grower to accept the yield reduction to produce the most profitable crop as a continuous monoculture. Cropping sequences containing forage legumes often compels the grower to operate a livestock enterprise if there is not a ready market for hay. Yield penalties of 15% or more with continuous no-tillage corn on poorly drained soils (Table 6.2) are excessive and may limit adoption of reduced-tillage systems.

Rotating tillage systems also serves to maintain corn yields on poorly drained soils; yield reductions are manifest for the second and succeeding years of no-tillage. Thus, there is little yield penalty for no-tillage corn following tilled corn. This relationship is illustrated by results from 4 years of two contiguous experiments on the Hoytville soil shown in Table 6.3 (Triplett and Van Doren, 1985). Experiment I presents individual years of the rotation study reported in Table 6.2. In Experiment II, no-tillage corn was planted into soil plowed for the previous year crop. All plowing was sometime before planting both experiments. Hoytville soil dries slowly and tilling wet soil creates clods that make immediate preparation of a suitable seedbed almost impossible.

A direct comparison of yields of the two separate experiments reported in Table 6.3 is not proper. Tillage treatment comparisons consist of percent yield change calculated as follows:

$$\frac{Y_{NT} - Y_{FP}}{Y_{FP}}$$

where Y_{NT} = no-tillage yield
Y_{FP} = Fall plowing yield

The change values calculated and their direction makes comparison possible. Yields from continuous, no-tillage corn in Experiment I averaged a significant 16% less than for fall plowing. In Experiment II, no-tillage corn that followed tilled corn averaged a nonsignificant 3% less than tilled continuous corn. Based on results of these studies, rotation of either crops or tillage reduced the adverse response to no-tillage on this soil. These relationships provide growers with important information in devising management strategies for reduced tillage on poorly drained soils.

A striking difference between the two soils compared in Table 6.2 is the degree of internal drainage. Water percolates through the Wooster soil much more rapidly than through the Hoytville soil. This leads to a question of whether improved drainage would serve to help maintain continuous, no-tillage corn yields on poorly drained soils. A corollary question involves determining the intensity of tillage required to maintain crop yields. In studies on the Hoytville soil, continuous corn planted on ridges constructed the previous season produced yields equal to continuous corn planted into fall-plowed soil. Thus, planting on low ridges as a means of improving drainage in the seedling zone also appeared to diminish the detrimental effect of reduced tillage for continuous corn. Ridge planting as described here is similar to the till plant system.

MULTIPLE CROPPING

Allocation of land use to one crop for an entire season is wasteful of land and climatic resources unless that crop is a full-season perennial. In the middle latitude humid regions of the North and South American continents, corn and soybeans are planted in early to midspring and mature in early fall. The cropping season occupies 5–6 months and management is directed toward maximum production of the single crop. Land is idle much of the year.

Two means are available to increase land productivity. The first is to ensure maximum production of the major crop, a method pursued in most temperate regions. An alternate means of increasing productivity is to develop cropping systems in which the land is used throughout the year. Multiple cropping employs this approach to intensify land use capability. With multiple cropping,

TABLE 6.3. Annual Corn Grain Yield from Various Tillage and Cropping History Combinations on Hoytville Silty Clay Loam Soil, Experiments I and II

	Experiment I						Experiment II					
	Tillage Not Rotated						Tillage Rotated					
	Continuous Corn			Corn–Oats–Hay			Continuous Corn			Corn-Hay		
Year	Fall Plow (mt/ha)	No-Tillage (mt/ha)	Difference[a] (%)	Fall Plow (mt/ha)	No-Tillage (mt/ha)	Difference[a] (%)	Fall Plow (mt/ha)	No-Tillage (mt/ha)	Difference[a] (%)	Fall Plow (mt/ha)	No-Tillage (mt/ha)	Difference[a] (%)
1968	5.9	4.9	−17[b]	5.9	6.3	+ 7[b]	7.8	7.5	−4	8.2	7.8	− 5
1969	5.9	4.7	−20[b]	6.0	6.0	0	7.5	7.1	−5	8.4	8.5	+ 1
1970	7.4	6.2	−16[b]	7.6	6.7	−12[b]	7.4	7.3	−1	7.4	7.9	+ 7
1971	10.5	9.3	−11[b]	11.0	10.8	−2	10.2	9.9	−3	11.4	10.1	−11[b]
Mean	7.4	6.3	−16[b]	7.6	7.4	−2	8.2	7.9	−3	8.8	8.6	− 2

[a] [(No-till) minus (fall plow)]/fall plow.
[b] Difference significant < 5% level of probability.

SOURCE: Adapted from Triplett and Van Doren (1985).

production potential of one crop may be sacrificed so that the second crop may contribute to total seasonal return. An evaluation of multiple cropping systems considers all crops within specified time periods.

Multiple cropping is common in areas where limited land resources are available to feed large populations. Subsistence farmers in tropical regions often plant several crops together or in succession. A crop with a short growing season (a vegetable) is interplanted in a crop (corn) requiring more time to mature. The short season crop completes its life cycle while the other becomes established. A third crop may be planted before the second matures. These systems strive for maximum use of land resources, and require careful planning and intensive hand labor.

There are constraints with full use of multiple cropping systems in temperature regions with mechanized agriculture. Freezing temperatures limit growth and alter the sequence of winter annual crops to occupy the land overwinter. Since the time interval between harvest and planting is crucial they must be mechanized. Availability of adequate moisture for both crops is essential in the planning.

Sequential cropping systems intensify crop production only in the time dimension (Andrews and Kassam, 1977). A second crop is planted following harvest of the preceding one: producing two crops during a single season is double cropping, while growing a third (often vegetables) is triple cropping.

A common sequence in low or middle latitudes involves a winter small grain crop [oats, wheat, or barley (*Hordeum vulgare* L.)], followed by soybeans planted in the stubble immediately after harvest. Small grains mature successively later at more northern latitudes, and the growing season for the second crop becomes shorter and impractical. The zone is not clearly defined, but is still moving northward as no-tillage management practices are improved. Use of early maturing small grains and soybean varieties, early harvest, and heat drying of the grain contribute to adoption. The current northern limit of wheat–soybean double cropping is approximately 41°N latitude. The western limit is determined by soil moisture availability and midsummer rainfalls. Soybean maturity is delayed by late planting so fall-seeded small grains do not follow well except in southeastern United States with a long frost-free season. Double-crop production is normally limited to three crops in 2 years in midwestern United States.

With good management, double-crop soybean yields have averaged 40–50% those of a full-season crop, whereas winter small grain yields are not affected. Productivity of multiple cropping systems is measured by calculating the land equivalent ratio (LER) of the crop involved (Andrews and Kassam, 1977):

$$\text{LER} = \frac{\text{multiple crop yield}}{\text{sole crop yield}}$$

Continuing, with a small grain yield of 1.0 plus a soybean yield of 0.4, LER = 1.4. Thus, land productivity per year is increased by 40%.

No-tillage practices are helpful in making double-cropping systems successful. By not tilling the soil, the succeeding crop can be planted more quickly. Seedbed preparation with plow-tillage normally takes a week or more. In the small grain–soybean sequence, soybean planting in the northern corn belt is delayed several weeks beyond the optimum date. Soybean yield potential declines rapidly with each day planting is delayed. This factor becomes more acute in more northern latitudes than further south. Multiple cropping of vegetables in Louisiana with no-tillage planting permitted four crops during the season instead of three in tilled soil (Standifer and Mohd noor, 1975).

Multiple-cropping systems can be profitable, rivaling or exceeding the net return of most single-cropping systems. The cost of land use represents a significant production expense, which is shared by both crops. Double-cropping profitability has led to increased wheat acreage in southeastern United States as the practice expands.

Several other crop sequences are in use in humid regions, including:

- Field peas (*Pisum sativum* subsp. *arvense* (L.) Poir.)–sorghum (*Sorghum bicolor* (L.) Monech.), sudangrass (*Sorghum bicolor* (L.) Monech), millets (*Pennisetum* and *Panicum* sp.).
- Canning peas (*Pisum sativum* L.)–sweet corn (Wisconsin).
- Sweet corn–sweet corn.
- Meadow–corn.
- Small grain–sunflower (*Helianthus annuus* L.).
- Small grain–buckwheat (*Fagopyrum esculentum* Monech.).
- Various vegetable crops planted in succession.

In southeastern United States the following are possible:

- Corn–soybeans (Florida, South Georgia).
- Small grain–grain sorghum.
- Soybeans–soybeans (Florida).

Triple cropping is possible with some vegetable crops. Other double-crop sequences increase as the tropics are approached.

In double-crop systems, the time interval between harvest of the first crop and emergence of the next allows use of nonselective herbicides for weed control. Weed control is, seemingly, less difficult in double-crop systems with delayed planting than for earlier planting.

Tillage and growing crops dry the soil. Plowing small grain stubble and disking to prepare a seedbed, accelerates soil moisture loss by evaporation. Replenishment is required for germination of a newly seeded crop. A soil untilled and mulch-covered conserves the moisture present. Moisture is a critical factor in the success of the crop that follows and, in areas depending solely on rainfall, double cropping should not be attempted if the soil is dry at the time of small grain harvest.

In *intercropping* systems, crop intensification is in both space and time dimensions, that is, two or more crops occupy the same field during a portion of the life cycle of each (Andrews and Kassam, 1977). Harvest requirements, differences in cultural practices, and the need for mechanization restrict the opportunities for many intercropping systems in temperate agricultures. Relay intercropping with wheat and soybeans are being developed (Jeffers 1984). Soybeans are planted in the growing wheat crop with a no-tillage planter, which minimizes damage to the wheat from traffic. The soybeans emerge and become established before wheat harvest. With the overlap of crops, the growing season is used more than with double cropping. Management of the two crops is more demanding than in sequential systems. Lodging of the small grain must be avoided. The number of herbicides to use in both crops is limited, which makes weed control more difficult.

Successful relay intercropping of wheat and soybeans offers several advantages over double cropping. Reduction in wheat yield, either from traffic or competition from soybean plants, is more than offset by increases in soybean production. LERs with relay intercropping have been has high as 1.6. Timely soybean planting and maturity reduce the danger of crop damage from early frosts. Timely maturity improves the potential for seeding a third crop following soybean harvest.

COVER CROPS

Most crop production systems, with grains or other annual crops, occupy the land for only a fraction of the year. When the principal crop is harvested, the soil may be bare, exposed to erosion, or occupied by weeds. In arid areas, volunteer crops and weeds are suppressed and in the presence of a caliche crust moisture is conserved. In humid areas, moisture conservation is equally important. Crops are often grown merely to provide cover between principal crops.

Cover crops represent a time-honored agricultural practice that was revived with no-tillage development. The cover crop protects the soil against wind and water erosion, a legume cover fixes nitrogen that will be available to the crop that follows. A cover crop provides forage for livestock, important in extending the grazing season. These reasons are expanded to include an anchored mulch for no-tillage systems.

Choice of Crop

Traditional factors governing selection of crop species for cover include ease of establishment, availability and cost of seed, ground cover provided, growth rate during the off season, feed and cover for wildlife, dependability (winter survival), and the prospect of the crop to become a weed. With legume cover crops, the amount of nitrogen fixed is very important. The most widely used cover crops are the small grains, rye (*Secale cereale* L.), wheat (Moschler et al., 1967), and oats. Legumes include hairy vetch (*Vicia villosa* Roth) and crimson clover (*Trifolium incarnatum* L.). Annual forage and turf grasses may be used in some areas (Mitchell and Teel, 1977).

Cover Crop Management

Rye, ryegrass (*Lolium multiflorum* L.), and wheat are often seeded broadcast as intercrops in corn and soybeans while the crop is maturing. They provide soil cover by the time the principal crop is harvested. Early establishment of the cover crop is especially desirable in areas with severe winters.

Seedings made following crop harvest establish more rapidly since a seedbed is prepared to provide seed–soil contact, but it also buries existing residue and increases the erosion hazard. Reduced-tillage seeding methods of cover crops are being developed. Seedings after crop harvest may not provide adequate time for establishment and soil cover in autumn before growth is arrested by cold weather. Weed control methods used during crop establishment are important. Herbicides, such as simazine in corn and orazylin in soybeans, may persist as harmful residues and reduce cover crop stands. Timing, herbicide rate, and growing season rainfall are factors that govern herbicide persistence.

Spring management practices influence the value and effectiveness of cover crops. Important functions of cover crops are to produce a desired amount of a persistent mulch, to improve soil tilth, and to provide nitrogen fertility. Furthermore, the cover crop should not interfere with the planting of the next crop. Time of planting following kill of the cover crop is important. Early kill limits dry matter production and the succulent herbage decomposes rapidly. Rye and wheat cover crops produce more persistent mulches than do many legume species, owing to higher cellulose content. Using the cover crop as forage, by either grazing or harvest, also reduces the cover provided.

Adverse Effects of Cover Crops

Delay in killing the cover crop increases dry matter production but has potentially adverse effects. A living cover crop uses stored soil moisture, which may be detrimental to the crop that follows. This was illustrated by studies reported by Sprague et al. (1963). In humid areas, rainfall ordinarily replenishes moisture used. In arid climates soil and crops are managed to store moisture and water used by the cover crop reduces grain yield potential. Thus, effective use of cover crops is limited primarily to humid regions. Older, more mature cover crop plants are more difficult to kill and require higher rates of herbicides.

Some pest problems can be affected by cover crops. Dense cover crop growth in early spring suppresses germination of seed of annual weed species.

The growing cover crop attracts some insect pests, such as army worms. The moths oviposit in the vegetation, and larvae that hatch in the dying cover crop move to feed on small crop plants as they emerge. Severe crop damage can be averted by timely insecticide application, but the potential for damage requires that producers monitor fields closely.

Evaluation

The value of cover crops for yield improvement has not been established for all reduced-tillage practices (Moschler et al., 1967; Van Doren and Triplett, 1973). On some soils, crop yields are decreased with no-tillage if the soil surface is bare. Cover crops clearly are needed for reduced-tillage crop production on these soils if the aboveground portion of the previous crop is removed, such as silage corn. Crops that provide minimal amounts of persistent residue also may be candidates for a cover crop. Yields are less affected by mulch cover levels on poorly drained soils that crack upon drying, and value of cover crops is less well defined. The suggestion that cover crops benefit wet soils by removing moisture in the spring is realistic but not fully quantified.

Cover crops represent one of several sources of mulch in no-tillage production systems. The amount of soil surface covered and persistence of the mulch represent the best measures of mulch quality and effectiveness. The residue from a corn crop yielding from 6 to 9 mt/ha of grain will provide about 95% cover at harvest. By the following spring the mulch has deteriorated in northern latitudes to about 70–80% of the soil surface. Coverage may be reduced to 50–65% by mid August by which time the crop canopy is complete. Rate of mulch deterioration is influenced by composition, weather factors, feeding by livestock, and soil microorganisms. Highly nitrogenous crops decompose more rapidly than do others due to the availability of nitrogen for microbiologs. Cover from a high-yielding corn crop provides adequate protection from erosion and moisture loss. As such, a cover crop would contribute little unless a legume is used. This is true in northern areas where cold weather arrests cover crop growth for most of the winter. In Ohio studies on well-drained and poorly drained soils, there was no positive effect of a rye cover crop on corn yield where residue was present from the previous crop. Both corn and soybeans served as previous crops in these multiyear studies (personal communication D. Eckert, Ohio State U., Columbus.).

Legume cover crops are sometimes considered as a source of nitrogen for the grain crop that follows. Based on the frequency of published reports on the subject, interest in legume cover crops declined when fertilizer nitrogen became inexpensive in the 1950s and revived about 1975 in response to more expensive energy supplies and increased nitrogen costs.

Mitchell and Teel (1977) in Delaware and Ebelhar et al. (1984) in Kentucky investigated the value of legume cover crops as a source of nitrogen for no-tillage corn production. Cover crops were interseeded after corn was mature or following corn harvest. The cover crop was killed with herbicides the next spring, and corn was planted into the mulch. In both studies, vetch provided the equivalent of 100 kg/ha nitrogen for the following corn crop. In New Jersey, Flannery (1981) reported that a vetch cover crop supplied the equivalent of 200 kg/ha nitrogen in corn harvested for silage. Silage harvest precedes grain harvest by several weeks and permits early establishment of the cover crop, increasing the potential for growth. Crimson clover also was an effective species in Delaware (Mitchell and Teel, 1977).

The amount of nitrogen fixed by the legume cover crop is determined by the amount of growth between its seeding and destruction for the following crop. Nitrogen fixation by legumes slows at temperatures below 10°C. The Delmarva peninsula (Mitchell and Teel, 1977) and Lexington, KY (Eberhar et al., 1984) both have climates with long periods of moderate temperatures between corn maturity and replanting, which permits considerable growth of the cover crop. In work in northern Ohio, Triplett (1962) reported that an alfalfa intercrop seeded in wide corn rows in June provided about 50 kg/ha N for the tilled corn that followed. Vetch was not reliable in these studies because of winter kill (unpublished). In the northern Ohio climate, growth of the legume essentially ceases by mid October and does not resume until mid or late April. Optimum planting time for corn in this area is about May 10; thus, the value of cover crops as a source of nitrogen is less than in areas with moderate climate.

Legume cover crops as a source of nitrogen often require a series of crop management compromises. Early destruction of the cover limits the amount of nitrogen fixed. Delaying crop planting to increase cover crop growth and nitrogen production reduces yield potential for some crops. In areas where moisture use by the cover crop is less important, the legume growing season (and nitrogen fixation) might be extended by interplanting the corn into the growing cover crop and delaying kill, perhaps until after corn emergence.

Finally, economy of the practice must be considered. A value can be placed on the nitrogen fixed by the legume by comparison with fertilizer N replacement. This must be offset by cost of cover crop establishment and kill. In the Kentucky studies, vetch was seeded at a 40 kg/ha rate. At current prices, seed cost largely offset the value of nitrogen produced. However, in the New Jersey studies (Flannery, 1981) a vetch cover crop for silage corn was the most profitable system.

MANAGEMENT OF MULCH COVER

Crop residue on the soil surface may have undesirable effects on crop growth and productivity. Under some circumstances of climate or cropping sequence, mulch removal may be the only means of maintaining crop yields with no-tillage planting. Elliott et al. (1978) point out that residues left on or near the soil surface frequently reduce growth and yield of the next crop as compared with those where residues are removed or when conventional tillage is used. Cereals seem particularly susceptible. In initial no-tillage seedings of wheat, yields decreased as residue amounts increased with crop rotation sequence. Growth appeared normal when residue amounts were low or the previous year's stubble was burned. With high residue levels, wheat emergence was poor, tillering was reduced and plants were spindly. Elliott et al. (1978) speculated that reduced yields might be associated with toxic compounds leached from crop residues and/or to microbial production of toxic compounds during residue decomposition. However, at present, it is very difficult to directly implicate phytotoxins as the cause.

In a later study in the same area, Cochran et al. (1982) seemingly overcame residue problems described by Elliott et al. (1978). At least part of the crop residue problem was mechanical. Disk drills used to direct-seed cereals often place seed in proximity of surface residues and in some cases may fold straw into the soil, enclosing seed in a straw envelope. This results in poor seed–soil contact and enhances exposure of seed to phytotoxins that may be present. In their 3-year study, there was no yield decrease with surface residues present. There was higher yield with residue present during a dry season because of moisture conservation (Cochran et al., 1982).

In the United Kingdom (UK) where reduced tillage is used for cereals, burning straw is common and is generally considered to be a prerequisite to successful direct drilling (no-tillage) (Soane et al., 1975). In the UK climate, where chronically wet soils are common, burning straw leads to increased surface tilth and soil surfaces are dried somewhat for several weeks after burning (Cannell et al., 1978). Problems associated with straw and other residue tend to be worse in wet years. Risks can be avoided by ensuring complete straw burn.

Excess straw poses a special problem in multiple cropping systems where soybeans follow small grains. Soybean planting date is delayed while the grain matures; planting proceeds immediately after harvest to ensure maximum productivity of the second crop. Straw from the freshly harvested grain has no time to decompose, and toxins that may be present in the straw have no chance to dissipate before seeding soybeans.

In an evaluation of various straw management alternatives for double cropping, Sanford (1982) reported that soybean yields were highest and most consistent when soybeans were no-tillage seeded into burned stubble. Disking to incorporate straw was a poor treatment; and planting into shredded straw reduced yields one year in three. Varying intensities of soybean leaf yellowing were noted in all three years where straw was not burned. The yellowing occurred on all treatments where straw was present, but was more intense where straw was left standing or shredded. Yellowing characteristically occurred soon after the first rain following emergence and persisted 3–4 weeks, followed by gradual disappearance.

Tillage dries the soil. In one year when rain did not occur for 10 days after planting, tilling burned stubble before planting reduced yields (Sanford, 1982). In areas where straw has value for other purposes such as for animal bedding, baling the straw for sale seems acceptable.

CHANGES IN TILLAGE MANAGEMENT

More than a half century of searching has produced an evolution of revolutionary changes in how field crops are grown. Shifts in tillage practices are gradual as improved techniques are discovered, refined, farm tested, and eventually adopted. "Conventional" Corn Belt tillage operations during the 1930s consisted of moldboard or disk plowing in fall or spring in preparation for planting a full-season row crop. As the soil warmed in spring, the field would be disked two or more times to provide a firm smooth seedbed. A field often might be worked again with a spike harrow or rotary hoe to break a soil crust or control emerging weeds before crop emergence. This was followed by from two to four cultivations before the crop canopy completed development in July. Check row hill planting permitted cross cultivation for better weed control than row planting but doubled the number of trips over the field.

Conventional tillage has changed. By 1980 "conventional" tillage had evolved to a moldboard plowing, a tandem hitch field cultivation, and smoothing to prepare the seedbed and possibly to incorporate herbicides before planting. During the period 1930–1980 the number of passes over a field to grow a crop of corn with "conventional" tillage had about halved. Weed control with herbicides has made these changes possible. No-tillage systems reduce the traffic further with less soil compaction and associated cost savings. An estimated 95% of the acreage of major row crops in the United States now receive one or more herbicide applications in place of postemergence cultivation.

In this discussion *plow-tillage* is defined as a system that employs a moldboard or disk plow to bury and incorporate plant residue in the soil (Chapter 1). *Surface-tillage* systems use cultivators, chisels, or disks for primary tillage to loosen the soil and retain plant residues on or near the surface. *No-tillage* represents a spray and plant system, reduced tillage includes both. Surface tillage involves a continuum of practices and demarcation between them is blurred although the results are similar. Selected herbicides may be used with any system.

Viewed from an historical perspective, farm operators in the United States are using much less tillage than 3 to 4 decades ago. Two-thirds of the row crop area is still plowed but during the last decade producers have moved rapidly to the adoption phase of some kind of reduced tillage. Only an estimated 5–10% of the total corn crop is produced with no-tillage; however, rapid expansion during the past decade in some areas indicates a well-developed trend that is likely to continue.

SELECTING TILLAGE SYSTEMS

No agricultural management practice is wholely beneficial under all situations. Compromise usually is essential because sites and situations are not predictable. Adoption of new practices requires weighing of benefits against undesirable effects. Tillage systems are site, soil, and crop specific, and their effects must be assessed for each locale. If tillage selections include only moldboard plowing and no-tillage, choices are relatively simple. These extremes, with the array of intermediate systems, permit reduction in tillage to meet most soil and crop management requirements. Furthermore, tillage choices should be made annually with some fields tilled in one season remaining untilled the next.

All aspects of reduced tillage must be considered before adoption. These include:

1. Elements of risk should not increase significantly. Agricultural disasters associated with weather and other uncontrollable circumstances must be reduced to a minimum. Crop failures associated with reduced-tillage systems often can be traced to the user's lack of knowledge. Growers must be willing to learn the new systems before adopting them on large acreages. Also, knowledge of how the systems perform locally, and possible causes of failure, must be available.

2. A profit margin is of prime importance to an economic system. A change to a new tillage system must be an improvement over that abandoned, whether on well-drained fertile soils or in problem areas. Yield penalties, if any, for reduced tillage should be offset by reduced costs of inputs. A complete assessment of this factor should include opportunities for management improvement. For example, if a yield penalty for delayed planting can be avoided by eliminating tillage operations, the net effect of adopting reduced tillage may be a yield improvement.

3. Systems must be matched to local soils, crops, pests, climate, equipment, and personnel regimes. General cultural practices must be adapted to localized situations, with on-site decisions essential. There are still gaps in knowledge that must be filled before a complete assessment of various tillage effects is available for different crops and localities. This process probably will continue for a few more decades before completion. As systems are refined and timing becomes more accurate, their adoption becomes more attractive.

Table 6.4 lists tillage effects of factors that influence crop productivity. Only plowing and no-tillage, the polar cases, are considered. Reduced-tillage systems have intermediate effects related to degree and depth of soil disturbances. The column "Effect on Crop" indicates positive, negative, or no direction. This effect can be so small as to be inconsequential or large enough to warrant a major consideration in adopting a specific system. Examples based on several factors in Table 6.4 may serve to illustrate this point.

Undecomposed organic matter remaining on the soil surface slows runoff, reduces evaporation, and increases rainfall infiltration. Slower runoff rates and protection of soil from raindrop impact reduces erosion (positive in all areas on sites where erosion is a problem). The mulch cover decreases diurnal temperature variation (slowing emergence in cold climates and beneficial to growth in tropical areas). Pathogens and insect pests may be increased because of mulch

TABLE 6.4. Comparison of Tillage Effects on Factors Influencing Crop Productivity

Factors	Plow-Tillage	No-Tillage	Effect on Crop[a]
Soil Factors			
Temperature	Warm days, cool nights	Little variation	±
Water intake	High following tillage, decreases with crusting	Lower initial rate. Maintained through season	+ 0
Surface-applied minerals	Mixed with soil to plow depth	Percolate slowly	0
Bulk density	Decreased by primary tillage	Little effect	$\underline{0}$
Compaction	Broken by tillage	Little disturbance	−
Aeration	Initial increase	Little affected	$\underline{0}$
Organic matter	Mixed with soil	Near surface	
distribution			±
decomposition rate	Rapid	Slower	±
Erosion hazard	Serious	Soil protected	+
Runoff rate	Rapid, when soil compact	Slowed by mulch	+
Evaporation	Rapid	Decreased by mulch	±
Biological Factors			
Weed control	Initially excellent	Reliance on herbicides	−
Disease organism	Innoculum buried (or incorporated)	At surface	−
Soil invertebrates			
beneficial	Disrupts life cycles	Little effect	+
destructive	Same as beneficial	Little effect	−
Machine Function			
Planting	Machine designed for loose soil	Special equipment for undisturbed soil	$\underline{0}$
Cultivation	Effective in loose soil root pruning	More difficult in mulch	$\underline{0}$
Traffic support (in wet soil)	Poor	Good	+
Management			
Timely operations	May be delayed	More timely to be reliable	+
Power demand	Increased by successive operations	Minimal	N/A
Labor demand	Increased by successive operations	Minimal	N/A
Dependability	Good	Can be erratic	

[a]Codes: +, positive benefit for no-tillage; 0, neutral or no effect; −, no-tillage detrimental. N/A not applicable.

(negative) and machine function in planting may be impaired (negative). Organic matter decomposition is slower in untilled soils. This is positive for long-term soil productivity but negative if elements released by organic decomposition constitute a major source of crop nutrition.

Soil bulk density, soil aeration, and compacted layers are modified by tillage. Compact soils with high bulk density restrict plant rooting and moisture movement only when bulk density exceeds certain limits ($\cong >1.6-1.7$ g/cm^3). Below this level, the effect generally is considered minimal. Compacted layers are formed in soils either naturally or by traffic. If tillage serves to break these layers (where present) and improves conditions for root development, the tillage effect is positive. If root development is not restricted, deep tillage usually has little effect on a crop. Air in unflooded soil usually supports respiration and root growth adequately. Under short-term flooding, O_2-laden water replaces air in the larger soil pores with little effect, but longer-term anaerobic conditions created in stagnant soil water can severely damage many crops. This can occur in both tilled and untilled soil but larger pores in tilled soil initially drain more rapidly, helping decrease the extent of anaerobic conditions short term. New channels open as dead roots decay in no-tillage improving drainage long term. Flooding and anaerobic conditions usually are hazards on relatively flat terrain and specific soils.

MATCHING SYSTEMS TO SITE CHARACTERISTICS

With the wide variety of soils, climatic regimes, and different crops that are grown, each carrying specific management needs, tillage systems selected should create best conditions for the crop at the location with a minimum of undesirable effects. Several different systems may provide adequate performance at a given location. Examples are now given of different sites and best tillage practices.

EXAMPLE 1. Site characteristics for this example include well-drained soils on slopes >5%. Soil erosion is a major problem with annual row crops, and the tilled, bare soil surface seals during the growing season, limiting rainfall infiltration. Mulch cover is necessary for no-tillage crop production. To sustain tilled systems, producers must include forages and small grains in the cropping sequence, which almost dictates a livestock enterprise.

No-tillage benefits include increased rainfall infiltration, and decreased erosion and evaporation. The net effects on corn production include higher yields, expanded land use capabilities, and more timely operations. In response to these positive benefits, no-tillage has been adopted most widely on such sites.

Geographical areas with soils and topography fitting the description include parts of Kentucky, West Virginia, eastern Ohio, Pennsylvania, Virginia, North

Carolina, and Maryland. Locally in this region, no-tillage may occupy more than half the corn acreage.

EXAMPLE 2. Site characteristics for this example include a poorly drained soil with high clay and high organic matter content. The soil cracks when dry and crop yields are not increased by mulch cover. Low soil temperature in early spring delays seedling emergence. Slopes are 0–2% and soil erosion is not a dominant problem. Yields of continuous, no-tillage corn have been lower than fall or winter moldboard plowing by 10–15% on this soil. Clayey soils dry slowly in spring, and plowing at planting time forms persistent clods that makes seedbed preparation difficult.

On such a soil the yield penalty for continuous no-tillage corn is unacceptable. Yet, research has identified means of mitigating this yield reduction with no-tillage, and other reduced-tillage systems also may perform satisfactorily. No-tillage corn yields are equivalent to fall plowing if corn follows some other crop or the soil was plowed for corn the previous season. If moldboard plowing is delayed until the planting season, no-tillage or reduced tillage (such as disk, plant) are better choices for timely crop establishment and avoid the rough seedbeds formed with deeper tillage.

A surface-tillage (till-plant) system is emerging as an excellent reduced-tillage practice for these poorly drained soils. Tillage moves the mulch from the seed row, facilitating planter function and rapid seedling emergence. The slight elevation of a ridge improves drainage in the seed zone. Continuous corn yields with the till-plant system have equalled fall plowing. Producers having soils described in this example should be equipped to employ one or more of several alternative tillage systems, depending on weather and previous cropping history.

EXAMPLE 3. Multiple cropping with soybeans may follow a small grain harvest on all soils. Delayed planting of the succeeding crop is a dominant factor influencing yield in multiple cropping systems. At small grain harvest, soil temperatures are adequate for seedling growth but inadequate soil moisture commonly limits seedling emergence. Tillage serves to further dry the soil, and surface- or no-tillage systems facilitate rapid planting with minimum moisture loss.

Small grain straw may interfere with seed placement and early growth of the second crop. Straw removal from the crop row zone is necessary and can be done by mechanical means or by burning. Choice of methods should consider the soil, value of straw, erosion hazard, and any restrictions on burning.

From the examples above, selection of tillage systems should be based on prevailing local conditions. A number of systems may create a suitable crop environment and result in comparable yields. All factors should be considered, in particular the timeliness of operations as well as operator skills in making systems perform properly.

PROGRESS IN SURFACE TILLAGE FOR ROW CROPS

Corn occupies a major share of the crop area in the warmer temperate humid regions, with rainfall exceeding 700 mm (28 in.) and a growing season of 3 months or more. No-tillage cultural management is popular and increasing for several reasons. Corn is a robust plant with wide adaptability and grows well in a monoculture. Selective herbicides are available to control a wide spectrum of competitive weeds. Specialized equipment is available for precision planting in untilled mulch-covered soil. It is a versatile crop easily adjusted in rotations with a variety of crops. Adapted techniques permit application of fertilizers and pesticides at planting time. Finally, it is a high-yielding nutritious crop attractively priced as a feed.

Several factors limit the expansion of no-tillage corn. A few weeds escape herbicides currently available. Included are the warm season grasses, johnsongrass and bermudagrass, and areas infested with these grasses should be avoided. Corn yield can be impaired on cold, poorly drained soils, but the adverse effects of poor drainage can be mitigated to some extent by tile draining, crop rotations, and certain types of tillage. Time of planting is determined largely by soil temperature in middle latitudes and by soil moisture in the tropics. Root diseases are suspected of being a contributor to lower yield on poorly drained soil and some insects are devastating if not controlled. Crop rotation reduces pathogen populations (Chapters 12 and 13).

Several forces function to expand the reduced-tillage corn area. Time and fuel conserved by eliminating operations increases cost efficiency. Land on slopes can be cropped more intensively without incurring unacceptable soil erosion. This provides the farmer with rolling terrain the option of expanding row crop production onto land currently having lower income potential. Finally, certain governments partially reimburse farmers who try an area of no-tillage as an introduction and incentive to acceptance.

Sorghum for grain and forage is normally grown in more arid areas than corn. Although corn is the preferred crop because of higher feed value and price, sorghum will tolerate unpredictable periods of drought with less precipitous decline in yield. Production areas overlap at about 635 mm (25 in.) rainfall in the northern production area to around 900 mm (35 in.) in the southern area. No-tillage with associated moisture conservation alters the risk of adaptability. Furthermore, sorghum tolerates delayed planting better than corn and fits well into multiple-cropping systems. In tropical areas two to three ratoon crops can be harvested where moisture is available and nitrogen is added between crops.

Both reduced- and no-tillage systems are used for sorghum production. Sorghum is grown in rows and equipment used to plant both corn and soybeans is suitable for sorghum. Sorghum normally follows other tilled crops in the cropping sequence, although some sorghum may be grown as a continuous monoculture. Many herbicides available for weed control in corn also can be used with sorghum. Thus, warm season perennial grass infestations represent the most important group of weeds that limit use of reduced tillage for sorghum. Volunteer sorghum is a serious weed problem for no-tillage sorghum in monoculture (Wiese et al. 1985). However, by including metolachlor in the herbicide mixture and treating seed with a safener (antidote to the herbicide), volunteer sorghum can be controlled while seeded plants are not harmed.

Since sorghum is produced in drier areas, soil moisture conservation represents the major reason for development and adoption of reduced-tillage practices for this crop. Much sorghum follows wheat in the cropping sequence and additional moisture conserved by good straw cover coupled with chemical control of weeds between crops can significantly increase sorghum yields (Wiese et al., 1985). Moisture conservation and rapid crop establishment are critical factors in successful multiple-cropping systems. Erosion control remains a reason for reduced tillage in humid areas.

There does not appear to be major restriction to the further development of reduced tillage for sorghum. The relationships between drainage levels, soil type, tillage system, and cropping sequence established for corn production have not been fully tested with sorghum. Details of other cultural practices also need further development.

The *soybean* production area rivals that for corn but, in general, tillage used for soybeans is more intense than for corn. Row crop planters can be used interchangeably and row widths of less than 25 cm (10 in.) are increasingly more common for soybean production. However, narrow row drills for use in untilled soils are still being perfected.

Soybeans are included in several cropping sequences owing to the wide range of maturities available among adapted cultivars. Soybeans are planted as a full-season crop and as a second crop following harvest of small grains. In double cropping no-tillage or some form of reduced tillage is more common than plowed systems.

Corn and soybeans complement each other in reduced-tillage production systems. Rotating corn and soybean reduces various disease and insect problems. Perennial grasses that limit reduced-tillage corn production are controlled by newly developed herbicides with soybeans. Conversely, perennial broadleaf weeds are more easily controlled in corn than in soybeans.

Small grains, wheat, oats, barley, and rye, are major crops in arid regions but are less important than corn or soybeans in humid regions. Their life cycles, however, include seasons that do not directly compete for land use with corn and soybeans. Because of this, small grains fit well into multiple-cropping systems and acreage is expanding in humid regions. Small grains also have large, vigorous seeds that emerge well from less than ideal seedbeds.

Reduced-tillage (disk-plant) practices are commonplace for both fall- and spring-seeded small grains in humid regions. No-tillage drills available for soybeans also seed small grains, whereas conventional drills may not penetrate untilled soil adequately. A few growers aerially seed small grains, including wheat, into maturing soybeans. Also, small grains are routinely included in sequence with other crops. Because of this, few clearly defined weed problems specifically associated with reduced-tillage small grain production have emerged in humid regions. Herbicides available for small grain weed control do not control perennial grasses.

Cotton requires a long growing season of about 5 months, intense sunlight, and high temperature to do well. Accordingly, it is normally grown in the lower latitudes and where adequate moisture is available. Seedlings begin slowly requiring 10–12 weeks to close the canopy for competitive weed control, about twice that for corn and soybeans. Cotton farmers use a minimum of 8–12 tillage operations annually, with weed control a major reason for tillage (McWhorter and Jordan, 1985). Reductions in tillage began in the United States when cotton production moved during the 1950s and 1960s to irrigated areas further west, where solation is more intense. No-tillage cotton came to follow no-tillage grain sorghum since cotton seedlings have a very tender hypocotyl and benefit from the protection of sorghum stubble, during its early stages, from abrasive blowing sand.

Reduced tillage for cotton production would be highly desirable. Many problems described for conventional tillage production of corn and soybeans are even more acute for cotton. Climate of the southeastern Cotton Belt is characterized by intense rainstorms and greater total annual rainfall than the Corn Belt. Because of this, soil erosion loss is two to three times greater than for similar tillage and cropping practices used in the Corn Belt. Many soils in the southeast are very fragile with shallow depths of topsoil. With the combination of fragile soils and increased erosion potential, row crop production cannot be sustained for long periods without severe losses of productive capacity of the land. Cropping systems utilizing reduced tillage and mulch cover would help reduce these problems. On many soils, a severe compaction problem develops with multiple field operations, making a good case for reduced tillage.

McWhorter and Jordan (1985) from a survey conducted in 1978, reported considerable research effort on development of reduced-tillage systems for cotton. Little of this work has been published because of negative results, and some experiments were abandoned because of poor weed control. Currently, adequate control of problem weeds is marginal with both herbicides and tillage, especially in the southeastern Cotton Belt. Any reductions in cultivation will likely increase weed competition. Availability of more effective herbicides could serve to accelerate tillage reduction for cotton production.

Reductions in the amount of tillage used for cotton, however, is taking place both in the southeast and the western irrigated areas. The least amount of tillage required has not yet been clearly defined. Row crop planters used for corn and soybeans also function well for reduced-tillage cotton. Weed control remains the key to successful reductions in tillage.

Tobacco is an intensively managed crop on well-drained soil, frequently located on modest slopes. Reduced tillage that maintains a mulch-covered soil lessens erosion, conserves moisture, and increases growth potential. Like tomatoes, tobacco seedlings are started in beds and transplanted to the field. Transplanters that perform well in untilled soil are available and have been used successfully in no-tillage tobacco production research studies. Acreage controls, maximum yield, and high quality dominate management decisions with tobacco.

Early tillage trials for tobacco production usually resulted in yield decreases in reduced-tillage systems. Successive trials have improved management techniques and narrowed the yield gap between tilled and untilled seedbeds. Efforts to refine procedures with reduced-tillage tobacco production continue (Worsham, 1985).

Weed control has been a major limitation in efforts to develop reduced-tillage methods for tobacco production, since few herbicides are approved for weed control in tobacco. Of these, some require incorporation to be effective and others do not provide season long control of a broad spectrum of weeds. Many of the herbicides must be used preemergence to weeds or the crop so that weed escapes cannot be easily controlled. Further herbicide development is required before reduced tillage systems with tobacco production will become practical.

Root crops offer special challenges for reduced-tillage production. Surgarbeet seeds have precise depth and planting requirements, but may be planted with row crop planters. Potato seed pieces require special planters that can be modified to function in untilled, mulch-covered soil. An erosion hazard may exist in production fields of both crops and moisture conserved by a mulch cover improves growing conditions for both crops without irrigation.

Herbicides available for potato culture include many of those used for weed control in soybeans. By limiting production to fields without perennial weed infestations the same chemicals apply to both. Sugarbeet herbicides offer versatility roughly equivalent to that for potatoes. Both crops follow other row crops in rotation, providing opportunities to deal with special weed problems.

Both sugarbeets and potatoes present special problems with reduced-tillage production. As root crops, the harvested part of the plant is located underground and shape or quality of the roots can be affected by soil conditions. Sugarbeets growing in untilled soil have an increased proportion of branched roots, but crop yields are roughly equivalent to those in tilled soil. Potato tubers are initiated on rhizomes of the mother plant. Expanding tubers that push aside the soil develop a green skin, when exposed to sunlight, which reduces product quality. In conventional tillage systems, postemergence cultivation keeps tubers covered. These special problems must be dealt with in reduced-tillage systems. Potatoes are being grown on a commercial basis in the northwest in a killed rye mulch. The mulch reduces wind erosion and damage to young plants by drifting sand particles. Herbicides effectively control weeds present in this area (Standifer and Beste, 1985).

Peanuts, a leguminous crop, are primarily produced in climatic areas where cotton is adapted. After the aboveground flowers are pollinated, the gescarpic plants "peg," pushing the developing fruit below ground. Harvesting equipment lifts the mature plants including fruits out of the soil. Because of these growth characteristics, peanuts are grown mainly on loam or sandy loam soils.

Traditionally, thorough tilling of soil has been recommended for peanut production for reasons of weed and disease control as well.

Research on no-tillage peanut production has been conducted in both North Carolina and Virginia. In the methods being developed, small grain cover crops are planted in the autumn and killed with herbicides the next spring. The peanuts are planted into the cover crop residue and productivity with no-tillage has been equal to conventional methods in these trials and no special disease or harvest problems have developed. No-tillage peanut production on a commercial scale is not common at this time (Worsham, 1985).

Sunflowers are grown primarily in drier areas, but are being evaluated in humid regions. Sunflower production research has included both reduced and no-tillage systems. At the present time, there is little or no reduced tillage used for sunflower production. Row crop planters function effectively to establish sunflowers. Herbicides available for weed control in sunflowers include some products used for soybeans, but are much more limited, and weed control currently represents a barrier to reduced-tillage sunflower production.

Reduced-tillage practices should be adaptable to a range of field and vegetable crops, including home gardens. Diverse production areas and crop requirements preclude blanket statements regarding adaptability of reduced tillage to vegetables.

SUMMARY

Reduced-tillage practices are dynamic and easily controlled in grain crop monocultures, making possible the solution of some major crop production problems. The basic requirements for crop growth have been described and satisfied in both tilled and untilled soil. Leaving the soil untilled and mulch covered improves moisture availability for the crop and reduces the amount of erosion. This, in turn, improves long-term soil productivity and increases land available for arable crops. Worker productivity is improved through fewer operations and cost structures are adjusted.

LITERATURE CITED

Andrews, D. J. and Kassam, A. H. 1977. The importance of multiple cropping in increasing world food supplies. *In* R. I. Papendick, P. A. Sanchez, and G. B. Triplett (eds.). *Multiple Cropping*. American Society of Agronomy, Madison, WI.

Cardwell, V. B. 1982. Fifty years of Minnesota corn production: sources of yield increase. *Agron. J.* **74**:984–990.

Cochran, V. L., L. F. Elliott, and R. I. Papendick. 1982. Effect of crop residue management and tillage on water use efficiency and yield of winter wheat. *Agron. J.* **74**:929–932.

Cannell, R. Q., D. B. Davies, D. Mackney, and J. D. Pidgeon. 1978. The suitability of soils for sequential direct drilling of combine harvested crops in Britain: A provisional classification. *Outlook on Agriculture* **9**:306–316.

Ebelhar, S. A., W. W. Frye, and R. L. Blevins. 1984. Nitrogen from legume cover crops for no-tillage corn. *Agron. J.* **76**:51–55.

Eckert, D. J. 1984. Tillage systems × planting date interactions in corn production. *Agron J.* **76**:580–582.

Elliott, L. F., T. M. McCalla, and A. Waiss, Jr. 1978. Phytotoxicity associated with residue management. *In* W. R. Oschwald (ed.). *Crop Residue Management Systems*. American Society of Agronomy, Madison, WI, pp. 131–146.

Flannery, R. L. 1981. Hairy vetch vs. rye cover crops for corn silage production using no-till. *Better Crops with Plant Food*. **66**:22–25.

Hodges, H. F., N. W. Buehring, R. E. Coats, John McMillan, N. C. Edwards, and C. Hovermale. 1983. The effect of planting date, row spacing and variety on soybean yield in Mississippi. Miss. Agr. & Forestry Exp. Sta. Bulletin 912.

Jeffers, D. L. 1984. A growth retardant improves performance of soybeans relay intercropped with winter wheat. *Crop Sci.* **24**:695–698,

Jordan, D. M. 1980, 1981, 1982, 1983. Ohio corn performance test. Agronomy Dept. Series 215, Ohio Agr. Res. & Dev. Ctr., Wooster, Ohio.

Kronstad, W. E., W. L. McCuistion, M. L. Swearingen, and C. O. Qualset. 1978. Crop selection for specific residue management systems. *In* W. R. Oschwald (ed.). *Crop Residue Management Systems*. American Society of Agronomy, Madison, WI.

McWhorter, C. G. and T. N. Jordan. 1985. Limited tillage in cotton production. *In* A. F. Wiese (ed.). *Weed Control in Limited-Tillage Systems*. Weed Sci. Soc. Amer., Monogr. Ser. No. 2. Champaign, Il.

Mitchell, W. H. and M. R. Teel. 1977. Winter-annual cover crops for no-tillage corn production. *Agron. J.* **69**:569–573.

Moschler, W. W., G. M. Shear, D. L. Hallock, R. D. Sears, and G. D. Jones. 1967. Winter cover crops for sod-planted corn: their selection and management. *Agron. J.* **59**:547–551.

Sanford, J. O. 1982. Straw management practices in soybean-wheat double cropping. *Agron. J.* **74**:1032–1035.

Soane, B. D., M. J. Buston, and J. D. Pidgeon. 1975. Soil moisture interaction in zero tillage for cereals and raspberries in Scotland. *Outlook on Agriculture* **8**:(special number).

Sprague, M. A., M. M. Hoover, Jr., M. J. Wright, H. A. MacDonald, B. A. Brown, A. M. Decker, J. B. Washko, K. E. Varney, and V. G. Sprague. 1963. Seedling management of grass-legume associations in the Northeast. New Jersey Agric. Exp. Stn. Bull. 804.

Standifer, L. C. and Mohd noor bin Ismael. 1975. A multiple cropping system for vegetable production under subtropical, high rainfall conditions. *J. Am. Soc. Hort. Sci.* **100**:503–506.

Standifer, Leon C. and C. Edward Beste. 1985. Weed control methods for vegetable production with limited tillage. *In* A. F. Wiese (ed.). *Weed Control in Limited-Tillage Systems*. Weed Sci. Soc. Amer. Monogr. Ser. No. 2, Champaign, IL.

Triplett, G. B., Jr. 1962. Intercrops in corn and soybean cropping systems. *Agron. J.* **54**:106–109.

Triplett, G. B., Jr., F. Haghiri, and D. M. Van Doren, Jr. 1979. Plowing effects on corn yield response to nitrogen following alfalfa. *Agron. J.* **71**:801–803.

Triplett, G. B., Jr. and D. M. Van Doren, Jr. 1985. An overview of the Ohio conservation tillage research. *In* F. D'Itri (ed.). *A Systems Approach to Conservation Tillage*. Lewis Publishers Inc., Chelsea, MI.

Van Doren, D. M., Jr. and G. B. Triplett, Jr. 1973. Mulch and tillage relationships in corn culture. *Soil Sci. Soc. Am. Proc.* **37**:766–769.

Van Doren, D. M., Jr., G. B. Triplett, Jr., and J. E. Henry. 1976. Influence of long term tillage, crop rotation and soil type combinations on corn yield. *Soil Sci. Soc. Am. J.* **40**:100–105.

Wiese, A. F., P. W. Unger and R. R. Allen. 1985. Reduced tillage in sorghum. *In* A. F. Wiese (ed.). *Weed Control in Limited-Tillage Systems*. Weed Sci. Soc. Amer. Monogr. Ser. No. 2, Champaign, IL.

Worsham, A. D. 1985. No-till tobacco *(Nicotiana tabacum)* and peanuts *(Arachis hypogaea)*. *In* A. F. Wiese (ed.). *Weed Control in Limited-Tillage Systems*. Weed Sci. Soc. Amer. Monogr. Ser. No. 2, Champaign, IL.

7

SUBSTITUTES FOR TILLAGE ON THE GREAT PLAINS

G. A. WICKS
Professor of Agronomy
University of Nebraska
West Central Research and Extension Center
North Platte, Nebraska

INTRODUCTION

The Great Plains extend from Texas into Canada. Included are parts of Texas, Oklahoma, Kansas, Nebraska, South Dakota, North Dakota, eastern Montana, Wyoming, Colorado, and New Mexico; and parts of Alberta, Manitoba, and Saskatchewan. Elevation on the east is 460–610 m gradually increasing westerly to 1370–1980 m.

Rainfall and temperature exert the greatest influence on crops grown on the Great Plains. Precipitation varies greatly across the region west to east from 300 to 800 mm in the southern Great Plains; from 300 to 600 mm in the central Great Plains; and from 300 to 500 mm in the northern Great Plains. Peak rainfall occurs during May, June, and July throughout the region. Thunderstorms of high intensity commonly occur for short duration. These intense storms cause considerable erosion if sufficient plant cover is not in place. Summer daytime temperatures are hot with the southern Great Plains having higher temperatures, hence, higher evaporation rates than further north. Winter temperatures range from warm to extreme cold and many degrees below freezing. The length of the growing season also dominates crop selection throughout the region. The region is characterized with an abundance of wind, a classic continental climate. Wind erosion is a constant threat throughout the year. Accordingly, crop production practices are developed around systems that hold soil and water erosion in check.

Many different crops are grown in the Great Plains. Many others have been tried and failed in this semiarid environment. Fallowing to conserve soil water is a common practice and has stabilized crop yields in many areas. Irrigation is common where water is available and where soils and climates are suitable. This presentation will focus on winter wheat (*Triticum aestivum* L.), spring wheat, corn (*Zea mays* L.), sorghum [*Sorghum bicolor* (L.) Moench], and soybean [*Glycine max* (L.) Merrill] production systems that return optimum yield and help reduce soil and water erosion.

Tillage with a plow and disk has been used almost exclusively for weed control and as a method of preparing a seedbed since native sod was first tilled during the late 1800s and early 1900s. Tillage destroyed vegetation and over time has caused severe erosion problems. As the organic matter was gradually depleted during the past 75 years it became necessary to add nitrogen fertilizer to maintain yield. During the last 20 years herbicides have become available that can replace all or part of the tillage operations for weed control. However, herbicides must be used intelligently, if they are to serve a useful purpose. One must know weeds present, herbicide capabilities, soil characteristics, moisture availability, and crop adaptation to use herbicides successfully in place of tillage.

All facets of crop management that affect herbicides must be incorporated into the system to develop economic weed control. These include amount and type of plant residue available, row spacing of the crop, seeding rate, date of planting, cultivar or hybrid selected, seed quality, fertilization, crop rotation, harvest efficiency and timeliness, and height of stubble.

Facilitating a shift in management techniques with the introduction of herbicides to replace tillage operations requires that, first, the herbicides be economical and dependable and, second, the soil and climate must be suitable for both the herbicide and the crop rotation selected. Third, there must be sufficient plant residue available from preceding crops. Advantages of leaving the crop residue on the soil surface include reduced wind velocity near the soil surface, cooler soil temperatures, less evaporation, reduced rate of water flow across the field, increased water infiltration into the soil, reduced runoff, snow trapped by standing residue, reduced stalk rot, reduced soil erosion, improved weed control, improved crop and forage yields, reduced reliance on insecticides, reduced losses from hail early in the growing season, lower fuel requirements, reduced soil compaction, reduced labor requirements, and lower equipment requirements.

Conservation tillage is any soil management practice that leaves the soil surface resistant to erosion and conserves soil moisture. At least 30% of the soil surface should be covered with plant residue after planting. In the Great Plains there are several terms that designate crop management practices intended to conserve the soil. These include stubble mulch, chemical fallow, ecofarming, lo-till, zero tillage (no-tillage) and ridge planting. These practices will be described in detail in succeeding paragraphs. Each system has been adapted to specific areas in the Great Plains and other similar areas of the world. One must evaluate the system and adapt those principles suitable for the geographic area.

STUBBLE MULCH

In wheat-fallow rotation areas it has been common to control weeds with sweep tillage from one wheat harvest until wheat is planted again (Fenster and Wicks, 1982; Unger et al., 1971). Depending on location, which determines planting and harvest dates, the fallow period extends about 14 months for winter wheat and 20 months for spring wheat. Downy brome (*Bromus tectorum* L.) has become a problem in the winter wheat–fallow rotation under stubble mulch. Many farmers have reverted to tillage methods that eliminate stubble in controlling this weed. Herbicides registered to selectively control downy brome in winter wheat are not satisfactory in most areas.

CHEMICAL FALLOW

Herbicides are used successfully in controlling downy brome along with other weeds during a fallow period. This practice, called chemical fallow, has increased winter wheat yields 10% (Wicks and Smika, 1973). A no-tillage drill should meet the following requirements: (a) row spacing of 25–36 cm, (b) at least 51 cm of clearance between ranks of openers, (c) at least 46 cm of vertical clearance between frame bottoms of shoe openers, (d) narrow shoe openers of 10 cm or less, and (e) equipped with press wheels to pack seed firmly into moist soil and anchor residues in the ridge (Fenster and Wicks, 1977). However, it has not been economic to use herbicides as the only means of weed control during the long fallow period. Sufficient crop residue must be produced to increase water storage and to reduce evaporation and soil crusting (Unger and Wiese, 1972). Crop yields may be reduced because many drills fail to place seed into moist soil for good germination and growth. Often, residual herbicides either may not last long enough or persist too long and injure the succeeding crop. Atrazine* has persisted excessively in high pH (above 7.8) or low organic matter (below 0.6%) soils, and guidelines have been suggested to identify problem soils (Smika and Sharman, 1982). These include Cadoma, Canyon, Dwyer, Heldt, Limon, Midway, Norka, Rosebud, and Plantner–Ascalon–Haxtun soil series. Nonresidual herbicides are too expensive for repeated use. Herbicides in the wheat–fallow rotation areas are used as fallow aids to tillage. The most effective herbicide timing is when weeds are most difficult to control

*Chemical names of herbicides and generic names are listed in Appendix A.

with tillage. For example, downy brome and volunteer wheat are very difficult to kill with sweep tillage in spring when soils are cool and wet. Herbicides can be selected that will effectively control these weeds, and tillage later in the summer aids in weed control and seedbed preparation. This tillage also reduces crop residue cover and encourages establishment of wheat with conventional drills.

ECOFARMING STRATEGIES

In the central Great Plains popular rotations have been winter wheat–sorghum or corn–fallow, Fig. 7.1. These rotations are well suited to herbicide use as a replacement for most tillage operations. Specific terms have been coined to describe this cropping system (Burnside et al., 1980). Ecofarming is defined as a system of controlling weeds and managing crop residues throughout a crop rotation with minimum use of tillage as a means to reduce soil erosion and production costs while implementing weed control, water infiltration, moisture conservation, and crop yield. Ecofallow is the period between wheat harvest and planting the next crop. Phillips (1964) in Kansas used both atrazine and sweep tillage following wheat harvest to control weeds until sorghum was

FIGURE 7.1. Sorghum planted into winter wheat stubble 10 months after planting. Wheat stubble received atrazine + paraquat after harvesting and glyphosate + alachlor at planting.

planted the following spring. This practice has developed into the most popular dryland cropping system throughout south central and southwest Nebraska and northern parts of northwest and north central Kansas.

Growing Winter Wheat

Many facets of crop management contribute to successful weed control and good crop yield. First, since winter wheat must compete with weeds, weed control begins during the fallow season prior to planting. Chisels, sweeps, and rodweeders are common implements used to control weeds and prepare the seedbed for winter wheat. The seedbed should be firm and the grain should be sown with a drill that places the seed firmly into moist soil. Row spacing (25–41 cm) should be as close as possible, depending on depth to moisture and amount of residue needed for erosion control. A wide selection of drills is available. Moderate seeding rates and narrow row spacing can reduce weed growth in the crop and in the stubble after wheat harvest (Vander Vorst et al., 1983). Hard rains before crop emergence can reduce stands of wheat by silt accumulating over the row in fields free of plant residue. These fields generally are planted with disk drills. In more arid areas shoe drills are needed to remove the dry surface soil and place the seed 2.5–5-cm deep into moist soil. Care must be taken not to cover semidwarf varieties with excessive soil as emergence may be impaired. Also some varieties require cool temperatures after harvest to break dormancy. The seed wheat should be an adapted variety able to compete with weeds (Challaiah et al. 1983). Fertilization should be according to soil test for optimum yields without lodging. Adequately fertilized winter wheat plants are vigorous and competitive to weeds both during growth and following harvest. Seeding rate is important to ensure proper tiller density. Wheat should be planted at the optimum date for the area to reduce damage from diseases and insects.

Winter wheat is a very competitive crop with most spring germinating weeds (Hoefer et al., 1981). Herbicide application for broadleaf control is not necessary in most years. Adequate stands of vigorously growing wheat are competitive to summer annual weeds. Weeds commonly remain small until the crop canopy is removed at harvest, then develop rapidly depending on availability of moisture and nutrients.

The wheat straw (including chaff) should be spread uniformily behind the combine or removed from the field. Straw may be chopped, but should not be left in windrows or piles, which interfere with herbicides reaching the soil or with planting and growth of the succeeding crop, Fig. 7.2. With increasing cutter bar size on combines, uniform straw distribution has become increasingly difficult. Headers with cutter bars greater than 6 m require special equipment to spread the straw uniformly. Volunteer wheat can be a most troublesome weed to combat. Combines must be properly adjusted and operated to reduce loss of grain. Variety, hybrid selection and proper fertilization also aid in preventing loss of grain.

FIGURE 7.2. Pile of straw left by stopping the combine in the field during harvest. Excess straw hinders planting and sorghum emergence.

Growing Corn and Sorghum

Herbicides should be applied to the wheat stubble within 30 days after harvest. Application should be delayed for several days after harvest to allow the stubble to settle and expose weeds. Paraquat and/or terbutryn are combined with atrazine to kill existing weeds. Terbutryn may aid atrazine in controlling late germinating weeds. Good coverage is essential. Grasses are difficult to kill if well rooted and well tillered. Respraying is sometimes necessary to kill weed escapes. Glyphosate is also used for controlling weeds after harvest.

The amount of atrazine used after wheat harvest depends on weeds to be controlled, soil type, organic matter, soil pH, amount of erosion, expected rainfall, expected soil persistence, and succeeding crops to be planted. Use rates generally range from 1 to 3.3 kg/ha. Some farmers in low rainfall areas may chose to plant wheat in place of corn or sorghum if precipitation has been insufficient to recharge the soil profile by planting time. Thus, the atrazine rate must be low enough to permit an option to plant winter wheat 12 months later. Should corn or sorghum be chosen, additional atrazine can be applied at planting time. If grasses, such as barnyardgrass [*Echinochloa crusgalli* (L.) Beauv.] or foxtail (*Setaria* spp.), are potential problems, then alachlor, cyanazine, metolachlor, or terbutryn alone or in combination should be used.

In some areas cyanazine and metolachlor may be applied up to 45 days ahead of planting; alachlor and terbutryn are more effective if applied within 5 days before or immediately after planting. Terbutryn is registered only for use on sorghum. The appropriate seed safener is required when alachlor or metolachlor is used on sorghum.

Volunteer wheat, Russian thistle (*Salsola kali* L.), and annual grasses may be present prior to planting. Glyphosate, paraquat, 2,4-D, and terbutryn may be used effectively prior to crop emergence. Best control is obtained when weeds are small.

Planting is a critical step in successful management of tillage systems. Adapted corn and sorghum hybrids should be selected. Since average soil temperature beneath a straw mulch may be as much as 10°C cooler (Doupnik and Boosalis, 1980) than a tandem disked seedbed in summer, this should be considered when selecting hybrids. Long-season hybrids should not be used where frost is likely in early fall, since delay in maturity reduces grain yield. Planter adjustment is important. Rolling coulters preceding the planter opener should be sharp and should cut the straw rather than push it into the soil. Since the ground is usually moist in spring, seed placement 2.5 cm deep is adequate. Crop injury has occurred when seed was inadequately covered and herbicides were applied preemergence. Corn has lodged when seed was not planted deep enough. If grain stubble was cut short and straw cover is heavy, opening devices should be used to part straw ahead of rolling coulters.

On the Great Plains weed control improves as residue amounts increase (Crutchfield and Wicks, 1983). Although there is risk of herbicides being intercepted by the residue, less herbicide is needed where more stubble is present. Optimum crop yields require 3000 kg/ha or more of residue under most conditions. Texas research (Unger and Wiese, 1979) has shown the merit of irrigating winter wheat to produce more straw. More crop residues reduced evaporation, increased water storage in the soil, and improved dryland sorghum yields.

There always is risk of poor weed control when herbicides are applied prior to or at planting time. Two management options remain if weeds need further control. First, herbicides may be applied postemergence depending on the crop and the weeds. Second, several cultivars are available that operate effectively in residue. Timely implementation of the weed control program is important to have alternative methods in place before the need arises.

No additional treatments following corn or sorghum harvest are necessary to control annual weeds until spring in the central and northern Great Plains. Three approaches have been successful in controlling weeds in sorghum or corn stubble while leaving sufficient residue on the soil surface for erosion control. First, no herbicide is applied and the land is tilled with sweeps to control weeds and prepare a seedbed for fall-planted wheat. Second, herbicides may be used to control all weeds. Volunteer corn and sorghum are difficult or expensive to control with presently available herbicides. Also, soil may become too hard for drills to penetrate if insufficient plant residue is not present.

Opportunity Farming

A third option is often called opportunity farming. Cyanazine is applied in spring to keep fields weed-free until mid-May. If sufficient soil moisture is available at sorghum planting time, sorghum will be planted and additional herbicides will be used to keep the crop weedfree, Fig. 7.3. If soil moisture is insufficient, the land will remain fallow and planted to wheat in the fall. Some farmers have harvested up to five sorghum crops before returning to winter wheat. Versatility is an important feature as the land is better protected from erosion as the fallow season is delayed. Also production income is increased with the use of sorghum.

Growing Soybeans

U.S. soybean hectarage has increased rapidly in the past decade, moving west into the eastern areas of the Great Plains. Soybean–winter wheat is a successful rotation where moisture is adequate, Fig. 7.4. Winter wheat is planted following soybean harvest and soybeans are planted 10 months after wheat harvest. Both crops are planted with no-tillage drills. Soil moisture may be inadequate and/or the growing season may be too short to double crop soybeans immediately after wheat in most areas of the Great Plains. Growing

FIGURE 7.3. Sorghum planted between rows with a no-tillage slot planter. Erosion potential is greatly reduced with the residue from previous sorghum and wheat crops.

FIGURE 7.4. Soybeans planted into winter wheat stubble 10 months after wheat harvest. Weeds were controlled with cyanazine + paraquat after harvest and metribuzin + metolachlor prior to planting soybeans.

soybeans in arid areas is risky at best because peak water usage by the soybean crop is in August, a time when rainfall is normally low. Risk is less when growing corn or sorghum than soybeans 10 months following wheat harvest.

The two crops aid weed control by alternating a summer and winter annual, and broadleaf and monocot crops dictate the use of different herbicides and different seasons. Similarly, downy brome has successfully been controlled by introducing corn or sorghum into the winter wheat-fallow rotation.

CONTINUOUS WINTER WHEAT

Growing continuous winter wheat is commonplace in the higher rainfall areas of central and southern Great Plains. Normally several tillage operations are used to control weeds and reduce plant residue so surface drills will function properly. Unfortunately, when wheat emergence is poor, due to untimely rainfall, lack of cover presents great potential for erosion. In Oklahoma (Stiegler et al., 1982), farmers apply herbicides after wheat harvest, eliminating most or all tillage between the harvest and replanting in the fall.

Wheat fields must be examined in both fall and spring for competing weeds. Timely applications of effective herbicides will control emerged weeds and provide preemergence control of those germinating later. Many of the weeds present in the stubble after harvest germinated while the crop was growing (Hoefer, et al., 1981). Selective herbicides applied in spring could reduce this pressure and eliminate weeds that may cause dockage at the elevator. Presently few herbicides with residual activity are registered for application on the growing wheat. Promising herbicides to control summer annual grasses include oryzalin, pendimethalin, chlorsulfuron, and metolachlor (Ghadiri et al., 1981). Controlling weeds before they produce seed, and become a serious problem, in a continuous wheat system should become an important management technique. Of these herbicides only a postemergence application of chlorsulfuron is effective on certain broadleaf weeds. Most broadleaf weeds, however, can be controlled with 2,4-D, bromoxynil, or dicamba when applied to wheat in the tillering stage. Chlorsulfuron or bromoxynil have an advantage over 2,4-D and dicamba in that they can be applied safely in the fall after the wheat has emerged. In this way much of the competition from winter annual broadleaf weeds, such as field pennycress (*Thaspi arvense* L.), is eliminated before their presence reduces yield. Further, 2,4-D may be employed as a harvest aid to reduce weed interference with combining.

Following harvest there are 45–120 days before winter wheat will be reseeded. These approaches are possible for herbicide weed control. First, residual herbicides can be used to provide control for extended periods of time. For example, cyanazine or terbutryn may be applied after harvest. Generally, weeds are present, so appropriate foliar-active herbicides should also be chosen. In some areas cyanazine may dissipate before the time to replant wheat, so an application of glyphosate may be needed to control newly emerged weeds. In Oklahoma low rates of atrazine provide longer residual control than cyanazine. Terbutryn has shorter residual activity than cyanazine and is preferred in areas of low rainfall or a short fallow period. Dicamba, 2,4-D, and chlorsulfuron effectively control broadleaf weeds during this fallow interval.

A second option is two or more applications of a nonresidual herbicide such as glyphosate. This has provided satisfactory weed control and is a feasible alternative providing the economics will permit.

A third option is to combine tillage and herbicides. Grain drills that will seed well through heavy wheat stubble may not be available, so some tillage may be necessary to reduce the residue prior to seeding. Disking may be the tillage used immediately after harvest or prior to fall wheat seeding. This could cover grain lost at harvest and increase the volunteer problem. Air seeders mounted on chisel plows, sweep plows, or rodweeders provide tillage at planting and control weed escapes. Also they incorporate less plant residue than a disk. Disadvantages include poor planting depth control of seed and residual straw left in the seed row. Tillage dries out the seedbed in some years and affects germination and emergence of wheat.

Winter annual grass weeds, such as downy brome, cheat (*Bromus secalinus* L.), and jointed goatgrass (*Alegilops cylindrica* Host), are difficult to control in continuous winter wheat as well as in the winter wheat-fallow rotation. Generally, if these weeds are present in surrounding waste areas, they will migrate into the field.

Wheat should be monitored closely for plant diseases. The standing stubble is also a haven for grasshoppers, which will quickly devour new seedlings if not controlled.

CONTINUOUS NO-TILLAGE

Zero tillage is a popular term describing a no-tillage system used in Canada that was developed for the semiarid prairie plains (Stobbe, 1979; Stobbe and Rourke, 1980). Several advantages can be listed for continuous zero-tillage cropping: (a) maximum protection against wind and water erosion; (b) standing stubble traps snow, improves soil moisture and winter survival, and increases yield; (c) soil salinity is minimized, since fallow has been eliminated and evaporation reduced; (d) potential of nitrogen leaching into the groundwater is reduced; (e) annual weed problems are decreased, since many species require seed to be incorporated into the soil for germination; (f) time and fuel are saved; (g) machinery costs are reduced; (h) this practice is well suited to many soil types; and (i) interseeding of wheat in spring provides an ideal seedbed. The limitations of the system consist of: (a) too great reliance on herbicides for weed control; (b) present herbicide cost exceeds that of tillage; (c) fertilizer application and placement is difficult; (d) heavy clay soils may be difficult to plant in wet years, since stubble reduces drying of the soil surface; (e) spreading straw and chaff is difficult; and f) new drills must be purchased or old drills modified for no-tillage planting.

RIDGE PLANTING

Ridge planting began in Nebraska during the late 1950s (Wittmuss, 1959). Different than listing, it consists of planting on top of the ridge of undisturbed ground that had been in a row crop the previous year. In principle, the objective is to move the corn or sorghum stalks and roots to the center between the rows. The idea was sound, but equipment, management problems, and tradition were difficult obstacles to overcome. Common problems were weeds, difficulty in staying on the ridge, poor stands, and difficulty in cultivation and furrow irrigation.

There are enough sound agronomic features to the system that it survived and is being adapted for use in the 1980s by many farmers. The advantages of ridge planting are (a) moisture retention by preservation of valuable plant residue. The soil does not need to be disked or plowed, which retains the plant residue on the soil surface thereby reducing runoff and evaporation. (b) Rows remain in the same place each year, so traffic patterns are established and soil compaction is maintained between the row. (c) Crop residue is kept on the soil surface as protection from wind and water. (d) The ridge provides an adequate seedbed, the soil is moist and has been measured up to 6°C warmer than the furrow in the spring (Buchele, et al., 1955). More rapid crop emergence in the seed row, where the residue is removed, provides a weed control advantage, since the cooler soil temperatures between the row slow weed germination and growth. Cultivation coverage is easier and more complete because of a greater height differential between crop and weeds. (e) During wet weather the ridge dries more quickly than level ground seedbeds. As the season progresses and dries, the furrows retain moisture longer. (f) Fuel and labor costs for seedbed preparation and planting are greatly reduced compared with flat seedbeds. Field operations, such as plowing, disking, fertilization, cultivation, and harvesting, use about 17% of the total energy consumed in production agriculture (Splinter, 1978); ridge planting eliminates several of these operations. (g) Weed seed on the soil surface are moved to between the row during planting (Wicks and Somerhalder, 1971), whereas, disking destroys the ridges and incorporates weed seed into the top few centimeters of soil. The crop is thus planted into the soil with the weed seed, both of which emerge together permitting greater weed control problems. Ridge planting is a control method for shattercane [*Sorghum bicolor* (L.) Moench] in Nebraska. EPTC + safener was the most used herbicide for shattercane until selected soil microbes shortened the longevity of EPTC (Obrigawitch et al., 1982). Shattercane infestations increase on fields with a history of EPTC use to levels attained when farmers first started using the chemical. Volunteer corn is a problem almost met by ridge planting. The kernels and ears are moved to the furrow where they are controlled by cultivation.

Winter annual weeds such as downy brome, tansy mustard [*Descurainia pinnate* (Walt.) Britt.], and field pennycress increase with ridge planting in Nebraska. A late March or early April application of atrazine at 1.1 kg/ha controls these weeds and early germinating broadleaf weeds such as kochia [*Kochia scoparia* (L.) Roth] and lambsquarters (*Chenopodium* spp.). However, there are increasing reports of kochia escaping atrazine even if applied at 2.2–3.3 kg/ha. An application of dicamba for 2,4-D ester prior to planting has been effective in eliminating weed escapes. A band application of alachlor + cyanazine suppresses shattercane at planting time. However, in severe weed infestations a broadcast application of appropriate herbicides should be used after planting or a higher rate of atrazine used ahead of planting.

Several equipment companies manufacture ridge planters (see Chapter 3). Basically a sweep or blade 25–30-cm wide or a large double-disk opener removes the previous crop residue from the row. Many standard planter units can be adapted to ridge planting. In some systems the planter is mounted behind a rotary tiller, which follows the sweeps. A couple of tines are used to surface-incorporate fertilizer, herbicide, and insecticide, if desired. In areas of limited rainfall adequate incorporation of herbicides provide more reliable weed control.

A stablizer fin helps keep the equipment on the row during planting and cultivation; straight rows are an advantage during cultivation.

SUMMARY

There are many cropping areas in the Great Plains where herbicides have replaced part or all tillage operations and more are in prospect. As better herbicides and equipment are developed, less reliance is placed on tillage as an integral part of crop production, providing the new practices are cost effective. Greater acceptance of reduced tillage systems will occur as herbicide use becomes more economical. Better and more dynamic control of all aspects of crop management is possible with herbicides in place of tillage.

Integrated management is the key to success in these cropping systems. Balanced attention is essential to application equipment; herbicide selection; soil type; climate; timeliness; selection of crop and cultivar; the succeeding crop; harvesting; and feasibility of mixing herbicides, insecticides, and fertilizers. Producers who adapt to the "new" agriculture will benefit from improved moisture management as their land becomes better protected from wind and water erosion. Improved yields with fewer inputs of energy and labor are possible.

LITERATURE CITED

Buchele, W.F., E.V. Collins, and W.G. Lovely. 1955. Ridge farming for soil and water control. *Agr. Eng.* **36**:324–329, 331.

Burnside, O.C., G.A. Wicks, and D.R. Carlson, 1980. Control of weeds in an oat (*Avena sativa*)-soybean (*Glycine max*) ecofarming rotation. *Weed Sci.* **28**:46–50.

Challaiah, R.E. Ramsel, G.A. Wicks, O.C. Burnside, and V.A. Johnson. 1983 Evaluation of the weed competitive ability of winter wheat cultivars. North Cent. Weed Control Proc. **38**:85–91.

Crutchfield, D.A. and G.A. Wicks. 1983. Effect on wheat mulch level on weed control in ecofarming corn production. *Abstr. Weed Sci. Soc. Am.* 1.

Doupnik, B., Jr., and M.G. Boosalis. 1980. Ecofallow-a reduced tillage system-and plant diseases. Plant Dis. **64**:31–35.

Fenster, C.R. and G.A. Wicks. 1977. Minimum tillage fallow systems for reducing wind erosion. *Trans. Am. Soc. Agric. Eng.* **20**:906–910.

Fenster, C.R. and G.A. Wicks. 1982. Fallow systems for wheat in western Nebraska. *Agron. J.* **74**:9–13.

Ghadiri, H., G.A. Wicks, C.R. Fenster, and O.C. Burnside. 1981. Control of weeds in winter wheat (*Triticum aestivum*) and untilled stubble with herbicides. *Weed Sci.* **29**:65–70.

Hoefer, R.H., G.A. Wicks, and O.C. Burnside. 1981. Grain yields, soil water storage and weed growth in a winter wheat-corn-fallow rotation. *Agron. J.* **73**:1066–1071.

Obrigawitch, T., R.G. Wilson, A.R. Martin, and F.W. Roeth. 1982. The influence of temperature, moisture, and prior EPTC application on the degradation of EPTC in soils. *Weed Sci.* **30**:175–181.

Phillips, W.M. 1964. A new technique of controlling weeds in sorghum in a wheat-sorghum-fallow rotation in the Great Plains. *Weeds* **12**:42–44.

Smika, D.E. and E.D. Sharman. 1981. Atrazine carryover and its soil factor relationships to no-tillage and minimum tillage fallow-winter wheat cropping in the Central Great Plains. Col. State Univ. Exp. Sta. Bull. 144.

Splinter, W.E. 1978. Energy requirements in managing food and fiber production. In *Renewable Resource Management for Forest and Agriculture.* Ed. J. Bethel and M. Massengale pp. 67–80.

Stiegler, J., R.Johnston, and H. Greer (eds). 1982. Lo-till farming. Okl. S. Univ. Coop. Ext. Serv.

Stobbe, E.H. 1979. Tillage practices on the Canadian prairie. *Outlook Agric.* **10**:21–26.

Stobbe, E. and D. Rourke, 1980. Tillage practices in winter wheat production. Proc. Tillage Symp., Kirkwood Inn, Bismark, ND. pp. 175–181.

Unger, P.W., R.R. Allen, and A.F. Wiese, 1971. Tillage and herbicides for surface residue maintenance, weed control and water conservation. *J. Soil Water Conserv.* **26**:147–150.

Unger, P.W. and A.F. Wiese, 1972. No-tillage research in the Panhandle of Texas. Proc. No-Tillage Systems Symp., Columbus, OH, pp. 103–107.

Unger, P.W. and A.F. Wiese, 1979. Managing irrigated winter wheat residues for water storage and subsequent dryland grain sorghum production. *Soil Sci. Soc. Am. J.* **43**:582–588.

Vander Vorst, P.B., G.A. Wicks, and O.C. Burnside. 1983. Weed control in a winter wheat-corn-ecofarming rotation. *Agron. J.* **75**:507–511.

Wicks, G.A. and D.E. Smika. 1973. Chemical fallow in a winter wheat-fallow rotation. *Weed Sci.* **21**:97–102.

Wicks, G.A. and B.R. Somerhalder. 1971. Effect of seedbed preparation for corn on distribution of weed seed. *Weed Sci.* **19**:666–668.

Wittmuss, H.D. 1959. Minimum tillage method can cut costs. *Nebr. Exp. Q.* **6**:3–5.

8

NO-TILLAGE PASTURE AND MEADOW IMPROVEMENT IN HUMID REGIONS

R. W. VAN KEUREN
Professor of Agronomy
Ohio State University
Ohio Agricultural Research and Development Center
Wooster, Ohio

INTRODUCTION

No-tillage is an effective method of seeding forages for pasture and meadow improvement. It virtually eliminates the erosion hazard, reduces equipment and energy requirements, and reliably provides good establishment when done carefully. Rolling land and steep slopes are typically kept in pasture and meadow all or most of the time to minimize erosion losses. Many sites also have shallow soils, or may be stony, wet, or have rough uneven terrain, and thus are not suitable for conventional tillage. These are often the most neglected areas on the farm. Thus, no-tillage offers an opportunity through changing species to bring nontillable land into production. No-tillage renovation also reduces the time in which a field is out of production during reestablishment compared with preparing a seedbed by plowing or disking.

Millions of hectares, much of it hilly land, are used for pasture and meadow in the humid regions throughout the world. Recent increases in row crop acreage, and urban and commercial use of prime agricultural land has intensified usage and improvement of forage crop land. Currently, there is also a renewed interest in forages for livestock feeding and as soil-conserving crops.

The acceptance of no-tillage techniques in recent years by producers has demonstrated the value of this practice. The development of effective seeding equipment and herbicides has made no-tillage pasture and meadow reseeding technically feasible. Originally of interest for sloping land, no-tillage forage establishment is being used more and more for all kinds of forage seedings, regardless of soil site.

A major advantage of no-tillage establishment is not having to prepare a seedbed by plowing or disking. This saves time, fuel, and need for certain equipment. Another major advantage, due to a surface mulch, is the conservation of moisture through reduction of surface runoff and evaporation (Phillips, 1981). The forage topgrowth must be reduced if moisture loss by transpiration is to be lessened. Surface soil moisture is particularly important early in establishment.

In contrast to annuals grown in a monoculture, such as corn and soybeans, the production, management, and utilization of forages is far more complex. Forages are usually mixtures of competing species, including grasses and legumes, with widely different growth characteristics. Most forages are perennials, involving the need for persistence for years, dormancy, and regrowth through vegetative regeneration. Forages are primarily utilized by livestock, and an adequate level of quality in terms of digestible energy and protein is needed. To obtain this quality, the plants are generally harvested repeatedly and at immature stages of growth. Much of the forage is harvested directly by grazing animals, and selective grazing, trampling, and other factors have a marked effect on the sward.

The objectives of this chapter are to present an overview of the principles and practices involved in no-tillage improvement of pastures and meadows, and to present research relating to no-tillage establishment of forages in the humid regions. Detailed differences do occur between regions in forage species adaptability, seeding dates, soil types, fertility needs, general management, and utilization, but similar principles and practices apply in most no-tillage systems.

Permanent pasture is defined as "a pasture of perennial or self-seeding annual plants kept for several years grazing" (CSSA, 1962). A meadow is "an area covered with grasses and legumes grown primarily for hay." Sod "usually refers to the top few inches of soil which is permeated and held together with grass roots, or grass and legume roots." Pasture renovation is defined as "the improvement of a pasture by partial or complete destruction of the sod, plus liming, fertilizing, and seeding as may be required to establish desirable forage plants" (CSSA, 1962). No-tillage pasture and meadow improvement or renovation is defined here as pasture or meadow improvement by drilling forage species into an undisturbed sod, with or without herbicide vegetation control, but including weed control, liming, and fertilizing as may be required to establish desirable forage plants.

EVALUATING THE NEED FOR IMPROVEMENT

The decision to improve a pasture or meadow is usually made because the forage stand has so deteriorated in yield and quality that it no longer meets the nutritional needs of the livestock or because the producer wants to expand his or her livestock numbers. The first step is to evaluate the field to be improved in terms of the species present, the soil capabilities for increased production, and needed improvements, including weed control and soil nutrient amendments. The field should also be evaulated in terms of livestock utilization, for example, will it be used as permanent pasture or meadow, or for winter grazing. Rough, hilly, stony, and shallow soil sites are best utilized as permanent pasture. Some fields can be utilized best as pasture and meadow in an integrated forage program. Fields used for winter grazing need to be well drained to minimize miring of the livestock and damage to the field and forage stand. In northern areas access to some kind of winter protection for livestock against winter storms is needed.

Species Replacement

Most permanent pastures have at least a remnant stand of desirable grasses and legumes, and the producer has the option of improving production by additions of soil fertility and grazing management. Over much of the northern humid region of the United States the major grass present in unimproved pastures is Kentucky bluegrass (*Poa pratensis* L.). The producer may, however, prefer to increase his or her production by reseeding to higher yielding grasses and legumes. For improved meadows and winter forage fields, more productive species should be used. Cool-season forages are grown from eastern Oklahoma to South Carolina and northward, as well as in the Pacific Northwest and irrigated areas in the northern Great Plains. These forages are predominately for dairy, sheep, and beef production. The major legumes are alfalfa (*Medicago sativa* L.), red clover (*Trifolium pratense* L.), birdsfoot trefoil (*Lotus corniculatus* L.), and white clover (*Trifolium repens* L.); the major seeded grasses include smooth bromegrass (*Bromus inermis* Leyss), orchardgrass (*Dactylis glomerata* L.), timothy (*Phleum pratense* L.), and tall fescue (*Festuca arundinacea* Schreb.). In wet areas reed canarygrass (*Phalaris arundinacea* L.) and creeping foxtail (*Alopecurus arundinaceus* Poir.) are useful. In most instances a mixture is desired, usually a grass and a legume. Complex mixtures of several species each of grasses and legumes are usually not recommended. In the Gulf Coast region warm-season perennial semitropical grasses are used, such as bermudagrass [*Cynodon dactylon* (L.) Pers.] and bahiagrass

(*Paspalum notatum* Fluegge), primarily for beef production. State and local agricultural extension service recommendations should be followed as to the species and cultivars best adapted.

Weed Problems

Prior to initiating pasture or meadow improvement is the best time to control weeds, particularly noxious and difficult to control weeds and encroaching brush. Major problem weeds include ironweed (*Vernonia* spp.), horsenettle (*Solanum carolinense* L.), wild *Allium* spp. (wild onion and wild garlic), Canada thistle [*Cirsium arvense* (L.) Scop.], and quackgrass [*Agropyron repens* (L.) Beauv.]. Control of such weeds should be initiated before seeding, preferably a year or more to allow sufficient time for several applications of herbicides if necessary. The use of herbicides is the only reliable method of control of noxious weeds in no-tillage systems. A 2,4-D* and dicamba mixture is effective on most of the common broadleaf weeds in pastures, but should be applied the season prior to seeding. Dicamba has a longer residual activity than the phenoxy herbicides, and the combination is more effective than 2,4-D alone. Picloram (Peters and Lowance, 1979) and triclopyr (Mann *et al.*, 1983) will provide better control of the more troublesome broadleaf weeds, such as ironweed. Some effective herbicides can be used prior to seeding but not without damage once the new forages are established (See Chapter 11.)

Soil Amendments

Low mineral fertility and pH are common causes of poor productivity of pastures and meadows as well as loss of stand (Van Keuren, 1976). Unimproved fields in most of the humid areas of the United States tend to be acid, low in phosphorus (P), and low to moderate in potassium (K). Liming is commonly needed to grow legumes. Commonly, for many years minerals have been removed from pastures as milk, meat, and bones, and by erosion with the sole source of restitution being spotty return of urine and manure by the grazing animals. Soil tests prior to seeding are essential to determine needs of lime, P, K, and, in some areas, other elements. Most sodseeding drills do not have fertilizer boxes, and the fertilizer may need to be applied separately, usually prior to seeding. Rapid incorporation of lime and fertilizer is not possible without tillage, but has generally not been a major constraint in no-tillage systems (Thomas et al., 1981). Plant roots feed near the soil surface–sod mulch interface to obtain surface-applied nutrients for satisfactory production, providing moisture is present. Excellent long-term production from alfalfa, red clover, and birdsfoot trefoil in mixtures with grasses have been obtained under no-tillage seeding and management on land with initial low pH and fertility (R. W. Van Keuren, Ohio Argic. Res. and Dev. Center,

*Chemical and generic names of herbicides are listed in Appendix A.

unpublished research). It is advisable, however, to apply needed lime preferably 6 months or more before seeding to permit its effect to penetrate to the root zone.

STAND ESTABLISHMENT

The objective in establishment is to provide the best conditions possible for rapid emergence and development of the seeded forages. We must observe the same principles and many of the same practices with no-tillage seedings as we do when seeding into a prepared seedbed. Important are optimum times of seeding, proper depth of seed placement, good seed–soil contact, and vegetation control. Although in no-tillage establishment we eliminate soil and wind erosion hazards, other problems need attention, primarily vegetation control and placement of seed.

Inherent in any tillage system is the control of existing vegetation. This is especially important with forages because most forage seeds are small and the seedlings are relatively weak and slow to develop. Competition must be kept to a minimum during the establishment period (Groya and Sheaffer, 1981). Removal of most of the top growth prior to seeding is also advisable for better control of seed placement. A large amount of plant material on the surface can interfere with penetration of the drill and obtaining uniform and shallow seed placement. Close grazing shortly before seeding is one method of removing surplus vegetation while utilizing the material.

Use of Herbicides

Herbicides are widely used in no-tillage systems for vegetation control, first proposed for pastures by Sprague (1952). Occasionally successful no-tillage seedings of legumes into cool season grasses are obtained without the use of herbicides (Kunelius et al., 1982; Taylor and Allinson, 1983). Some research indicates that herbicides are not needed for thin weak sods, but do enhance legume establishment in dense rapidly growing grass sods (Taylor et al., 1969). Herbicide treatment of the grass has also shown more rapid development of the legume component (Van Keuren and Triplett, 1970), and an increase in forage yield (Linscott and Vaughan, 1982). Some researchers believe that a herbicide is always advisable to ensure successful seedings (Myers and Triplett, 1973). Given the unpredictable hazards of weather, insects, disease, plant competition, and other problems, a herbicide can be regarded as one of several steps to ensure greater success in no-tillage seedings. The additional cost is minimal, considering the costs of lime, fertilizer, seed, and lost time and yield should the seeding fail.

The presence of dense growing vegetation in no-tillage seedings provides a favorable environment for insects or disease. Among the problems that have been experienced are field crickets (*Gryllus* spp.) (Hoveland et al., 1972);

snails and slugs (*Mollusca* spp.) (Byers, 1979; Kalmbacher et al., 1979; Welty et al., 1981; and nematodes and fungal diseases (Sheaffer et al., 1982). The use of insecticides, nematicides, and fungicides has improved legume establishment (Hoveland et al., 1972; 1981a; Sheaffer et al., 1982). (See Chapters 12 and 13.) Desiccation of the sward with paraquat, followed by burning, controlled snails (*Polygyra cereolus* Muhlfield) (Kalmbacher et al., 1979). However, burning is often not possible because of weather and environmental restrictions. Spraying a sod with glyphosate several weeks prior to seeding allowed sufficient time for grass and weed desiccation to provide a less favorable environment for slugs and allowed good legume seedling development (Welty et al., 1981). The use of granular carbofuran, 2, 3-dihydro-2, 2-dimethyl-7-benzofuranyl methylcarbamate, an insecticide, together with paraquat improved the establishment of "Tillman" ladino clover into tall fescue sod and in some cases was essential for establishment (Rogers et al., 1984).

In general, herbicides for no-tillage pasture and meadow improvement can be divided into four groups: (1) target herbicides for controlling certain plants, for example, removing broadleaf weeds from grass with 2,4-D and dicamba; (2) short-term "burn-off" or desiccant herbicides, such as paraquat, that kill the topgrowth only; (3) general control or broad spectrum herbicides, such as glyphosate, for controlling most vegetation, enabling producers to completely change the composition of the pasture or meadow; and (4) herbicides, such as pronamide, for control or temporary growth suppression of grasses, depending on the rate used (Triplett et al., 1977). In using a herbicide, the producer should read the label and follow prescribed usage carefully. The herbicide must have clearance for use in no-tillage systems. Some herbicides that have been used experimentally with satisfactory results have not been labeled for no-tillage. Very likely future developments in herbicides will provide even better and more selective materials than are currently available.

Time and Rate of Seeding

Timeliness of seeding is important (Fribourg and Strand, 1973). Seeding time needs to coincide with periods of adequate moisture and favorable temperature (Groya and Sheaffer, 1981; Taylor et al., 1972; Strand and Fribourg, 1973.) Early spring is generally the season when best moisture conditions prevail. Early spring planting is also important to provide time for new seedlings to become well established before the onset of summer heat, drought, and excessive weed competition. Fall seedings must be made sufficiently early to allow adequate development before cold temperatures prevail. Taylor et al. (1969) reported early spring seedings of alfalfa and white clover into Kentucky bluegrass sod the most successful, followed by late summer seedings, poorest establishment was in midsummer. Studies in North Carolina showed that excellent stands and yields of alfalfa were obtained by fall seeding into paraquat-sprayed tall fescue sod, but poor stands from spring seedings (Mueller and Chamblee, 1984). An advantage of no-tillage fall seedings compared with conventional seedings is the protection afforded legume seedlings by a sod mulch against soil heaving (Rogers et al., 1983). With early spring seedings, some pasture or hay will usually be available the seeding year. With fall seedings, it is possible to graze the old pasture through early summer, followed by a late summer seeding sufficiently early to ensure good fall growth before winter.

In general, the seeding rates used with plowed or disked seedbeds are adequate with no-tillage establishment. In one no-tillage study, increasing the seeding rates of birdsfoot trefoil, alfalfa, and crownvetch (*Coronilla varia* L.), resulted in higher initial seedling counts in Kentucky bluegrass pastures compared with conventional seeding rates, but generally not higher in seeding-year legume yield or total dry matter production (Clark et al., 1975). Increased seeding rates may increase establishment-year legume yield where high grass competition occurs, more so with alfalfa than with red clover (Sheaffer and Swanson, 1982).

Placement of seed in the sod requires special equipment that can penetrate the matted growth at the soil surface. The sodseeding drill must open the sod to enable the seed to reach the soil, yet not place the seed too deep. Shallow seed placement together with good seed–soil contact is essential for rapid germination and emergence of the seeded forage (Taylor et al., 1969). A description of sodseeding drill requirements is given in Chapter 3.

Seedling Management

Following pasture and meadow reseeding there is need for controlled grazing or mowing to reduce the competiton from recovery regrowth of the old vegetation or new weeds (Norman and Green, 1957; Robinson and Cross, 1960). Using paraquat, for example, a rapid burn-down of the existing vegetation is followed within a few weeks by partial or complete recovery. In addition, fertilizer applied during renovation enhances the growth and competitiveness of the old sward that survives. This regrowth, of course, is important because it produces livestock feed and reduces the time the field is out of production. The vegetative regrowth can be grazed by cattle if care is taken to remove the top growth rapidly by using brief heavy stocking, that is, use the animals much like a mower and with the least amount of treading damage. If legumes have been interseeded into an existing sod, it is not advisable to graze with sheep until the plants are well established, since sheep selectively graze the legume seedlings and severely injure the stand. Clipping or removal as harvested forage is an alternative to increase light penetration and improvement of interseeded legumes (Barnhart and Wedin, 1978). In years after establishment, it is necessary to provide good fertilizer management to maintain vigor and productivity.

Integrated Forage Management

Pasture and meadow improvement implies increased production and quality. Since forages are utilized only by livestock, improvement will have a major influence on the livestock program. Livestock numbers and their management should be adjusted to utilize the additional feed. The producer initially considers reseeding to meet his or her livestock needs. Occasionally a producer fails to adjust his or her livestock program to utilize the increased forage available and thereby realizes little return for his or her effort. Renovated pastures often too soon revert to the original low productivity owing to poor grazing practices and maintenance fertilization. Continuous overgrazing together with failure to fertilize are major causes of loss of stand and low productivity. Grasses will remain vigorous in pastures and meadows for many years with good fertility and harvest management. The perennial legumes are shorter-lived than grasses and will need to be either reseeded periodically or allowed to set seed to maintain a stand if they have the potential to reseed.

A producer needs to view his or her pastures and meadows as a total integrated forage system. Harvest of ungrazed forage as hay or silage from pastures during periods of rapid growth is advisable. Conversely, meadow aftermath may be grazed to supplement pastures during periods of low production. Rotational grazing, especially of the tall-growing forage grasses and legumes, allows better control of grazing, forage utilization, and persistence of the legumes, compared with continuous grazing.

Another aspect of forage management is weed control. Noxious weeds must be controlled prior to reseeding as discussed earlier. Frequently these weeds encroach into pastures and meadows if soil fertility and good pasture management are not maintained. Ironweed, Canada thistle, and horsenettle encroach into pasture, since they are favored by selective grazing. Clipping to prevent seed formation and occasional use of herbicides may be necessary. Occasional harvest of pastures for hay or silage is one method of clipping such weeds. Wick applicators for control of tall-growing weeds provide one method of using herbicides without injury to the forages, especially the legumes.

PASTURE AND MEADOW IMPROVEMENT

Pastures and meadow improvement include the following situations: (1) upgrading the productivity of thin weedy permanent pastures; (2) introducing into meadows improved legumes and grasses (including complete replacement of the species); (3) establishment of perennial legumes into cool season grasses; (4) introducing winter annual legumes, annual ryegrass (*Lolium multiflorum* Lam.), and small grains into subtropical perennial grasses; (5) establishment of warm-season annual grasses such as sudangrass and sorghum–sudangrass (*Sorghum* spp.), millets (*Setaria* spp.), and annual root crops or rape (*Brassica* spp.); and (6) establishment of warm-season perennials such as switchgrass (*Panicum virgatum* L.) and big bluestem (*Andropogon gerardii* Vitm.). All are well suited to no-tillage techniques.

Permanent Pastures

Unimproved permanent pastures are generally thin weedy stands with low productivity because of low fertility and overgrazing. The most common species on the better soils in northern humid regions of the United States is Kentucky bluegrass, although a number of other grasses, legumes, and weeds may be present. Such pastures can be improved by correcting the basic fertility deficiencies, controlling weeds, and better grazing management. Such a program, however, will be slow with only a moderate increase in productivity. Site limitations, such as shallow, droughty soils, may preclude marked changes in production and such investment improvements may be all that can be justified. On better sites, productivity can be increased further and more rapidly by introducing more productive grass–legume mixtures. Most no-tillage forage seedings in recent years have been confined to permanent pastures on deeper soils.

A series of steps should be followed to ensure a successful surface tillage or no-tillage renovation of permanent pasture (Graber, 1936; Myers and Triplett, 1973; Sprague et al., 1962a; Wilkinson, 1976). Successful seedings are the result of following these steps: clearing of brush, liming, controlling weeds, and removing vegetative growth prior to seeding; reducing competition by close grazing or with herbicides applied at or prior to seeding; placing the seed properly with a sodseeding drill to penetrate the sod; applying adequate fertilizer at seeding time and seeding at favorable times of the year. Successful maintenance of stand requires good grazing management, continued weed control, and periodic applications of needed fertilizer and lime. With this procedure the residual grasses in the original sod are not killed and also will respond to improved growing conditions.

Paraquat is an economical and effective herbicide to temporarily control vegetation by "burning-off" or desiccating the aboveground growth (Allen, 1966), taking only 2 or 3 days (Davies and Davies, 1981). Most established perennial species will recover in several weeks. This rapid temporary control enables the new seeding to become established more easily than with no treatment. The degree of sod suppression increases with increased rates of paraquat from 0.28 to 1.12 kg/ha (Decker and Dudley, 1976) but varies markedly with grass species and season. Orchardgrass is more difficult to suppress than Kentucky bluegrass or tall fescue, and spring growth is more difficult than summer growth. Split applications of paraquat gives more effective control than a single application. If the perennial broadleaf weeds are not killed prior to using paraquat, they will recover rapidly following suppression of the sward and compete strongly with the new seeding (Triplett et al., 1975). Occasionally warm-season annual grasses such as crabgrass (*Digitaria* spp.), foxtail (*Setaria* spp.), and barnyardgrass [*Echinochloa crusgalli* (L.) Beauv. var. *crusgalli*] also will invade paraquat-sprayed areas, but are not a major problem and can be controlled largely by grazing or clipping.

On steep hills, shallow soils (0–30 cm) over bedrock, or stony sites, a combination of chemical renovation and broadcast seed has been used successfully to establish birdsfoot trefoil where drilling was difficult or impossible (Winch and Watkin, 1980). In these Ontario studies, Canada bluegrass (*Poa compressa* L.) pasture productivity was markedly increased by controlling perennial weeds and brush, applying P and K, suppressing the existing sod with paraquat or dalapon, and broadcasting adapted birdsfoot trefoil varieties ("Empire," "Leo," or "Carroll").

Reseeding Meadows

Occasionally a producer wants to eliminate all existing vegetation in a field and establish a different forage mixture. In most instances grass is the major component of the sward to be controlled, although a wide range of broadleaf weeds may be present. The broad spectrum herbicide glyphosate gives good control of mixtures of species. Several herbicides, herbicide combinations, and application rates and dates have been evaluated for such no-tillage situations.

Of the currently labeled materials, glyphosate is inactivated rapidly in most soils (Sprankle et al., 1975). It has been used in a spray-and-plant procedure under some conditions, but is much more effective if seeding is delayed 3–4 weeks after spraying depending on the species to be sown (Welty et al., 1981). For best control the vegetation to be treated should be growing actively (Campbell, 1976); grasses should be 20 cm in height and in the 3–4 leaf stage. Delay permits desiccation of the killed sward, better light penetration, a less favorable environment for insects and diseases and greater ease in seeding. A simultaneous spray and plant procedure in some instances has had a phytotoxic effect on the seeded forage depending on rate and exposure of the seed to the herbicide (Campbell, 1976; Salazar and Appleby, 1982a; Segura, et al., 1978). Delayed seeding, following spraying, reduces or largely eliminates this hazard. Because it is normally advisable to seed as early as possible in the spring, timing becomes important. Delays encountered in allowing the resident vegetation to grow actively before spraying are followed by a further delay of seeding. Late summer seeding, thus, is frequently more advisable, or a late fall herbicide application may be followed by spring seeding.

A split application of paraquat (fall and spring), or fall-applied pronamide plus spring-applied paraquat, have also given reasonably good control of mixed grass sods in New Hampshire (Mueller-Warrant and Koch, 1983).

Seeding Legumes into Sod

Frequently an existing sward has a good stand of grass with little or no legume, and the producer wishes to interseed a legume. The value of legumes in terms of yield contribution, as a nitrogen source for the companion grass, and for their value as livestock feed has been long recognized. Forage legumes in the humid regions are generally short-lived and not as persistent as grasses. Periodic reseeding of the legume is usually necessary to maintain a desirable mixture. All of the legumes commonly used in the northern United States can be seeded successfully with no-tillage into established grass stands.

Although legumes can be seeded successfully into actively growing sod without a grass-suppressing herbicide (Taylor and Allinson, 1983), better stands have been obtained when such a herbicide has been used (Linscott and

Vaughan, 1982; Nichols and Peters, 1983; Olsen et al., 1981; Sprague et al., 1962a; Van Keuren and Triplett, 1970. A herbicide promotes more rapid development of the legume seedlings, survival, and their contribution to the total yield (Linscott and Vaughan, 1982; Mueller-Warrant and Koch, 1980).

Legumes differ in case of establishment in grass swards, with red clover relatively easy to establish, followed by alfalfa, with trefoil the most difficult (Decker and Dudley, 1976; Kunelius and Campbell, 1984; Martin and Marten, 1977; Van Keuren and Triplett, 1970). This differential between legumes is related to seedling vigor. There is also a difference between established grasses on the ease of legume establishment; seedling alfalfa competing more successfully with Kentucky bluegrass than with smooth bromegrass (Groya and Sheaffer, 1981). Time of seeding is also important, influenced mostly by temperature and the moisture availability. Early spring is a favorable period for interseeding legumes into temperate grasses (Taylor et al., 1969), with late summer an acceptable alternative (Fribourg and Strand, 1973).

Paraquat is used widely with no-tillage seeding of legumes (Decker et al., 1969; Taylor et al., 1969, 1972; Triplett et al., 1975; Van Keuren and Triplett, 1970, 1972; Watkin et al, 1971), and is currently labeled for this use. It can be sprayed simultaneously with the seeding, and banded over the seed to reduce the amount of herbicide needed (Decker et al., 1969; Taylor et al., 1969, 1972; Van Keuren and Triplett, 1972), although this procedure requires special mounting of the spray nozzles on the drill. Taylor et al. (1979) reported that paraquat banded over the seeded row improved legume stands only when the Kentucky bluegrass sod was dense and growing rapidly. Spraying paraquat on legume seeds at seeding time has no effect on germination or subsequent development, but can severely affect grass seed germination (Appleby and Brenchley, 1968). Delaying the interval between spraying and drilling reduces the effect on grass seedling establishment (Davies and Davies, 1981). Early spring seeding of red clover into tall fescue without a herbicide has been successful when the pasture was being utilized for winter forage and was very closely grazed at the time of seeding (R. W. Van Keuren, Ohio Agric. Res. and Dev. Center, unpublished data). The spring regrowth of established plants in such fields was noted to be slower than in pastures not subjected to winter utilization.

The amount of grass suppression by paraquat and the concomitant increase in legume stand results from increasing the rate applied (Decker and Dudley, 1976). A slight suppression of Kentucky bluegrass–timothy sod with a spring application of 0.28 kg/ha; a severe suppression at 0.56 kg/ha (Triplett et al., 1975) and nearly complete kill at 1.12 kg/ha (Decker and Dudley, 1976) were reflected in better legume establishment. Tall fescue is suppressed about 4 weeks with 0.3 kg/ha of paraquat, 7 weeks with 0.6 kg/ha, and 8 weeks with 0.9 kg/ha (Olsen et al., 1981). Grass suppression by paraquat at rates normally used is relatively short-lived (3–4 weeks), usually sufficient time for legume seedlings to become established, yet permitting good recovery of the grass sod for grazing later in the seeding year.

Occasionally increased weed encroachment occurs following suppression of a grass sod by reducing grass competition to broadleaf weeds, that is, dandelions (*Taraxacum officinale* Wiggers) (Triplett *et al.,* 1975; Kunelius *et al.,* 1982), or seedling annual weedy grasses (R. W. Van Keuren, Ohio Agri. Res. and Dev. Center, unpublished data). The use of 2,4-D prior to seeding will eliminate most broadleaf weeds present. Injury to seedling legumes sometimes occurs from phenoxy herbicide applications made at or shortly before seeding (G. B. Triplett, Jr. and D. K. Myers, Ohio State U., unpublished data). Injury is minimized if seedings are delayed for 7–10 days following the first 10–15 mm rainfall after herbicide application. Pound (1981) confirmed that dry soil conditions following herbicide application and seeding was the major cause of seedling injury. Birdsfoot trefoil seedlings are more sensitive to these herbicides than alfalfa. A study in Mississippi on overseeding legumes into bermudagrass [*Cynodon dactylon* (L.) Pers.] sod, showed that dicamba at 0.56 kg/ha or 2,4-D + dicamba at 1.68 + 0.56 kg/ha should be applied at least 30 days prior to overseeding to avoid residual activity on overseeded legumes (Griffin et al., 1984). Their results indicated that crimson (*Trifolium incarnatum* L.) and white clover generally were less sensitive to residual 2,4-D + dicamba than to dicamba alone.

Several herbicides suppress grass growth to varying degrees, depending on the rate used, and there is also a difference between grass species in their relative tolerance to these chemicals. Dalapon effectively suppresses grasses for legume establishment in sods (Harrington and Washko, 1962; Parsons and Davis, 1964; Sprague, 1956; Sprague et al., 1962b) and appears to compare favorably with paraquat and glyphosate for this purpose. Rates of 0.5, 1.0, and 5.0 kg/ha of paraquat, glyphosate, and dalapon, respectively, are the minimum applications for effective control of most cool-season grasses for establishment of birdsfoot trefoil (Linscott and Vaughan, 1982). For interseeding birdsfoot trefoil into Kentucky bluegrass and perennial ryegrass, 3.0 kg/ha of dalapon was sufficient. Dinoseb and paraquat are equally effective in suppressing tall fescue top growth the first week after application, but effects of dinoseb at 2.8–8.4 kg/ha do not persist, limiting its usefulness for legume establishment (Olsen et al., 1981). Mefluidide at 0.9–2.7 kg/ha suppresses grass growth, but not as effectively as paraquat or glyphosate. Glyphosate delays foliar desiccation until after the first week,and then is effective for good legume establishment. The herbicides in the above studies were broadcast sprayed immediately following no-tillage drilling in mid-March in this southern Illinois study. Furthermore, glyphosate at 0.6 kg/ha was sufficient to give good establishment of alfalfa and red clover in tall fescue (Olsen et al., 1981), in smooth bromegrass, and in a mixture of smooth bromegrass, quackgrass, and Kentucky bluegrass (Sheaffer and Swanson, 1982). At 1.7 and 1.8 kg/ha, glyphosate resulted in almost complete control of the grasses. Dalapon at 5.6 and 9.0 kg/ha with April seedings of alfalfa into a mixed smooth bromegrass, Kentucky bluegrass, and quackgrass sod or with glyphosate at 0.8 and 1.1 kg/ha with late May seedings gave the greatest seeding year legume yields

(Martin *et al.*, 1983). These researchers suggested that the herbicide and herbicide rate used can be selected for grass suppression and legume establishment based on the extent of grass vegetative development.

Glyphosate at 2.2 kg/ha resulted in better establishment of "Viking" birdsfoot trefoil in orchardgrass than paraquat at 1.1 kg/ha (Nichols and Peters, 1983). Very little birdsfoot trefoil was established by direct seeding into orchardgrass without the herbicide.

Glyphosate applied at seeding time in spring at 0.84 kg/ha on Kentucky bluegrass–smooth bromegrass in Minnesota studies killed about one-half the Kentucky bluegrass and severely retarded the remainder, with the smooth bromegrass recovering fully in about 6 weeks. This treatment enabled good establishment of alfalfa, red clover, and birdsfoot trefoil (West et al., 1980). However, variability in results can occur from year to year between grass species and glyphosate rates (Linscott and Vaughan, 1982). This was observed with birdsfoot trefoil seeded into several kinds of sod immediately following spraying. Birdsfoot trefoil has a weak seedling, compared with red clover and alfalfa, and thus may be more sensitive to factors affecting germination and early seedling development, which vary from year to year and between different seedings, owing to temperature, moisture, depth of planting, degree of suppression of competition, and presence of insects and disease. As an example, sod-seeding perennial legumes into tall fescue was not dependable in Alabama trials (Hoveland et al., 1981a). Striped field crickets were a major cause of clover failure in autumn seedings, resulting in late winter seedings being more reliable. Broadcast seeding resulted in stands almost equal to drilling, and the use of herbicides to suppress tall fescue competition was important to legume establishment.

Generally legumes establish successfully with no-tillage drilling into sods sprayed with glyphosate shortly before seeding or on the same date (West *et al.*, 1980; Olsen *et al.*, 1981: Groya and Sheaffer, 1981; Linscott and Vaughan, 1982; and Sheaffer and Swanson, 1982). However, injury can occur when legume seed is sown on bare soil before or after spraying with glyphosate or surface-sown on sprayed sod (Campbell,1974, 1976). There was little or no effect on germination, but seedlings became deformed and growth was restricted. Light rates of glyphosate, 0.5–1.5 kg/ha, and delayed sowing, 20 days following spraying, reduced or eliminated the effect on alfalfa (Campbell, 1976). A Michigan study showed further that under conditions where alfalfa seeds imbibe glyphosate, seedling growth may be inhibited (Moshier and Penner, 1978). A very high rate of glyphosate (3.4 kg/ha) reduced germination and growth of alfalfa and red clover when seed was sown on sprayed soil surface (Salazar and Appleby, 1982b). Glyphosate, however, is rapidly inactivated by soil contact, especially soils high in clay and organic matter contents (Sprankle *et al.*, 1975).

In view of the above research, delaying seeding until after spraying to avoid contact of glyphosate with seed that may not have been covered in drilling appears advisable. Because of potential problems with insects, snails, and

FIGURE 8.1. Spraying a pasture field with paraquat preliminary to no-tillage seeding red clover into a tall fescue sod. (Ohio Agr. Res.and Dev. Center photograph.)

FIGURE 8.2. No-tillage seeding of a grass–legume mixture into a glyphosate-killed sod. (Ohio Agr. Res.and Dev. Center photograph.)

slugs, which harbor in green vegetation, delay of seeding until desiccation of the vegetation also reduces the chance of such damage occurring. Of the above herbicides, however, only paraquat and glyphosate are currently registered and labeled for use in forage establishment. Paraquat appears to be the most useful chemical for suppressing grass growth for no-tillage seeding of forage legumes (Fig. 8.1); glyphosate appears to be best for situations where it is desirable to severely reduce or eliminate the grass stand (Fig. 8.2).

Broadcast Seeding

An age-old no-tillage method of seeding forages is to broadcast or overseed in very early spring on frozen ground. In northern areas repeated freezing and thawing of the soil in early spring enhances the probability of the seed being covered by soil. The method is used most widely to introduce or increase legumes in a thin grass sward. White clover, for example, can be established successfully in an untilled Kentucky bluegrass sod by broadcasting in late winter or early spring (Taylor et al., 1972). Double the normal seeding rate is normally required and better stands are obtained by drilling into the soil (0.6 cm) than from surface sowing. Generally, good to satisfactory stands of alfalfa, red clover, birdsfoot trefoil, and other legumes were obtained by March frost seedings in Iowa (Shaller and George, 1978) and the best stands were obtained in orchardgrass, followed by smooth bromegrass and the poorest in Kentucky bluegrass. The major reasons for poor establishment from surface seedings include poor germination, too heavy accumulation of plant residue on the surface, inability of sprouting seed to reach and penetrate the soil, desiccation, insect damage, and weed competition. Greatest losses occur in summer and least in winter (Campbell and Swain, 1973). Greater assurance of successful seedings are obtained by drilling than by broadcasting.

Natural Reseeding

Many legumes, particularly annual legumes, can be maintained by encouraging natural reseeding. Under close grazing subclover (*Trifolium subterraneum* L.), rose clover (*T. hirtum* All.), Korean lespedeza (*Lespedeza stipulacea* Maxim.), and white clover will set sufficient seed to maintain satisfactory stands. Subclover is used extensively in Australia, the Mediterranean, western Oregon, and eastern Texas. Rose clover is widely used for range in California. Korean lespedeza is limited primarily to east central United States. White clover is widespread throughout humid regions. Two winter annual clovers used widely in the Gulf Coast region, crimson (*Trifolium incarnatum* L.) and arrowleaf (*T. vesiculosum* Savi), require special management to maintain stands by natural reseeding, primarily by limiting the grazing during flowering in the spring to allow seed set. Natural reseeding also contributes to maintenance of stands of red clover and birdsfoot trefoil. Birdsfoot trefoil, indeterminate in growth habit, can be deferred grazed to allow seed to set and still maintain good quality forage. Red clover in a late summer and early fall deferred grazing program provides an opportunity to set seed, which aids in stand maintenance (R. W. Van Keuren, Ohio Agric. Res. and Dev. Center unpublished research). A high percentage of hard seed, common to some of these legumes, aids stand maintenance year to year by building a supply of viable seed in the soil.

Establishment in Quackgrass

A special problem in no-tillage forage seeding is the presence of quackgrass. This aggressive,strongly rhizomatous grass is a common invader throughout northern United States and Canada and is difficult to control. Putting a pasture or meadow through corn (*Zea mays* L. subsp. *mays*) is a convenient method of controlling this weed with the herbicides available, but this is frequently not possible in many no-tillage forage situations.

Glyphosate has made control of quackgrass much easier than previously. Complete eradication generally is not possible, but there is practical control of quackgrass in a forage stand. Best control of quackgrass is obtained by treating the plants when actively growing in the jointing to early heading stage rather than at earlier or later vegetative stages (Friesen, 1977). The major objective is to obtain good translocation of the herbicide to the rhizomes to kill the buds. Glyphosate is not translocated to the rhizomes when the plant is in the 2 leaf-stage, but is increasingly translocated at the 3 and 4 leaf stage (Rioux et al., 1974). Spring and fall treatments at 1.1 and 2.2 kg/ha were equally effective in New Hampshire and there appears to be no advantage to split fall and spring applications (Mueller-Warrant and Koch, 1983). Frost increases glyphosate activity (Davis et al., 1978, 1979), and in the fall relatively low rates, 0.56 and 1.12 kg/ha, gave better control than in spring (Ivany, 1981). At 2.24 kg/ha, no seasonal effect was noted.

A problem associated with seeding forages into quackgrass is the allelopathic effect of dead quackgrass on the seeded forages, especially the legumes (Kommedahl et al., 1959; Welbank, 1963). There is indirect evidence that germination and early growth of alfalfa is inhibited by the rapid death and decomposition of quackgrass following spring glyphosate treatment (Mueller-Warrant and Koch, 1980). A May 1 glyphosate application, compared with May 15, significantly increased the growth of alfalfa seeded into quackgrass (Mueller-Warrant and Koch, 1983). Leaching of the allelochemicals may be associated with rainfall.

No-Tillage Seeding into Small Grain Stubble

Forages can be successfully no-tillage seeded into small grain stubble (Vough et al., 1981). This practice is compatible with crop rotations used by many livestock producers. The producer can optimize grain and straw yield, and subsequently establish legumes or grass–legume mixtures. As with pasture renovation, certain steps should be followed to ensure maximum success (Myers, 1981). These include liming and fertilization, removing excess straw, controlling weeds as needed, and seeding in late summer with a sodseeding drill. If broadleaf weeds are present, an application of 2,4-D several weeks prior to seeding is advisable, preferably after a rain to reduce herbicide toxicity to the legume. At planting time 0.28 kg/ha of paraquat is applied to control volunteer grain and grassy weeds. If quackgrass or other perennial weedy grasses are present, glyphosate may be used.

No-Tillage Seeding into Subtropical Perennial Grasses

Subtropical perennial grasses are grown widely across the southern United States, primarily along the Gulf Coast from eastern Texas to Georgia and Florida. Bermudagrass, bahiagrass, and dallisgrass (*Paspalum dilatatum* Poir.), predominate, with digitgrass (*Digitaria decumbens* Stent.), limpograss [*Hemarthria altissima* (Poir.) Stapf. et C. E. Hubb.], and stargrasses (*Cynodon* spp.), of importance in south Florida. This is primarily a beef-cattle-producing area, with some dairy cattle. The subtropical grasses are dormant in much of the region for 5–6 months during late fall, winter, and early spring. No-tillage seeding of cool season legumes and annual grasses is a widespread practice to provide grazing during this period. Excellent forage production and animal performance are obtained extending the grazing season appreciably (Anthony et al., 1971; Hoveland, *et al.*, 1977; Knight, 1970; Knight *et al.*, 1975). Cool season legumes seeded into the pastures also provide substantial amounts of nitrogen to the sward (Palmertree, 1977). The most common species used are annual ryegrass, winter oats (*Avena sativa* L.), wheat (*Triticum aestivum* L.), and rye (*Secale cereale* L.), plus crimson, arrowleaf, white, and subterranean clovers or mixtures of both grass and clover. In general, the presence of legumes provides better seasonal distribution of yield than grasses alone. The warm-season perennial grass sods (bermudagrass or bahiagrass) seeded to winter annuals come close to an ideal year-around grazing system in this region. Further north the more winter hardy species are also useful.

Winter oats, wheat, and annual ryegrass have long been fall-seeded into dallisgrass–bermudagrass sods in the lower South without a herbicide (Dudley and Wise, 1953). Moisture is the most critical factor influencing successful seedings, and sodseeding equipment is necessary to penetrate the sod in placing the seed. The small grains should be placed about 5 cm deep in light soils and annual ryegrass should be placed about 2.5 cm. No-tillage seedings have the advantages of lower establishment cost and better sod conditions in wet weather to support livestock compared with a plowed or disked seedbed. Delayed seeding until late September or early October is recommended to avoid dry weather. The pastures should be closely grazed prior to seeding, and the grazing animals should be removed at seeding time (Coats, 1957). In Mississippi annual ryegrass is seeded in early October in the northern part of the state and in late October in the south. Needed fertilizers should be applied at seeding, and additonal nitrogen in mid-February to optimize yield. Seeding oats, rye, or wheat at 112 kg/ha is suggested, and/or annual ryegrass at 28 kg/ha. The grasses do not provide grazing until mid-January in Mississippi (McKie, 1977). Competition from the winter grass or cereal growth delays spring production of the perennial grass. Because of this, no more than two-thirds of the pasture should be overseeded.

Disking bermudagrass sod once or twice and broadcasting annual ryegrass gives slightly higher yields than drilling with a Midland Zip® seeder (Watson et al., 1975). However, the use of paraquat permits seeding in the bermudagrass sod up to six weeks earlier and provides fast and thorough desiccation of bahiagrass and dallisgrass sods; residual desiccation effects are evident for 15–18 days. The use of paraquat results in significantly higher dry matter yields at the first harvest, compared with no herbicide, and this early growth comes at a critical time in livestock production. Paraquat rates from 0.28 to 1.12 kg/ha are equally effective. Both "Gulf" ryegrass and "Redland" red clover can be successfully no-tillage seeded into paraquat-treated "Pensacola" bahiagrass and "Alicia" bermudagrass to extend forage productivity and improve forage quality (Montgomery et al., 1983). Glyphosate on bahiagrass markedly increases rye and annual ryegrass production in Alabama compared with no herbicide (Hoveland *et al.*, 1981b). There were no differences between glyphosate applied at 0.28, 0.56, and 1.12 kg/ha and its use on bermudagrass did not affect annual grass production.

Temperature and moisture are critical factors for germination of winter legumes. Crimson and subclover are sown from the middle of September until November, depending on location and usage; arrowleaf from August until November. Planting in November or later, however, results in slower germination and poorer seedling growth and survival, and delayed and reduced forage production (Evers, 1980). "Yuchi" arrowleaf clover germinates at low temperatures, with seedlings continuing to emerge through December and January in Alabama when moisture is adequate (Hoveland, *et al.*, 1969). Crimson clover is less likely to germinate successfully during this period. A dry autumn followed by a cold winter may result in poor stands of crimson clover, but arrowleaf frequently continues to germinate and grow. Nitrogen fertilization at seeding time reduces both the stand and subsequent yield of crimson clover seeded into "Coastal" bermudagrass, although it increases the total forage yield (Knight, 1967). Winter annual clovers are fall-seeded with cereal grains or annual ryegrass in the lower South from September 1 to October 15, depending on the latitude.

Close grazing or mowing late in summer is necessary prior to seeding. If summer growth is removed, disking of bermudagrass is not beneficial to crimson clover establishment (Knight, 1967). Satisfactory stands of all species are ensured by proper seeding into a weed-free sod that has been limed and fertilized as indicated by soil tests and crop needs. Shallow placement of the seed (5–10 mm) in the soil gives the best chance for seedling emergence and development. Seedling survival of arrowleaf clover and *Rhizobium* nodulation was markedly influenced by seeding depth, with 13 mm optimum (Rich et al., 1983). Drilling subclover in 13 cm rows gave slightly higher yields than drilling in 25 cm rows or broadcasting in a Texas study (Evers, 1982).

Among the winter legumes, crimson clover is seeded alone at 22–34 kg/ha, with the higher rates more desirable; with cereal grain or annual ryegrass the rate may be reduced to 17–22 kg/ha. With arrowleaf clover, 6–10 kg/ha are used in pure stands, and 5–8 kg/ha with annual ryegrass or cereal grains. With subclover 12–15 kg/ha are adequate; although with overseeding aerially or broadcasting into rough, unprepared sites, rates up to 20 kg/ha are used to compensate for poor seed placement and germination. No large increase in yield of "Mt. Barker" subclover above 13 kg/ha was obtained in eastern Texas, using rates from 4 to 36 kg/ha (Evers, 1982). The use of glyphosate increased the stand of "Yuchi" arrowleaf clover in bahiagrass in Alabama compared with no herbicide and increased clover production in spring (Hoveland et al., 1981b). Glyphosate was not effective on bermudagrass and paraquat was not effective in suppressing bahiagrass sufficiently to increase the clover stand over no herbicide.

In Florida, interseeding "Apollo" alfalfa, "Pennscott" red clover, and Tillman ladino clover in November and December into semidormant bahiagrass increased winter pasture production (Kalmbacher *et al.*, 1980). Disking the bahiagrass sod once and broadcasting the legume seed resulted in better stands and yields than sodseeding with a Midland Zip® seeder. Use of herbicides with disking and broadcasting resulted in little further yield increase. Paraquat with sodseeding resulted in better legume stands and forage yield compared with sodseeding alone. A combination of paraquat, burning, and seeding with the sodseeder gave the best results, but likely not the most economical.

December seedings of "Kenland" red clover, Yuchi arrowleaf clover, and "Nolins" white clover in limpograss were equally successful in Florida by broadcasting on undisturbed sod or on disked sod, or drilling into disked sod (Ruelke and Quesenberry, 1981). The greatest yields were obtained from the red clover utilized as a winter annual.

Guidelines have been outlined for establishing clovers and grasses in existing sods in most areas of Mississippi (Watson et al., 1975), and these have wide application for the region. These steps are: (1) weaken existing grass sod by hard, close grazing or mowing (dense, vigorous sods may be further weakened by tillage and/or herbicides such as paraquat); (2) apply needed lime and fertilizer for the seeded crop; (3) plant high-quality seed of adapted species at the optimum time; (4) observe renovated fields frequently and control harmful insects; and (5) manage grazing to favor growth, development, and production of the introduced species.

Warm-Season Perennial Grasses

Warm-season perennial grasses are used as summer forage in central United States to supplement cool-season pastures and as wildlife cover (Rountree et al., 1980). These grasses include switchgrass, indiangrass [*Sorghastrum nutans* (L.) Nash], big bluestem, and little bluestem [*Schizachyrium scoparium* (Michx.) Nash]. Seedling establishment is relatively slow and weed competition is a major factor (Bryan and McMurphy, 1968; Martin et al., 1982) and a long period of weed suppression is needed. Important factors in establishment and maintenance include proper planting time, use of adapted varieties, weed control, time and frequency of mowing or grazing, and adequate fertility with proper timing and rate of nitrogen (Warnes and Newell, 1971). Limited information is currently available on no-tillage establishment of warm-season grasses. Spring seedings with paraquat at 0.3 kg/ha and atrazine at 2.2 kg/ha or glyphosate at 2.2 kg/ha applied at seeding time gave successful establishment of switchgrass into a predominantly Kentucky bluegrass sod in Nebraska (Samson and Moser, 1982). In this study warm-season perennial grass remnants, primarily big bluestem, rapidly increased in response to the atrazine and glyphosate application. Atrazine, however, has not been labeled nationwide for use on these perennial grasses but has been approved in some states. Both are effective in suppressing the old sod, but atrazine provided season-long control of annual weeds and white clover, resulting in a more vigorous stand of the grasses by the end of the first growing season than with paraquat plus glyphosate. Atrazine generally suppresses Kentucky bluegrass very well (Samson and Moser, 1982), and at 2.2 kg/ha was effective for big bluestem establishment. Indiangrass was found to be sensitive at this rate (Martin et al., 1982), but at 1.1 kg/ha was satisfactory for indiangrass and little bluestem. Reduced rates of atrazine are indicated on the label for soils in low organic matter and caution is advised when seeding warm-season perennials under these conditions. In general, a combination of paraquat to initially control the vegetation and atrazine for long-lasting control appears to be a promising program for no-tillage establishment.

Annual Forages

A number of annual forages can be no-tillage seeded, including sudangrass [*Sorghum sudanense* (Piper) Stapf], sorghum–sudangrass hybrids (*Sorghum* spp.), and pearl millet [*Pennisetum americanum* (L.) Leeke], for summer pasture, hay, and silage throughout the humid regions and under irrigation; small grains for fall and early spring pasture in the northern states and winter grazing in the Gulf Coast region; annual ryegrass as winter pasture in the lower South; and root crops and rape for late summer and fall pasture and fodder in the northern states. The fast-growing taller summer annuals are also useful as smother crops to further weaken and kill weedy species. Small grains and annual ryegrass for winter pasture in the south has been discussed earlier. All the annual forages can successfully be seeded into pastures or meadows by removing the existing topgrowth by close grazing or mowing, use of a herbicide to control regrowth, followed by sodseeding to get proper seed placement (Sprague *et al.*, 1962a, b). Compared with the perennial grasses and legumes, these crops have larger seeds and are thus generally easier to seed and to establish. Paraquat can be used successfully to fall seed small grains into bluegrass pasture (Van Keuren and Triplett, 1972).

Successful seedings of sudangrass into Kentucky bluegrass have been reported (Sprague *et al.*, 1962b) providing competition from a partially killed sward is not great. For maximum season-long growth of the seeded annual, the sward must be suppressed longer than for seedling establishment only. No-tillage herbicide systems for forage sorghum and sorghum–sudangrass hybrids have been developed that provide for season-long control of sod and weed competition (Gallaher and Cummins, 1976). Combinations of herbicides and specific herbicides used depend on the weeds present. Gallaher and Cummins used 3.4 kg/ha propazine and 0.56 kg/ha paraquat with a surfactant. Atrazine has been labeled to include forage-type sorghum and sorghum–sudangrass hybrids in addition to corn. It should be noted that the sorghums and sudangrass are more sensitive to atrazine than is corn, and should not be seeded with atrazine on sands, sandy loams, or sandy clay loams. Attention should be given to the carryover of atrazine and its effect on crops that follow such as cereal grains. Limited information is available on the use of paraquat and glyphosate for no-tillage seeding of annual crops but, as with perennial forages, these herbicides may be useful with these crops also.

Root crops such as turnips and foliar crops such as rape have limited use in the United States as supplemental catch crops for grazing or soiling. These crops are successfully established in sod without tillage in the United Kingdom (Evans, 1973; Jeater and McIlvenny, 1968; Toosey, 1971) and in New Zealand (Whittles and Williams, 1968). Drilling kale into paraquat-sprayed sod gave similar results to drilling into a plowed and prepared seedbed (Jeater and McIlvenny, 1968). Similar yields of turnips and rutabagas were obtained from using paraquat and dicamba compared with traditional cultivation; rape and choumoellier returned higher yields with traditional cultivation (Whittles and Williams, 1968). No-tillage sodseeding results in better livestock utilization due to firmer soil conditions and footing compared with a plowed seedbed (Warboys and Nuttall, 1968). Removal of excess surface plant material gave better seedling emergence and establishment than with a trashy surface and was important to permit the herbicide to function better (Squires and Elliott, 1972, 1975).

In the United States turnips and rape were successfully no-tillage established in early August in pasture sods using paraquat at 1.12 kg/ha or glyphosate at 1.12 kg/ha (Jung et al., 1983). Phosphorus fertilizer increased yield in seedings made in infertile hill-land pasture in this Pennsylvania study. A subsequent study showed that it was necessary to control competition of the sod because forage turnips and forage rape are poor competitors, but that it was not necessary to kill the sod to obtain good *Brassica* stands (Jung et al., 1984).

LOOKING AHEAD

Currently, we have the technology of herbicides, sodseeding equipment, and forage species, as well as the scientific knowledge to adapt no-tillage methods for establishment of pastures and meadows in farm situations where tilled seedbeds are now used. In addition the no-tillage systems enable producers to establish forages on sites that cannot be tilled because of slope, shallow soils, stoniness, poor drainage, or rough and uneven terrain. Where it is impossible to use ground-spraying and sodseeding equipment, aircraft can be used, although additional information is needed on the use of herbicides under such conditions.

In the future the greatest advancements will be in the development of better herbicides and their delivery to "fine-tune" vegetation control. This will involve greater selectivity, for example, eliminating one perennial grass from a mixture, such as tall fescue from orchardgrass, bermudagrass from bahiagrass, a grass from a legume, or a particular noxious weed from forage mixtures. We will be able to suppress certain species for specific periods of time to enable the interseeded species to establish well, restoring the vigor of the resident sod after establishment to maximize yield. Increasing use of plant growth regulators will be used to manipulate sward growth, increase or decrease vegetative regeneration, inhibit flowering maturity, and suppress one competitive species to favor another to provide the desired grass–legume mixture. Season-long suppression of perennial species will permit production of an annual crop in a "sleepy-sod" concept, which would permit the use of more supplemental forages in rotation. Early spring perennial production could be followed by a warm-season annual to maximize summer yield followed by regrowth of the perennials in the fall. This would have useful application as a soil conservation measure, especially on sloping land. A major problem is labeling of useful herbicides for the many forage species and conditions encountered, especially where relatively small acreages are involved.

Millions of hectares of land suitable for pasture and meadow remain unimproved throughout the humid regions of the world. The potential for no-tillage methods of improvement, conservatively estimated in excess of 400%, is virtually untapped. The current limitations are the demand for, and profitability of, animal production.

SUMMARY

The no-tillage system provides an efficient and successful procedure for improving pastures and meadows (Table 8.1). It is adaptable to all kinds of land capability situations from rough hill land to rotation cropland. It can be adapted for all kinds of forage species including perennial grasses and legumes, annual legumes, cereal grasses, warm-season annuals, and root crops. Although originally considered primarily as a method for improving land subject to erosion and not easily cultivable, it is equally as attractive for forages in rotation. A unique use of the no-tillage system is the seeding of legumes into established grasslands, enabling producers to upgrade periodically the legume stand, which typically is shorter lived than the companion grasses. No-tillage reestablishment must of necessity be integrated with maintenance management practices to extend desirable mixtures and long-term production. The production, management, and utilization of forages is profoundly complex because they generally are perennials involving long-term persistence and regrowth through vegetative regeneration, usually grown in mixtures of competing species and are often harvested frequently at immature stages of growth. In addition, the grazing animal has a profound effect on the sward. All of these factors affect the performance of the established forages as well as the frequency of reseeding or interseeding.

TABLE 8.1. Summary Outline of No-Tillage Procedures for Seeding Forage Legumes and Grasses in Middle Latitude Humid Regions

A. *General Procedure for All Seedings*

1. Select site and determine lime and fertilizer needs.
2. Apply needed lime, preferably 6 months to a year prior to seeding.
3. Use control measures for noxious weeds, brush, and trees.
4. Apply needed corrective fertilizer.
5. Follow-up forage management and periodic maintenance fertilization.

Note: The above steps without reseeding will result in slow but marked improvement at least cost.

B. *Permanent Pasture*

Purpose: Improve pastures by seeding legume mixture into existing bluegrass or mixed grass sward.

Procedure:

1. In early spring or late summer apply 2,4-D 7–10 days prior to seeding for weed control. In late summer graze close prior to seeding.
2. Apply paraquat at or shortly before seeding for suppressing existing vegetation.
3. Drill shallow at 6–13 mm.

Results: More rapid improvement than A, 1–5 and higher yield potential.

C. *Seeding Meadows and Pastures*

Purpose: Eliminate all vegetation and seed a new mixture.

Procedure:

1. In early spring or late summer seedings allow grass in old sod to reach 20 cm in height and 3–4 leaf stage.
2. Apply glyphosate; for early spring seedings, apply in late fall.
3. Delay several weeks before seeding to allow old vegetation to desiccate.
4. Two to three applications of paraquat may be substituted for glyphosate.

Results: Complete change in forage species, yield potential greater, quicker, and longer than B, but more expensive.

D. *Seeding Legumes into Grass Sod*

Purpose: Introduce or reseed perennial legumes into perennial cool-season grasses.

TABLE 8.1. *(Continued)*

Procedure:
1. Apply 2,4-D 7–10 days prior to seeding if broadleaf weeds present; for noxious weeds use other herbicides previous season.
2. Apply paraquat at or shortly before seeding. Glyphosate may be substituted for paraquat as species warrant.
3. Drill seed shallow (6–13 mm).
4. Use insecticide and/or nematicide if needed.

Results: Improve forage quality and supply nitrogen to companion grass.

E. *Seeding Legumes or Legume–Grass into Quackgrass*

Purpose: Establish forages in quackgrass-infested fields.

Procedure:
1. Allow quackgrass to reach 20 cm in height and 3–4 leaf stage.
2. Apply glyphosate to vigorously growing vegetation.
3. Delay several weeks before seeding to allow desiccation of old vegetation.
4. Drill shallow.
5. Use insecticide and/or nematicide as needed.

Results: Replace quackgrass with desired forages without excessive cost or time delay.

F. *Establishment in Small Grain Stubble*

Purpose: Forage establishment when small grains are in the rotation.

Procedure:
1. After grain harvest, remove straw, leaving a short stubble.
2. Delay seeding several weeks after grain harvest to allow weeds and volunteer grain to grow.
3. If broadleaf weeds are present, apply 2,4-D.
4. Apply paraquat at or shortly before seeding.
5. If quackgrass is present, apply glyphosate as in E above instead of paraquat.
6. In the upper Midwest no-tillage drill alfalfa in early August.

Results: Allows maximum production from small grain followed by rapid forage establishment.

G. *Cool-Season Legumes and Annual Grasses into Subtropical Perennial Grasses*

Purpose: Seed perennial or winter annual clovers and grasses into bermudagrass, bahiagrass, and dallisgrass pastures.

Procedure:
1. Closely graze pasture prior to seeding.
2. Seed in late September or October in Gulf region, earlier in Maryland, Delaware, and Virginia.
3. Use paraquat for earlier seeding than if a herbicide is not used.
4. Sod-seed legumes at 5–10 mm, small grains at 5.0 cm, and annual ryegrass at 2.5 cm depth. Nitrogen fertilization needed if grasses only are used.
5. Crimson, arrowleaf, and subclover can be managed to reseed, arrowleaf is less dependable.

Results: To provide winter-grazing from winter-dormant perennial pasture.

TABLE 8.1. *(Continued)*

H. *Warm-Season Perennial Grasses*

Purpose: Establish switchgrass or big bluestem for supplemental pasture and wildlife cover.

Procedure:

1. Apply paraquat at 0.3 kg/ha and atrazine at 2.2 kg/ha or atrazine alone for switchgrass or big bluestem on silt or clay loam soils with $<1\%$ organic matter.
2. Glyphosate can be substituted for grasses not atrazine-tolerant.

Results: Rapid establishment of the warm-season perennial grasses.

I. *Annual Forages*

Purpose: Seed supplementary or emergency forages.

Procedure:

Forage sorghums and sudangrass:

1. Atrazine can be used preplant before sod vegetation is more than 15–20 cm tall. Rate depends on soil type, but cannot be used on light sandy soils.
2. Propazine and paraquat also can be used.

Small grains and root crops:

Paraquat or glyphosate to control sod vegetation. Use glyphosate if wish to kill sod.

Results: Establishment of supplemental forages without tillage.

LITERATURE CITED

Allen, H. P. 1966. The role of paraquat as an aid to the renewal of grassland. In Proc. X Int. Grassl. Congr. Helsinki, Finland, pp. 326–330.

Anthony, W. B., C. S. Hoveland, E. L. Mayton, and H. E. Burgess. 1971. Rye-ryegrass-Yuchi arrowleaf clover for production of slaughter cattle. Auburn Univ. (Ala.) Agric. Exp. Sta. Circ. 182.

Appleby, A.P. and R. G. Brenchley. 1968. Influence of paraquat on seed germination. *Weed Sci.* **16**:484–485.

Barnhart, S. K. and W. F. Wedin. 1978. Post-interseeding management of *Poa* and *Bromus*-dominated swards. Agronomy Abstr. **1978**:91.

Bryan, G. G. and W. E. McMurphy. 1968. Competition and fertilization as influences on grass seedlings. *J. Range Manage.* **21**:98–101.

Byers, R. A. 1979. Arthropod and mollusc pests of no-till forages. In Invited Papers and Abstracts, Amer. Soc. Agron., N. E. Br., pp. 13–15.

Campbell, M. H. 1974. Effects of glyphosate on the germination and establishment of surface-sown pasture species. *Aust. J. Exp. Agric. Anim. Husb.* **14**:557–560.

Campbell, M. H. 1976. Effect of timing of glyphosate and 2,2-DPA application on establishment of surface-sown pasture species. *Aust. J. Exp. Agric. Anim. Husb.* **16**:491–499.

Campbell, M. H. and F. G. Swain. 1973. Factors causing losses during establishment of surface-sown pastures. *J. Range Manage.* **26**:355–359.

Clark, D. W., W. F. Wedin, G. Ayres, and L. Gittins. 1975. Interseeding legumes following strip tillage. Proc. No-tillage Forage Symposium. Oct. 14-16, 1975. Ohio State Univ. and Ohio Agric. Res. and Dev. Center, Wooster, Ohio, pp. 72–81.

Coats, R. E. 1957. Sod seeding-brown loam tests. Mississippi Agric. Exp. Stn. Bull. 554.

Crop Science Society of America. Committee on Crop Terminology. 1962. Summary of terms compiled by Committee on Crop Terminology. *Crop Sci.* **2**:85–86.

Davies, W. I. C. and J. Davies. 1981. Varying the time of spraying with paraquat or glyphosate before direct drilling of grass and clover seeds with and without calcium peroxide. *Grass Forage Sci.* **36**:65–69.

Davis, H. E., R. S. Fawcett, and R. G. Harvey. 1978. Effect of fall frost on the activity of glyphosate on alfalfa (*Medicago sativa*) and quackgrass (*Agropyron repens*). *Weed Sci.* **26**: 41–45.

Davis, H. E., R. S. Fawcett, and R. G. Harvey. 1979. Effects of frost and maturity on glyphosate phytotoxicity, uptake, and translocation. *Weed Sci.* **27**:110–114.

Decker, A. M. and R. F. Dudley. 1976. Minimum tillage establishment of five forage species using five sod-seeding units and two herbicides. *In* J. Luchok (ed.). *Hill Lands*. Proc. of an Int. Symposium, Morgantown, W. Va. Oct. 3–9, 1976. West Virginia University Books, Morgantown, WV, pp. 140–146.

Decker, A. M., H. J. Retzer, M. L. Sarna, and H. D. Kerr. 1969. Permanent pastures improved with sod-seeding and fertilization. *Agron. J.* **61**:243–247.

Dudley, R. F. and L. N. Wise. 1953. Seeding in permanent pasture for supplementary winter grazing. Mississippi Agric. Expt. Sta. Bull. 505.

Evans, T. 1973. New approaches to increasing fodder production. Swedes and turnips in upland situations. *Outlook in Agric.* **7**:171–174.

Evers, G. W. 1980. Germination of cool season annual clovers. *Agron. J.* **72**:537–540.

Evers, G. W. 1982. Subterranean clover seeding rates. Texas Agric. Exp. Stn. Forage Research in Texas, 1982.

Fribourg, H. A. and R. H. Strand. 1973. Influence of seeding dates and methods on establishment of small-seeded legumes. *Agron. J.* **65**:804–807.

Friesen, H. A. 1977. Effect of growth stage on quackgrass control with glyphosate. *Weed Sci. Soc. Amer. Abstr.*:3.

Gallaher R. N. and D. G. Cummins. 1976. Year-round forage production utilizing multiple cropping-minimum tillage management. *Georgia Agric. Res.* **17**(3):12–14.

Graber, L. F. 1936. Renovating bluegrass pastures. Wisconsin Agric. Exp. Stn. Circ. 277.

Griffin, K. L., V. H. Watson, W. E. Knight, and A. W. Cole. 1984. Forage legume response to dicamba and 2,4-D applications. *Agron. J.* **76**:487–490.

Groya, F. L. and C. C. Sheaffer. 1981. Establishment of sod-seeded alfalfa at various levels of soil moisture and grass competition. *Agron. J.* **73**:560–565.

Harrington, J. D. and J. B. Washko. 1962. Sod subjugation with herbicides for pasture renovation. Pennsylvania Agric. Exp. Stn. Bull. 694.

Hoveland, C. S., M. W. Alison, Jr., R. F. McCormick, Jr., W. B. Webster, V. H. Calvert II, J. T. Easor, M. E. Ruf, W. A. Griffey, H. E. Burgess, L. A. Smith, and H. W. Grimes, Jr. 1981a. Seeding legumes into tall fescue sod. Auburn Univ. (Alabama) Agric. Exp. Stn. Bull. 531.

Hoveland, C. S., W. B. Anthony, E. L. Mayton, and H. E. Burgess. 1972. Pastures for beef cattle in the Piedmont. Auburn Univ. (Alabama) Agric. Exp. Stn. Circ. 196.

Hoveland, C. S., R. F. McCormick, Jr., J. A. Little, G. V. Granade, and J. G. Starling, 1981b. Growth suppressant chemicals for establishment of winter annual forages on bahia and bermudagrass sods. Auburn Univ. (Alabama) Agric. Exp. Stn. Bull. 533.

Hoveland, C. S., W. B. Anthony, J. A. McGuire, and J. G. Starling. 1977. Overseeding winter annual forages on Coastal bermudagrass sod for beef cows and calves. Auburn Univ. (Alabama) Exp. Stn. Bull. 496.

Hoveland, C. S., E. L. Carder, G. A. Buchanan, E. M. Evans, W. B. Anthony, E. L. Mayton, and H. E. Burgess. 1969. Yuchi arrowleaf clover. Auburn Univ. (Alabama) Agric. Exp. Stn. Bull. 396.

Ivany, J. A. 1981. Quackgrass (*Agropyron repens*) control with fall-applied glyphosate and other herbicides. *Weed Sci.* **29**:382–386.

Jeater, R. S. L. and H. C. McIlvenny. 1968. Direct drilling of kale. *Weed Res.* **8**:145–148.

Jung, G. A., R. E. Kocher, and A. Glica. 1984. Minimum-tillage forage turnip and rape production on hill land as influenced by sod suppression and fertilizer. *Agron. J.* **76**:404–408.

Jung, G. A., W. L. McClellan, R. A. Byers, R. E. Kocher, L. D. Hoffman, and J. J. Donley. 1983. Conservation tillage for forage Brassicas. *J. Soil Water Cons.* **38**:227–230.

Kalmbacher, R. S., D. R. Minnick, and F. G. Martin. 1979. Destruction of sod-seeded legume seedlings by the snail (*Polygyra cereolus*). *Agron. J.* **71**:365–368.

Kalmbacher, R. S., P. Mislevy, and F. G. Martin. 1980. Sod-seeding bahiagrass in winter with three temperate legumes. *Agron. J.* **72**:114–118.

Knight, W. E. 1967. Effect of seeding rate, fall disking, and nitrogen level on stand establishment of crimson clover in a grass sod. *Agron. J.* **59**:33–36.

Knight, W. E. 1970. Productivity of crimson and arrowleaf clovers grown in a Coastal bermudagrass sod. *Agron. J.* **62**:773–775.

Knight, W. E., V. H. Watson, and T. Kight. 1975. Research reveals value of renewed interest in legumes. *Mississippi Agric. and Forestry Exp. Stn. Res. Rep.* **1**:19.

Kommedahl, T., J. B. Kotheimer, and J. V. Bernadini. 1959. The effects of quackgrass on germination and seedling development of certain crop plants. *Weeds* **7**:1–12.

Kunelius, H. T. and A. J. Campbell, 1984. Performance of sod-seeded temperate legumes in grass dominant swards. *Can. J. Plant Sci.* **64**:643–650.

Kunelius, H. T., A. J. Campbell, K. B. McRae, and J. A. Ivany. 1982. Effects of vegetation suppression and drilling techniques on the establishment and growth of sod-seeded alfalfa and birdsfoot trefoil in grass dominant swards. *Can. J. Plant Sci.* **62**:667–675.

Linscott, D. L. and R. H. Vaughan. 1982. Influence of herbicides on direct-seeding establishment of birdsfoot trefoil (*Lotus corniculatus*) into grass swards. *Weed Sci.* **30**:567–571.

Mann, R. K., S. W. Rosser, and W. W. Witt. 1983. Biology and control of tall ironweed (*Vernonia altissima*). *Weed Sci.* **31**:324–328.

Martin, N. P. and G. C. Marten. 1977. Methods of interseeding legumes in three humid perennial grass pastures. Amer. Soc. Agron. Abstr. 102.

Martin, A. R., R. S. Moomaw, and K. P. Vogel. 1982. Warm-season grass establishment with atrazine. *Agron. J.* **74**:916–920.

Martin, N. P., C. C. Sheaffer, D. L. Wyse, and D. A. Schriever. 1983. Herbicide and planting date influence establishment of sod-seeded alfalfa. *Agron. J.* **75**:951–955.

McKie, J. W. 1977. Overseeding and sod seeding permanent summer pastures. Mississippi Coop. Ext. Serv. Info. Sheet 829.

Montgomery, C. R., B. D. Nelson, M. Allen, L. Mason, and R. P. Mowers. 1983. Evaluation of Pensacola bahiagrass and Alicia bermudagrass with and without interplanted ryegrass and red clover. Louisiana Agric. Exp. Stn. Bu. 748.

Moshier, L. and D. Penner. 1978. Use of glyphosate in sod seeding alfalfa (*Medicago sativa*) establishment. *Weed Sci.* **26**:163–166.

Mueller, J. P. and D. S. Chamblee. 1984. Sod-seeding of ladino clover and alfalfa as influenced by seed placement, seeding date, and grass suppression. *Agron. J.* **76**:284–289.

Mueller-Warrant, G. W. and D. W. Koch. 1980. Establishment of alfalfa by conventional and minimum-tillage seeding techniques in a quackgrass dominant sward. *Agron. J.* **72**:884–889.

Mueller-Warrant, G. W. and D. W. Koch. 1983. Fall and spring herbicide treatment for minimum-tillage seeding of alfalfa (*Medicago sativa*). *Weed Sci.* **31**:391–395.

Myers, D. K. and G. B. Triplett. 1973. No-tillage pasture renovation. Ohio Coop. Ext. Serv. Agron. Tips F-7.

Myers, D. K. 1981. No-tillage legume seeding following small grain. Ohio Coop. Ext. Serv. Agron. Tips F-10.

Nichols, R. L. and R. A. Peters. 1983. Sod-seeding birdsfoot trefoil in established orchardgrass with overall or banded herbicides. *Can. J. Plant Sci.* **63**:267–276.

Norman, M. J. T. and J. O. Green. 1957. The renovation of downland permanent pastures. *J. Brit. Grassl. Soc.* **12**:74–80.

Olsen, F. J., J. H. Jones, and J. J. Patterson. 1981. Sod-seeding forage legumes in a tall fescue sward. *Agron. J.* **73**:1032–1036.

Palmertree, H. D. 1977. Cool season legumes. Mississippi Coop. Ext. Serv. Info. Sheet 910.

Parsons, J. L. and R. R. Davis. 1964. Establishment and management of birdsfoot trefoil in Ohio. Ohio Agric. Exp. Stn. Res. Bull. 967.

Peters, E. J. and S. A. Lowance. 1979. Herbicides for renovation of pastures and control of tall ironweed (*Vernonia altissima* Nutt). *Weed Sci.* **27**:342–345.

Phillips, R. E. 1981. Soil moisture. In No-tillage research: Research reports and reviews. Kentucky Agric. Exp. Stn. Special Pub. pp. 23–42.

Pound, W. E. 1981. Factors influencing the longevity of 2,4-D in pasture renovation. M.S. Thesis, The Ohio State University.

Rich, P. A., E. C. Holt, and R. W. Weaver, 1983. Establishment and nodulation of arrowleaf clover. *Agron. J.* **75**:83–86.

Rioux, R., J. D. Bandeen, and G. W. Anderson. 1974. Effects of growth stage on translocation of glyphosate in quackgrass. *Can. J. Plant Sci.* **54**:397–401.

Robinson, G. S. and M. W. Cross. 1960. Improvement of some New Zealand grasslands by oversowing and overdrilling. In C. L. Skidmore (ed.), Proc. Eighth Int. Grassl. Cong., Hurley, England, pp. 402–405.

Rogers, D. D., D. S. Chamblee, J. P. Mueller, and W. V. Campbell. 1983. Conventional versus no-till seeding for establishment of ladino clover as influenced by insects and grass suppression. In Agronomy Abstracts, American Society Agronomy, p. 118.

Rogers, D. D., D. S. Chamblee, J. P. Mueller, and W. V. Campbell. 1984. Fall sod-seeding of ladino clover into tall fescue as influenced by time of seeding and grass and insect suppression. *Agron. J.* **76**:1041–1046.

Rountree, B. H., A. J. Bjugstad, and H. N. Wheaton. 1980. Big bluestem, switchgrass, and indiangrass. Missouri Coop. Ext. Serv. Sci. & Tech. Guide.

Ruelke, O. C. and K. H. Quesenberry. 1981. Topseeding winter clover on limpograss, potentials and problems. *Soil and Crop Sci. Soc. of Florida* **40**:162–164.

Salazar, L. C.and A. P. Appleby. 1982a. Germination and growth of grasses and legumes from seeds treated with glyphosate and paraquat. *Weed Sci.* **30**:235–237.

Salazar, L. C. and A. P. Appleby. 1982b. Herbicidal activity of glyphosate in soil. *Weed Sci.* **30**:463–466.

Samson, J. F. and L. E. Moser. 1982. Sod-seeding perennial grasses into eastern Nebraska pastures. *Agron. J.* **74**:1055–1060.

Schaller, F. W. and J. R. George. 1978. Improving pasture by frost seeding. Iowa Coop. Ext. Serv. Pm-856.

Segura, J., S. W. Bingham, and C. L. Foy. 1978. Phytotoxicity of glyphosate to Italian ryegrass (*Lolium multiforum*) and red clover (*Trifolium pratense*). *Weed Sci.* **26**:32–36.

Sheaffer, C. C., D. L. Rabas, F. I. Frosheiser, and D. L. Nelson. 1982. Nematodes and fungicides improve legume establishment. *Agron. J.* **74**:536–540.

Sheaffer, C. H. and D. R. Swanson. 1982. Seeding rates and grass suppression for sod-seeded red clover and alfalfa. *Agron. J.* **74**:355–358.

Sprague, M. A. 1952. The substitution of chemicals for tillage in pasture renovation. *Agron. J.* **44**:405–409.

Sprague, M. A. 1956. The use of herbicides in pasture renovation. In Proc. Weed Soc. Amer. New York, NY.

Sprague, M. A., R. D. Ilnicki, R. J. Aldrich, A. H. Kates, T. O. Evrard, and R. W. Chase. 1962a. Pasture improvement and seedbed preparation with herbicides. New Jersey Agric. Exp. Stn. Bull. 803.

Sprague, M. A., R. D. Ilnicki, R. W. Chase, and A. H. Kates. 1962b. Growth of forage seedlings in competition with partially killed grass sods. *Crop Sci.* **2**:52–55.

Sprankle, P., W. F. Meggitt, and D. Penner. 1975. Rapid inactivation of glyphosate in the soil. *Weed Sci.* **23**:224–228.

Squires, N. R. W. and J. G. Elliott. 1972. Surface organic matter in relation to the establishment of fodder crops in killed sward. Proc. 11th Brit. Weed Control Conference pp. 342–347.

Squires, N. R. W. and J. G. Elliott. 1975. The establishment of grass and fodder crops after sward destruction by herbicides. *J. Brit. Grassl. Soc.* **30**:31–40.

Strand, R. H. and H. A. Fribourg. 1973. Relationships between seeding dates and environmental variables, seeding methods, and establishment of small-seeded legumes. *Agron. J.* **65**: 807–810.

Taylor, R. W. and D. W. Allinson. 1983. Legume establishment in grass sods using minimum-tillage seeding techniques without herbicide application: Forage yield and quality. *Agron. J.* **75**:167–172.

Taylor, T. H., J. S. Foote, J. H. Snyder, E. M. Smith, and W. C. Templeton, Jr. 1972. Legume seedling stands resulting from winter and spring sowings in Kentucky bluegrass (*Poa pratensis* L.) sod. *Agron. J.* **64**:535–538.

Taylor, T. H., E. M. Smith, and W. C. Templeton, Jr. 1969. Use of minimum tillage and herbicides for establishing legumes in Kentucky bluegrass (*Poa pratensis* L.) swards. *Agron. J.* **61**: 761–766.

Taylor, T. H., W. F. Wedin, and W. C. Templeton, Jr. 1979. Stand establishment and renovation of old sods for forage. In R. C. Buckner and L. P. Bush (ed.) Tall Fescue. Am. Soc. Agron. Monograph. Amer. Soc. of Agron., Inc., Madison, Wisconsin, pp. 155–170.

Thomas G. W., K. L. Wells, and L. Murdock. 1981. Fertilization and liming. In No-tillage research: Research reports and reviews. Kentucky Agric. Exp. Stn. Special Pub., pp. 43–54.

Toosey, R. D. 1971. Direct drilling of fodder crops. *Agric. Dev. and Advisory Serv. Rev.* **3**:121–133.

Triplett, G. B., Jr., R. W. Van Keuren, and V. H. Watson. 1975. The role of herbicides in pasture renovation. In Proc. No-tillage Forage Symposium. Ohio State Univ. and Ohio Agric. Res. and Dev. Center, Wooster, Ohio, pp. 29–41.

Triplett, G. B., Jr., R. W. Van Keuren, and J. D. Walker. 1977. Influence of 2,4-D, pronanide, and simazine on dry matter production and botanical composition of an alfalfa-grass sward. Crop Sci. 17:61–65.

Van Keuren, R. W. 1976. Hill land improvement in eastern United States. In J. Luchok (ed.). *Hill Lands*. Proc. of an Intern. Symp. Morgantown, W. Va., 3–9 Oct. West Virginia University Books. Morgantown, WV, pp. 77–90.

Van Keuren, R. W. and G. B. Triplett, Jr. 1970. Seeding legumes into established grass swards. In

M. J. T. Norman (ed.), *Proc. XI Int. Grassl. Cong.*, University of Queensland Press, St. Lucia, Queensland, pp. 131–134.

Van Keuren, R. W. and G. B. Triplett, Jr. 1972. No-tillage pasture renovation. *In* Proc. No-tillage Systems Symposium. Ohio State University and Ohio Agric. Res. and Dev. Center, Wooster, Ohio, pp. 69–80.

Vough, L. R., A. M. Decker, and R. F. Dudley. 1981. Influence of pesticide, fertilizer, row spacings, and seedings rates on no-tillage establishment of alfalfa. In Proc. *XIV Int. Grassl. Cong.*, Lexington, Kentucky, U.S.A. Western Press, Boulder, Colorado, pp. 547–550.

Warboys, I. B. and M. Nutall. 1968. The direct drilling of green fodder crops in West Wales. Proc. 9th Brit. Weed Control Conference.

Warnes, D. D. and L. C. Newell. 1971. Establishment and yield response of warm-season grass strains to fertilization. *J. Range Manage.* **22**:235–240.

Watkin, E. M., J. E. Winch, and G. W. Anderson. 1971. Establishment of birdsfoot trefoil in paraquat treated roughland pastures. *Can. J. Plant Sci.* **51**:163–166.

Watson, V. H., W. E. Knight, and R. E. Coats. 1975. Pasture renovation and overseeding programs for the lower South. In Proc. No-tillage Forage Symposium. Ohio State Univ. and Ohio Agric. Res. and Dev. Center, Wooster, Ohio, pp. 97–104.

Welbank, P. J. 1963. Toxin production during decay of *Agropyron repens* (couch grass) and other species. *Weed Res.* **3**:205–214.

Welty, L. E., R. L. Anderson, R. H. Delaney, and P. F. Hensleigh. 1981. Glyphosate timing effects on establishment of sod-seeded legumes and grasses. *Agron. J.* **73**:813–817.

West, C. P., N. P. Martin, and G. C. Marten. 1980. Nitrogen and rhizobium effects on establishment of legumes via strip tillage. *Agron. J.* **72**:620–624.

Whittles, J. G. and P. P. Williams. 1968. Studies of direct-drilled Brassicas in the Waikato and Central Plateau Region. In Proc. 21st N. Z. Weed Pest Conference, pp. 159–162.

Wilkinson, S. R. 1976. Principles of forage and pasture renovation with reduced tillage systems. *Bull. Entomological Soc. Amer.* **22**:294–295.

Winch, J. E. and E. M. Watkin. 1980. Trefoil establishment on roughland pasture. Ontario Ministry of Agric. Agdex 136/22.

9

INTEGRATED MANAGEMENT SYSTEMS FOR IMPROVEMENT OF RANGELAND

CHARLES J. SCIFRES
THOMAS M. O'CONNOR PROFESSOR
Department of Range Science
Texas A&M University
College Station, Texas

INTRODUCTION

Definitions of rangeland are often hinged explicitly on a lack of suitability for cultivation (Stoddard et al., 1975), which, unfortunately, imposes the connotation of low productivity; and imparts the broad misconception that these lands are used for livestock grazing "because they're good for nothing else." More precise definitions are couched in terms of best land use based on present societal needs and preferences as well as land capability. Considerable land area still used as rangeland in the Great Plains and south to the Texas coast is indeed capable of producing crops. As Heady (1975) stated, "varying economic and social pressures result in continual land-use changes. Thus, rangeland areas and rangeland products exist as a part of man's total land-based system, but there are indistinct and changing boundaries among various uses of the land."

Rangeland includes prairies, shrublands, savannahs, wet meadows, grazeable deserts, coastal marshes, and other ecosystems that provide grazing for domestic animals; habitat for wildlife; a source of recreation; and numerous other amenities and products for mankind. The most satisfactory definition of rangeland for the following discussion is simply "native pasture on natural grazing land" (Dyksterhuis 1955). Rangelands occupy roughly one-half of the earth's surface; more than 340 million hectares of the United States is classified as rangeland (Office of Technology Assessment 1982).

Range management is a land management discipline that skillfully applies an organized body of knowledge to a renewable natural resource (Heady, 1975). Basic to effective management of rangland is a clear understanding of its classification into units called range sites (Dyksterhuis, 1958). These complexes, the result of specific edaphic–topographic–climatic interactions, form specific environments that vary widely in the kinds and amounts of vegetation that they are capable of supporting. In addition, different range sites may respond in quite different ways to the same range management practice. However, expected vegetation change on any specific site is predictable based on documented successional patterns. A series of vegetational assemblages (plant communities), which serve as indicators of the state of ecological health of the rangeland based on the natural production capability, may successively occupy the same site. A quantitative assessment of ecological progression on a range site is referred to as "range condition." Moreover, the direction of change in ecological health, toward improved condition or retrogression, may also be quantified and is referred to as "range condition trend." The ultimate vegetational assemblage of a site, based on natural potential, is referred to as "climax" or "potential" vegetation.

Range managers must understand the nature and sequencing of plant groupings (plant succession) on a range site, as well as the forces that direct change in the vegetation, if they are to effectively manage that site (Bonham, 1983). Furthermore, application of range management methods must be carefully planned and executed with cognizance of the ecological nature of rangeland if maximum sustained advantage of the differential production capabilities of range sites is to be attained.

Rangeland ecosystems, then, are managed by procedures derived from long-standing ecological principles. These management principles focus on expediting secondary plant succession; a striking contrast to agronomic management practies that are designed to retard succession largely for the sake of perpetuating monocultures. Thus, range improvement plans are developed to improve range condition and/or facilitate more efficient utilization of the range. Economic success from application of range improvement practices depends largely on the ability of the manager to match the appropriate practice to ecological potential of the site. Because the "crop" (natural forage) is improved relatively slowly and must be harvested by herbivores for conversion into a marketable product, range improvement practices should be implemented with a relatively long planning horizon. Whitson and Scifres (1980) suggested that most management programs should be develped with a planning horizon of at least 15–20 years.

Rangeland ecosystems are, for the most part, innately fragile—easily damaged and slow to recover. Unfortunately, by conservative estimate, more than one-half of the U.S. rangelands are producing less than half their potential. Mismanagement, continual overgrazing by livestock is the most often cited example, has caused the original vegetation to be displaced entirely or largely replaced by plant cover of lower value to range animals, particularly to cattle.

The primary objectives of this chapter are to (1) provide a brief description of selected methods to vegetation manipulation for range improvement and expected generalized responses to their application and (2) present a framework for application of improvement practices in integrated vegetation management systems to improve efficiency of range management. Since it is beyond the scope of this chapter to entertain range management and range improvements in detail, and excellent coverage of these subjects is available (Heady, 1975; Soddart et al., 1975; Vallentine 1972), range improvement via brush management, grazing management, and revegetation will be emphasized.

BRUSH–HERBACEOUS STAND INTERACTIONS

The term "brush" as used herein refers to a growth of shrubs or small trees usually of a type undesirable to livestock and/or unsuitable for timer management. However, the definition should be moderated to include the thought that brush in certain amounts may be useful (such as for watershed protection) or can be managed for useful purposes (such as for wildlife habitat) (Scifres, 1980).

That woody plants compete with herbaceous plants for environmental resources is generally accepted, but the specific outcomes of such interactions are not well documented and the interactive mechanism(s) are not fully understood. However, an appreciation of the interactions of woody plants (many of which are required components of quality habitat for important game animals on rangeland) with herbaceous forages (which are essential for livestock production) is critical to the planning of effective brush management systems. Presented in most simple form, the function for decreasing herbaceous production with increasing canopy cover of woody plants (Dahl et al., 1978; Grelen and Lowry, 1978; Halls and Schuster, 1965; Jameson, 1967; Koshi, 1953) is a curvilinear relationship as generalized for warm-season grasses in Figure 9.1. As woody plant coverage increases, the botanical composition also shifts with sun-loving species replaced by shade-tolerant grasses of relatively low vigor.

A greater proportion of the herbaceous production may be sacrificed as brush cover increases when environmental resources such as rainfall are limiting (Scifres et al., 1983). Conversely, the greatest absolute positive grass production response to brush removal usually occurs during "wet" years (Scifres and Polk, 1974; Scifres et al., 1977).

There are certain cases when the brush–herbaceous production relationship is not so straightforward as described for warm-season grasses. For example, the presence of light-to-moderate woody plant canopy cover may promote

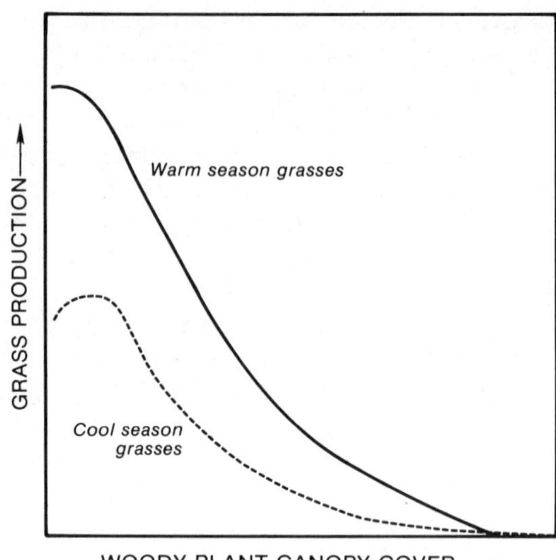

FIGURE 9.1. Generalized production response of warm and cool season grasses to increasing woody plant cover in south Texas [adapted from Scifres et al. (1982, 1983c)].

the abundance of cool-season grasses (Fig. 9.1). Production of Texas wintergrass (*Stipa leucotricha*) increased as huisache (*Acacia farnesiana*) canopy cover increased to roughly 20% in a study on the Texas Coastal Prairie; and Texas wintergrass production was not reduced until brush cover exceeded 60% (Scifres et al., 1982). Other cool-season and/or shade-tolerant grasses such as species of *Phalaris, Limnodia*, and *Vulpia* may also grow beneath the shrubs (Drawe et al., 1978). As a result, yearlong grass production (warm season + cool season) may be greater where scattered huisache plants exist than on brush-free areas (Scifres et al., 1982). A similar relationship exists between honey mesquite (*Prosopis glandulosa* var. *glandulosa*) and cool-season grasses in semiarid areas (Brock et al., 1978). Apparently, the woody plant cover ameliorates the environment, probably by reducing air temperatures and light levels, such that growth and development of cool-season species are favored.

Grass growth may be precluded by some woody plant growth forms. In contrast to the relatively open growth forms assumed by huisache and honey mesquite, mixed-brush (*Prosopis*–*Acacia*) of south Texas forms dense assemblages in which plants grow extremely close and branch heavily near the ground. Native grasses typical of brush-free areas may occupy only the interstitial zones separating the brush mottes and to the immediate driplines of the woody plants. Scifres et al. (1983c) reported that only bunch cutgrass (*Leersia monandra*), a low statured species, persisted from the canopy edges to about 1.5 m inward from the brush driplines. No herbaceous vegetation occurred in the center of the brush stands. Similar relationships have been reported with lotebush (*Ziziphus obtusifolia*) and grasses on semiarid rangelands in north Texas (Foster et al., 1983). These competitive effects appear to be expressed primarily in terms of plant needs for space, at least until moisture becomes limiting.

Excessive cover of woody vegetation on rangeland reduces livestock production not only by constraining forage production, but by reducing forage quality (largely by negatively influencing botanical composition of herbaceous stands) and availability, and by causing bodily injury to and/or poisoning of livestock. Also, heavy brush cover reduces management efficiency by increasing costs of handling and caring for livestock. More than 38 million hectares of rangeland in Texas alone is producing less than its potential because of excessive woody plant cover (Scifres, 1980). Thomas (1970) estimated that effective brush management could double, perhaps triple, livestock production on western rangelands.

ORIGIN OF THE WOODY PLANT PROBLEM

Even though excessive brush cover poses serious management problems, it should be recognized that woody plants are natural components of rangeland vegetation of the southwestern United States. Relatively few of the problem woody species are introduced, and many of the most serious present-day problems were once associated with pristine grasslands. As an example, remains of honey mesquite from archeological digs along the Frio River of Texas have been dated to 1300 BC (Hester, 1980). The diaries of explorers and travelers through Texas in the 1700s refer often to presence of specific woody plants that are now considered to be major problems (Inglis, 1964).

Most range managers and ecologists agree that the contemporary "brush problem" is the result of dramatic increases in the density and stature of woody plants which originally occurred as scattered individuals or as small and localized aggregates of species. The increase in abundance of woody plants largely occurred in the last 150 years or so (Bogusch, 1952, Buffington and Herbel, 1965; Gadzia and Ludwig, 1983; Gardner, 1951; Hester, 1980; Inglis, 1964; Johnston, 1962; Scifres, 1980; York and Dick-Peddie, 1969). Pristine grasslands were once stable ecosystems in dynamic equilibrium with environmental influences (notably fire, herbivory which had coevolved with the vegetation, and drought) (Bray, 1901). The shift from grassland to brushland is attributed to drastic reduction (even cessation) of naturally occurring prairie fires, continual grazing overuse by domestic livestock (not only because of increased livestock numbers, but as the result of restricting the animals to specific areas by fencing), and the compounding of these influences by periodic drought (Scifres et al., 1983a). Contemporary brushlands have apparently reached a state of equilibrium with this "new" set of disturbances. Moreover, because most brush management methods do not remove entire woody populations, because the seeds of many species may persist in the soil for years, and because of the presence of woody plants on adjacent areas, those areas improved by brush removal are constantly disposed to revert to woody plant communities. This is further justification for planning brush management programs over relatively long-term periods.

WOODY PLANT MANAGEMENT

The philosophical approach of research directed toward coping with the brush problem has undergone considerable evolutionary change during the past 40 years. Research in the 1940s was based on a perceived need to *eradicate* woody plants (Fisher et al., 1946), an overly optimistic if not impossible goal. The term "brush control," as a philosophical replacement for "brush eradication," rapidly gained popularity and was almost universally used by the mid-1950s. However, the objective of brush control, applied in the purest sense, still carries the connoted desire to kill 100% of the targeted woody plant stand.

Considerable information has been accrued, especially during the past 10 years, relative to the potential values of woody plants. In addition, the necessity of economic justification for employing control methods has become inescapable. These factors provided impetus for emergence of the *brush management* concept. The idea of "brush management" was originally presented in the mid-1960s (Box and Powell, 1965). However, the approach has only recently been formalized (Scifres, 1980); and it is apparently gaining acceptance by researchers and range livestock producers. Appreciation of the concept has been enhanced by the increasing values of rangeland for uses other than solely as grazing by domestic livestock, especially its importance as wildlife habitat. Thus, many livestock producers now have an economic reason for managing use of all range vegetation, not just the grass. "Although the concept of brush management on rangeland is not new, the time appears right for promoting its general acceptance" (Scifres, 1978).

A significant amount of theory and technology applicable to improvement of brush-dominated rangeland has resulted from research efforts during the last four decades. Available technologies may be classified as chemical, prescribed burning, biological control including selective grazing, or mechanical methods. All brush management methods, except some mechanical practices which precede artificial revegetation, depend on the processes of secondary succession for vegetation improvement. Thus, the no-tillage concept is pervasive to discussion of brush management procedures.

Chemical Brush Management

Scifres (1980) classified herbicides for range improvement based on their primary route of entry into plants following broadcast application. Herbicides are either applied to the foliage of the target species or to the soil where they are taken up by the woody plants' roots. Some herbicides such as picloram* are included in both categories, depending on target species, application rate and formulation used. Phenoxy herbicides, especially 2,4-D, 2,4,5-T and silvex have formed the basis for most chemical brush management efforts for the past 45 years (Figurc 9.2). These foliar-active compounds have been applied alone or in various mixtures with dicamba or picloram (Scifres and Hoffman, 1972) with ground or aerial equipment for control of an array of troublesome species. Although the rangeland uses of 2,4,5-T and silvex have been cancelled, research results and user experience with these compounds form a valuable basis of reference for developing new herbicides.

Dicamba may be applied alone or in combination with picloram for control of honey mesquite and certain associated woody species (Scifres and Hoffman, 1972). Dicamba also effectively controls many species of herbaceous weeds on rangeland, including certain troublesome perennial species (Scifres, 1980).

Introduction of picloram (Hamaker et al., 1963) not only offered potential for broadening the spectrum of species controlled by phenoxy herbicides, but soil applications effectively control species, such as junipers (*Juniperus* spp.), which are not highly susceptible to foliar-active herbicides. A related herbicide, triclopyr, was registered in 1985 for brush management on rangeland. Its projected use pattern is much as for 2,4,5-T including application in combination with picloram.

Clopyralid is structurally similar to picloram and appears promising for improved control of honey mesquite. Clopyralid at 0.6 kg/ha killed 60–68% of the honey mesquite in experiments in north Texas, compared to 21–30% mortality from the same rate in 1:1 mixture of 2,4,5-T and picloram (Jacoby et al., 1981). The experimental herbicide is also effective when combined with picloram.

Adaptation of the photosynthesis inhibitor tebuthiuron for range improvement (Fig. 9.2) added a new dimension to brush control. Tebuthiuron, in pellet form, is active against a broad spectrum of woody plants (Scifres et al., 1979, 1981a), and its application is not so constrained by state of phenological development of the target species as with most foliar sprays (Scifres, 1980). However, maximum brush control usually requires at least two growing seasons following tebuthiuron application. In contrast, defoliation is complete within 30 days following foliar application of herbicides such as dicamba or picloram. Therefore, a positive forage response usually does not occur until the second growing season following tebuthiuron application (Scifres and Mutz, 1978; Scifres et al., 1981b) but a flush of grass growth may occur by summer or fall after application of conventional sprays in the spring.

Hexazinone (Fig. 9.2) is the most recent herbicide introduction for brush control on rangeland. Hexazinone formulations evaluated incude wettable powder, liquid, and large particles referred to as "gridballs." Hexazinone

*Common and chemical names of herbicides are listed in Appendix A.

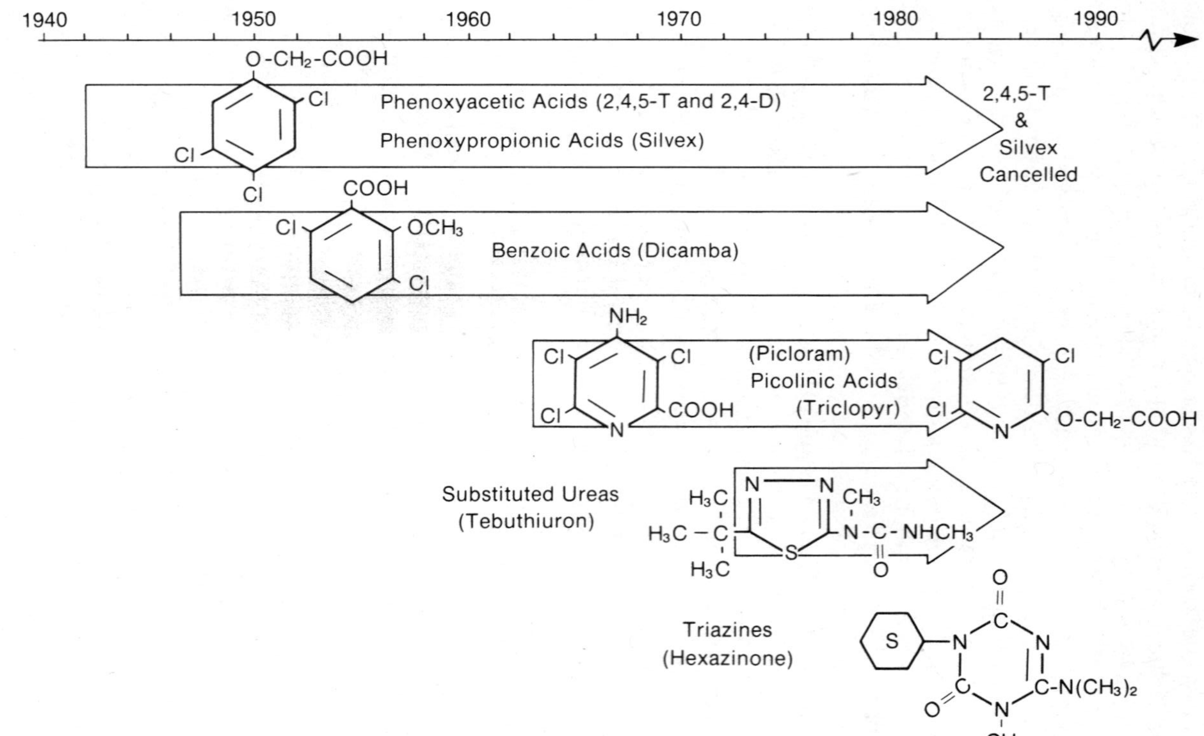

FIGURE 9.2. Primary herbicide families used for brush management on rangeland and the relative time periods of their availabilities.

liquid is used primarily for individual-plant treatment of woody plants on rangeland. Application of the large particles in grid patterns (1.5- or 3-m spacings) at 2 or 4 kg/ha effectively controlled post oak (*Quercus stellata*) and blackjack oak (*Q. marilandica*) in east central Texas (Scifres, 1982). Several other species were moderately susceptible to the treatment. The successful application of dry herbicide particles in widely spaced grids was first described in 1978 with the herbicide karbutilate (Scifres et al., 1978). The application method facilitates effective use of soil-active herbicides that are nonselective when applied broadcast. Thus it may allow adaptation of additional herbicides now used for industrial weed control to selective use for brush management on rangeland.

Herbicides for brush management on rangeland have become increasingly sophisticated, relative to chemistry, formulations available, and methods of use, since the 1960s (Fig. 9.2). Thus, technical understanding and skills of the range manager must also be increased for effective management application of these compounds.

There are several advantages to the use of herbicides for brush management, including:

1. A variety of application methods are available to meet specific treatment needs (Scifres, 1980). These vary from techniques for treatment of individual plants in relatively thin stands to aerial application with either fixed- or rotary-winged aircraft for rapid treatment of dense stands on large areas.
2. Aerial application is rapid, efficient, and independent of terrain and woody plant growth type.
3. There is no soil disturbance, except minor perturbations that might result from use of ground equipment, associated with herbicide application.
4. Standing woody debris remaining after herbicide application is often adequate to serve as screening cover for wildlife.
5. Applied under optimum environmental conditions and at the correct rate(s) and formulation, herbicides are extremely effective for selective control of woody plants.
6. Soil-applied herbicides such as tebuthiuron, picloram, and hexazinone may be applied during relatively broad periods each year and effectively control many species of woody plants. Application as relatively large dry particles reduces potential for displacement by wind to adjacent ecosystems.
7. Associated herbaceous weeds may be effectively controlled to further reduce competition with desirable grasses.

Some disadvantages associated with herbicide use include:

1. Desirable forbs (especially legumes) are usually reduced, at least for the growing season of application, in the native forage complex.
2. Application of herbicide sprays must be carefully timed to coincide with the appropriate phenological stage of woody plant development. Thus, the

time period for most effective use of foliar-active herbicides is usually restricted to a few weeks each spring (Scifres, 1973) and/or, with some species, to a short period each fall (Scifres, 1980). Since spray applications must also be restricted to certain environmental conditions (especially relative to wind speed and air temperatures), the time available for application in any given year may be seriously restricted.

3. The spectrum of species controlled in mixed brush stands may not be adequate to allow maximum forage release. Moreover, control of some woody species may allow resistant species to increase and form more difficult-to-manage stands than was the original brush cover.

Prescribed Burning

The use of controlled fire to achieve stated range management objectives has achieved renewed acceptance by range researchers and producers during the last decade (Scifres, 1980). Because of a concentrated effort by relatively few dedicated researchers, prescribed burning has assumed its rightful place as a range management technology. The recent review of fire ecology and prescribed burning applications by Wright and Bailey (1982) is recommended to the reader with interests in details of fire applications for natural resource management.

Effective use of prescribed burning requires development of skills largely accrued by experience (Scifres, 1980; Wright and Bailey, 1982). A number of generalized "prescriptions" have been developed for various purposes, but no two fires are exactly alike and no two areas to be burned are identical. Therefore, the most important skill is understanding the principles underlying effective burning so that they may be successfully applied to a specific area. Prescribed burning appears to have most utility for brush management as a secondary treatment to prolong improvement of rangeland initially treated by chemical or mechanical methods. Used in this context, prescribed burning offers several advantages not characteristic of other methods including:

1. Prescribed burning is relatively inexpensive. Although costs vary with objective of the burn and local variations in terrain and fine fuel characteristics, most prescribed burns for range improvement can be installed for \$10.00–\$15.00 per hectare. Primary costs are associated with fire guard installation, deferment from grazing to build fuel and, subsequent to burning, for forage plant recovery, and labor to install and monitor the burn (Scifres, 1980).

2. Prescribed burning may extend the treatment lives of costly, initial brush management practices resulting in an improved economic outcome for the land manager (Garoian et al., 1984; Whitson and Scifres, 1980). In some cases, prescribed burning may essentially salvage costly investments in tame pasture establishment by suppressing invading woody plants (Hamilton and Scifres, 1982).

3. Forage quality may be improved by burning, a change not generally

possible with other brush management methods. Young, actively growing tissues that emerge following burning usually contain greater relative amounts of crude protein, phosphorus, and other nutrients than do tissues of the same forage species on unburned areas (Scifres, 1980).

4. Some species, not grazed or utilized relatively little by livestock on unburned areas, may be readily accepted by grazing animals and furnish highly nutritious forage after burning (McAtee et al., 1979; McGinty et al., 1983; Oefinger and Scifres, 1977; Wright 1974). Improvement in patability and forage quality may be reflected as increased weaning weights of calves from burned pastures (Mutz et al., 1984; McGinty and Smeins, 1983).

5. Desirable broadleaved species, selectively removed by herbicide treatments, may recover to their original position in the native forage complex following prescribed burning (Scifres, 1975, 1980).

6. Prescribed burning may reduce the incidence of external parasites, such as Gulf Coast tick (*Amblyomma maculatum*), at least for one growing season (Oldham et al., 1982; Wright 1974). However, there is argument that mortality of the ticks caused directly by the fire and by creating unfavorable microenvironmental conditons may be offset by subsequent concentration of hosts on burned areas (Minshull and Nerval, 1982).

7. Quality and availability of browse for wildlife and livestock may be improved by prescribed burning (Rasmussen et al., 1983).

Primary limitations to prescribed burning for brush management include:

1. Fine fuel load and continuity under heavy brush cover is often not adequate for installation of effective prescribed burns.

2. Many woody species rapidly replace their topgrowth following top removal. Therefore, prescribed burning may offer only short-term woody plant suppression (Hamilton et al., 1981).

3. Successful prescribed burns must be timed to take advantage of environmental conditions, especially moisture. Burning under marginal moisture conditions followed by a prolonged dry period may necessitate inordinate extension of grazing deferment because of slow recovery of desirable species.

Mechanical Brush Management

An array of methods have been developed for mechanical control of woody plants. These methods may be categorized as those that remove the topgrowth only and those designed to remove entire plants (Scifres, 1980). Simple top removal methods include shredding, roller chopping (roller cutting), and crushing. Methods designed to remove entire plants include power grubbing, root plowing, dozing, chaining, and cabling (Scifres, 1980). Detailed descriptions of each of these methods and discussion of applications are available elsewhere (Scifres, 1980; Vallentine, 1971).

In general, simple top removal methods are less costly and cause less soil disturbance than those designed for complete plant removal but are characterized by a relatively short treatment life because of the regrowth potential of woody plants (Hamilton et al., 1981). Energy-intensive methods such as dozing or root plowing cause maximum soil disturbance and are relatively expensive, but are highly effective for brush removal. These methods are often used in preparation for range seeding.

There have been some advancements in increasing energy efficiency of mechanical methods. For example, low-energy power grubbers have increased cost effectiveness for removal of species such as huisache (Bontrager et al., 1979) and honey mesquite (Wiedemann et al., 1977; McFarland and Ueckert, 1982). McFarland and Ueckert (1982) reported total costs of $3.05 per hectare for controlling sparse stands (37 plants per hectare), $12.50 per hectare for removing moderately dense stands (185 plants per hectare), and $78.13 per hectare for treating dense stands (1048 plants per hectare) of honey mesquite with a rear-mounted, low-energy grubber.

BIOLOGICAL BRUSH MANAGEMENT-SELECTIVE GRAZING

Biological brush management methods use natural enemies (plant-feeding and disease-causing organisms) to reduce or eliminate the economic impact of an undesirable species or plant stand. Direct action of the biological agent (removal of photosynthetic area, damage to reproductive structures) and/or indirect action (promotion of conditions that favor secondary infections, reduced vigor) are involved in the general mode of action of biocontrol methods. There have been striking successes from biocontrol methods; use of the moth *Cactoblastis cactorum* on pricklypear in Australia and leaf beetles (*Chrysolina* sp.) for control of Klamathweed (*Hypericum perforatum*) are often cited examples. Insects such as conchuela (*Cholorochroa ligata*) and a seed bettle (*Algarobius prosopis*) may significantly reduce the production of viable seed by honey mesquite (Smith and Ueckert, 1974). Twig girdlers (*Oncideres rhodosticta*) (Polk and Ueckert, 1974) may significantly reduce photosynthetic area of mesquite in some years. However, research is still needed to develop programs that effectively use these and other similar organisms in a management context.

The negative impact of some plant species may be greatly reduced, in some cases, by using domestic livestock as the "biocontrol agent." Various species of grazing animals, perhaps even individuals within species, prefer certain growth forms of vegetation, species of plant within growth forms, and certain parts of individual plants. However, grazing preferences relative to growth forms may be generalized. Grass is preferred by cattle, forbs are preferred by sheep, and browse is preferred by goats. Overlap in grazing preference is largely regulated by availability and diversity among and within the various growth forms, season of year, and plant growth stage. For example, cattle will browse the new buds, succulent leaves, and young twigs of many woody plants in the spring.

They also may browse certain species again in the fall when the availability and/or quality of herbaceous species is low; or small amounts of certain browse species may occur in their diets throughout the year. However, cattle normally do not use enough browse during any season to be considered a control agent (Kirby and Stuth, 1982; Scifres, 1981a).

The preference of goats for woody plants is so strong that they represent the most effective biological brush management agent among various species of domestic livestock (Merrill and Taylor 1976). Stocking density of goats, length of grazing period, and season of grazing for effective brush suppression depends on the botanical composition and intensity of the brush problem. However, one goat per 0.8–1.2 ha yearlong for 5 years reduced cover of a stand of mixed brush on the Edwards Plateau by 83% (Merill and Taylor 1976).

Two approaches are used for "goating" as a brush management method. The goats may be left in the target pasture yearlong; or the goats may be used in a given pasture for a shorter period and then moved to another area. Twelve to twenty goats per hectare may be required to achieve the desired brush control level with short-term grazing (30 days or less). Goats are particularly effective for sprout control following brush management practices that have removed the mature brush growth. Goats have provided effective control of some species following chaining, cabling, chopping or bulldozing of brush (Scifres, 1981a).

Primary considerations when incorporating use of goats in a brush management program include the additional costs to management for facilities to contain, house, and protect the goats from predators (Scifres 1980, 1981a; Vallentine 1971). Cool-season browse supplied for goats may be a limiting factor in areas where most woody plants are deciduous. Also, the general diet preference of goats overlaps considerably with that of white-tailed deer (*Odeocoilus virginianus*) raising the concern for potential competition between the two species.

Both Spanish and Angora goats are used for brush control. Spanish goats consume more browse than Angoras, but the latter species offers potential increased income to the land manager from the sale of mohair.

Integrated Brush Management Systems

Each brush control method has applications for which it has proven a certain level of biological effectiveness. However, each method is characterized by limitations that prevent complete effectiveness. Moreover, rising costs of energy, heavy equipment, and herbicides now preclude use of some practices by many ranch firms. Economic efficiency of brush management may be increased by (1) improving biological effectiveness of practices without increasing costs and/or (2) increasing the effective life of the practice.

Combinations of methods have been investigated in efforts to develp complementary treatment sequences, especially use of low-cost measures that extend the effective treatment life of more costly practices. Some of these combinations, chaining following aerial spraying of honey mesquite, for example (Scifres, 1973), have proven highly effective. Aerial application of 0.6 kg/ha of 2,4,5-T for honey mesquite control in north Texas has an average treatment life of 7 years. Chaining 2–3 years following spraying may extend range improvement effectiveness for 10–15 years.

Because of their importance to brush management systems, some workers have viewed the treatment combinations themselves as systems. Although treatment combinations or sequences may be used as components for brush management systems, they provide only one element of brush management systems (Scifres, 1981b). "Development of effective brush management systems necessitates that all processes required for designing, implementing, and maintaining range improvement . . . from the development of objectives for a given management unit to assessment of economic feasibility . . . be given appropriate consideration" (Welch and Scifres, 1980). Integrated brush management systems (IBMS) development is a rational, decision-making process that seeks to optimize sustained yield of all rangeland products (Scifres et al. 1983a,b). Yield optimization should maximize economic returns over a prede termined planning horizon.

The IBMS concept reduces dependence on any given brush management method or combination of methods by emphasizing an appropriate sequencing of a series of alternatives (Scifres, 1981b). The alternatives are selected such that the unique strengths of one method, which may be in terms of biological efficacy or economic efficiency, compensates for characteristic weaknesses of the other(s) (Scifres, 1980; Scifres, et al., 1983a).

The initial step in the IBMS planning process is the setting of goals and objectives for "best use" of the management unit that is targeted for improvement. Best use of the targeted management unit is developed in terms of overall enterprise objectives. The production goals must be reasonable and attainable in an appropriate time frame or planning horizon.

Realistic goal setting is obviously not possible without reliable estimates of production potential of the range resource for various uses. Most importantly, present production level must be compared against production capability as the result of implementing applicable range improvement practices. This estimate of potential production change is essential for development of economic projections that allow comparison of alternative production strategies. These comparisons are quantified based on procedures developed from ecological principles described earlier (Whitson et al., 1979).

Effective grazing management is an essential component of IBMS (Scifres et al., 1983a). Brush management systems offer only the basis for increasing output, whereas grazing management system determines efficiency of translating that increased yield (herbage) into merchandisable product. No specific grazing management system may necessarily be considered universally superior. The grazing management strategy must be developed to best fit the specific resource and management needs. IBMS may incorporate an existing grazing management strategy, the existing grazing management system may be modified as a result of IBMS development, or the grazing management strategy may be developed as a component of IBMS.

Available brush management alternatives that are applicable to the specific problem vegetation and that have characteristics allowing achievement of the stated management goals must be identified. This evaluation includes consideration of established as well as "new" technologies. Development of IBMS has resulted in improved applications for long-standing practices such as prescribed burning (Scifres et al., 1983b).

Prescribed burning and herbicide application have proven to be synergistic against several brush problems when applied in the IBMS context. Prescribed burning has been used as a complementary treatment by (1) extending the effective life of range improvement following herbicide treatment, (2) increasing herbicide effectiveness on a specific target species, and (3) broadening the spectrum of species controlled by the herbicide. In all cases, the use of prescribed burning allows decreased use of herbicide either by eliminating the need for multiple applications or by reducing the rate of herbicide required for control of the undesirable species.

Management of areas supporting excessive cover of Macartney rose (*Rosa bracteata*) provides an example of the use of prescribed burning to extend the treatment life of herbicide application. Macartney rose is an evergreen vine that may render potentially productive rangeland unsuitable for livestock. The major conventional treatment in the early 1970s was to broadcast apply three, successive, annual applications of 2,4-D at 1.1–2.2 kg/ha (Scifres, 1980). This treatment was only partially effective so that it was necessary to repeat the program about every 10 years. The repeated herbicide application essentially removed all desirable broadleaves from the range ecosystem. Substitution of a single application of 2,4,5-T + picloram at 1.1 kg/ha for the multiple 2,4-D application effectively improves Macartney rose-dominated rangeland if the herbicide treatment is followed in 18–24 months by prescribed burning. Prescribed burns may then be applied at about 3-year intervals after the initial burning (Scifres, 1975) and, within a sound grazing management system, will perpetuate range improvement (Garoian et al. 1984). Based on results from 1972 to 1982, this approach when applied in the IBMS context, increased internal rate of return to management by 11% compared to the standard herbicide treatment (Table 9.1).

Similar results have been obtained with herbicide-prescribed burning systems for improvement of south Texas rangeland dominated by a running mesquite (*Prosopis* spp.) complex (Scifres et al., 1983b). Conventional treatment of running mesquite has been two or three successive applications of 0.85 kg/ha of 2,4,5-T (Scifres, 1980). A single application of 1.1 kg/ha of 2,4,5-T + picloram followed by prescribed burning the second year after spraying and prescribed burning periodically thereafter effectively controlled the running mesquite. In this case, the researchers believe that after the installation of the first two burns at 3-year intervals, a burning frequency of 5 years will maintain range improvement (Scifres et al., 1983b).

TABLE 9.1. Annual Rate of Return of Selected Treatments for Improvement of Macartney Rose-Dominated Rangeland on the Texas Coastal Prairie

Treatment	Rate of Return (%)[a]
Three, successive, annual broadcast applications of 2,4-D (1.1 kg/ha)	5.0
Single aerial spray of 2,4-5-T + picloram (0.6 kg/ha of each herbicide in mixture)	6.3
Single aerial spray of 2,4,5-T + picloram (0.6 kg/ha of each herbicide in mixture) followed by prescribed burning	16.1

[a]Based on evaluations for the period 1972–1982.

Use of prescribed burning to reduce herbicide application rate for effective brush management may also be exemplified with Macartney rose. Gordon et al. (1982) found that application of picloram pellets at 1.1 kg/ha (active ingredient) directly into the ash immediately following burning resulted in more effective Macartney rose control than twice that amount of herbicide applied to unburned areas. The workers hypothesized that regrowth of Macartney rose following burning placed a greater water-use demand on the soil system than did undisturbed growth; and that greater water uptake resulted in a concomitant increase in herbicide uptake. It may also be that the young, rapidly growing top growth is physiologically more susceptible than the original growth to picloram. Similar results have occurred from field experiments where the same treatments were applied to common goldenweed (*Isocoma coronopifolia*) (Mayeux and Hamilton, 1983). Preliminary results of Ueckert (1983) indicate potential for decreasing application rate by as much as 50% and significantly increasing control level when 2,4,5-T + picloram or picloram alone are applied to recently burned pricklypear (*Opuntia* spp.), compared to control of unburned pricklypear with the herbicide. In other cases, however, prescribed burning prior to herbicide application has decreased effectiveness of control. For example, Ueckert et al. (1983) found that prescribed burning in late February in west Texas reduced the control of rayless goldenrod (*Isocoma wrightii*) by subsequent applicaiton of picloram pellets.

Prescribed burning may be applied to broaden the spectrum of species controlled by herbicide application, as was demonstrated in use of tebuthiuron–fire systems for management of whitebrush (*Aloysia lycioides*)-dominated areas in south Texas. Tebuthiuron pellets applied in the spring or fall at 1.1–2.2 kg/ha effectively control whitebrush, but the treatment does not control several associated species such as honey mesquite, lime pricklyash (*Zanthoxylum fagara*), or pricklypear (Scifres et al., 1979). Prescribed burning 2–3 years after aerial application of tebuthiuron pellets removed standing dead whitebrush stems and improve botanical composition and production of native range forage stands, compared to areas treated with the herbicide only (Scifres et al., 1983b). Moreover, burning at 3–5-year intervals maintains the herbicide-treated areas brush free, whereas adjacent unburned areas may be invaded by honey mesquite and other woody plants.

The positive influence of prescribed burning on herbicide effectiveness was purposely emphasized with these examples. Based on information accrued to date, herbicide-prescribed burning interactions may be categorized as one of three types (Table 9.2). In most cases, the combination of treatments appears to be additive. The herbicide conditions the vegetation by removing the woody plant canopy cover and, in concert with proper grazing management, releases herbaceous species to form a continuous load of fine fuel adequate for installation of effective burns. Herbicide treatment thins the woody plant stand such that burning does not need to be applied as frequently or as severely as on areas not pretreated. However, prescribed burning functions as expected based on results with similar vegetation types that were not pretreated with herbicide. Where burning as the initial treatment allows reduction in herbicide dosage for effective brush control (i.e., control exceeds the "expected" level or that level on unburned rangeland), the treatment combination may be deemed synergistic. However, as exemplified by the rayless goldenrod example (Ueckert et al., 1983), fire-conditioned interactions may not always be synergistic or even additive (Table 9.2). Much more research is needed in this area. There, conceivably, could exist herbicide-conditioned systems that are synergistic or antagonistic, or additive fire-conditioned systems. However, the classification scheme presented here hopefully will form a basis for categorizing treatment interactions for IBMS.

GRAZING MANAGEMENT STRATEGIES

Grazing management systems may be categorized as one herd–one pasture; one herd–two pastures; one herd–multiple pastures; or multiple herds–multiple pastures. The one herd–one pasture strategy, called continuous grazing, is generally viewed as the least satisfactory approach to grazing management, since there are no grazing deferments to allow the forages to reinstate vegetative vigor following top removal or to set seed. If the range is overstocked (i.e., numbers of livestock exceed carrying capacity), damage to the vegetation can be expected.

One herd–two pastures systems allows for some of the simplest approaches to planned grazing. Two pastures may be grazed in various sequences including "4-months on, 4-months off," a graze–defer sequence of 3-6-3-3-6-3 months, or a graze 6 months–defer 6 months rotation. The sequences are developed so that deferment occurs one year in the fall followed by deferment the next year in the spring. If stocked properly, improvement in range condition occurs rapidly with even simple systems, compared to overstocking on a continuous basis, and brush management systems are easily integrated into the grazing management strategy (Scifres, 1980). Brush management strategies, such as herbicide application, may be scheduled for the rest periods so that one pasture is treated each year. The longer deferment periods (i.e., 6 months) are prefera-

TABLE 9.2. Qualitative Summary of the Interactive Nature of Herbicide-Prescribed Burning Combinations Applied for Brush Management

Type of interaction	Function		Result	Examples
	Herbicide	Prescribed Burning		
Herbicide conditioned (additive)	Removes woody plant canopy; reduces density of live woody plants; releases herbaceous species to develop continuous load of fine fuel	Removes dead standing debris; suppresses sprout regrowth on surviving plants; suppresses re-invasion of woody plants; reinstates forbs to herbaceous stands	Two or more herbicide applications replaced by prescribed burning	Macartney rose (Scifres, 1975); whitebrush, running mesquite (Scifres et al., 1983b)
Fire conditioned (synergistic)	Toxicant for control of postburn brush growth	Reduces large volume of mature and/or woody growth to uniform and/or rapidly developing vegetative tissues. Reduces surface litter and mulch, hence increases deposition of herbicide on soil surface	Dosage of herbicide for effective control reduced	Macartney rose (Gordon et al., 1982); common goldenweed (Mayeux and Hamilton 1983); pricklypear (Ueckert et al., 1983).
Fire conditioned (antagonistic)	As above	As above	Herbicide effectiveness decreased	Rayless goldenrod (Ueckert et al., 1983).

ble treatment times because of the opportunity for extended rest following treatment. Treatment of pastures during the rest period not only ensures some grazing deferment following application of the brush management practice, but spreads treatment costs over 2 years. This also allows more time for addition of livestock to takc advantage of improved vegetation conditions than when both pastures are treated during the same year.

One herd–multiple pastures may be developed with a relatively few pastures (three or four), but most recent interest has focused on systems that normally use eight or more pastures, referred to as short-duration grazing systems (SDG). These systems impose heavy grazing pressure on each pasture for a relatively short time, followed by a relatively long period of grazing deferment. The concept is premised on uniformily and quickly harvesting the available forage, and induce utilization of species that are normally less preferred as well as those that are most preferred by the livestock. Livestock producers have shown considerable interest in in SDG, but much research is yet to be accomplished to answer certain questions related to livestock performance and rate and direction of vegetation change.

Because of the long periods of grazing deferment throughout the year, SDG is highly amenable to integration with brush management practices. Since there are a relatively large number of pastures in SDG systems, additional flexibility in scheduling deferments following installation of brush management treatments is possible. In extreme cases, an entire pasture may essentially be omitted from the system to allow additional deferment with little or no adjustment in the grazing scheme.

A common multiple-herd–multiple pasture strategy is the three herd–four pasture or Merrill grazing system. This system schedules 12 months grazing followed by 4 months rest for each pasture. Although less flexible than SDG relative to integration of brush management into the grazing schedule, the Merrill system is highly compatable with most range improvement practices.

Major improvements in range condition and livestock performance are possible through use of grazing systems, but integration of grazing management with brush management offers potential for greater improvement than from either of the practices applied singly. Deferment following brush management is most effective when the rest period is scheduled to allow the more desirable species to set seed before grazing is resumed (Scifres, 1975). Theoretically, this practice should be followed each year until desirable forage species occupy most of the area previously occupied by brush. However, such a tightly designed approach is not usually possible. The most satisfactory approach is to utilize the deferment system built into a sound grazing management strategy and plan for improvement to occur progressively over a number of years, with particular care given to selecting stocking rates in years of above normal precipitation.

Welch and Scifres (1980) presented an example of the integration of brush management into grazing management systems. They used honey mesquite as the target species and assumed that the IBMS utilized aerial spraying–

chaining-prescribed burning. The following paragraphs present the rationale used for selection of each treatment.

1. *Chemical.* Aerial application of herbicides such as 2,4,5-T, 2,4,5-T + picloram, or 2,4,5-T dicamba (usually at 0.6 kg/ha total herbicide with the herbicide mixtures in 1:1 ratios) will effectively defoliate the honey mesquite and, on the average, will kill about 25–40% of the plants (Scifres 1973; Scifres and Hoffman 1972). Aerial spraying typically releases range forage as well as improves the load and continuity of the herbaceous plant stand for use as fine fuel. Herbicide applications must be carefully timed for maximum effectiveness since honey mesquite is most susceptible to foliar sprays for only 40–90 days after bud break in the spring (Scifres, 1973). Surviving honey mesquite plants sprout from their branches and basal bud zones, and usually require subsequent treatment for maximum control. The spray treatment leaves the trunks and stems standing, and usually reduces the abundance and diversity of forbs. It is also highly desirable to apply the chemical prior to or coinciding with a normal grazing deferment period to expedite improvement in the forage stand.

2. *Mechanical.* Chaining following aerial spraying is an excellent method for uprooting and knocking down the stand of honey mesquite trunks (Scifres, 1973). Applied when soil-water content is adequate to allow maximum uprooting, a high percentage of the surviving trunks, the source of new sprouts, will be dislodged (Scifres, 1980).

3. *Prescribed burning.* After the fine fuel load has been developed as a result of woody plant removal, by aerial spraying and chaining, and grazing management, prescribed burning may be used to suppress surviving honey mesquite sprouts and reduce cactus populations. Fire also removes a considerable amount of the woody debris left by chaining, improves grazing distribution, increases the palatability of forage plants, and improves botanical composition of the forage stand (Scifres, 1980).

Each of the three range-improvement methods—chemical, mechanical and prescribed burning—were chosen for a specific purpose, to complement the action of the other methods and to avoid repeating the use of herbicides or mechanical methods as outlined earlier in the discussion of IBMS. The timing of brush management treatment installation might be accomplished as in Table 9.3 with the two pasture–one herd system. Calendar dates are used for ease of presentation only with full realization that grazing periods must be adjusted for factors such as rainfall.

Pasture 2 in the hypothetical program is scheduled for deferment from June 1 through September 30. Therefore, that pasture would be the logical choice to receive the first aerial spray, since the herbicide application would be followed by a 4-month-deferment period. After 18–24 months, the herbicide application should have ample time to achieve maximum effectiveness. Thus, chaining might be applied from October 1 to January 31 depending on soil-water conditions, the second regular deferment period following aerial spraying or

TABLE 9.3. Theoretical Treatment Sequence for Integrating a Chemical–Mechanical–Fire Management System for Honey Mesquite Within Two Pasture–One Herd Grazing Systems

Time Period	Pastures[a]		Year		Brush Management Treatment	
	1	2				
February 1–May 31	D	G	1			
June 1–September 30	G	D	—		Spray pasture 2	
October 1–January 31	D	G	—		12 months	
February 1–May 31	G	D	2			
June 1–September 30	D	G	—	24 months	Spray pasture 1	
October 1–January 31	G	D	—			
February 1–May 31	D	G	3			
June 1–September 30	G	D	—		Chain pasture 2	18 months
October 1–January 31	D	G	—	8 months		
February 1–May 31	G	D	4		Burn pasture 2	
June 1–September 30	D	G	—		Chain pasture 1	
October 1–January 31	G	D	—	8 months		
February 1–May 31	D	G	5		Burn pasture 1	

[a]D = deferred from grazing; G = pasture grazed.
SOURCE: Welch and Scifres (1980).

from June 1 to September 30, the third deferment period after the herbicidal treatment of pasture two (Table 9.3). The latter date would allow a time lapse of 24 months after spraying before chaining, a desirable time interval, but may present problems because of normally dry soil conditions.

Chaining the pasture during a scheduled deferment period allows further improvement of fine fuel continuity and load, essential for effective application of prescribed burning. The prescribed burns may be applied during the dormant season, just ahead of spring greenup of forage species, and allow some deferment before grazing.

Pasture 1 may be aerially sprayed the year following treatment of pasture 2 (Table 9.3). Using the same treatment sequence as described for pasture 1, the brush management treatment sequence is applied in 48 months. After the initial burns, prescribed burning may be used as needed (probably at 5–7-year intervals) to suppress invading honey mesquite seedlings and surviving sprouts, as well as pricklypear and associated brush species.

The same system may be used in a three pasture–two herd grazing management system. The primary difference is that prescribed burning may not be applied until 27 months after chaining in the three pasture–two herd system compared to a time lapse of only 8 months in the former grazing system (Table 9.4).

The aerial spray–chain sequence was chosen for presentation because of its established effectiveness (Scifres, 1980). Burning techniques have been developed specifically for chained rangeland (Wright, 1974). However, there are no documented attempts to meld these methods into a brush management system in concert with grazing management. The proposed system should be viewed with an attitude of flexibility in timing treatment application. It theoretically offers the following advantages:

1. Only a single spray treatment is applied. This approach should be of economic advantage compared to using sprays repeatedly—even at only 5–7-year intervals.
2. Chaining is one of the most economically efficient mechanical brush management methods available (Whitson and Scifres, 1980). It complements the action of the herbicide and improves the rangeland for application of prescribed burning.
3. Prescribed burning is a relatively low-cost method for maintaining range improvement (Garoian et al. 1984; Whitson and Scifres 1980).
4. Treatment costs are spread over a series of years. This, of course, is of lesser advantage in inflationary times than others; but is still a positive attribute if cash flow is a limitation to installation of range improvements.

RANGE SEEDING

The preceding discussions have emphasized improvement of range forage stands of native species, via "natural seeding," mostly with methods that do not cause significant soil disturbance. However, in some cases, "artificial seedings" may be the most effective means for establishing a forage base. This may be accomplished by seeding a mixture of native species or by establishing a single-species stand that complements use of native forage stands.

The "ideal" seedbed for grass establishment is very near that for establishment of an agronomic crop. The seedbed is free from competitive vegetation, well pulverized on the surface, and firm within and below the seeding zone (Vallentine, 1971). Grass seedbeds may be improved if there are moderate amounts of plant residue on the surface to aid in trapping moisture, reduce

TABLE 9.4. Theoretical Treatment Sequence for Integrating a Chemical–Mechanical–Fire Management System for Honey Mesquite Within Three Pasture–Two Herd Grazing Systems

	Pastures[a]				
Time Period	1	2	3	Brush Management Treatment	
March 1–May 31	G	G	D		
June 1–August 31	D	G	G	Spray pasture 1	
September 1–November 30	G	D	G		
December 1–February 28	G	G	D		
March 1–May 31	D	G	G		18 months
June 1–August 31	G	D	G	Spray pasture 2	
September 1–November 30	G	G	D		
December 1–February 28	D	G	G	Chain pasture 1	
March 1–May 31	G	D	G		
June 1–August 31	G	G	D	Spray pasture 3	
September 1–November 30	D	G	G		
December 1–February 28	G	D	G	Chain pasture 2	
March 1–May 31	G	G	D		
June 1–August 31	D	G	G		27 months
September 1–November 30	G	D	G		
December 1–February 28	G	G	D	Chain pasture 3	
March 1–May 31	D	G	G	Burn pasture 1	
⋮		⋮		⋮	

[a]D = deferred from grazing; G = pasture grazed.

SOURCE: Welch and Scifres (1980).

evaporative losses, and protect the young seedlings from drying winds and blowing soil particles.

Stuth and Dahl (1974) illustrated the importance of seedbed preparation to development of grass stands on rangeland when seeding a mixture of native grasses. The more intense the seedbed preparation, the greater the establishment of seeded species (Fig. 9.3). However, they found that root plowing and roller chopping (moderate intensity preparation) allowed a lesser cost per unit increase in grazing capacity than did the more intensive operations.

Methods of seedbed preparation vary a great deal among geographic regions, among ranchers within regions, and among locations on the same ranch because of differences in range sites, original vegetation cover, and objectives for the conversion. Therefore, only a few examples will be contrasted here.

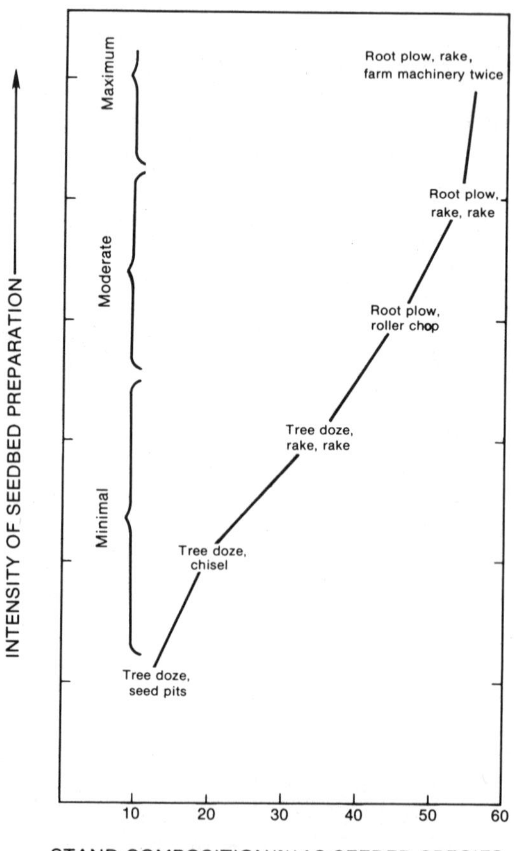

FIGURE 9.3. Generalized relationship of stand success measured as composition by seeded species (native grasses) with intensity of seedbed preparation on rangeland [adapted from Stuth and Dahl (1974)].

Where heavy brush cover exists on the area to be seeded, a sequence of practices may be required including chaining two-ways; raking, stacking, and burning the debris; root plowing, root raking, stacking, and burning; or root plowing, roller chopping, and seeding. Seeding may be accomplished with a grass drill, or the seed may be broadcast on the surface. If broadcast seeding is used, the seed may be quickly covered by dragging a light chain over the surface. Dragging for seed coverage, however, leaves a relatively loose, "fluffy" seedbed. Therefore, in many cases the soil surface is firmed with an implement such as the "cultipacker." Even with this maximum tillage effort, the seedings may not develop satisfactory stands. Grass seeds are usually planted about 1–5 mm deep and grass seedlings are extremely fragile. Stand establishment may be limited by insufficient moisture in the seed zone for germination and seedling emergence, soil crusting, seedling mortality from dessicating winds, and damage by rodents, lagomorphs and insects.

Brush control and seeding operations may be conducted simultaneously. Exhaust-driven seed blowers on crawler tractors may be used to disseminate the seed during the root plowing operation, or the seed may be aerially applied to the disturbed area. These approaches reduce the number and intensity of operations for seeding but also, because of seedbed limitations, may reduce the probability of stand success. Seeding depth also varies greatly from place to place in the treated area. Most of the seeds land on the soil surface so that they must be covered by the action of rainfall. Others may fall in soil cuts several inches deep and be buried so deeply that seedling emergence is not possible. Also, dense stands of weeds usually develop following soil disturbance such as root plowing. Although the weed cover offers protection to the emerging grass seedling, it also competes vigorously for soil water. In many cases, it is advisable to kill the weed cover with a selective herbicide to reduce competition for water and form a protective cover of standing dead plant material.

Methods such as chemical seedbed preparation require least tillage. These methods are discussed in detail by Vallentine (1971), and include applying herbicide in the spring, followed by drilling of grass seed. One of the most efficient processes involves seeding directly into strips that have been treated with herbicide. The herbicide application and seeding are completed in a single operation. Contact herbicides such as paraquat have been used effectively for establishing perennial grasses on annual grasslands (Vallentine 1971). Paraquat is extremely effective and has no measurable soil residual activity. Vegetation removal in the herbicide-treated bands promotes seedling establishment, and vegetation between seeded rows apparently creates a favorable microclimate for the new seedlings.

Range seeding as a range improvement alternative must be scrutinized carefully and not only from the standpoint of biological success but relative to economic risk and effectiveness. Data of Whitson and Scifres (1980) offer a basis for economic comparison of alternatives for improvement of honey mesquite-dominated rangelands in Texas. Although the absolute values are outdated (the study was based on 1978 data including cattle prices of $0.97/kg), relative comparisons are valid. If projected annual rate of return is used as the selection criterion, aerially spraying or chaining would be deemed preferable to root plowing and seeding [in the example, to a single species stand of kleingrass (*Panicum coloratum*) (Table 9.5)]. If investment capital were limited, the seeding project would be even less attractive since, for example, cost for conversion to kleingrass was estimated at about 10 times for the least expensive spray treatment (Whitson and Scifres, 1980). However, from the viewpoint of increasing annual cash flow and total weaned calf production, the range seeding was the superior alternative (Table 9.5).

In many cases, the range forage base is represented by a mixture of artificially seeded pastures and native range. In south Texas, for example, common buffelgrass (*Cenchrus ciliaris*), an introduced, warm-season, perennial grass, is used extensively for tame pasture and for seeding following root plowing. Crude protein content of buffelgrass exceeds that of many native species during the 260–280-day growing season. Seedings of buffelgrass may be used in rotation with native range during selected periods, such as the breeding season and/or during lactation, when a relatively high nutritional plane is required. However, since the carrying capacity of buffelgrass pasture in good condition may double that of native range, care must be exercised to balance pasture sizes to ensure adequate forage year around.

The emphasis on grass seedings in this section does not imply that forbs and shrubs should not be utilized for artificial revegetation of rangeland. A balanced forage supply on rangeland consists of an approximate mix of grasses, forbs, and browse. Browse plants, especially evergreen species, are a good source of crude protein, phosphorus, and carotene. Shrubs such as fourwinged saltbush (*Atriplex canescens*) have successfully been used to revegetate sites that are too dry or saline to allow establishment of many commercially available herbaceous species (McArthur et al. 1978; Ueckert et al. 1982).

TABLE 9.5. Projected Annual Rates of Return, Increased Cash Flow, and Total Weaned Calf Production Over a 20-Year Planning Horizon for Improvement of Deep Soils Having Dense Honey Mesquite Cover in the Rolling Red Plains, Texas[a]

Treatment	Annual Rate of Return (%)	Increased Cash Flow ($/ha)	Total Weaned Calf Production (kg/ha/year)
Aerial spray (2,4,5-T)	15.4	3.26	20.2
Aerial spray (2,4,5-T + picloram)	10.1	3.09	20.7
Chain	10.5	3.21	20.6
Root plow, seed to kleingrass	4.5	10.72	61.8

[a] All values based on 1978 dollars and cattle price of $0.97/kg.

SOURCE: Whitson and Scifres (1980).

McFarland and Ueckert (1982) reported aboveground standing crops of fourwing saltbush of 1124–2328 kg/ha at 5 months after direct seeding on leveled oil well slush pits on rangeland in western Texas. Standing crops at 16 months after seeding exceeded 16,000 kg/ha (D.N. Ueckert, personal communications). However, browsing by deer and lagomorphs can be devastating to shrub plantings. Furthermore, establishment of acceptable, productive stands of shrubs in rangeland plantings appears to require planting in pure stands, as opposed to plantings in mixtures with grasses, and possibly suppression of competing vegetation. Preemergency herbicides may prove useful for controlling competing vegetation in shrub plantings (Ueckert, personal communication); (Whisenant et al. 1982).

VERTICAL INTEGRATION OF RANGE IMPROVEMENTS: THE SAGEBRUSH TO WHEATGRASS EXAMPLE

Western rangelands are among the greatest natural resources of the nation. Presently, some 8.4 million cows reside an average of 3 months annually on public lands (Young and Evans, 1980). Concentration of domestic herbivores has caused shrubs, such as big sagebrush (*Artemisia tridentata*), to increase in density at the expense of herbaceous species. Once established, the life expectancy of big sagebrush may exceed 150 years (Ferguson 1964).

Current research is emphasizing conversion of degraded sagebrush rangeland to productive, stable sites dominated by perennial grasses. This has been approached by considering a range of technologies that may be appropriate for integration into effective systems. System effectiveness is determined by economic evaluation, vegetation response and red-meat production, and their influence on a large number of interacting environmental parameters. This approach applies conventional integrated pest management (IPM) strategies to a range improvement problem.

Vertically integrated weed control systems include combinations of 2,4-D application for brush management and atrazine for herbaceous fallows to reduce competition and allow desirable forage species to become established (J.A. Young, personal communication). Crested wheatgrass (*Agropyron desertorum*) is seeded with a rangeland drill (Evans and Young, 1977).

Where herbaceous vegetation does not constitute a serious competition hazard, the sagebrush may be sprayed and perennial grasses seeded shortly thereafter. A special rangeland drill (Young and McKenzie, 1982) with modified openers is used to make furrows when seeding to provide a more favorable microenvironment for the developing seedlings.

Seriously degraded range sites may be plowed with a special brushland plow and seeded with the rangeland drill. In some cases, prescribed burning may be used to remove the brush pior to seeding with a rangeland drill. On sites with sufficient remnant perennial grasses (one plant per square meter), the prescribed burn may not be seeded, since the destruction of the woody vegetation by the fire stimulates herbaceous vegetation succession, the direction of which is shaped by a grazing management system.

Vertically integrated weed control–revegetation systems are presently in the developmental stages. The alternative technologies are being evaluated on a large-scale basis by J.A. Young, R.A. Evans, and coworkers in Nevada. All range improvement techniques employed for the sagebrush conversion to wheatgrass involve a period of deferment from grazing to allow the desirable forage to respond to the environmental potential released by the weed control treatments. The vertically integrated weed control systems involve a 3–4-year period from initiation to completion. The steps include herbicide application, fallow season, seedling year of the established forage, and grazing when the perennial grasses are well established. In the first year of use, grazing is usually deferred until after the wheatgrass has set seed.

Degraded sagebrush rangelands may produce 60–170 kg/ha of herbage in "good" years and less than 60 kg/ha in dry years (J.A. Young, personal communication). Improved sagebrush rangelands produce 550–2,250 kg/ha of higher-quality herbage during "average to good" years, and produce some forage even in extremely dry years. Approximately 0.5 million ha of the 23 million ha of sagebrush in Nevada alone have been seeded to crested wheatgrass. The increased production and quality of forage translates to more than greater livestock-carrying capacities. The higher-quality forage during breeding seasons leads to improved calving percentages and increased livestock production efficiency. The sagebrush to wheatgrass conversion serves as only one example, but the principles of application of these systems may well be applicable to a high proportion of semiarid rangelands.

CONCLUSIONS

Rangeland is a critically important component of human's total land-based, life support system. Rangeland ecosystems are managed by procedures derived from long-standing ecological principles; and, because of its expanse, nature and use, rangeland is managed extensively and generally without tillage. Most range management practices are designed to expedite secondary plant successon and/or improve utilization of the rangeland by herbivores that convert primary "crop" (natural vegetation) to marketable products.

Practices such as brush management and grazing management often offer the most effective approaches to improving production on rangeland. Therefore, the "no-till" philosophy is inherent in the general foundation of range management doctrine and pervasive to technology development for range improvement. However, in some cases, the range has become so depleted that artificial revegetation is required to restore production in a reasonable length of time. In these cases, initial range restoration is then followed by conventional range management practices.

Range-improvement methods have traditionally been applied as isolated practices with their implementation planned independently, one of the other. However, because of the multiple uses of rangeland, which contribute to the ultimate economic success of the ranch firm, range managers are now viewing range-improvement practices in the context of overall range resource use planning. Moreover, recent research has demonstrated that a well-planned sequencing of brush control methods increases economic, ecological, and management benefits at the ranch firm level. These factors have provided the impetus for emergence of concepts such as the integrated brush management systems (IBMS) approach to range resource management. Such integrated management systems hold promise for meeting the continually increasing societal demands on natural resources.

ACKNOWLEDGMENTS

I am grateful to Julia Scifres for typing and other preparations of this manuscript; and to D.N. Ueckert, J.W. Stuth, J.A. Young, R.A. Evans, and T.G. Welch for their constructive criticisms.

LITERATURE CITED

Bogusch, E.R. 1952. Brush invasion in the Rio Grande Plains of Texas. *Texas J. Sci.* **4**:85–91.

Bonham, C.D. 1983. Range vegetation classification. *Rangelands* **5**(1):19–21.

Bontrager, O.E., C.J. Scifres, and D.L. Drawe. 1979. Huisache control by power grubbing. *J. Range Manage.* **32**:185–188.

Box, T.W. and J. Powell. 1985. Brush management techniques for improved forage values in south Texas. *Trans. North. Amer. Wildl. and Nat. Resour. Contr.* **38**:285–294.

Bray, W.L. 1901. The ecological relations of the vegetation of western Texas. *Bot. Gaz.* **32**: 99–123; **32**:195–207; **33**:262–291.

Brock, J.L., R.H. Haas, and J.C. Shaver. 1978. Zonation of herbaceous vegetation associated with honey mesquite in north-central Texas. *Proc. Internat. Rangeland Congr.* **1**:187–189.

Buffington, L.C. and C.H. Herbel. 1965. Vegetation changes in semi-desert grassland rangeland from 1858 to 1963. *Ecol. Monogr.* **35**:139–164.

Dahl, B.E., R.E. Sosebee, J.P. Goen, and C.S. Brumley. 1978. Will mesquite control with 2,4,5-T enhance grass production? *J. Range Manage.* **31**:129–131.

Drawe, D.L., A.D. Chamrad, and T.W. Box. 1978. Plant communities of the *Welder Wildlife Refuge.* Welder Wildl. Found. Contrib. No. 5, Series B. (Rev.).

Dyksterhuis, E.J. 1955. What is range management? *J. Range Manage.* **8**:193–196.

Dyksterhuis, E.J. 1958. Range conservation as based on sites and condition classes. *J. Soil Water Conserv.* **13**:151–155.

Evans, R.A. and J.A. Young. 1977. Weed control-revegetation systems for big sagebrush-downy brome rangelands. *J. Range Manage.* **30**:331–336.

Ferguson, C.W. 1964. Annual rings in big sagebrush, *Artemisia tridentata.* Papers of Lab. of Treerings Res., Univ. of Arizona Press, Tucson.

Fisher, C.E., J.L. Fultz, and H. Hopp, 1946. Factors affecting action of oils and water soluble chemicals in mesquite eradication. *Ecol. Mongr.* **16**:109–126.

Foster, M.A., C.J. Scifres, and P.W. Jacoby. 1984. Herbaceous vegetation-lotebush (*Ziziphus obtusifolia* (T.& G.) Gray var. *obtusifolia*) interactions in North Texas. *J. Range Manage.* **37**:317–320.

Gadzia, J.S.G. and J.A. Ludwig. 1983. Mesquite age and size in relation to dunes and artifacts. *Southw. Nat.* **28**:89–94.

Gardner, J.L. 1951. Vegetation of the croesotebush area of the Rio Grande Valley in New Mexico. *Ecol. Monogr.* **21**:379–403.

Garoian, L.R., J.R. Conner, and C.J. Scifres. 1984. Economic evaluation of fire-based systems for Macartney rose-dominated rangeland. *J. Range Manage.* **37**:111–115.

Gordon, R.A., C.J. Scifres, and J.L. Mutz. 1982. Integration of burning and picloram pellets for Macartney rose control. *J. Range Manage.* **35**:427–430.

Grelan, H.E. and R.E. Lohrey. 1978. Herbage yield related to basal area and rainfall in a thinned longleaf plantation. So. Forest. Exp. Sta. Res. Note 50-232.

Halls, L.K. and J.L. Schuster. 1965. Tree-herbage relations in pine-hardwood forests of Texas. *J. For.* **63**:282–283.

Hamaker, J.W., H. Johnston, R.T. Martin, and C.T. Redemann. 1963. A picolinic acid derivative: A plant growth regulator. *Science* **141**:363.

Hamilton, W.T., L.M. Kitchen, and C.J. Scifres. 1981. Height replacement of selected woody plants following burning or shredding. Tex. Agr. Exp. Sta. Bull. 1361.

Hamilton, W.T. and C.J. Scifres. 1982. Prescribed burning during winter for maintenance of buffelgrass. *J. Range Manage.* **35**:8–12.

Heady, H.F. 1975. *Rangeland Management.* McGraw-Hill, New York.

Hester, T.R. 1980. *Digging into South Texas Prehistory.* Corona Press, San Antonio, TX.

Inglis, J.M. 1964. A history of the vegetation on the Rio Grande Plains. Texas Parks and Wildlife Dept. Bull. 45.

Jacoby, P.W., C.H. Meadors, and M.A. Foster. 1981. Control of honey mesquite (*Prosopis juliflora* var. *glandulosa*) with 3,6-dichloropicolinic acid. *Weed Sci.* **29**:376–378.

Jameson, D.A. 1967. The relationship of tree overstory and herbaceous understory vegetation. *J. Range Manage.* **20**:247–249.

Johnston, M.C. 1962. Past and present grasslands of southern Texas and northern Mexico. *Ecology* **44**:456–466.

Kirby, D.R. and J.W. Stuth. 1982. Botanical composition of cattle diets grazing brush managed pastures in east-central Texas. *J. Range Manage.* **35**:434–436.

Koshi, P.T. 1953. An evaluation of forage production under various densities of oak woodland. M.S. Thesis, Dep. Range and Forestry, Texas A&M College, College Station.

Mayeux, H.S., Jr. and W.T. Hamilton. 1983. Response of common goldenweed *(Isocoma coronopifolia)* and buffelgrass *(Cenchrus ciliaris)* to fire and soil-applied herbicides. Weed Sci. **31**:355–360.

McArthur, E.D., A.P. Plummer, G.A. Van Epps, D.C. Freeman, and K.R. Jorgensen. 1978. Producing fourwing saltbush seed in seed orchards. Proc. Internat. Rangeland Congr. **1**:406–410.

McAtee, J.W., C.J. Scifres, and D.L. Drawe. 1979. Digestible energy and protein content of gulf cordgrass following burning or shredding. J. Range Manage. **32**:376–378.

McFarland, M.L. and D.N. Ueckert. 1982. Honey mesquite control: Use of a three-point-hitch mounted, hydraulically assisted grubber. Texas Agr. Exp. Sta. Prog. Rep. 3981, *In* Brush Manage. and Range Improvement Res. 1980–1981, Texas Agri. Exp. Sta. Consol. Prog. Rep. 3968–4014, pp 48–50.

McFarland, M.L. and D.N. Ueckert. 1982. Rehabilitation of rangeland disturbed by energy exploration in West Texas. Texas Agri. Exp. Sta. Prog. Rep. 3983 *In* Brush Manage. and Range Improvement Res. 1980–81, Texas Agri. Exp. Sta. Consol. Prog. Rep. 3968–4014, pp 53–56.

McGinty, A., F.E. Smeins, and L.B. Merrill. 1983. Influence of spring burning on cattle diets and performance on the Edwards Plateau. J. Range Manage. **36**:175–178.

Merrill, L.B. and C.A. Taylor. 1976. Take note of the versatile goat. Rangeman's J. **3**:74–76.

Minshull, J.I. and R.A.I. Norval. 1982. Factors influencing the spatial distribution of *Rhipicephalus appendiculatus* in Kyle Recreational Park, Zimbabwe. W. Afr. J. Wildl. Res. **12**: 118–123.

Mutz, J.L., T.G. Greene, C.J. Schifres, and B.H. Koerth. 1984. Prescribed burning for improvement of Pan American balsamscale range. Texas Agr. Exp. Sta. Bull. 1492.

Oefinger, R.D. and C.J. Scifres. 1977. Gulf cordgrass production, utilization, and nutritional value following burning. Texas Agr. Exp. Sta. Bull. 1176.

Office of Technology Assessment. 1982. Impacts of technology on U.S. cropland and rangeland productivity. Congress of the United States, U.S. Govt. Print. Office.

Oldham, T.W., C.J. Scifres, and D.L. Drawe, 1982. Gulf coast tick response to presribed burning on the coastal prairie. Texas Agr. Exp. Sta. Prog. Rep. 3933, in Brush Management and Range Improvement Res. 1980–81, Cons. Prog. Rep. 3968–4014, pp. 83–86.

Polk, K.L. and D.N. Ueckert. 1973. Biology and ecology of a mesquite twig girdler, *Oncideres rhodosticta*, in West Texas. *Ann. Entomol. Soc. Amer.* **66**:411–417.

Rasmussen, G.A., C.J. Scifres, and D.L. Drawe, 1983. Huisache growth, browse quality, and use following burning. *J. Range Manage.* **36**:337–342.

Scifres, C.J. 1973. *Mesquite, Distribution, Ecology, Uses, Control, Economics.* Texas Agr. Exp. Sta. Res. Monogr. 1.

Scifres, C.J. 1975. Systems for improving Macartney-rose infested coastal prairie rangeland. Texas Agr. Exp. Sta. Bull. 1225.

Scifres, C.J. 1978. Range vegetation management with herbicides and alternate methods: An overview and perspective. Symp. Use of Herbicides in Forestry, U.S. Dep. Agr.-U.S. Environ. Prot. Agency, Arlington, VA (Feb. 21–22):151–165.

Scifres, C.J. 1980. *Brush Management. Principles and Practices for Texas and the Southwest.* Texas A&M Univ. Press, College Station.

Scifres, C.J. 1981a. Selective grazing as a weed control method. *CRC Handbk. Pest. Manage. in Agri.* **11**:369–376.

Scifres, C.J. 1981b. Integrated brush management systems (IBMS) for rangeland: Concepts and application. *Internat. Stockmen's School, Beef Cattle Sci. Handbk.* **18**:74–79.

Scifres, C.J. 1982. Woody plant control in the Post Oak Savannah of Texas with hexazinone. *J. Range Manage.* **35**:401–404.

Scifres, C.J., G.P. Durham, and J.L. Mutz. 1977. Range forage production and consumption following aerial spraying of mixed brush. *Weed Sci.* **25**:48–54.

Scifres, C.J., W.T. Hamilton, J.M. Inglis, and J.R. Conner. 1983a. Development of integrated brush management systems (IBMS): Decision-making processes. Brush Manage. Symp., Soc. Range Manage, Albuquerque, New Mexico. 97–104.

Scifres, C.J. and G.O. Hoffman. 1972. Comparative susceptibility of honey mesquite to dicamba and 2,4,5-T. *J. Range Manage.* **25**:143–145.

Scifres, C.J. and J.L. Mutz. 1978. Herbaceous vegetation changes following applications of tebuthiuron for brush control. *J. Range Manage.* **31**:375–378.

Scifres, C.J., J.L. Mutz, and W.T. Hamilton. 1979. Control of mixed brush with tebuthiuron. *J. Range Manage.* **32**:155–158.

Scifres, C.J., J.L. Mutz, and C.H. Meadors. 1978. Response of range vegetation to grid placement and aerial application of karbutilate. *Weed Sci.* **26**:139–144.

Scifres, C.J., J.L. Mutz, R.P. Smith, and G.A. Rasmussen. 1983b. Integrated brush management systems (IBMS): Concepts and case studies with running mesquite and whitebrush. Texas Agr. Exp. Sta. Bull. 1450.

Scifres, C.J., J.L. Mutz, R.E. Whitson, and D.L. Drawe. 1982. Interrelationships of huisache canopy cover with range forage on the coastal prairie. *J. Range Manage.* **35**:558–562.

Scifres, C.J., J.L. Mutz, R.E. Whitson, and D.L. Drawe, 1983c. Mixed-brush canopy cover-rainfall interrelationships with native grass production. *Weed Sci.* **31**:1–4.

Scifres, C.J. and D.B. Polk, Jr. 1974. Vegetation response following spraying a light infestation of honey mesquite. *J. Range Manage.* **27**:462–465.

Scifres, C.J., J.W. Stuth, and R.W. Bovey. 1981a. Control of oaks (*Quercus* spp.) and associated woody species on rangeland with tebuthiuron. *Weed Sci.* **29**:270–275.

Scifres, C.J., J.W. Stuth, D.R. Kirby, and R.F. Angell. 1981b. Forage and livestock production following oak (*Quercus* spp.) control with tebuthiuron. *Weed Sci.* **29**:535–539.

Smith, L.L. and D.N. Ueckert. 1974. Influence of insects on mesquite seed production. *J. Range Manage.* **17**:61–65.

Stoddard, L.A., A.D. Smith, and T.W. Box. 1975. *Range Management.* McGraw-Hill, New York.

Stuth, J.W. and B.E. Dahl. 1974. Evaluation of rangeland seedings following mechanical brush control in Texas. *J. Range Manage.* **27**:146–149.

Thomas, G.W. 1970. The western range and the livestock industry it supports. In *Range Research and Range Problems.* Crop Sci. Soc. Amer., Madison, WI.

Ueckert, D.N. 1983. Personal communication. Texas Agr. Res. and Ext. Center, San Angelo, TX 76901.

Ueckert, D.N., J.L. Petersen, M.L. McFarland, and R.L. Potter. 1982. Fourwing saltbush revegetation trials in West Texas. Texas Agri. Exp. Sta. Prog. Rep. 3985. In *Brush Management and Range Improvement Research 1980-81*, Texas Agri. Exp. Sta. Consol. Prog. Rep. 3968-4014, pp. 59-62.

Ueckert, D.N., S.G. Whisenant, and G.W. Sultemeier. 1983. Control of rayless goldenrod (*Isocoma wrightii*) with soil-applied herbicides. *Weed Sci.* **31**:143–147.

Vallentine, J.F. 1971. *Range Development and Improvements.* Brigham Young Univ. Press, Provo, Utah.

Weidemann, H.T., B.T. Cross, and C.E. Fisher, 1977. Low-energy grubber for controlling brush. *Trans. Amer. Soc. Agr. Engr.* **20**:210–213

Welch, T.G. and C.J. Scifres. 1980. Integrating range improvement with grazing systems. *Proc. Symp. Grazing Manage. Systems for Southwest Rangelands*, Albuquerque, New Mexico, pp. 91–94.

Whisenant, S.G., D.N. Ueckert, and J.E. Huston. 1982. Forage shrub revegetation trials in west central Texas. Texas Agric. Exp. Sta. Prog. Rep. 3984. In *Brush Management and Range Improvement Research 1980-81*, Texas Agric. Exp. Sta. Consol. Prog. Rep. 2968-4014, pp. 56–59.

Whitson, R.E., W.T. Hamilton, and C.J. Scifres. 1979. Techniques and considerations for economic analysis of brush control alternatives. Texas Agr. Exp. Sta., Dep. Tech. Rep. 79-1.

Whitson, R.E. and C.J. Scifres. 1980. Economic comparisons of alternatives for improving honey mesquite-infested rangeland. Texas Agr. Exp. Sta. Bull. 1307.

Wright, H.A. 1974. Range burning. *J. Range Manage.* **27**:5–11.

Wright, H.A. and A.W. Bailey, 1982. *Fire Ecology. United States and Southern Canada.* Wiley, New York.

York, J.C. and W.A. Dick-Peddie. 1969. Vegetation changes in southern New Mexico during the past 100 years. In *Arid Lands in Perspective.* Univ. Arizona Press, Tucson.

Young, J.A. and D. McKenzie, 1982. Rangeland drill. *Rangelands* **4**(3):108–113.

Young, J.A. and R.A. Evans. (Ed.) 1980. Physical, biological, and cultural resources of the Gund research and demonstration ranch, Nevada. U.S. Dep. Agr., Agri. Res. and Manual W-11.

10

NO-TILLAGE AND SURFACE-TILLAGE SYSTEMS TO ALLEVIATE SOIL-RELATED CONSTRAINTS IN THE TROPICS

R. LAL
Soil Physicist
International Institute of Tropical Agriculture
Ibadan, Nigeria

INTRODUCTION

Recognition of the difficulties of protecting bare soil from erosion and high temperatures led to the method of "shifting cultivation" and "bush fallow rotation" widely practiced by subsistence farmers throughout the tropics. By this method, fertility lost during the cropping phase of each cycle is replenished by fallowing with natural vegetation. If the fallow cycle is sufficiently long, the fertility so restored permits cropping through another cycle without commercial inputs. During the cropping phase, a considerable number of woody perennials are retained by incomplete clearing, soil is minimally disturbed, and the residue from leaf litter and weed regrowth is retained as a mulch. The system is ecologically stable provided the fallow period is long enough to restore soil physical, chemical, and biological properties. Reduced mechanical soil manipulation is, therefore, an integral part of traditional farming in the tropics. It is because of the minimum disturbance of the soil and vegetative cover that erosion and irreversible degradation of soil quality are not serious problems under shifting cultivation.

The often quoted "high risks" of erosion in the tropics are only partly due to an inherent susceptibility of these soils to erosion. Tropical rains have higher intensity, bigger drop size, and more energy load than equivalent rains in temperate regions. The chemical fertility of most tropical soils is concentrated in the superficial surface horizon and the aboveground biomass. The loss of this thin surface layer leads to a very rapid decline in soil productivity. There are also differing microclimatic conditions at the beginning of the growing season between temperate and tropical regions. In spring, soils in the temperate region are wet and cold at below-optimal soil temperature. The opposite is true in the tropics where soils are too hot and too dry in spring following a long dry period. Soil management in the tropics strives to lower soil temperature and to conserve moisture in the root zone.

Attempts at replacement of traditional farming systems with those that utilize modern machinery and agrochemicals have often been unsuccessful because of the ecological instability and because modern systems are not adapted to the needs and resources of the people who use it. Rather than replacing this subsistence, but ecologically sound, traditional agriculture, it is logical to improve its productivity by appropriate inputs of modern technology. In addition to increasing production and land use intensity, an improved system should also (a) maintain favorable soil structure and curtail erosion, (b) use minimal agrochemicals and other fossil-fuel-based inputs, and (c) maintain high levels of production per unit area.

Adjusted no-tillage systems meet these requirements in the tropics. The features embodied in them are (a) sowing of crops directly into untilled or least-tilled soil, (b) prevention of soil erosion and suppression of weed growth with crop residue mulch, and (c) maintenance of favorable soil structure through biological rather than mechanical means. No-tillage requires fewer capital-intensive inputs, minimizes soil disturbance with soil cover, improves soil structure, and facilitates intensive land use by prolonging soil productivity. This chapter, a state-of-art review of no-tillage farming research in the tropics, describes the potential and limitations of this system for diverse soils and ecologies. On the basis of existing information, research and development priorities for adaptation of this system to a wide range of crops and soils are discussed.

HISTORICAL DEVELOPMENT

Ironic as it is, the current trends in scientific temperate region agriculture are toward decreasing the intensity and frequency of mechanical tillage in favor of reduced or no-tillage systems, while mechanized tillage operations of plowing and harrowing are recommended for intensifying tropical agriculture. Attempts should be made to improve the traditional agriculture rather than replace it by systems developed elsewhere. The importance of retaining the basic ingredients of minimal soil disturbance and use of crop residue mulch to improve and intensify land use in the tropics was recognized by Abeyratne (1956). His work maintained the productivity of upland crops for fragile, easily compacted and eroded Alfisols in the dry zone of Sri Lanka. While realizing the importance of mulch cover and of no soil disturbance, Abeyratne and contemporary workers were also aware of the severe problem of controlling weed growth on untilled soil without using herbicides. The yield reduction of 36% in uncultivated upland rice (*Oryza sativa* L.) compared with plowed Sri Lanka Alfisols was attributed to severe weed competition. In Central Africa, Muller and Bilerling (1953) reported work on Oxisols in Belgian Congo (now Zaire) that more yield of upland rice on unplowed than on plowed land was due to less soil erosion, no crusting, and slight or no drought stress. Ospina and Nel (1957) reported from Colombia that plowing often resulted in accelerated erosion.

The development of paraquat and other growth regulators in the mid-1960s facilitated weed control without plowing and led to the widespread use of no-tillage and mulch farming in commercial tropical agriculture. In Ghana, Kannegieter (1967, 1969) and Gordon (1968) successfully demonstrated the use of paraquat for suppressing cover crops and weeds for growing maize (*Zea mays* L.) without plowing. Many researchers in Asia concluded that, as in the management of uplands, excessive presowing tillage and puddling may not be essential for lowland rice cultivation provided weeds are efficiently controlled (Bhan and Tilak, 1964; Mabbayad, 1967; Mittra and Pieris, 1968; Moomaw et al., 1968). Although crop yields on unplowed soils were generally lower than on conventionally plowed lands, these studies indicated the potential of the no-tillage system in using these otherwise difficult-to-manage soils in harsh climatic environments.

MULCH FARMING IN THE TROPICS

The value of crop residue mulch on the soil surface to moderate the effects of harsh tropical climate has long been recognized. Experiments conducted for diverse tropical soils and ecologies indicate that mulches control weeds and soil erosion, improve soil organic matter content, and improve crop yields (Abu-Zied, 1973; Ashrif and Thornton, 1965; Chaudhary and Prihar, 1974; Jurion and Henry, 1969; Lal, 1975; Lawes, 1962). The effectiveness of mulch in preventing soil erosion is widely documented (Lal, 1976a). Mulches also decrease maximum soil temperatures (Lal, 1973; Ghuman and Lal, 1982; Lal, 1978) and improve soil structure and water retention capacity (Lal et al., 1980). In spite of the favorable influence of mulch materials on soil properties, it is difficult to generalize its effects on crop yields. Crop yield response to mulching depends on a multitude of interacting factors, for example, the nature of mulch material, soil properties, crop characteristics, and prevailing micro- and meso-climates (Bates, 1975; Chaudhary and Prihar, 1974; Lawson and Lal, 1979;

Okigbo, 1973; Panikar and Walunjkar, 1976). Even the biotic environments are modified by the mulch. For example, Yu et al. (1981) observed that fungi, actinomycetes, ammonifying bacteria, N-fixing bacteria, and potassium bacteria in mulched plots as 58.3%, 25.8%, 47.3%, 56.3%, and 56.1% more, respectively, than in the unmulched control.

Although many researchers strongly believe mulch to be the critical factor in intensifying land use and increasing crop production in the tropics, it is more economical to produce mulch *in situ* than to bring mulch from elsewhere. In this connection the adoption of no-tillage farming on tropical soils is logistically a better alternative than plowing.

SOIL AND ECOLOGICAL FACTORS OF NO-TILLAGE IN THE TROPICS

Soil properties that favor no-tillage farming are (i) a coarse-textured surface horizon or self-mulching clayey soils with high initial porosity, (ii) low susceptibility to soil compaction and crusting, (iii) good internal drainage, (iv) high biological activity of earthworms and other soil fauna, and (v) friable consistency over a wide range of soil moisture contents. These conditions minimize the necessity for mechanical soil manipulation in seedbed preparation. The no-tillage system is generally not adaptable to soils that do not fulfill these requirements.

Alfisols in the Humid and Subhumid Tropics

Alfisols are formed in the humid, subhumid, and semiarid tropics of West Africa, south Asia, northeast Thailand, northeast Brazil, and northern Australia. Terra Roxa soils of the Amazon Bazin are also Alfisols. In West Africa, most Alfisols contain predominantly low-activity clays with kaolinite as the most abundant clay mineral. Therefore, these soils are characterized by low CEC (cation exchange capacity) and low available water and nutrient retention capacity, and are structurally unstable. Cultivated Alfisols slake readily, develop crusts, are easily compacted, and are susceptible to accelerated soil erosion. The merits of a no-tillage system for Alfisols are discussed below.

Soil Physical and Chemical Properties

Field experiments conducted on a wide range of Alfisols in West Africa and elsewhere have indicated that the infiltration rate of these soils can drop more than 50% under conventional methods of plowing and harrowing within 1 year of returning bush fallow land to arable farming (Wilkinson and Aina, 1977). Cultivated Alfisols are easily compacted, and the severity of compaction increases with the frequency and intensity of mechanized tillage. Baffoe-Bonnie and Quansah (1975) observed that on an Alfisol in the subhumid region of Kumasi, Ghana, reduced tillage operations caused the least compaction, maintained high porosity, and had the lowest soil and water losses. In an identical environment of southwest Nigeria, Lal (1974, 1976b) and Aina (1979a) also observed more favorable physical and chemical soil properties with no-tillage than conventionally plowed systems.

Table 10.1 compares the effects of three tillage systems and a fallow on physical and chemical properties of an Alfisol near Ife, in southwest Nigeria. The physical and chemical properties of fallow and no-tillage systems were superior and more favorable to crop production than plow and disk-harrow treatments. The deterioration in soil physical and chemical properties was greatest in the system that involved both primary and secondary tillage operations, that is, plowing followed by disk harrowing. Similar results have been reported in Africa by Kannegieter (1969), Ajunwon et al. (1978), Juo and Lal (1978), Agboola (1981), and Olaniyan (1983); in Australia by Melville (1978) and McCowen et al. (1985) in northern Australia; and in Sri Lanka by Wijewardene (1980). These favorable effects of the no-tillage system on soil physical properties are partly attributed to the prevalence of biotic activity of earthworms and other soil fauna (Lal, 1976; Aina, 1979a).

Hydrological Properties

Opara-Nadi and Lal (1983) observed significant differences in moisture retention capacities in the 0–10-cm- and 10–20-cm-depth layers due to tillage treatments. Soil from the no-tillage treatments had higher moisture retention than that from plowed plots in the pF range <1.6. Ghuman and Lal (1984)

TABLE 10.1. Effects of Three Tillage Methods and of Fallowing on Physical and Chemical Properties of an Alfisol, Near Ife, Nigeria

Property	Fallow	No-Till	Plow	Plow and Disk Harrow	$LSD_{0.05}$
pH	5.3	5.0	4.5	5.0	0.2
Ca^{2+} (meq/100 g)	3.20	1.75	1.22	1.35	0.27
K^+ (meq/100 g)	0.37	0.19	0.14	0.12	0.03
NO_3-N (ppm)	22.4	16.6	9.4	10.2	4.5
CEC (meq/100 g)	10.3	6.6	5.9	5.5	0.9
Bulk density (g/cm^3)	1.15	1.42	1.48	1.51	0.03
Total porosity (%)	55.9	49.2	46.0	44.1	2.3
Air-filled porosity at 60 cm H_2O suction %)	16.9	10.9	8.2	6.8	1.2
Water-stable aggregates >2.36 mm (%)	79	48	23	15	5
Saturated hydraulic conductivity (cm/hr)	115	44	20	15	4

SOURCE: Aina (1979a).

reported the mean infiltration rate on plots under no-tillage and plow-tillage for 12 consecutive years ranged from 4.6 to 42.9 cm/hr in the latter compared with 15 to 120 cm/hr in the former.

In contrast to Alfisols in Nigeria, no-tillage experiments conducted at Buoake, Ivory Coast, with similar soils and rainfall conditions, indicated that soil physical and hydrological properties were improved by plowing (Buanec, 1974). The no-tillage treatment used in Ivory Coast, however, did not maintain a crop residue mulch.

Soil and Water Conservation

Favorable soil physical properties in an untilled soil maintain high rates of water infiltration. Coupled with the buffering effects of a mulch against raindrop impact, a no-tillage system is an effective erosion deterrent on slopes up to 15% provided there is an adequate quantity of mulch (Lal, 1976c). In this study, runoff was reduced 80% and soil loss 99% by no-tillage compared to plowing. Similar observations were made in Ghana by Baffoe-Bonnie and Quansah (1975) and Bonsu and Obeng (1979). Where this system is employed, other erosion control measures such as terraces and diversion channels are often unnecessary.

Soil Temperature

Adverse effects of drought stress are accentuated by high soil temperatures. Crops grown with methods of seedbed preparation that expose soil surface to high insolation suffer more from supraoptimal soil temperatures than those grown on an untilled seedbed with a residue mulch (Maurya and Lal, 1981). Many studies in Africa (Lal, 1976a; Aina, 1979a; Nangju, 1979) and northern Australia (McCowen et al., 1985) have shown temperatures of the surface horizon of untilled mulched soil to be 8–16°C lower than that of a plowed or ridged soil surface.

Crop Response

Different crops respond differently to no-tillage. Response depends on initial soil conditions, quantity and quality of the surface mulch, effective rooting depth, and prevalent micro- and mesoclimate.

MAIZE AND SORGHUM (*SORGHUM VULGARE PERS.)*

Yield. Much research from West Africa, southern Asia, and northern Australia describe the response of maize to methods of seedbed preparation. Kannegieter (1967, 1969) reported from Ghana that no-tillage maize sown directly through killed kudzu (*Pueraria thumbergiana* Sieb. and Eucc., Benth.) yielded more than double that on land that was plowed and then mulched with the slashed kudzu residue. In the absence of an adequate quantity of mulch and when the soil is initially compacted, maize grain yield on an unplowed soil can be less. For example, on an Alfisol near Kumasi, Ghana, maize yield was significantly less on an unplowed than on a 22- or 37-cm-deep plowed seedbed (Ofori and Nandy, 1969). Initial soil conditions and the presence of residue mulch are important in determining the response of maize to a no-tillage system.

In Southwest Nigeria, Aina (1982) obtained similar or more maize grain yield from mulched no-tillage than from mulched ridged or mounded seedbeds. In a long-term study at IITA (International Institute of Tropical Agriculture), Lal (1982) compared maize grain yield on no-tillage and conventionally plowed plots for 24 consecutive seasons (two seasons per year). Over and above the seasonal fluctuations in yield, due to variations in rainfall, maize grain yield from no-tillage plots was greater (Fig. 10.1). This was particularly so in those seasons that experienced short duration drought. These data also confirm that in the absence of mechanized operations, leading to soil compaction, a sustained and economic maize grain yield can be obtained with no-tillage systems for maize. On these coarse-textured sandy soils, equivalent grain yields can be obtained even without adopting the elaborate and capital-intensive systems of chiseling, primary and secondary tillage, plowing at the end of the previous rainy season, or contour ridges (Table 10.2). In fact, the lowest yield was obtained from the most elaborate ridged treatment. The removal of crop residue mulch from no-tillage plots also reduced yield and

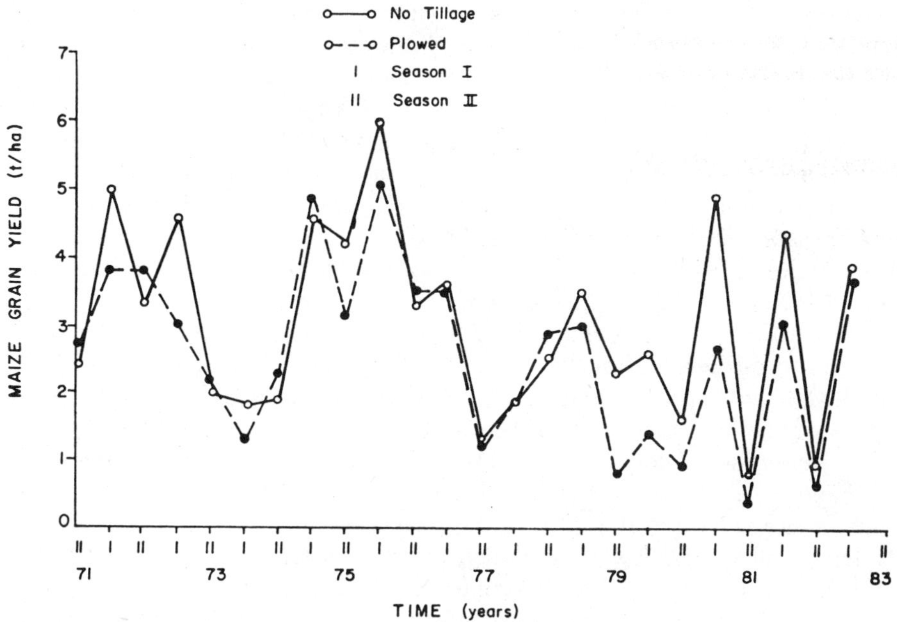

FIGURE 10.1. Maize grain yield on an Alfisol for 24 consecutive crops grown with no-tillage and conventional plowing methods of seedbed preparation.

TABLE 10.2. Effects of Different Methods of Seedbed Preparation on Maize Grain Yield From an Alfisol in Western Nigeria

Tillage Treatment	Infiltration Rate (cm/3 hr)	Maize Grain Yield (mt/ha)
No-till with residue mulch	168	4.4
No-till with chiseling in the dry season	80	4.2
Molboard plowing, two harrowing	68	3.7
Disk plowing, rotovation	71	4.6
No-till without residue mulch	138	3.7
Plowing at the rain end, two harrowings at seeding	74	3.8
Plowing, two harrowing with mulch	79	4.4
Plowing, two harrowing and ridging	59	3.4
$LSD_{0.05}$	50	1.0

SOURCE: Unpublished data (Lal, 1983).

mechanized tillage and frequent traffic caused soil compaction. With severe compaction, grain yield with a no-tillage system was superior to that on a plowed seedbed (Couper et al., 1979).

Drought Stress. High maize grain yield without tillage on soils of low available water-holding capacity is partly due to more favorable soil moisture and temperature regimes in untilled soil. Lal et al. (1978a) monitored diurnal fluctuations in leaf water potential of maize grown with no-tillage and plowed systems for variable irrigation cycles. Although leaf water potential decreased with increase in days after irrigation, the rate of decrease was slower with no-tillage than plowing, implying that both degree and duration of soil moisture stress were affected by tillage methods. Decreased drought stress suffered by no-tillage maize is partly due to a root system development in an undisturbed soil (Fig. 10.2). Channels created by earthworms and decomposing roots provide continuous and preferential pathways for root development and water transmission (Ghuman and Lal, 1983). Opara-Nadi and Lal (1983) established an empirical relationship between the maize grain yield and available water content for the two tillage systems. Their data show that the maize grain yields increased at the rate of 0.45 t/ha for each 1% increase in the available water capacity in the range of 20–39%. In this experiment, the maximum available water storage capacity was 38.1%, 33.6%, 32.6%, and 29.1% for the no-tillage without mulch and conventional tillage without mulch, respectively. Lal et al. (1978) also computed the water use efficiency (WUE) of maize and found it to be more for no-tillage than a conventionally plowed system.

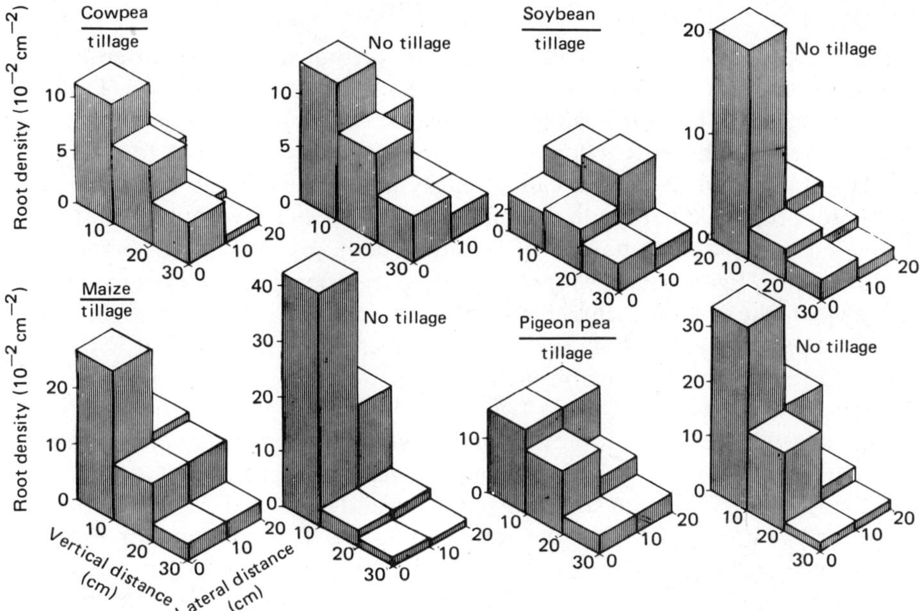

FIGURE 10.2. Root system of crops grown with conventional plowing and no-tillage methods of seedbed preparation on an Alfisol (Maurya and Lal, 1978).

Fertilizer Requirements. Fertilizer response of maize in a no-tillage system can be different than in a plowed system. Losses of nitrogen by ammonia volatilization from surface-applied fertilizer in a no-tillage system can be more than from a plowed seedbed with the fertilizer incorporated (Acquaye and Cunningham, 1965) (see Chapter 5). Kang et al. (1980) observed that maize grain yield from no-tillage unfertilized plots was significantly lower than from plowed treatments. Differences in maize grain yield among tillage treatments were less when fertilizer was applied (Kang and Messan 1983). The higher response to fertilizer in no-tillage than in plowed plots is partly due to the immobilization of available N by the microorganisms in plots that receive mulch material with a wide C:N ratio. This response can be less and even negative when the experiments are conducted for a sufficiently long period to allow the available soil N to attain stability.

For these Alfisols with low activity clays, P placement in the soil has no advantage over broadcast application. Kang and Yunusa (1977) showed that although P movement was slow with no-tillage methods, P availability was equally effective in plowed and no-tillage systems. Juo and Lal (1978) and Armon et al. (1980) also observed significant P movement up to 30 cm depth in no-tillage plots within 6 years (see Chapter 4).

COWPEA *VIGNA UNGUICULATA* (L.) WALP

Crops that do not leave adequate amounts of mulch respond differently to no-tillage systems than those that do. For example, growing of cowpea following cowpea with the no-tillage may deteriorate soil properties because of the lack of adequate mulch. Cowpea grown in a rotation with a cereal, for example, maize or sorghum, however, can be easily adapted to the no-tillage system. Literature reported here indicate that cowpea responds favorably to no-tillage provided an adequate quantity of residue from a previous crop is available.

In Nigeria, Ezedinma (1964) observed no significant differences in cowpea grain yield when planted on a freshly plowed ridge/furrow system or on ridges formed the year before. Lal (1976b) reported cowpea yields of 600–770 kg/ha with no-tillage in comparison with 470–610 kg/ha on plowed seedbeds. Nangju (1979) observed that in two of three years, cowpea grown with no-tillage outyielded those on plowed and ridged treatments. He observed that percentage emergence was significantly lower on ridges (83%) and plowed seedbeds (89%) than with strip tillage (97%) and no-tillage (98%) (Nangju et al., 1975). Furthermore, on the ridged plots, it took 12 days for the first emergence of cowpea, compared with 3 days in strip or no-tillage seedbeds.

Cowpea grown on ridges and plowed seedbeds suffer more from drought stress and high soil temperature than those in untilled soil. Lal et al. (1978a) measured the grain yield and water use efficiency of cowpea on no-tillage and plowed plots under different irrigation frequencies. For irrigation every 4, 8, and 12 days, water use efficiency was significantly greater with no-tillage than on plowed seedbeds. Aina (1979b, 1982) observed less drought stress and more cowpea grain yield with a no-tillage mulch than on ridged or mounded seedbeds. While evaluating the effects of tillage and mulching treatments on an Alfisol, Lal (1979) observed that cowpea growth on ridges was suppressed by very high soil temperature and reported a linear decrease in the leaf area index with an increase in the maximum soil temperature. In the first growing season from April to July in western Nigeria, the maximum temperature at 1500 hours was highest on ridges compared with flat and mulched seedbeds. Lowest yield was obtained on ridges and soybean [*Glycene max* (L.) Merrill] responded similarly (Table 10.3). In the second growing season from August to November, with frequent rains, with a mulch, yields were affected more by fungal diseases than other treatments.

SOYBEAN

Soybean is even more sensitive to fluctuations in soil temperature and moisture than maize or cowpea. Soybean germination and seedling establishment are adversely affected by supraoptimal temperatures and drought stress commonly experienced during April–May in West Africa. Of all the grain crops, soybean should respond most favorably to a no-tillage mulch.

Nangju et al. (1975) observed that percentage emergence of soybean was 0.9%, 33.4%, 50.7%, and 53.9% for ridges, plowed, strip tillage, and no-

TABLE 10.3. Effects of Mulching and Methods of Seedbed Preparation on Grain Yield of Soybean and Cowpea in 1977 on an Alfisol

	Cowpea (kg/ha)		Soybean (kg/ha)	
Treatment	First Season	Second Season	First Season	Second Season
Black plastic mulch	640	600	1930	1300
Clear plastic mulch	670	760	1230	1320
Straw mulch	730	380	1730	1570
Ridges	460	450	230	1410
Plowed	620	600	1600	1500
Aluminum foil mulch	850	650	2100	1460
$LSD_{0.05}$	260	380	670	360

SOURCE: Lal (1979).

tillage treatments, respectively. The days to first emergence in the same order of treatments were 12, 6, 5, and 5, respectively. Nangju (1979) obtained satisfactory soybean yields on no-tillage and strip tillage seedbeds, and a complete failure of a 1975 soybean crop sown on ridges.

CASSAVA (*MANIHOT ESCULENTA* CRANTZ) AND OTHER ROOT CROPS

Tubers require more "root room" for adequate development than do the feeder roots of grain crops. In addition to soil depth and drainage, to which most tropical root crops are extremely sensitive, response of root crops to tillage systems also depends on their physiology. Tuber shape and rate of development greatly influence yield response to tillage methods. Also important are methods of propagation and establishment, most of which are vegetatively propagated. Seed-sets and root cuttings require favorable soil moisture and temperature regimes for good stand establishment. In general, root and tubers do not produce satisfactory yields on compacted, untilled, or shallow soils.

In northern India, Verma (1973) observed an increase in size and yield of cassava tuber from earthing up twice during the initial stages of development. In the humid tropics of southern India, however, Thamburaj et al. (1980) reported that no-tillage and mulching of cassava outyielded conventional plowing and harrowing. In Ghana, Ofori (1983) reported significant yield increases in cassava tubers when a compacted Alfisol was plowed compared with the traditional method. Because cassava tubers develop in a horizontal direction, Ofori observed no improvement in tuber yield by 30-cm over 22-cm plowing depths. In western Nigeria, Aina (1982) also reported lower cassava tuber yield on an untilled than on plowed and ridged soil. Lal (1980) obtained equivalent tuber yield from large-scale no-tillage and plowed watersheds of 4–5 ha each. In addition to the adverse effects of compaction, Akobundu (1983) reported that low cassava yield on untilled plots was also due to severe weed competition. Uncontrolled weed growth caused a yield reduction of 71% in no-tillage and 54% in plowed and harrowed treatments.

Lal (1975) reported the effects of methods of seedbed preparation and of mulching on tuber yield of yam (*Dioscorea* spp). On a shallow Alfisol, mulching of ridges and mounds significantly improved plant establishment. There was an eightfold increase in yield by mulching a flat seedbed for the early planted yams as compared to no yield on unmulched heaps and ridges. If adequately mulched, ridges and mounds can be useful for production of tropical root crops, particularly for shallow soils underlain with gravelly horizons. Maurya and Lal (1980) also reported a satisfactory yield of sweet potato [*Ipomoea batatis* (L.) Lam.] on an untilled and mulched Alfisol.

Ultisols

Ultisols are more abundant than Alfisols in the humid tropics of Central Africa (Ivory Coast and Sierre Leone), southeast Nigeria, Cameroon, southeast Asia, and the Amazon Basin. Recent studies have shown that with appropriate land use and balanced fertilizer inputs continuous food crop production is achievable (Sanchez et al., 1982). Because Ultisols have lower mineral fertility than Alfisols and are acidic, the choice of crops to be grown is more limited for adaptation to a no-tillage system. Regardless of tillage method, maize does not grow well on an unlimed Ultisol. Even if lime were locally available, surface application in a no-tillage system would not be as effective in neutralizing subsoil acidity as its incorporation by plowing and harrowing. A more practical alternative would be choice of a crop that can be grown without liming. Although problems of acidity and nutrient imbalance are more severe than the limitations imposed by physical characteristics, erosion, compaction, crusting, drought stress, and high soil temperatures are also important limitations on most Ultisols in the humid tropics (Pla Sentis et al., 1979).

Crop Response to No-Tillage

Few experiments have been reported on the development of appropriate tillage systems for Ultisols. With their high chemical fertility, Alfisols are generally the first choice for arable land use. In the Amazon Basin, for example, Terra Roxa and annually renewable flood-plain soils (Inceptisols) are intensively used for food crop production, rather than nutrient-deficient Ultisols and Oxisols. The no-tillage system is suitable for some crops on a wide range of Ultisols regardless of their chemical constraints.

Grain Crops

With adequate liming, no-tillage maize outyielded plow tillage of an Ultisol in Liberia (Lal and Dinkins, 1979) and southeastern Nigeria (Rodriguez and Lal, 1979a). In Liberia, maize grain yield was 2.8 and 1.3 mt/ha with no-tillage and plow tillage, respectively. Soil inversion by plowing brought highly acidic subsoil to the surface resulting in unfavorable physical and chemical environments for root growth and a maize grain yield of 3.8 mt/ha in a limed no-tillage system compared with 2.6 mt/ha in limed plowed plots. Maurya and Lal (1979c) reported slightly more maize grain yield from plowed than an untilled Ultisol in eastern Nigeria. This superior yield was attributed to a deeper root system development when lime was incorporated by plowing (Fig. 10.3). Hayward et al. (1980) reported that a common practice in many Central American countries is to grow maize with no-tillage on hilly areas of El Salvador, Venezuela, Costa Rica, Nicaragua, and Ecuador.

No-tillage on acidic soils is better suited to those crops that are tolerant of low soil pH. Lal and Dinkins (1979) reported upland rice yields of 2.9 and 1.9 mt/ha for no-tillage and plowed ultisol in Liberia. Similarly, Maurya and Lal (1979c) reported 80% more cowpea yield from an untilled than plowed Ultisol.

Crop response to tillage systems is greatly influenced by mulching. In addition to physical effects, mulching also alters the nutrient regime and biotic environments. Ogunremi et al. (1984) reported lower grain yields of upland

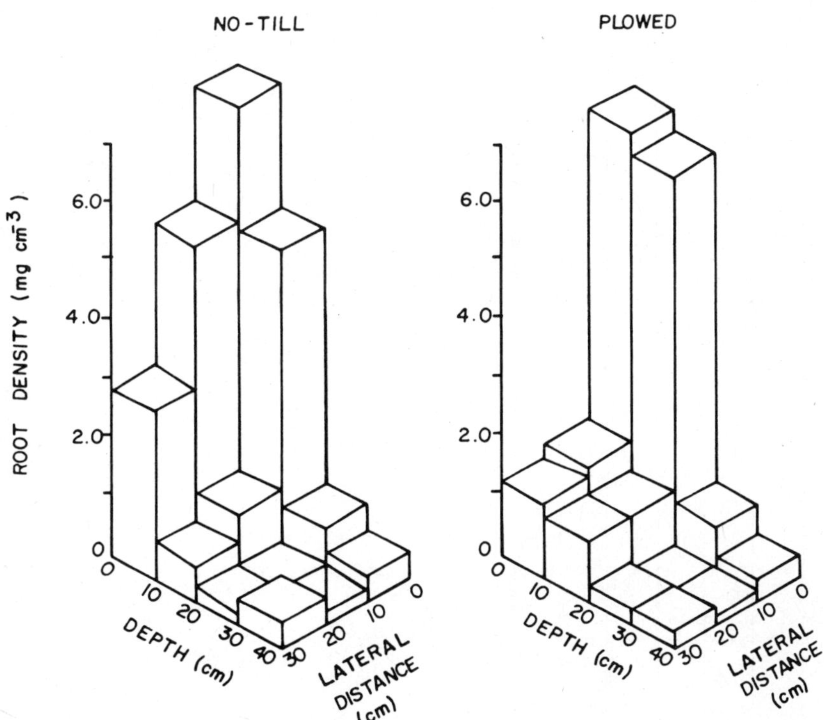

FIGURE 10.3. Effects of tillage methods on root development on an Ultisol in eastern Nigeria (Maurya and Lal, 1979).

rice with no-tillage than plowed treatments without a mulch (Fig. 10.4). With mulching, however, equivalent yield was obtained in no-tillage and plowed plots. Soil compaction caused by mechanized operations, however, may eventually cause severe yield reductions in no-tillage upland rice.

Mechanized farm operations cause severe compaction on most tropical soils, and the no-tillage system is not readily adapted on compacted soils. Perez Escolar and Ortiz Lugo (1973) reported from Puerto Rico that sugarcane

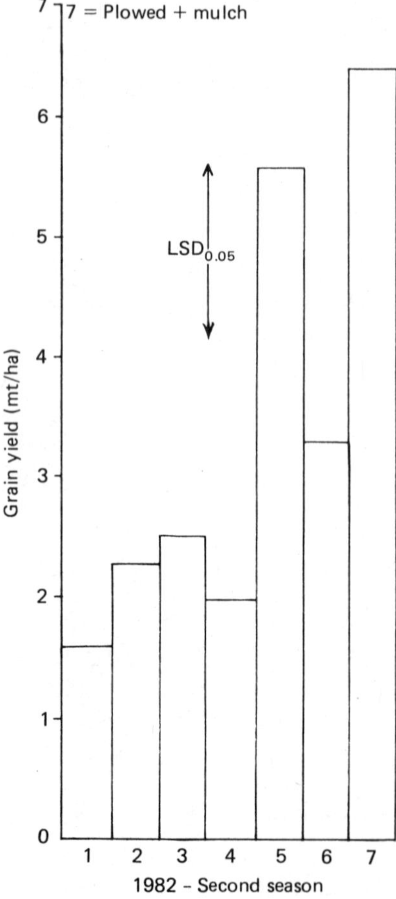

FIGURE 10.4. Grain yield of upland rice on an Ultisol in eastern Nigeria as influenced by methods of seedbed preparation and mulching (Ogunremi et al., 1986).

(*Saccharum officinarum* L.) yield from compacted Ultisols was improved by deep plowing. This mechanical ameliorative measure causes soil erosion (Jackson, 1979). Agronomic systems should be developed that cause the least soil compaction. Compacted soils are preferably restored by fallowing rather than by deep plowing.

Root Crops

Tropical root crops, for example, cassava and yam, are more adapted than grain crops to acidic soils in humid environments. Okigbo (1979) reported no significant differences in cassava tuber yield in relation to frequency and intensity of preplant cultivation. Similar observations were reported by Maurya and Lal (1979). Akobundu (1983), however, reported significantly higher cassava tuber yield in ridged and plowed than no-tillage plots: the mean tuber yield 13.5, 13.4, and 8.3 mt/ha, respectively. As in his findings on an Alfisol, yield reductions on no-tillage plots were attributed to severe weed infestations (Akobundu, 1983).

In a mulching and tillage study, Maduakor et al. (1983) showed significant and positive effects of mulch but no effect of tillage methods on tuber yield of yam on an Ultisol in eastern Nigeria. The mean tuber yield on a ridged seedbed was 14.1 mt/ha compared with 14.8 mt/ha on a flat unridged seedbed. For comparatively deep soils of sandy loam texture, tillage has little effect on tuber development of acid tolerant tropical root crops and there are no obvious advantages of ridging over flat planting. However, mulch does improve soil moisture and temperature regimes and tuber yield.

Oxisols

Oxisols are similar in some ways to Ultisols in chemical and physical soil properties. In central Brazil and the Amazon Basin they are highly susceptible to water erosion and have low available water holding capacity (Brazil, 1980). Crops often suffer from drought even a few days after heavy rains. Important objectives of seedbed preparation, therefore, are to conserve soil and water resources, increase water use efficiency, and sustain economic productivity.

On a Brazilian Oxisol, Sidiras et al. (1982, 1983) observed significant improvements in water-stable aggregates and soil-water retention at 0.06, 0.33, and 1 bar suctions after 4 years of continuous no-tillage farming (Table 10.4). Improvements in soil moisture retention in the no-tillage system were more pronounced in the top 20-cm layer than in the soil beneath. In addition to improvements in soil structure and infiltration rates, the presence of crop residue mulch on no-tillage plots significantly decreased runoff and soil losses from an Oxisol on 4% slope. In comparison with conventional tillage, grain yield of soybean was improved by 4% with chisel plow and by 7% with the no-tillage system (Sidiras et al., 1982). In another study in Typic Haplorthox, Kemper and Derpsch (1981) reported soybean grain yields of 1434, 1505, and 1987 kg/ha on conventional tillage, minimum tillage, and no-tillage seedbeds, respectively.

TABLE 10.4. Soil-Water Retention (%) at Different Suctions After 4 Years of Conventional Tillage, Chisel Plow, and No-Tillage on an Oxisol in Brazil

	Suction (bars)		
Tillage Treatments	0.06	0.33	1
		3–10-cm Depth	
Conventional tillage	36.4	30.8	29.0
Chisel plow	39.2[a]	34.6[a]	32.6[a]
Direct drill	44.7[a]	40.3[a]	38.2[a]
		11–20-cm Depth	
Conventional tillage	38.9	33.2	31.4
Chisel plow	42.6[a]	38.0[a]	35.9[a]
Direct drill	43.3[a]	39.9[a]	38.1[a]

a = Significant at 1% level of probability.

SOURCE: Sidiras et al. (1982).

Tropical root crops respond differently to no-tillage than grain crops. In addition to the problem of weed control, additional soil volume (or root room) is required for proper development of bulky tuberous roots. Development of root tubers can be better in plowed than untilled Oxisol. Plowing is not so essential for recently developed Inceptisols with favorable subsoil properties. Ezumah (1983) observed the effects of tillage methods on cassava in an Oxisol in Zaire and reported inferior plant stands for no-tillage compared to tilled and ridged plots. The cassava tuber yield was 7.3, 5.9, and 4.0 mt/ha for plowed, ridged, and no-tillage treatments in an Oxisol, compared with 18.9, 16.6, and 16.4 mt/ha, respectively for a sandy loam Inceptisol.

SOIL CHARACTERISTICS UNSUITABLE FOR NO-TILLAGE

A no-tillage system is generally ineffective on soils with degraded antecedent soil physical conditions. Some important factors to be considered while making a choice of an appropriate method of seedbed preparation on physically deteriorated soils are:

Compaction. Crop establishment with no-tillage is generally unsatisfactory on soils that have a compacted surface horizon.

Microrelief. An uneven ground surface adversely affects the seeding process with no-tillage. There are more pest and rodent problems on an untilled and mulched soil than in plowed and clean seedbed.

Residue mulch. A no-tillage system cannot be adopted successfully if there is an inadequate amount of residue mulch on the soil surface.

Seed–soil contact. One of the reasons for poor crop stand on an untilled soil is improper equipment for seeding through a layer of crop residue mulch. Lack of seed–soil contact such as that caused by the smearing of a wet clayey soil during seeding also results in a poor stand and low yields.

Soil and climate characteristics of the semiarid regions render surface soil conditions unsuitable for direct adaptation of the no-tillage system. In general, the longer the duration of the dry season, the lesser is the amount of crop residue mulch on the surface at the beginning of the monsoon to protect soil from raindrop impact and to provide favorable soil temperature and moisture regimes. Whatever residue remains is either grazed or burned. Uncontrolled grazing also compacts the soil surface and necessitates some mechanical tillage to loosen the seedbed.

Predominant upland soils of the semiarid and arid tropics are Alfisols, Inceptisols, and Vertisols. Charreau (1977) described the following soils in the semiarid regions of West Africa:

1. Ustropepts are found mainly on ancient eolian deposits with a mean annual rainfall of 500–900 mm.
2. Ustalfs are commonly developed on gneiss and sandstone parent materials and have well-developed profiles.
3. Alfic Eutrustox are found in southern Senegal.
4. Vertisols and other soils containing 2:1 clay minerals are developed on basic parent materials.
5. Hydromorphic soils (Tropaquents, Tropaquepts, and aquic subgroups of Tropuldalfs and Tropuldults) are widespread in the area.
6. Iron pan soils are commonly found in regions with rainfall of 500–900 mm, and are of negligible agronomic importance.

Tillage needs for this wide range of soils in West Africa and elsewhere in the semiarid tropics are difficult to generalize. Soils that contain at least 20% clay with some proportion of expanding clay minerals (high-activity clays) are "structurally active" because of their swell–shrink characteristics. Some form of no-tillage or reduced tillage system can be used for these soils. Those soils that contain less than 20% clay or nonexpanding minerals have negligible swell–shrink characteristics and are "structurally inactive." For these soils, some mechanical tillage is necessary to obtain an adequate seedbed (Charreau, 1977). Only Vertisols and some hydromorphic and ferrallitic soils are structurally active. Most soils of semiarid West Africa are structurally inert and have (i) weak structure with low porosity, (ii) low infiltration rate and low available water capacity, and (iii) are easily compacted.

Tillage Systems for Structurally Inert Soils

The objectives of seedbed preparation for coarse-textured soils containing predominantly low-activity clays are to improve (i) porosity and relative proportion of macropores, (ii) root growth in subsoil, (iii) water infiltration capacity, and (iv) water storage within the soil profile. Different methods of seedbed preparation are recommended for different agroecological environments.

Semiarid West Africa

Intensive soil surface management studies have been conducted by IRAT in Francophone West Africa (Charreau, 1972; Charreau and Nicou, 1971; Chopart and Nicou, 1976; Nicou, 1974a, 1974b; Nicou and Chopart, 1979; Poulain and Tourte, 1970; Chopart, 1981). Most studies indicate the benefits of deep tillage and soil inversion in terms of decreasing losses due to water runoff and conserving more water for plant growth. Data from Senegal (Nicou, 1977) indicate a significant increase in available water in the soil profile and of yield of millet [*Pennisetum typhoides* (Burm) Staph & C.E. Hubb.] and groundnuts (*Arachis hypogaea* L.) from plowing compared with the no-tillage system. However, it is important to realize that rather than using crop residue in the no-tillage treatment as mulch, it was burned. Charreau (1977) summarized 17 years of experiments and indicated the beneficial effects of plowing on grain yield of Pearlmillet, sorghum, maize, lowland rice, cotton (*Gossypium hirsutum* L.), and groundnut (Table 10.5). Nicou (1979) showed from his studies at Bambey that the root weight of groundnut is increased linearly with total porosity and that the pod yield increases linearly with total root weight as shown by the following empirical equations:

$$\text{Root weight (mg)} = 148\ (\text{total porosity, \%}) - 5600, \quad r = 0.91^{**}$$
$$\text{Pod yield (kg/ha)} = 1.37\ (\text{Root weight, mg}) + 233, \quad r = 0.82^{**}$$

Plowing increased total porosity and thus the groundnut yield. Nicou (1974a), however, indicated that plowing brings about only transient improvements and that soil structure is easily degraded during the cropping phase. Nicou (1979), Chopart (1981), and Chopart et al. (1981), while summarizing the results of tillage studies conducted in Senegal, Togo, and Ivory Coast, supported the conclusion of Charreau (1977) that structurally inert soils of the Sahel region benefit markedly from mechanical tillage. Results of experiments by Nur and Gasim (1974) in Sudan Gezira also support Charreau's conclusion that intensive tillage increased groundnut pod yield by 72%. No-tillage system, however, is feasible for Alfisols in the subhumid regions of Ivory Coast (Table 10.6).

**Means that correlation coefficient is significant at 1 percent level of probability.

TABLE 10.5. Mean Effects of Plowing on Yields of Crops in the West African Dry Tropical Area (IRAT Experiments; Soil Surface Layers, Sandy to Coarse Loamy), 1952–1969

	Vegetation Removed before Plowing						Vegetation Incorporated by Plowing					
	Number of Experimental Observations			Yield on Unplowed Controls	Increases from Plowing		Number of Experimental Observations			Yield on Unplowed Controls	Increases from Plowing	
Crops	Total	Beneficial	%	(kg/ha)	kg/ha	%	Total	Beneficial	%	(kg/ha)	kg	%
Pearl millet (grain)	22	21	95	1,245	+256	+21	5	4	80	971	+365	+38
Sorghum (grain)	46	39	85	1,874	+536	+29	2	2	100	2039	+532	+26
Maize (grain)	6	6	100	2,093	+568	+27	12	10	83	1474	+970	+66
Rainfed rice (paddy)	11	11	100	966	+1515	+157	1	1	100	1547	+705	+46
Cotton (seed)	7	7	100	1,629	+433	+27	12	10	83	1240	+423	+34
Groundnut (pods)	31	27	87	1,412	+274	+19	113	81	71	1661	+119	+7

SOURCE: Charreau (1977).

TABLE 10.6. Tillage Effects on Crop Yields for Different Soils of the Semiarid Regions of West Africa

	Plowing (kg/ha)	Minimum Tillage (kg/ha)
Senegal		
Groundnuts	2029	1536
Millet	1635	1546
Maize	3014	1515
Rice (rainfed)	3417	1765
Togo		
Maize	1991	608
Sorghum	3392	1259
Groundnuts	1111	666
Ivory Coast		
Maize		
Class 1 soils[a]		
Variety IRAT 81	4355	4510
CJB	3125	2920
Class ⅔ soils[b]		
IRAT 81	3340	2740
CJB	2180	1800
Rainfed rice		
Class 1 soils		
Variety IRAT 13	3240	3340
Moroberekan	1700	1850
Class ⅔ soils		
Variety IRAT 13	2200	1900
Moroberekan	1240	1560

[a] Class 1 soils: soils with good water retention properties.
[b] Class ⅔ soils: soils with gravel content, poor water retention properties.
SOURCE: Nicou (1979).

Southern Africa

Soils of the semiarid regions in Botswana are mostly sandy and are underlain by a thick ferruginous concretionary horizon. These soils are also structurally inert, and lack of moisture is the most important factor limiting crop yield. Studies by Willcocks (1979a, 1979b, 1981) indicate the importance of reduction in soil bulk density by mechanical tillage and by periodic subsoiling operations. Whiteman (1975) showed that moisture conservation by fallowing in alternate years increased crop yield. He demonstrated that chemical weed control during fallowing was more effective in soil-moisture conservation than plowing and mechanical weed control measures. Sorghum grain yields were significantly higher after bare fallow than after cover crop, weed, fallow or maize.

Soils of the Semiarid Tropics in India

Studies conducted at ICRISAT (Charreau, 1977) and elsewhere in India support the conclusion that structurally inert soils require mechanical ameliorative measures to improve soil and water conservation and crop yield. In Gujrat, Borole et al. (1972) reported yield data from a 23-year tillage study showing that deep plowing consistently produced higher yields of pearl millet than shallow plowing or no plowing. From a 10-year tillage study conducted near Banglore, Gidnavar et al. (1972) reported a low percentage of water-stable aggregates in soils receiving minimum tillage treatment. Subramanian et al. (1975) reported that for a red sandy loam soil near Madras deep tillage (20−45 cm) improved soil porosity, saturated hydraulic conductivity, and structural stability. In the subtropical regions of northern India, plowing also improved the infiltration in an Inceptisol and increased grain yield of pearl millet by 29% over the no-tillage treatment (Malik et al., 1973). This study also showed that chiseling in the row zone was as effective as plowing.

In most studies reported in this section, the beneficial effects of plowing over untilled soil as measured in terms of grain yields are undisputably impressive. Nevertheless, these results should be evaluated in view of the following:

1. Crop residue mulch was not used in most of these studies since it was not available, was deliberately removed, or was burned. Wherever the residue mulch was used, its advantages in soil and water conservation and in improving crop yields were substantial in comparison with the unmulched no-tillage technique commonly used as a standard control.
2. Soil physical conditions at sowing time were unfavorable. Because the no-tillage system maintains a status quo, the results would have been different if steps were taken to improve soil physical conditions by fallowing or by using deep-rooted cover crops prior to implementing the no-tillage system.
3. Amelioration in soil structure caused by mechanical tillage was temporary, and the improved structure was easily degraded (Nicou, 1974a). Tillage and fertilization, although temporarily increasing crop yields, did not bring about a complete restoration of degraded lands unless improvements in organic matter were also brought about (Diatta, 1974).
4. Chemical weed control during fallowing was as effective in moisture conservation in Botswana as were mechanical tillage and ridging (Whiteman, 1975). The benefits of plowing were, therefore, limited to plowing's role as a weed control measure.
5. The effects of plowing the entire field were similar to those of chiseling in the row zone alone.

It is important to realize, therefore, that the solution to bringing about permanent improvements in soil structure and water transmission characteristics of these soils cannot be through mechanical tillage alone. Steps must be taken to improve soil organic matter content by improved and more efficient fallowing, controlled grazing, and prohibiting the widespread use of fire as methods of land preparation. Once structural improvements are brought about through improvement in soil organic matter content, mechanical tillage may not be as necessary as it is now for these degraded soils.

Tillage Methods for Structurally Active Soils

Tillage response of structurally active soils, those that contain expanding lattice (high-activity) clay minerals and with clay contents exceeding 20%, is generally similar to that of Alfisols in the humid and subhumid regions. These soils have more favorable structure, and if managed properly, maintain better soil moisture retention and transmission properties than structurally inert soils.

Alfisols, Inceptisols, and Aridisols

Guinea Savanna Zone of Nigeria

The Nigerian savanna extends from 7°N to 13°N with corresponding range of rainfall from about 1200 to 1500 mm per annum spread over 7 months in the south to 3 months in the north. The region is dominated by ferruginous soils. Soils of the northern savanna are affected by aeolian materials and of the southern Niger–Benue depression by being formed over sandstone (Jones and Wild, 1975). Loess plain soils have a high proportion of silt and fine sand, and "cap" or crust due to raindrop impact. Most soils are dominated by low-activity clays with CEC ranging from 1 to 10 meq/100 g and have low levels of soil organic matter content (about 1.0%). The bulk density of uncultivated soils is usually 1.4 g/cm^3. They have weakly developed structure. Risk of erosion, by both wind and water, is high. These ferruginous soils cover 60% of the region.

Lawes (1961, 1962, 1966) showed that the use of crop residue mulch is extremely effective in preventing formation of "cap," in maintaining a high infiltration rate, and in preventing runoff of rainfall (Table 10.7). Mechanically disturbed bare soil may lose 70% of the total rainfall; and even with cultivation at fortnightly intervals, 50% may be lost. In his study, the mean sorghum grain yield in plots mulched with groundnut shell was 29% higher than the unmulched control.

While researchers recognize the beneficial effects of residue mulch, it is difficult to procure amounts adequate to be effective for soil and water conservation. Because of the difficulties of obtaining it in the semiarid regions, erosion control is sought through mechanical means of ridging and rough plowing. In this context Kowal (1970) reported that ridges made up and down the slope encourage rather than control gully erosion. Bababe (1977) confirmed Kowal's findings and observed the least runoff and soil erosion from cropped lands with a flat seedbed. The maximum runoff and soil erosion was observed from land ridged at 1 m intervals. Even tied ridges were not effective in preventing runoff and soil erosion. In accordance with Lawes's data (Table

TABLE 10.7. Relative Efficiency of Soil Surface Treatments in Conserving Rainfall

Treatment	Percentage of Rainfall Penetrating Soil Surface	Maximum Infiltration Rate (mm/hr)
Bare soil		
Undisturbed	31	12.7
Hoed at fortnightly intervals	49	22.9
Hoed after every storm	57	30.5
Mulches		
Dead grass	90	>127
Groundnut shells	89	>127
Sorghum stalks	98	>127

SOURCE: Lawes (1961).

10.7), runoff and soil erosion would have been negligible if crops sown on flat land were also mulched at 4–6 mt/ha of straw.

Killed weeds can be a good source of mulch, and the channels created by their decaying roots usually improve water infiltration. Dunham (1979) and Dunham and Aremu (1979) reported significantly higher infiltration rate with no-tillage (54 mm/hr) than with plowing and harrowing (18 mm/hr). Similarly, the percentage of water stable aggregates in the surface horizon of no-tillage plots were 38%, 42%, and 57% compared with 15%, 19%, and 33% of plowed plots with size ranges of > 2 mm, 1–2 mm, and 0.5–1 mm, respectively. Although grain yields of maize, sorghum, and seed cotton were lower with no-tillage than plowed treatments, yields with no-tillage were improved by retaining the crop residue mulch and supplementing nutrient supply with additional applications of N and P (Dunham, 1982a).

Dunham (1982b) concluded that soil physical properties of the surface layer were superior in no-tillage, manual tillage, and ox-driven tillage methods to those cultivated with tractor-driven implements (Table 10.8). In fact, percentage of water stable aggregates and infiltration rates decreased with increases in frequency and intensity of tractor-driven tillage operations. The relative yield for the zero-tillage treatment increased in the third and fourth years, and was more than that of subsoiled, moldboard plowed, and all ridged treatments. Dunham (1982b) concluded that the most effective way of maintaining surface soil stability and a high infiltration rate is through reducing soil disturbance to the minimum, and by improving soil organic matter content. No-tillage farming, if done properly, with adequate weed control, and liberal use of residue mulch, is potentially useful in the savanna zone of Nigeria and similar soils elsewhere.

In similar ecological regions of northern Ghana, Takayi (1970) observed only a slight improvement in maize grain yield in ridged plots over the untilled

TABLE 10.8. Effects of Tillage Methods on Soil Physical Properties and Crop Response at Samaru, Nigeria

Treatment	Bulk density (g/cm^3) 0–6 cm	Percentage of Water Stable Aggregates		Infiltration Rate (mm/hr)	Relative Yield[a]				
					Second Year	Third Year		Fourth Year	
		<2 mm	<0.5 mm		M	M	C	M	C
Manual tillage	1.24	28	36	90a	100	100	100	100	100
Ox tillage	1.29	25	39	49a	98	102	106	147	105
Tractor tillage	1.39	12	25	30bc	97	106	85	127	86
Zero tillage	1.51	38	57	67a	84	77	76	137	109
Tractor ridge/splitting	1.36	15	33	22bc	98	96	85	125	85
Disk harrowing (flat)	1.45	—	—	50ab	102	108	92	150	88
Disk harowing (ridged)	1.33	—	—	16c	96	106	85	132	82
Disk plowing (ridged)	1.43	9	24	24bc	97	112	87	114	82
Moldboard plowing (ridged)	1.38	—	—	32bc	97	107	78	125	92
Subsoiling (flat)	1.41	12	18	34c	94	104	85	117	87
Manual tillage yield (kg/ha)	—	—	—	—	5130	2900	1720	2120	1290
$LSD_{0.05}$	0.08	6	6	—	—	14	13	—	—

Figures followed by the same letters are statistically identical.
[a]M = maize; C = cotton.

SOURCE: Dunham (1982b).

control. The differences would have been greater in favor of no-tillage had weed control been adequate. Leyenaar and Hunter (1977) concluded that maize grown on ridges was adversely affected by drought stress, high soil temperatures, and severe lodging; yield was the highest on a flat seedbed.

SEMIARID REGIONS IN EAST AFRICA

Relatively few tillage experiments on arable lands have been done in East Africa, although the importance of mulching and no-tillage for perennial crops has long been realized (Fuggles-Couchman, 1934; Jameson, 1970). Pereira and Jones (1954a, 1954b) observed that total pore space, percolation rate, and percentage of water stable aggregates were less in a clean weeded bare soil than in that protected by slashed or tall weed cover (Table 10.9). Disking and plowing lowered total porosity, noncapillary pore space, and percolation rate compared with hand hoeing. Pereira and Jones (1954b) concluded that bare soil subjected to tropical storms declines rapidly and substantially in its ability to transmit rainfall. Mulching, by whatever means, should be beneficial. Their results did not support need for mechanical tillage because the excellent structural conditions developed under a grass mulch approached optimum.

Northwood and Macartney (1971) on three soils in Tanzania—a Ferrallitic (Alfisol) at Kongwa, an Inceptisol at Sambwa, and an Andisol at W. Kilimanjaro—compared the effects of depth and width of cultivation on maize grain yield (Table 10.10). On the structurally inert Alfisol at Kongwa, strip tillage (20 cm wide, 9 cm deep) was superior. An increase in width of the tilled strip beyond 20 cm was not beneficial. On a well-structured Andisol, there was no benefit from tilling more than 2.5-cm-wide and 5-cm-deep strip, sufficient for seed placement.

Compacted soils of low organic matter content of the semiarid tropics may require both primary and secondary tillage operations. For example, Macartney et al. (1971) observed a complete failure of no-tillage maize on a compacted Alfisol with a bulk density of 1.52 g/cm^3. The initial high infiltration rate, caused by plowing, disking, and harrowing, decreased drastically over 2 or 3 cropping seasons due to progressive soil structural deterioration and surface sealing. Macartney et al. (1971) concluded that soil tillage be limited to the row (seed) zone only. They further recommended that dead weed and crop residue mulch be retained between the rows to improve water infiltration by preventing surface sealing. This is similar to the "zonal tillage" concept of "no-tillage" farming in the United States.

At Morogoro, Tanzania, Huxley, (1975, 1979, 1982), highlighted the merits of zero- and minimum-tillage in a 3 year study (Table 10.11). The benefits in yield of maize brought about by plowing over the no-tillage were smaller than those obtained in other experiments by mulching or N fertilizer. Khatibu and Huxley (1979) found the effects of rates of N applications on the grain yield of cowpea were not greatly affected by no-tillage. Although the cowpea vegetative growth was inhibited among tillage treatments to some degree initially, but diminished by harvest. Also at Morogoro, Maseri and Jana (1979) found grain

TABLE 10.9. Structure Effects of Weed-Control Cultivations on a Lateritic Soil

	Measurements on Undisturbed Soil Cores				Measurements on Disturbed Soil				
	Pore space (% of total soil volume)		Percolation (cm/hr)		Dry-sieving fractions (% of whole soil)		Water-stable aggregates greater than 0.5 mm (% of whole soil)		
	Total Pore Space	Pores Drained at 50 cm Water Tension	Under 1 cm Static Head	Artificial Rain at 15 cm/hr	Clods over 1.25 cm Diameter	Soil Passing 0.5 mm Diameter Sieve	After Vacuum Wetting	After Immersion Wetting	After Rain-Impact Wetting
W_0 clean weeded	63.7	19.9	24.9	11.2	31.3	44.2	41.2	24.8	27.9
W_1 slashed weeds	65.1	23.0	26.2	12.7	33.6	46.8	41.9	27.9	30.8
W_2 tall weeds	65.8	24.4	39.6	13.0	19.3	45.3	40.3	28.9	31.2
$LSD_{0.05}$	0.7	1.6	14.2	1.3	3.7	4.3	3.9	5.3	3.1

SOURCE: Pereira and Jones (1954).

TABLE 10.10. Effects of the Width of Strip Cultivation on Maize Grain Yield for Two Soils in Tanzania (kg/ha)

Depth of Cultivation (cm)	Width of Cultivation (cm) 2.5	10	20	30	45	Mean
	(a) An Alfisol at Kongwa					
5	526	745	1120	1075	1081	909
9	656	909	1247	1242	1231	1057
Mean	591	857	1184	1159	1156	
	(b) An Andisol at W. Kilimanjaro					
5	1516	1317	1581	1424	1419	1452
9	1475	1592	1568	1547	1343	1504
Mean						

SE	Width	Depth	CV (%)
(a) Kongwa	56	35	6.5
(b) W. Kilimanjaro	NS	NS	14

SOURCE: Northwood and Macartney (1971).

TABLE 10.11. Effects of Cultivation on Maize/Legume Yields at Morogoro

Treatment	Maize/Legume Yield (mt/ha) Zero-Tillage	Cultivated First Year	Cultivated) Every Year	$LSD_{0.05}$
Maize	2.14	2.71	2.98	0.21
Cowpea	0.47	0.59	—	0.04
Soybean	0.11	0.13	0.15	0.02

SOURCE: Huxley (1979).

yields of maize and soybean with mulch and zero-tillage treatments were equivalent or better than with plowing (Table 10.12). Soil moisture reserves were also high in mulch and no-tillage treatments throughout the growing season.

The potential of no-tillage farming in soil and water conservation and crop yield for arable land use in East Africa has been summarized by Dihenge (1979) and Huxley (1982). They concluded that the no-tillage method with crop residue mulch can be used to advantage for many soils and crops. In an economic appraisal Dihenge (1979) reported that the cost required to produce 1 kg of maize grain to be 1.25 times higher with plowing than with no-tillage. While evaluating the economic feasibility of no-tillage system, it is important to consider whether to improve the indigenous farming systems by incorporating

TABLE 10.12. Effect of Different Tillage Methods on Grain Yield of Maize and Soybean a Morogoro, Tanzania

Tillage treatment	Maize (kg/ha)	Soybean (kg/ha)
Conventional tillage	6370	1648
Conventional tillage + mulch	6740	2031
Strip tillage	5850	1386
Zero tillage	7270	1826
$LSD_{0.05}$	1238	524

SOURCE: Maseri and Jana (1979).

a no-tillage component into them or to introduce new systems. The results obtained to date in East and Central Africa (Morel and Quantin, 1972) indicate that, in the long run, mechanized-intensive cultivation of these ferrallitic soils result in deterioration of soil properties, and that the yields obtained by no-tillage and mulch systems are promising indeed.

TROPICAL AUSTRALIA

Some soils and environmental constraints to crop production in the Northern Territory and in the semiarid regions of northeastern Queensland, Australia, are similar to those of Alfisols and Ferruginous tropical soils of semiarid tropical West Africa (Bauer, 1983; Williams et al., 1983). A no-tillage system with residue mulch has been very encouraging on some Alfisols at Katherine, Northern Territory (McCowen et al., 1980a, 1980b, 1985). They concluded that if weed and pasture regrowth are adequately controlled by preplant application of glyphosate or other systemic herbicides, no-tillage maize and sorghum has generally outyielded the conventional plow-based method of seedbed preparation. Mulch in the interrow zone decreased the maximum soil temperature by 8−16°C beneath the mulch and by 4−5°C in the rows themselves. Soil moisture under the mulch was also more favorable than in plowed bare soil. Seedling establishment of both maize and sorghum was more than 40% greater with no tillage. Grain yield of maize was 5.6 mt/ha with plowing and 7.1 mt/ha with no-tillage. Stover yield of sorghum was 20% higher with no-tillage.

Reestablishment of pasture following no-tillage maize or sorghum production is not a serious problem. Peake et al. (1983) and Fisher and Phillips (1970) showed that reestablishment of *Alysicarpus vaginalis* (DC) Prodv. and *Stylosanthes humatas* and *S. guianensis* (Schum & Thonn.) was excellent after a maize crop treated with the herbicide atrazine.

Experiments conducted on some Alfisols and Vertisols in the drier regions of western Australia have shown structural stability of no-tillage soil better than that of plowed land (Hamblin, 1980). However, structural improvements with no-tillage in soils of the dry regions take place at a slower rate than in the wet tropics. In Fiji, Chandra (1977) demonstrated the merits of no-tillage and minimum tillage systems for sorghum and other grain crops in the semiarid regions.

Vertisols

Vertisols are developed in the semiarid tropics, and are extensively found in West Africa, India, the Caribbean, and tropical Australia. These soils contain high amounts of the swelling clays predominant in montmorillonitic clay minerals. These structurally "active" soils, under favorable conditions, exhibit self-mulching characteristics.

In the subtropical regions of Israel, Stibbe and Ariel (1970) compared sorghum grain yields under plowed and no-tillage methods of seedbed preparation and observed that yield response to tillage methods depended on rainfall amount and distribution. Yield in 1967 with sufficient and well-distributed rains was suppressed 40% by no-tillage. Yield in 1968 with low rainfall was 10–16% lower in a plowed than no-tillage system. These soils contain about 35% clay in the surface horizon. In a similar study on a Vertisol containing 64% clay, Stibbe and Ariel (1970) reported significantly lower cotton yield for no-tillage than plow tillage in 1969, a wet season, than in 1970 (Table 10.13). Increase in cotton yield during 1969 with subsoiling was due to a traffic-induced pan at 30 cm depth. In 1970, however, subsoiling produced lower yields than the no-tillage treatment. These authors concluded that if the plow pan does not exist and a soil moisture deficit is expected, no-tillage can be practiced on summer crops seeded directly through the winter wheat stubbles. In droughty years, no-tillage gave equal or better yields than plow tillage. In drought situations, economic evaluation of different tillage systems should be considered.

In Ghana, Mante (1979) investigated the influence of depth of plowing on physical properties of a Vertisol and sorghum grain yield (Table 10.14). Shallow cultivation of 7.5 cm depth gave low bulk density and favorable moisture content along with low power requirement. No differences in sorghum grain

TABLE 10.13. Yield Response to No-Tillage on Two Vertisols in Israel

Tillage Treatment	Depth of Tillage (cm)	Sorghum Yield (kg/ha)		Seed Cotton Yield (kg/ha)	
		1967	1968	1969	1970
Plowed	40	4386	2464	1630	1497
Plowed	25	4167	2303	1457	1430
Subsoil	40	3980	2711	1562	1181
No-tillage		2266	2728	1384	1235
Total rainfall (mm)		609	461	1088	619
$LSD_{0.05}$		1031	477	139	228

SOURCE: Stibbe and Ariel (1970).

TABLE 10.14. Effects of Different Depths of Cultivation of a Vertisol on Soil Properties and Sorghum Grain Yield in Ghana

Depth of Cultivation (cm)	Volumetric Moisture Content (%)		Soil Bulk Density (g/cm³)		Grain Yield (tonne/ha)
	With Mulch	Without Mulch	Cropped	Fallowed	
30	0.44a	0.46a	1.14	1.02	1.32
15	0.47b	0.43b	1.07	1.02	0.96
7.5	0.49b	0.47a	1.07	1.15	1.13
LSD (.05)	—	—	—	—	0.71

Figures followed by the same letter are statistically similar.
SOURCE: Mante (1979).

yield were evident. With self-mulching Vertisols, small advantages in grain yield from shallow cultivation may not be worth the extra effort.

That minimum cultivation of Vertisols can produce satisfactory yields is also supported by research in India. Srivastava and Singh (1972) found no significant differences in seed cotton yield between moldboard plowing plus two harrowings, cultivation to 15 cm depth, and strip tillage in the row zone only. These conclusions were also supported by the watershed management experiments conducted at the International Crops Research Institute for the Semi-Arid Tropics (ICRISAT), at Hyderabad, India.

THE CASE FOR RIDGES

Graded-ridge furrow system, once installed, can be used for many years provided the drainage channels are maintained and the weeds are adequately controlled (Kampen et al., 1981). In the savanna regions of Nigeria and elsewhere in semiarid tropical Africa, crops are traditionally grown on ridges, on small hillocks, or on mounds (Kowal and Stockinger, 1973). Ridge cropping is widely practiced on a range of soils (sandy to clayey), crops, rainfall regimes, and topography. Crops are grown on ridges or hillocks on shallow soils to increase the effective rooting volume; on poorly drained soils to grow upland crops; on nutrient-deficient soil to heap up the fertile ash-rich top soil; in manually powered farming to save labor by tilling only half of the land; on steep slopes to provide drainage channels up and down the slope for disposal of surplus water; and on sloping lands to conserve water and control erosion. On all soils, ridging facilitates easy harvesting of roots and tuberous crops.

This system of land preparation has evolved as an integral component of subsistence farming throughout tropical Africa and is well adapted for small-size, low-input, subsistence farming. However, the ridge–furrow system often breaks down on large-size, machine-powered, commercial farms. Ridges erode on structurally unstable soils, mechanical weed control is difficult, and crops suffer from high soil temperatures and low soil moisture, because ridges tend to dry rapidly, and crops often lodge. If rainfall exceeds the capacity of the ridge–furrow system to store surplus water, gully erosion is often severe on sloping ridged land.

Tie-ridging and ridging with the addition of cross ties in the furrows is an improvement over the traditional ridge–furrow system to hold surplus water and allow more time for infiltration into the soil. Tie-ridging is similar to "basin-listing" in the semiarid regions of southeastern United States. The tie-ridge system was initially developed in semiarid regions of Tanzania, where it proved an effective water-conserving technique. Prentice (1946) compared yields of cotton, sorghum, and maize on tied-ridged and flat seedbeds from 1939 to 1946. With the exception of cotton in 1942, yields were more from tie-ridge than from flat seedbeds. Yield benefits in the tie-ridged system were greater in years of partial drought.

It was the crop failure due to drought stress and soil degradation that led to the abandonment of the "Groundnut Scheme" on some 100,000 acres at Kongwa, in semiarid Tanzania (Wood, 1950). Pereira et al. (1958) observed that neither broad beds nor tied-ridges provided satisfactory erosion control, although tied-ridges prevented gully erosion (Pereria et al., 1967). Macartney et al. (1971) at Kongwa reported no beneficial effect of tie-ridged system on maize grain yield. On a similar soil in India, Ali and Prasad (1974) observed no effect of ridging on moisture conservation and on grain yield of pearl millet.

This apparent contradiction regarding the beneficial effects of tie-ridge systems can be resolved in terms of soil properties. It seems to be effective in water conservation on some soils but not others. The Dagg and Macartney (1968) data (Table 10.15) show that tie-ridged land produced significantly more maize grain yield only on a red soil (Alfisol) but not on a Vertisol or an Andisol. The ridge–furrow system reduced surface runoff and conserved more water in the root zone on structurally unstable red soil only. With soils of high waterholding capacity (Vertisol) or high permeability (Andisol), the beneficial effects of a tied-ridge system are negligible.

TABLE 10.15. Maize Grain Yield from a Mechanized Tie-Ridge System for Three Soils in Tanzania[a]

	Maize Grain Yield (kg/ha)			
Soil	Tie-ridged	Ridge	Flat	Means
Black soil (Vertisol)	3274	3251	3085	3204
Red soil (Alfisol)	3433	3029	2628	2930
Ash soil (Andisol)	1763	1113	1815	1543
Mean	2824	2465	2389	

[a] $LSD_{0.05}$ tillage, 600; soils, 347.

SOURCE: Dagg and Macartney (1968).

These conclusions are confirmed by Honisch (1974), with maize, sorghum, and millet on structurally inert soil in Zambia. On a structurally unstable soil with no water-stable aggregates, > 0.5 mm tied-ridges produced the highest grain yield of maize and sorghum. On inert soils in the Sahel region of Upper Volta, the tie-ridge system has also proved beneficial in moisture conservation (Table 10.16). In this study, the beneficial effects of tie-ridge systems were more pronounced in seasons of frequent droughts during the growth cycle and for shallow soils of low infiltration.

In addition to soil characteristics, the effectiveness of ridge planting depends on duration, intensity, and time of occurrence of drought stress. In the Eastern Province of Uganda, Walton (1962) observed that the ridge system had no beneficial effects on yield of an early planting on high-yielding productive soils where severe drought stress does not occur during the critical flowering stage. On the contrary, ridge planting can be extremely beneficial on marginal lands of low-yielding potential where the frequency of drought during critical stages is high.

It is apparent that ridge planting is a risk-avoidance system for small subsistence agriculture. It is effective on marginal lands of low-yield capacity, for structurally inert soils of low infiltration rate, and for regions prone to frequent drought stress during critical stages of growth. It is not an effective system for commercial agriculture or for high-yielding productive soils of good infiltration rate and favorable soil structure. Ridge planting or mechanical tillage provides a transient beneficial effect by improving infiltration. For most soils flat cultivation methods with mulch in the interrow zone and zonal tillage in the seed zone to alleviate soil compaction are preferable. A combination of ridge and no-tillage systems also can be used because ridges made once can be used for many cropping seasons (IITA, 1981).

TABLE 10.16. Maize Grain Yield in a Tie-Ridge System in Upper Volta[a]

	Maize Grain Yield (kg/ha)		
	Management level		
Ridge System	Low	High	Mean
1. No earthing-up	1040	1480	1260
2. Earthing-up at 30 days after planting (DAP)	990	1470	1230
3. Earthing up at 30 DAP and tying the ridges every other furrow	1840	2540	2190
4. Earthing up at 30 DAP and tying-all ridges	2040	3280	2660
Mean	1480	2190	1840

[a] $LSD_{0.05}$, management, 701; ridges, 416.

SOURCE: IITA (1981).

NO-TILLAGE SYSTEMS FOR UPLAND SOILS

Structurally active soils in the humid and subhumid tropics respond favorably to no-tillage provided they are not initially compacted and have adequate crop residue mulch. No-tillage farming is not directly applicable to compacted and eroded soils, or to those that are structurally inert. In these situations special soil and crop management systems are needed to achieve the benefits of no-tillage. The principles of the no-tillage system are scale-neutral. The technique can be adapted both for small-size subsistence farmers and large-size commercial agriculture. The special requirements for a no-tillage system to be effective are:

1. Adequate quantity of crop residue mulch.
2. Good initial soil structure with a predominance of interconnected macropores.
3. Bioactive soils with an abundance of soil fauna.
4. Good internal drainage.

If the no-tillage system is to be used for soils that do not meet these requirements, special techniques are needed to achieve these favorable conditions. A wide range of agronomic systems has been developed to improve soil structure and provide a renewable source of crop residue mulch. Rather than imposing a no-tillage system on structurally degraded and compacted soil, steps should be taken to restore its structure so that it can respond to no-tillage.

Structural Restoration of Compacted Lands

Improvements in soil organic matter usually result in structural amelioration of a degraded soil. Fallowing is an effective system of increasing soil organic matter, bringing about improvements in soil structure and in alleviation of soil compaction. Regeneration with natural vegetation cover is slow and inefficient and generally requires many years for complete recovery. Improved fallow with specially planted grass or legume species or woody perennials is more efficient in restoring the soil structure. Addition of manures and organic matter may be required to establish good grass or a legume cover (Pereira and Beckley, 1952).

The beneficial effects of grass and leguminous covers are well recognized for semiarid regions of Kenya and elsewhere in East Africa. Pereira et al. (1954) and Pereria (1956) investigated various lengths of fallowing with grass and legume mixtures on water-stable aggregates, total porosity, infiltration and percolation rates, and yield of arable crops. Their data confirmed the important soil improvements achieved during a fallow period under grass, and also that structural improvements are rapidly lost during the arable cultivation. Soil improvements were more rapid with grass than legume covers. From their experiments on the abandoned land of the "Groundnut Scheme," Pereira et al. (1958) confirmed that fallowing with teff grass [*Eragrostis abyssinica* (Jacq.) Link.] improved rainfall infiltration and decreased runoff and erosion. Establishment of grass sods can be a problem on plowed bare soil and is generally better with no-tillage or by undergrowing through standing crops in a relay system than by seeding after mechanical tillage (Peers, 1962). Improvements in soil structure and fertility by fallowing with grass and legume covers were also demonstrated in Uganda (Jameson and Kerkham, 1960), Nigeria (Dennison, 1959; Vine, 1953), Central Africa (Jurion and Henry, 1969), and Malaysia (Weng et al., 1979).

In northern Nigeria, Wilkinson (1975) observed that infiltration rates increased during the fallow period. Similar to the findings of Pereira in East Africa, Wilkinson confirmed that the equilibrium infiltration rate increased with duration of the fallow period as by the following equation:

$$\text{Equilibrium infiltration rate (cm/hr)} = 0.76 + 1.982(\text{Fallow Period, yr})^{1/2}$$

However, most of the increase in infiltration rate was lost by the end of the first cropping season. For example, the infiltration rate of 5.84 cm/hr at the end of the first cropping season declined to 2.34 cm/hr at peak rainy season and to 0.97 cm/hr at the end of the cropping cycle. Similar data were obtained in western Nigeria by Wilkinson and Aina (1976), when high infiltration rate obtained under prolonged bush fallow declined rapidly with arable cropping.

Lal et al. (1979) reported significant improvement in infiltration rate of an eroded and compacted Alfisol in western Nigeria (Table 10.17). There were also differences in infiltration rate among various grass species. Substantial improvements in water infiltration are brought by earthworms and other soil fauna. Juo and Lal (1977) demonstrated that deep-rooted woody perennials, for example, *Cajanus cajan* (L.) Mills and *Leucaena leucocephala* (C. Lam) deWit, are more effective in improving infiltration and macrochannels than

TABLE 10.17. Effects of Cover Crops on Infiltration Characteristics With and Without Worm Activity (Wilson et al., 1982).

	Accumulative Infiltration (cm/3 hr)		Equilibrium Infiltration Rate (cm/3 hr)	
Cover Crop	With Worm Activity	Without Worm Activity	With Worm Activity	Without Worm Activity
Brachiaria	490	64	75	19
Centrosema	220	72	30	18
Pueraria	270	76	90	16
Stylosanthes	390	74	60	16

SOURCE: Wilson et al. (1982).

shallow-rooted annuals. Similar positive effects of L. *glauca* were observed in East African soils by Pereira et al. (1954). Mehanni (1974) reported from Australia that fallowing with deep-rooted plants had the same effect on structure as deep ripping. The choice of an appropriate fallow crop depends on many factors, and may vary among soils and ecological regions. Economic considerations are also important, because farmers do not want to spend time growing an uneconomic cover crop if climate permits the growth of a food crop (Vine, 1953).

Management of Planted Fallows

Pereira et al. (1954) and Wilkinson (1975) showed that the benefits of fallowing are lost by the end of first cropping when primary and secondary tillage operations are used. Many studies have shown structural properties are at the optimum when the soil is protected by vegetation cover and any disturbance alters them adversely. It is therefore important that improvements in soil structure and fertility brought about by fallowing be maintained by the tillage system which follows. Many experiments conducted at the International Institute of Tropical Agriculture (IITA) (Lal, 1983; Wijewardene, 1980) and elsewhere in the tropics (Ogborn, 1982) have demonstrated that deterioration of soil structure and infiltration rate during the cropping phase can be drastically reduced by managing the fallow cover through a no-tillage system. Cover crops and pastures are a source of producing *in situ* mulch for no-tillage farming (McCowen et al., 1980; Wilson and Akapa, 1983).

The adoption of no-tillage on land under fallow can be done by one of the following commonly used practices.

Killed Sod as Residue Mulch

Contact and systemic herbicides have facilitated the suppression of luxurious vegetative growth. The destruction of pastures by herbicides or mechanical means is a good substitute for plowing (Hood et al., 1963). In Ghana, Kannegieter (1967, 1969) successfully demonstrated the use of a herbicide for growing maize through a killed *Pueraria* sod. The yield of maize grown through an undisturbed *in situ Pueraria* mulch was significantly more than when plowed under or used as mulch after plowing. Wilson (1979) discussed the use of legume covers for no-tillage production of food grains and recommended the inclusion of cover crops in the rotation every 3 or 4 years. In addition to soil physical properties, cover crops suppress weed growth and control nematodes.

Lal et al. (1979) and Wilson et al. (1982) compared grain yield of maize seeded through a range of suppressed grass and legume covers (Table 10.18). With chemical spray in 1976, high grain yields were obtained following *Psophocarpus*, *Stylosanthes*, and *Pueraria*. Among the mowed treatments, the maximum yield was obtained with *Stylosanthes*. In 1977, the highest maize grain yield was obtained in *Stylosanthes* irrespective of the method of suppres-

TABLE 10.18. Effect of Cover Crop and Method of Supression on Grain Yield of Subsequent Maize Crops

	1976 (mt/ha)		1977 (mt/ha)	
Cover Crop	Sprayed	Mowed	Sprayed	Mowed
Grasses				
Brachiaria	0.3	0.1	2.0	0.3
Paspalum	0.1	0.0	0.2	0.0
Cynodon	0.1	0.0	0.0	0.0
Legumes				
Pueraria	3.2	2.0	1.9	0.3
Stylosanthes	3.1	3.0	2.8	2.0
Stizolobium	2.1	1.0	0.9	0.8
Psophocarpus	3.6	2.2	2.2	1.0
Centrosema	2.2	1.8	1.8	0.9
Control	1.8	0.8	1.3	0.3
$LSD_{0.05}$				
Cover crop	0.7		0.8	
Suppression	0.3		0.5	

SOURCE: Wilson et al. (1982).

sion. Although improvements in soil structure and infiltration rate were more in grass than legume covers, low maize grain yield in grass covers was attributed to difficulties in effective suppression.

The beneficial effects of grass and legume mulch on crop yield have been demonstrated for an Ultisol in the Upper Amazon Basin in Peru by Wade and Sanchez (1983). Their data show that incorporation of grass and legume mulch may, for some crops, be more beneficial than leaving it on the surface. McCowen et al. (1985) found that maize and sorghum yields on chemically suppressed legume pastures without tillage on Alfisols in Northern Territory, Australia were superior to those on plowed soil.

Live Mulch

A live mulch system is based on the principles of mixed cropping (Lal, 1975). A low-growing crop, preferably a legume, is grown specifically as a cover crop. The objective is to establish a fast-growing, easily controlled perennial legume that can readily smother other weeds while aggressively covering the entire ground surface. These legumes are used as a perennial cover even during the cropping phase so that soil structure, infiltration rate, and chemical fertility are preserved. A small strip is opened, with or without using herbicides, to seed a seasonal food crop through it. The system works if the fallow crop is not competitive for light, moisture, and nutrients. Table 10.18 describes such a system. Mechanically mowed *Stylosanthes* was effectively controlled, but other grasses and legumes were not and drastically suppressed maize grain yield.

Some of the climber legumes, *Psophocarpus* and *Centrosema*, suffocated young maize.

Akobundu (1980, 1982) compared maize grain yield using three legume covers as live mulch with that obtained with no-tillage and plowing. With effective weed control and proper suppression of live mulch, grain yields were equivalent to those for no-tillage and a plowed system.

The advantage of live mulch lies in providing a continuous ground cover and in maintaining a favorable soil structure. Lal et al. (1978b) and Akobundu (1980) reported significantly more earthworm activity under a live mulch than in a no-tillage system using the previous crop residue as a mulch. Although the live mulch system of planted fallow crops is dependent on the availability of appropriate herbicides, this system has potential in both tropical and temperate environments (Miller and Bell, 1982). Its success depends to a great extent on the choice of appropriate legume covers that can be used as live mulch with a minimal dependence on herbicides.

Agroforestry and Alley Cropping

While the live mulch system is based on retaining a low-growing legume and growing a grain crop through it, agroforestry, the practice of growing deep-rooted perennial woody shrubs in association with seasonal food crops, is essentially "mixed cropping." This system of orderly mixed cropping with selected perennial shrubs and legume trees is an attempt to minimize the environmental degradation of intensive arable land use and to preserve and enhance soil structure and chemical fertility (Kang et al., 1981; Juo and Lal, 1977; Mongi and Huxley, 1979; Pereira et al., 1954; Vergara, 1982). In agroforestry, arable crops are grown between rows of specially planted woody shrubs or trees, which are regularly pruned during the cropping season to prevent shading, to provide mulch, and to reduce water use. Many researchers have shown that food and crops interplanted with nitrogen-fixing legume trees are generally more productive than a monoculture and cause less soil degradation. Either an alternate row or alternate strip system of tree arrangements can be used, and the trees are pruned at an appropriate height to promote rapid regeneration.

Commonly used legume species grown in association with food crops include *Sesbania*, *Gliricidia*, *Leucaena*, *Parkia*, *Ramon*, *Calliandra*, *Albizia*, *Acacia*, *Cassia*, *Pithecellobium*, *Mimosa*, *Prosopis*, and *Samanea*. The choice of an appropriate specie depends on soil, climate, and seasonal crops to be grown. The desirable species should have rapid growth, nitrogen-fixing capability, and a multipurpose nature (Vergara, 1982). In addition, it is important to consider its compatibility with food crops since its root system should not be competitive.

The practice of using prunings as a mulch for no-tillage food crop production between the rows of trees retains the benefits of no-tillage and of tree fallow to soil structure and mineral fertility during the cropping phase. Attempts should be made to balance agroforestry with the no-tillage to attain the maximum environmental benefit of this integrated land use system.

Crop Residue Management and Agronomic Packages

The benefits of no-tillage agriculture are partly due to the physical, chemical, and biological effects of mulch and partly to the maintenance of pore channels created by decaying roots and burrowing soil fauna. Cropping systems should therefore be developed to ensure an adequate quantity of surface mulch for the no-tillage system to be effective. Other uses (forage, fuel, fencing) must not be competitive. A high-residue-producing crop must be grown once during every cropping phase.

Agronomic and cultural practices required to seed through crop residue without plowing are different from those required for a mechanically prepared seedbed. The success of no-tillage depends on integrating the whole system which differs between soils and crops and necessitates local adaptive research for specific agroecological environments. The ecological limits of no-tillage farming then can be extended by development of appropriate packages of agronomic practices.

No-tillage farming must fit the overall framework of the soil's constraints and its potential, crop management requirements, and the socioeconomic constraints and aspirations of the community. A no-tillage system with a mulch of cover crops (dead or alive), woody perennials, or previous crop residue provides the framework for an economically viable and socially acceptable farm system. Other components will include fertilizer rates and modes of application, seeding methods, pest control, and rotations.

SOIL SUITABILITY FOR NO-TILLAGE

The review presented above indicates that a no-tillage system for upland crops is soil and crop specific, and is more applicable for some environments than for others. Lal (1983) developed a numerical rating system to assess soil's suitability for no-tillage systems. This rating method, tentative and preliminary as it is, is based on a few contrasting sites of experimentation and considers properties such as erosivity, erodability, soil loss tolerance, rooting depth, available water holding capacity, cation exchange capacity, pH, and others.

Figures 10.5 and 10.6 relate tillage and seedbed requirements to soil properties and constraints. Special attention is given to germination and seedling establishment, runoff and erosion, and soil structure. It is important to note that soils with similar physical characteristics may respond differently to tillage methods depending on the prevailing soil-moisture regime. Soils with similar moisture regimes may require different tillage because of variations in their physical properties. Friable, coarse-textured, self-mulching, and structurally active soils are likely to respond better to no-tillage or reduced tillage than soils with massive structure or which are easily compactable.

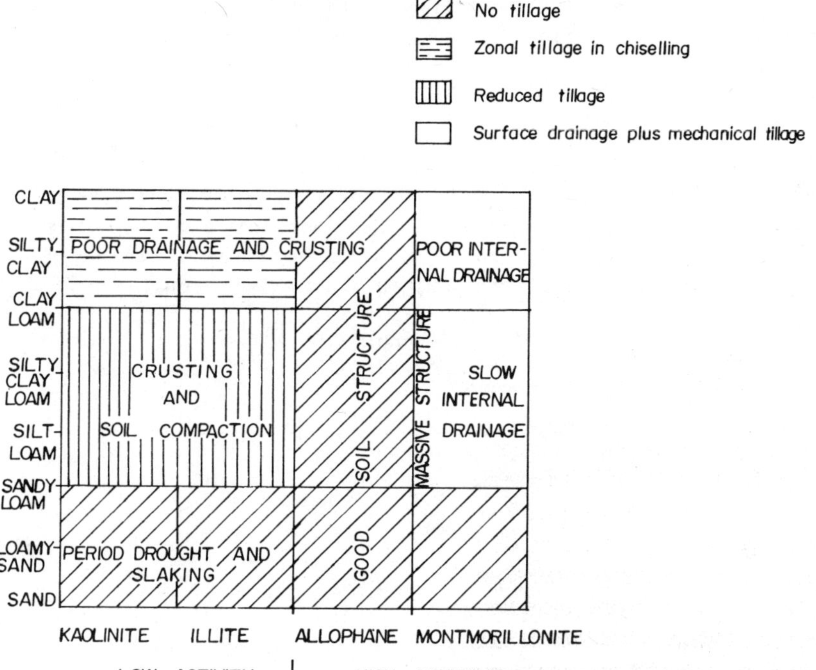

FIGURE 10.5. Soil suitability for different tillage systems in relation to texture and mineralogical composition.

No-tillage does not mean merely elimination of mechanical tillage. It is but one component of the overall cropping system and embodies specific requirements for mulch management, seeding equipment, weed control, crop rotation sequence, insect and pathogen control, methods and rates of application of soil amendments, seeding rate, and plant establishment. All these requirements vary with different soils, crops, and agroecologies. The ecological limits of a no-tillage system can even be extended by developing an improved package of cultural practices specifically to alleviate soil and climatic constraints. This requires location-specific research and should be extrapolated to another region with caution.

TILLAGE SYSTEMS FOR RICE

Rice, being a semiaquatic plant, grows better in hydromorphic than in upland environments. Soil and environmental conditions in lowland rice culture are therefore different than those of uplands. For example, soil erosion is not a constraint in leveled and bunded paddies. Drought stress, as it occurs in upland culture, is not a severe constraint, yet water management is necessary for successful rice culture. Aquatic weeds are different from upland weeds.

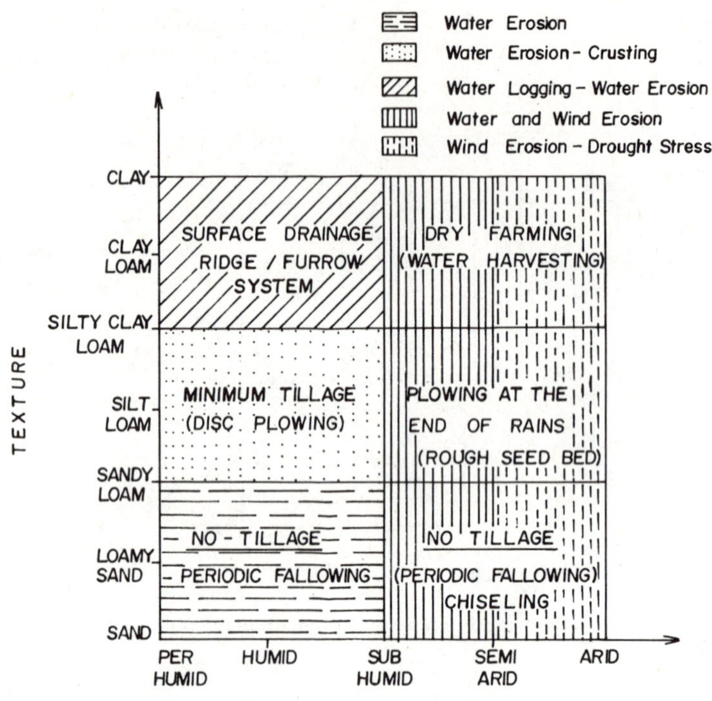

FIGURE 10.6. Soil suitability for different tillage systems in the tropics in relation to soil texture and moisture regimes.

The objectives of seedbed preparation for lowland rice production are to (1) minimize seepage losses and percolation rate, (2) control weeds, (3) reduce labor and turn around time to facilitate multiple cropping, and (4) preserve soil structure to enable satisfactory production of an upland crop following rice.

The traditional method of seedbed preparation for rice consists of dry and wet plowing. The latter, "puddling," decreases seepage losses and retains water, controls weeds, and facilitates seedling transplant. Puddling, plowing the soil when at a saturation point, is a process of destroying the soil structure (Koenigs, 1961, 1963). The macropores (transmission pores) are virtually eliminated, and high proportion of micropores (retention pores) facilitates more water retention in the puddled soil. Sanchez (1973a; 1973b) observed that drainage rates of a montmorillonitic clay were about a 1000 times slower in a puddled than in a granulated soil. The puddled layer itself, and not the creation of a "plow pan," decreases the percolation rate in a paddy field. Puddling lowers the risk of drought stress in rainfed rice culture. By curtailing percolation losses, puddling drastically reduces the rate of crop-water use (De Datta et al., 1973).

Despite its merits, puddling is laborious, requires high energy inputs, the preparation time between harvest and planting the next crop is long, considerable water is required for wet plowing, and the destruction of soil structure adversely affects productivity of the following upland crop.

No-Tillage Farming of Rice

In Asia, the tillage requirements for paddy culture have been investigated for more than three decades (Allen and Haynes, 1953; Ashby 1949; Bhan and Tilak, 1964; Coleman, 1949; Lovely, 1961). During the last two decades, the availability of herbicides to control weeds economically has facilitated the reduction of tillage operations without adversely affecting crop yields. Many experiments conducted on soils of heavy texture have shown that satisfactory yields of paddy rice are obtainable provided weeds are controlled. In the Philippines, Mabbayad (1967) and Mabbayad et al. (1969) reported that yields from plots without tillage were as good as those from plowed and harrowed ones. Test data of minimal tillage of transplanted rice (Mabbayad and Buenacosa, 1967) show no significant difference in rice grain yield between untilled and conventionally tilled seedbeds.

In another study, Mabbayad et al. (1969) investigated the effects of different rates of herbicide application and tillage methods on rice grain yield and observed satisfactory yields on untilled plots where weeds were efficiently controlled. Moomaw et al. (1968) and De Datta (1974) also concluded that the minimum tillage techniques using paraquat and other chemicals successfully reduce tillage operations, and even make no-tillage feasible for transplanted rice. Similar investigations in Sri Lanka have shown satisfactory rice grain yield from untilled paddy. Mittra and Pieris (1968) evaluated the effects of two rates of N on rice grain yield for transplanted and direct seeded rice for no-tillage and conventional puddling methods of seedbed preparation. The water requirement of conventionally puddled plots was 2.4 times more than that of untilled paddy. For the third consecutive crop, a maximum grain yield of 4800 kg/ha was obtained in a transplanted no-tillage plot and the minimum 3578 kg/ha for a plowed treatment (Table 10.19). Promising results of rice yield with minimum and no-tillage techniques were also reported from Sri Lanka by Seth et al. (1971). Wang (1975) investigated the effects of soil pulverization on rice in Taiwan and concluded that plowing beyond 15 cm depth had no beneficial effect on yield.

The effects of puddling on rice grain yield depend on soil properties and the prevailing climate. From his experiments on a montmorillonitic heavy-textured Typic Dystrandept in the Philippines, Sanchez (1973a, 1973b) observed no significant differences in grain or straw yield between a puddled and a granulated soil, and concluded that the advantages of puddling, if any,

TABLE 10.19. Effects of Zero-Tillage and Conventional Plowing on Grain Yield for Direct Seeded and Transplanted Rice for Two Levels of N Application in Wet Zone of Sri Lanka

Tillage	Seeding Method	N Level (kg/ha)	Rice Grain Yield (kg/ha)		
			First Crop	Second Crop	Third Crop
Zero tillage	Transplanted	44	2937ab	3475a	4522ab
Zero tillage	Transplanted	66	—	3623a	4798a
Conventional tillage	Transplanted	44	2794ab	2788a	3578d
Conventional tillage	Transplanted	66	—	3382a	2902cd
Zero tillage	Direct seeded	44	2424b	2821a	4063bcd
Zero tillage	Direct seeded	66	—	3507a	4347abc
Conventional tillage	Direct seeded	44	3289a	2944a	3901d
Conventional tillage	Direct seeded	66	—	3495a	4415abc

Figures followed by same letters are statistically similar.

SOURCE: Mittra and Pieris (1968).

are related to decreasing water losses and not to increasing nutrient supply. P uptake by rice seedlings was similar between the two tillage treatments (Sanchez and Briones, 1973). Conversely, many earlier investigations (De Datta et al., 1966; Koenigs 1961, 1963) supported the belief that puddling increased P availability. Beneficial effects, if any, of puddling on nutrient availability depend on soil properties.

Effects of plowing on nitrogen economy and water use efficiency of rice paddy were investigated by De Datta and Kerim (1974). Contrary to the data of Mitra and Pieris (1968) and of Sanchez (1973a, 1973b), water economy of rice was found to be 2.5 times higher in puddled than in the granulated soil. Because of the larger losses of water, nitrogen losses were also greater in granulated soil. Consequently, the plants grown in granulated soil had less nitrogen and lower grain yield.

In Malaysia, Brown and Quantrill (1973), investigating the effects of conventional puddling and minimum tillage for three rates of nitrogen and for three contrasting soils, observed no significant differences in grain yield between two tillage methods. In Surinam, Scheltema (1974) observed that differences in rice grain yield among tillage treatments were smaller at 180 kg than at 120 kg urea per ha. With dry/wet tillage there is a risk of excessive vegetative growth at the expense of the grain yield. He concluded that no-tillage in flooded rice is practicable but weed control is more troublesome than with plowing.

Tillage experiments conducted in western Nigeria also indicate the effects of soil properties on rice response to tillage methods. For soils of heavy texture, no-tillage can be adopted successfully for both seeded and transplanted rice (Maurya and Lal, 1979b; Rodriguez and Lal, 1979b). Table 10.20 shows no significant effect of planting techniques or of tillage methods on rice grain yield for a heavy-textured soil. No-tillage system is therefore equally effective on clayey soils. For coarse-textured sandy soils, however, leaching losses of added inorganic fertilizers can be substantial in an unpuddled and flooded paddy. Nutrient imbalances and toxicity due to anaerobic conditions may also be high for sandy soils owing to their low buffering capacity. The effects of anaerobic conditions can be made even more severe by the decomposition of crop residues under flooded conditions, and this can adversely affect both seedling growth and crop establishment. Thus, rice yield with no-tillage can be less than with conventional tillage involving both dry and wet plowing.

For highly permeable sandy soils, losses of water by percolation and of nitrogen by leaching can be drastically curtailed by compacting the soil. Grain yields have been satisfactory from a compacted compared with no-tillage or conventionally puddled paddy. Transplanting in compacted and unpuddled soil, however, is difficult for manual operations.

An analysis of the research indicates that for clayey soils with less leaching losses of water and plant nutrients, no-tillage or minimal tillage system produces satisfactory rice grain yields. Effective weed control for the no-tillage system, however, is more troublesome. For sandy or highly granulated soils,

TABLE 10.20. Rice Grain Yield on a Heavy Textured Soil as Affected by Tillage, Planting Methods and Fertilizer[a]

Planting Method		Rice Grain Yield (mt/ha)				
		N_1P_1	N_1P_2	N_2P_1	N_2P_2	Mean
Transplanting	No-tillage	6.1	6.1	5.6	5.1	5.7
Transplanting	Conventional tillage	6.5	5.8	5.6	4.8	5.7
Direct sowing	No-tillage	6.0	6.7	5.9	5.4	6.0
Direct sowing	Conventional tillage	5.7	5.5	6.1	5.0	5.6
Mean yield	1. No-tillage	5.8				
	2. Conventional tillage	5.6				
	3. Transplanted	5.7				
	4. Direct sowing	5.8				

[a] $LSD_{0.05}$: Planting methods, no significance; tillage systems, no significance; fertilizer, 0.56. N_1 = 60 kg/ha N; N_2 = 120 kg/ha N; N_1 = 13 kg/ha P; P_2 = 26 kg/ha P.

SOURCE: Rodriguez and Lal (1979b).

low grain yields with the no-tillage system are partly due to high leaching losses of nitrogenous fertilizer. When at appropriate soil moisture, even compaction of a sandy soil can have beneficial effects on rice grain yield.

BIOTIC FACTORS AND PEST INCIDENCE

Crop residue retained on the soil surface drastically alters microclimate at the soil surface. Incidence of weeds, diseases, and insect infestations can be different when a mulch is present or absent (see Chapters 12 and 13). The source of infection varies as well as the altered soil temperature and moisture regimes, affecting propagation of insects and pathogens. Residue mulch also suppresses weed growth (Lal, 1975) (see Chapter 11).

In the tropics, earthworm activity is greater in no-tillage mulched than in plowed soil. Lal (1976b) measured four to five times the activity in no-tillage than plowed plots (Table 10.21). Lal and De Vleeschauwer (1982) reported that worm casts contained 1.5–2.3 times more organic matter; 1.2–1.8 times more nitrogen; 1.3–1.6 times more Bray-P; 2.1–3.2 times more exchangeable Ca^{2+}; 2.5–3.8 times more exchangeable Mg^{2+}; 2.2–3.1 times more exchangeable K^+; and 1.2–1.4 times more exchangeable Na than the top 10 cm of soil. Furthermore, the nutrient status of both casts and soil was better in no-tillage than in plowed treatments. Earthworm casts are extremely stable to raindrop impact, and possess a high water retention capacity (De Vleeschauwer and Lal, 1981; Lal and Akinremi, 1983). In a grassland site in Orissa, India, Dash and Patra (1979) reported that about 75 mt/ha/yr of worm casts were produced, and that casts contained 0.47% N compared with 0.35% in the surrounding soil.

TABLE 10.21. Effects of Tillage Treatments on Earthworm Activity Under Different Crop Rotations

	Casts/m^2/year		Equivalent Weight (mt/ha)	
Crop Sequence	No-Till	Plowed	No-Till	Plowed
Maize–maize	1060	90	41.3	3.5
Maize–cowpea	1220	372	47.6	14.5
Pigeon pea–maize	464	100	18.1	3.9
Soybean–soybean	42	3	1.6	0.1
Cowpea–cowpea	28	36	1.1	1.4
Mean	563	120	22.0	4.7

SOURCE: Lal (1976b).

Caveness (1979) reported that in western Nigeria nematode populations [*Pratylenchus sefaensis* Fortuner, 1973 and *P. brachyurus* (Godfrey 1929) T. Goodey, 1951] were four and five times more numerous on maize in plowed than in no-tillage soils after 7 and 13 consecutive cropping seasons, respectively. At lower population densities, *Helicotylenchus pseudorobustus* (Steiner, 1914) Golden, 1956 and *Meloidogyne incorgnita* (Kofoid & White 1919) Chitwood 1949 juveniles were more numerous in no-tillage than in tilled soils.

Pests and their influence on crop production in no-tillage systems is a controversial issue, and the magnitude of effects varies among soils, crops, and prevailing climate (see Chapter 12). Rijn (1983) and Shenk and Saunders (1983a; 1983b) have reviewed the incidence of pests in relation to tillage methods for tropical environments. Shenk and Saunders (1983b) observed that on an Ultisol in Costa Rica insect pest damage was reduced and production increased in no-tillage systems as compared to plowed fields. *Spodoptera frugiperda* (J. E. Smith) and *Diabrotica balteata* Lec. colonization incidence was significantly greater in plowed fields. Shenk and Suanders also observed that the soil inhabiting pests reduced plant population and vigor more in plowed than in no-tillage plots.

ARE HERBICIDES INDISPENSABLE?

Small landholders in the tropics generally practice no-tillage and reduced tillage without benefit of herbicides. Weed control is achieved with manual systems of slash, burn, and hoeing. Attempts at modernization are generally based on a system of mechanical soil manipulation. Mechanization is expensive, and in view of the harsh climatic environments, destructive, because it accelerates soil erosion and processes of soil degradation. Attempts to modernize tropical agriculture would be well served if the drudgery of manual weed control could be alleviated by the introduction of an herbicide-based technology while keeping intact the reduced-tillage component. The arguments generally put forth are: (1) herbicides are expensive, (2) chemicals are not easily available, (3) they are potentially hazardous, (4) they can pollute environments, and (5) the technology is too sophisticated for small landholders.

Many reports have indicated that while the use of herbicides is desirable for efficient and timely weed control, it is not indispensible. For example, crop residue mulch will drastically suppress weed growth. Weed growth in mulched plots was one-third of that in plowed bare soil. Tables 10.22 and 10.23 show that timely manual weeding is as effective as herbicides. Drudgery can be greatly reduced if manual weeding is done with improved tools and at a stage when weed growth is not excessive. Herbicide use can also be decreased by appropriate crop rotations, and aggressive cover crops to smother other weeds

TABLE 10.22. Effects of Methods of Weed Control and Seedbed Preparation on Cowpea Yield

Tillage Treatment	Weed Control	Cowpea Grain Yield (kg/ha)
1. No-tillage	Manual slashing three days before seeding	1323ab
2. No-tillage	Manual slashing immediately before seeding	1470ab
3. No-tillage	Paraquat	1188ab
4. Conventional plowing	With residue mulch	1161ab
5. Conventional plowing	—	1021bc
6. No-tillage	Weeds cut and removed	620c

Figures followed by same letters are statistically similar.
SOURCE: Kamara (1980).

TABLE 10.23. Effects of Tillage Methods and Weed Control on Upland Rice

Tillage Method	Weed Control	Rice Grain Yield (kg/ha)
Conventional plowing	Manual	1524
Conventional plowing	Stam F-34 T	1425
No-tillage	Manual	1433
No-tillage	Stam F-34T	1560
	No-till mean	1499a
	Plow-tillage mean	1497a
	Manual weeding mean	1479a
	Chemical weed control	1517a

Figures followed by same letters are statistically similar.
SOURCE: Nyoka (1980).

but, in turn, that can be adequately controlled by mechanical rather than chemical means. For example, the data in Table 10.18, show that the maize grain yield following mechanically mowed *Stylosanthes* was equal to that chemically sprayed. There is a need for selection of suitable cover crops for different agroecologies of the tropics. *Stizolobium aterrimum* Piper & Tracy (Syn. *Mucana pruriens* DC. var. Utilis Wall) dies at the end of a long dry season and grain crops can be seeded through its mulch without a chemical spray (Wilson, 1979).

SUMMARY

The no-tillage system of crop production is rapidly gaining popularity in the tropics and has been widely researched for 15 years. A wide range of crops can be grown successfully without tillage on structurally active Alfisols, Ultisols, Oxisols, Andisols, and Inceptisols. Some form of reduced or zonal tillage is also feasible on self-mulching Vertisols. However, structurally inert Alfisols and Ferruginous soils in the semiarid regions of West Africa respond more favorably to deep tillage. Where applicable, no-tillage with a mulch has the advantages of improved soil and water conservation, soil moisture and temperature regimes, soil organic matter, and stable crop yields.

No-tillage or reduced-tillage systems are not easily adapted to soils with compacted surface horizons, uneven ground surface, or inadequate mulch. Amelioration of soil physical characteristics by fallowing with grass, legume, or woody perennials prior to adopting the no-tillage system is advisable for degraded and compacted soils. Alternates for managing a planted fallow for no-tillage farming include a grain crop in a chemically killed sod, live mulch, or agroforestry.

The tie-ridge system is an effective alternative method of improving water infiltration only on structurally inert soils.

No-tillage farming can be exended to a wide range of soils and crops by choosing or adjusting agronomic systems of fertilizer application, crop rotations and crop combinations, cover crops, pest control, and seeding equipment to fit the site and situation. Research is needed to develop numerical rating methods for assessing soils suitability for reduced tillage on the basis of soil properties, potential erosion risks, and resource optimization.

On clayey soils of low permeability, the no-tillage system gives satisfactory yield of paddy rice provided weeds can be controlled effectively. Unpuddled granulated or coarse-textured soils are less well suited. The turnaround time for double cropping is reduced with the no-tillage system. Most upland crops grow better on well-structured soils than those whose structure has been destroyed by dry and wet tillage. Coarse-textured sandy soils respond more favorably to compaction than to puddling.

Reduced tillage in one form or another is widely practiced by subsistence farmers in the tropics without herbicides. Economically viable and sustained crop yields are obtainable from a range of tropical soils with no-tillage farming. If soils are susceptible to erosion and easily degraded, a no-tillage system may be the only choice to manage them effectively. Technology should be developed to extend the ecological limits of its use.

LITERATURE CITED

Abeyratne, E.L.F. 1956. Dryland farming in Ceylon. *Trop. Agric. (Ceylone)* **112**(3):191–229.

Abu-Zeid, M. O. 1973. Continuous cropping in areas of shifting cultivation in southern Sudan. *Trop. Agr.* **50**:285–290.

Acquaye, D. K. and R. K. Cunningham. 1965. Losses of N by ammonia volatilization from surface fertilized tropical soils. *Trop. Agric.* **42**:281–292.

Agboola, A. A. 1981. The effects of different soil tillage and management practices on the physical and chemical properties of soil and maize yield in a rain forest zone of western Nigeria. *Agron. J.* **73**:247–251.

Aina, P. O. 1979a. Soil changes resulting from longterm management practices in western Nigeria. *Soil Sci. Soc. Am. J.* **43**:173–177.

Aina, P. O. 1979b. Tillage, seedbed configuration and mulching: a preliminary report on effects on soil and crops. *Ife. J. Agric.* **1**:26–35.

Aina, P. O. 1982. Soil and crop response to tillage and seedbed configuration in southwestern Nigeria. Proc. 9th ISTRO Conf., Osijek, Yugoslavia, pp. 72–78.

Ajunwon, S. O., B. A. Olunuga, and R. Lal. 1978. Applicability of no-tillage techniques in the forest and savannah regions of western Nigeria. *Nigerian J. Sci.* **12**:247–286.

Akobundu, I. O. 1980. Live mulch: a new approach to weed control and crop production in the tropics. Proc. 1980 British Crop Protection Conf., pp. 377–380.

Akobundu, I. O. 1982. Live mulch crop production in the tropics. *World Crops*:125–126; 144–145.

Akobundu, I. O. 1983. Weed control in no-tillage cassava in the subhumid and humid tropics. *In* I. O. Akobundu and A. E. Deutsch (eds.) *No-tillage Crop Production in the Tropics*. IPPC Document 46-B-83, Oregon State Univ., Corvallis, OR, pp. 119–126.

Ali, M. and R. Prasad. 1974. Effects of mulches and type of seedbed on pearl millet under semi-arid conditions. *Experimental Agriculture* **10**:263–272.

Allen, E. F. and D. W. F. Haynes. 1953. A review of investigations into the mechanical cultivation and harvesting of wet padi with special reference to the latter. *Malaya Agric. J.* **36**:61.

Armon, M. N., R. Lal, and M. Obi. 1981. Effects of tillage systems on properties of and Alfisol in southwest Nigeria. *Ife J. Agric.* **3**:1–13.

Ashby, H. K. 1949. Dry padi mechanical cultivation experiments, Kalantan, season 1950–51. *Malaya Agric. J.* **32**:32–41.

Ashrif, M. I. and I. Thornton. 1965. Effects of grass mulch on groundnuts in the Gambia. *Expl. Agri.* **1**:145–152.

Bababe, B. 1977. Tillage practices in relation to erosion in the savanna soils of Nigeria. Dept. of Soil Sci., B. Sc. Diss. A. B. U., Zaria, Nigeria.

Baffoe-Bonnie, E. and C. Quansah. 1975. The effect of tillage on soil and water loss. Ghana J. Agric. Sci. **8**:191–195.

Bates, L. 1975. Soil husbandry. *Tea in East Africa* **15**:18–26.

Bauer, F. A. 1983. The northwestern environment. Conf. Proc. "Agro-Research for Australian Semi-Arid Tropics" 21–25 March, 1983, Darwin, Australia.

Bhan, V. M. and K.B.V.R. Tilak. 1964. Recent trend in tillage studies. *The Allahabad Farmer* **38**:45–48.

Bonsu, M. and H. B. Obeng. 1979. Effects of cultural practices on soil erosion and maize production in the semi-deciduous rain forest and forest-savanna transition zones of Ghana. In R. Lal and D. J. Greenland (eds.). *Soil Physical Properties and Crop Production In The Tropics*. Wiley, U.K., pp. 509–519.

Borole, E. D., R. M. Patel, and J. R. Patel. 1972. The longterm effect of ploughing treatments under no-manure practice—a statistical assessment. *B. A. College of Agriculture, Magazine* **24**:67–76.

Brazil. 1980. Fundacão Instituto Agronomico Do Parana, Direct drilling in Parana State. IAPAR.

Brown, I. A. and R. A. Quantrill. 1973. The role of minimal tillage in rice in Asia, with particular reference to Japan. *Outlook Agric.* **7**:179–183.

Buanec, B. Le. 1974. Observations on soil profile loosening in ferrallitic soils: Effects on soil characteristics and growth of annual crops. *Agronomie Tropicale* **29**:1079–1099.

Caveness, F. E., Nematode populations under a no-tillage soil management regime. In R. Lal (ed.). *Soil Tillage and Crop Production*. IITA Proc. Series 21, IITA, Ibadan, Nigeria, pp. 133–145.

Chandra, S. 1977. Minimal tillage practices: Possible applications in Fiji agriculture. *Fiji Agricultural Journal* **39**:39–46.

Charreau, C. 1972. Problèmes posés par l' utilisation agricole des sols tropicaux par des cultures annuelles. *L' Agr. Trop.* **27**:905–929. (1972).

Charreau, C. 1977. Some controversial technical aspects of farming systems in semi-arid West Africa. In G. H. Cannell (ed.). *Proc. Int. Symp. on Rainfed Agric. in Semi-Arid Regions*. Univ. of California, Riverside, CA, pp. 313–360.

Charreau, C. and R. Nicou. 1971. L' amelioration du profil cultural daus les sols sableux et sablo argileux de la zone tropicale seche Ouest Africaine et ses incidences agronomiques. L' Agri. Trop. **26**:209–255; **26**:565–631; **26**:903–978; **26**:1184–1247.

Chaudhary, M. R. and S. S. Prihar. 1974. Root development and growth response of corn following mulching, cultivation, or interrow compaction. *Agron. J.* **66**:350–355.

Chopart, J. L. 1981. Le travail du sol au Senegal. Institut Senegalais De Recherches Agricoles (I.S.R.A.), Bambey, Senegal.

Chopart, J. L., J. M. Kalms, J. Marquette, and R. Nicour. 1981. Comparison de differentes techniques de travail du sol en trois ecologies de l' Agrique de l' Quest. Institut De Recherches Agronomiques Tropicales et al Des Cultures Vivriers, Montipellier, France.

Chopart, J. L. and R. Nicou. 1976. Influence du labour sur le developpement radiculaire de différentes plantes cultiveés au Sénégal—Conséquences sur leur alimentation hydrique. *L' Agr. Torp.* **31**:7–28.

Coleman, P. G. 1949. Investigations into the mechanical cultivation of wet padi in Wellesley Province during the season 1948–49. *Malaya Afric. J.* **32**:185.

Couper, D. C., R. Lal, and S. L. Classen. 1979. Mechanized no-till maize production on an Alfisol in tropical Africa. In R. Lal (ed.). *Soil Tillage and Crop Production*, IITA Proc. Series 2, IITA, Ibadan, Nigeria, pp. 147–160.

Dagg, M. and J. C. Macartney. 1968. The agronomic efficiency of N.I.A.E. mechanized tie ridge system of cultivation. *Expl. Agric.* **4**:279–294.

Dash, M. C. and U. C. Patra. 1979. Worm cast production and nitrogen contribution to soil by a tropical earthworm population from a grassland site in Orissa, India. *Renue d' Ecologie et de Biologie du sol* **16**:79–83.

De Datta, S. K. 1974. Weed control in rice: present status and future challenge. *Phil. Weed Sci. Bull.* **1**:1–16.

De Datta, S. K. and Kerim, N.S.A.A.A. 1974. Water and nitrogen economy of rainfed rice as affected by soil puddling. *Soil Sci. Soc. Amer. Proc.* **38**:515–518.

De Datta, S. K., H. K. Krupp, E. I. Alvarez, and S. C. Modgal. 1973. Water management practices in flooded tropical rice. In *Water Management Philippine, Irrigiation Systems: Research and Operations*. IRRI, Los Banos. Philippines, pp. 1–18.

De Datta, S. K., J. C. Moomaw, V. V. Rocho, and G. V. Simsiman. 1966. Phosphorus supplying capacity of lowland rice soils. *Soil Sci. Soc. Amer. Proc.* **30**:613–617.

Dennison, E. B. 1959. The maintenance of soil fertility in the southern Guinea Savanna Zone of northern Nigeria. *Trop. Agric.* **36**:171–178.

De Vleeschauwer, D. and R. Lal. 1981. Properties of worm casts in some tropical soils. *Soil Sci.* **132**:175–181.

Diatta, S. 1974. Effect of cultivation on the plateau soils of continential Casamance. Results of two-year experiments. *Agronomie Tropicale* **30**:344–351.

Dihenge, H. 1979. Tanzania's experience in appropriate technology for tillage operations. Proc. "Appropriate Tillage Workshop", IAR, Zaria, Nigeria, Commonwealth Secretariat, pp. 117–125.

Dunham, R. J. 1979. Cultivation experiments with zero tillage Proc. "Appropriate Tillage Workshop" 16–20 Jan., 1979 IAR Zaria, Nigeria, pp. 59–61 (1979).

Dunham, R. J. 1982a. No-tillage crop production research at Samaru. Inst. for Agric. Research, Samaru, Zaria.

Dunham, R. J. 1982b. Soil management research in the Nigerian Savanna. Institute of Agric. Research, Sumaru Zaria.

Dunham, R. J. and A. J. Aremu. 1979. Soil conditions under conventional and zero tillage at Samaru. Soil Sci. Soc. Nigerian Conference, Kano.

Ezedinma, F. O. C. 1964. Effect of preparatory cultivation on the general performance and yield of cowpea. *Nigerian Agric. J.* **1**:21–25.

Ezumah, H. C. 1983. Agronomic considerations of no-till farming. In I. O. Akobundu and A. E. Deutsch, eds.). *No-tillage Crop Production in the Tropics*. IPPC Document 46-B-83, Oregon State Univ., Corvallis, OR, pp. 102–110.

Fisher, M. J. and L. J. Phillips. 1970. *Australian J. Exp. Agric. Animal Husbandry* **10**:755–762.

Fuggles-Couchman, H. 1934. Green manure and cultivation trials at Morogoro, Eastern Province, Tanganyike Territory. *East African Agric. J.* **5**:208–210.

Ghuman, B. S. and R. Lal. 1982. Temperature regime of a tropical soil in relation to surface conditions and air temperature and its fourier analysis. *Soil Sci.* **134**:133–140.

Ghuman, B. S. and R. Lal. 1984. Water percolation in a tropical Alfisol under conventional plowing and no-till systems of management. *Soil Till. Res.* **4**:263–276.

Gidnavar, V. S., S. K. Gumaste, and K. Krishnamurthy. 1972. Effect of tillage on soil aggregation under rainfed conditions. In "Proc. Seminar on Drought." *Research Series, Univ. of Agric. Sciences, Hebbal 14*:313–318.

Gordon, J. 1968. A minimum cultivation trial with maize using paraquat herbicide. *The Ghana Farmer* **12**:2–5.

Hamblin, A. P. 1980. Changes in aggregate stability and associated organic matter properties after direct drilling and ploughing on some Australian soils. *Aust. J. Soil Res.* **18**:27–36.

Hayward, D. M., T. L. Wiles, and G. A. Watson. 1980. Progress in the development of no-tillage systems for maize and soybean in the tropics. *Outlook Agric.* **10**:255–261.

Honisch, O. 1974. Water conservation in 3 grain crops in the Zambezi valley. *Expl. Agric.* **10**:1–8.

Hood, A.E.M., H. R. Jameson, and R. Cotterell. 1963. Destruction of pasture by paraquat as a substitute for ploughing. *Nature* **197**:748.

Huxley, P. A. 1975. Zero-cultivation studies at Morogoro. In R. A. Luse and K. O. Rachie (eds.).

Proc. of IITA Collaborators Meeting in Grain Legume Improvement IITA, Ibadan, pp. 152–154.

Huxley, P. A. 1979. Zero-tillage at Morogoro, Tanzania. In R. Lal (ed.). *Soil Tillage and Crop Production*. IITA Proc. Series 2, IITA, Ibadan, Nigeria, pp. 259–270.

Huxley, P. A. 1982. Zero/minimum tillage in the tropics: some comments and suggestions. Kenya Soil & Water Cons. Workshop, 10–12 March 1982, Nairobi, Kenya.

IITA. 1981. Role of tied ridges in maize production in Upper Volta, *Research Highlights 1981*, IITA, Ibadan, Nigeria, pp. 7–10.

Jackson, P. T. 1979. Minimum tillage of sugarcane in St. Kitts. *Agric. Engineer* **34**(3):84.

Jameson, P. D. (ed.). 1970. *Agriculture in Uganda*. Oxford University Press.

Jameson, J. D. and R. K. Kerkham. 1960. The maintenance of soil fertility in Uganda. I. Soil fertility experiment at Serere, *Emp. J. Exp. Agr.* **28**:179–192.

Jones, M. J. and A. Wild. 1975. Soils of the west African Savanna: The Maintenance and Improvement of their Fertility. Commonwealth Agricultural Bureaux, Harpenden, U.K.

Juo, A.S.R. and R. Lal. 1977. The effect of fallow and continuous cultivation on chemical and physical properties of an Alfisol in Western Nigeria. *Plant Soils* **47**:567–584.

Juo, A.S.R. and R. Lal. 1978. Nutrient profile in a tropical Alfisol under conventional and no-till systems. *Soil Sci.* **127**:168–173.

Jurion, F. and J. Henry. 1969. Can primitive farming be modernized? I.N.E.A.C. Hors Series, Belgium.

Kamara, C. S. 1980. Effects of mulch tillage techniques in Sierra Leone on cowpea growth and yield. *Trop. Trop. Grain Legume Bulletin* **19**:10–13.

Kampen, J., J. Hari Krishna, and P. Pathak. 1981. Rainy season cropping on deep Vertisols in the semi-arid tropics: effects on hydrology and soil erosion. In R. Lal and E. W. Russell (eds.). *Tropical Agricultural Hydrology*. Wiley, U.K., pp. 257–272.

Kang, B. T. and A. D. Messan. 1983. Fertilizer management for no-tillage crop production. In I. O. Akobundu and A. E. Deutsch (eds.). *No-Tillage Crop Production In The Tropics*. IPPC Document 46-B-83, Oregon State Univ., Corvallis, OR, pp. 111–118.

Kang, B. T., K. Moody, and J. O. Adesina, 1980. Effects of fertilizer and weeding in no-tillage and tilled maize. *Fertilizer Research* **1**:87–93.

Kang, B. T., G. F. Wilson, and L. Sipkens. 1981. Alley cropping maize (*Zea mays* L.) and Leucaena in southern Nigeria. *Plant Soil* **63**:165–179.

Kang, B. T. and M. Yunusa. 1977. Effects of tillage methods and P fertilization on maize in the humid tropics. *Agron. J.* **69**:291–294.

Kannegieter, A. 1967. Zero cultivation and other methods of reclaiming Pueraria fallowed land for food crop cultivation in the forest zone of Ghana. Trop. Agriculturist **123**:51–73.

Kannegieter, A. 1969. The combination of a short term pueraria fallow, zero cultivation and fertilizer application: Its effect on a fallowing maize crop. *Tropical Agriculturist* **125**:1–18.

Kemper, B. and R. Derpsch. 1981. Results of studies made in 1978 and 1979 to control erosion by cover crops and no-tillage techniques in Parana, Brazil. *Soil Till. Res.* **1**:253–267.

Khatibu, A. I. and P. A. Huxley, 1979. Effect of zero cultivation on the growth, nodulation and yield of cowpea at Morogoro, Tanzania. In R. Lal (ed.). *Soil Tillage and Crop Production*. IITA Proc. Series 2, IITA, Ibadan, Nigeria, pp. 271–280.

Koenigs, F.F.R. 1961. The mechanical stability of clay soils as influenced by moisture conditions and some other factors. Centre for Publication and Documentation No. 677, Wogeningen.

Koenigs, F.F.R. 1963. The puddling of clay soils. *Neth. J. Agric. Sci.* **11**:145–156.

Kowal, J. 1970. The hydrology of a small catchment basin at Samaru, Nigeria IV. Assessment of soil erosion under varied land management and vegetation cover. *Niger. Agric. J.* **7**:134–147.

Kowal, J. and K. Stockinger. 1973. Usefulness of ridge cultivation in Nigerian agriculture. *J. Soil Water Cons.* **28**:136–137.

Lal, R. 1973. Soil temperature, soil moisture and maize yield from mulched and unmulched tropical soils. *Plant Soil* **40**:321–331.

Lal, R. 1975. Role of mulching techniques in tropical soil and water management. IITA Tech. Bull. 1.

Lal, R. 1976a. Soil erosion on Alfisols in western Nigeria. II. Effects of mulch rates. *Geoderma* **16**:377–387.

Lal, R. 1976b. No-tillage effects on soil properties under different crops in western Nigeria. *Proc. Soil Sci. Soc. Amer.* **40**:762–768.

Lal, R. 1976c. Soil erosion on Alfisols in western Nigeria. I. Effects of slope, crop rotation and residue management. Geoderma 16:363–375.

Lal, R. 1978. Influence of within and between row mulching on soil temperature, soil moisture, root development and yield of maize in a tropical soil. *Field Crops Res.* **1**:127–139.

Lal, R. 1979. Soil and micro-climatic considerations for developing tillage systems in the tropics. In R. Lal (ed.). *Soil Tillage and Crop Production*. IITA Proc. Series 2, IITA, Ibadan, Nigeria, pp. 48–62.

Lal, R. 1982. Effects of 10 years of no-tillage and conventional plowing on maize yield and properties of a tropical soil. Proc. 9th ISTRO Conf., Osijek, Yugoslavia; pp. 111–117.

Lal, R. 1983. No-till Farming. IITA Tech. Bull. 2, Ibadan, Nigeria.

Lal, R. and O. Akinremi. 1983. Physical properties of earthworm casts and surface soil as influenced by management. *Soil Sci.* **135**:114–122.

Lal, R. and D. De Vleeschauwer. 1982. Influence of tillage methods and fertilizer application on chemical properties of worm castings in a tropical soil. *Soil Till. Research* **2**:37–52.

Lal, R., D. De Vleeschauwer, and R. M. Nganje. 1980. Changes in properties of a newly cleared Alfisol as affected by mulching. *Soil Sci. Soc. Am. J.* **44**:827–833.

Lal, R. and E. L. Dinkins. 1979. Tillage systems and crop production on an Ultisol in Liberia. In R. Lal (ed.). *Soil Tillage and Crop Production*. IITA. Proc. Series 2, IITA,Ibadan, Nigeria, pp. 221.

Lal, R., P. P. Maurya, and S. Osei-Yeboah. 1978a. Effects of no-tillage and plowing on efficiency of water use in maize and cowpea. *Expl. Agric.* **14**:113–120.

Lal, R., G. F. Wilson, and B. N. Okigbo. 1979. Changes in properties of an Alfisol produced by various crop covers. *Soil Sci.* **127**:377–382.

Lal, R., G. F. Wilson and B. N. Okigbo, 1978b. No-tillage farming after various grasses and leguminous cover crops in tropical Alfisol. I. Crop performance. *Field Crops Res.* **1**:71–84.

Lawes, D. A. 1961. Rainfall conservation and the yield of cotton in Northern Nigeria. *Emp. J. Expl. Agric.* **29**:307–318.

Lawes, D. A. 1962. Influence of rainfall conservation on the fertility of the loess plain soils of northern Nigeria. Samaru Research Bull. 24, IAR, Samaru, Nigeria.

Lawes, D. A. 1966. Rainfall conservation and the yields of sorghum and groundnuts in Northern Nigeria. *Expl. Agric.* **2**:139–146.

Lawson, T. L. and R. Lal. 1979. Response of maize to surface and buried straw mulch on a tropical Alfisol. In R. Lal (ed.). *Soil Tillage and Crop Production* IITA Proc. Series 2, IITA, Ibadan, Nigeria, pp. 63–74.

Leyenaar, P. and R. B. Hunter. 1977. The effect of seedbed preparation on maize production in the Coastal savanna zone of Ghana. *Ghana J. Agric. Sci.* **10**:47–51.

Lovely, W. G. 1961. Mechanical considerations. North Central Weed Control Conf., 18th Annual Report, pp. 130–131.

Mabbayad, B. B. 1967. Tillage techniques and planting methods for lowland rice. M. Sc. Dissertation, College of Agric., Univ. Philippines, Los Banos.

Mabbayad, B. B. and I. A. Buenacosa. 1967. Test on "Minimal Tillage" of transplanted rice. *The Phil. Agr.* **51**:541–551.

Mabbayad, B. B., B. N. Emerson, and E. L. Aragon. 1969. Further tests on "minimal tillage" and rates of nitrogen application on transplanted rice. *The Phil. Agr.* **53**:200–210.

Macartney, J. C., P. J. Northwood, M. Dagg, and R. Dawson. 1971. The effect of different cultivation techniques on soil moisture conservation and the establishment and yield of maize at Kongwa, Central Tanzania. *Trop. Agric.* **48**:9–17.

Maduakor, H. O., R. Lal, and O. Opara-Nadi. 1983. Effects of methods of seedbed preparation and mulching on the growth and yield of white ham on an Ultisol in southeast Nigeria. Field Crops Res. **9**:119–130.

Malik, A. S., V. Kumarand, and M. K. Moolani. 1973. Dryland research in Northwest India I. Effect of variable pre-planting tillage on soil moisture, growth and yield of pearl millet. *Agronomy Journal* **65**:12–14.

Mante, E.F.G. 1979. The effect of tillage depth and other farming practices on some physical properties of a vertisol and on the yield of sorghum. Proc. "Appropriate Tillage Workshop, 16–20 Jan. 1979. IAR, Zaria, Nigeria, Commonwealth Secretariat, U.K., pp. 95–106.

Maseri, S. T. and R. K. Jana. 1979. Influence of tillage and crop combinations on maize and soybean in the semi-arid regions of Tanzania. In R. Lal (ed.). *Soil Tillage and Crop Production*. IITA Proc. Series 2, IITA, Ibadan, Nigeria, pp. 281–293.

Maurya, P. R. and R. Lal. 1979a. Effects of straw mulch and soil moisture regimes on upland rice growth and production. In R. Lal, (ed.). *Soil Tillage and Crop Production*. IITA Proc. Series 2, IITA Ibadan, Nigeria, pp. 326–335.

Maurya, P. R. and R. Lal. 1979b. Influence of tillage and seeding methods on flooded rice. In R. Lal (ed.). *Soil Tillage and Crop Production*. IITA Proc. Series 2, IITA Ibadan, Nigeria, pp. 337–347.

Maurya, P. R. and R. Lal. 1979c. No-tillage system for crop production on an Ultisol in Eastern Nigeria. In R. Lal (ed.). *Soil Tillage and Crop Production*. IITA Proc. Series 2, IITA, Ibadan, Nigeria, pp. 207–200.

Maurya, P. R. and R. Lal. 1980. Effects of no-tillage and plowing on roots of maize and leguminous crops. *Expl. Agric.* **16**:185–193.

Maurya, P. R. and R. Lal. 1981. Effects of different mulch materials on soil and on root growth and yield of maize and cowpea. *Field Crops Res.* **4**:33–45.

McCowen, R. L., R. K. Jones, and D. C. I. Peake. 1980a. A ley farming system for the semi-arid tropics. In I. M. Wood (ed.). *Proc. Aust. Agron. Cont., Qld. Agric. Call.*, Lawes, pp. 188–189.

McCowen, R. L., R. K. Jones, and D. C. I. Peake. 1980b. Short-term benefits of zero-tillage in tropical grain production. In I. M. Wood (ed.). *Proc. Aust. Agron. Conf. Qld. Agric. Coll.*, pp. 220–221.

McCowen, R. L., R. K. Jones and D. C. I. Peake. 1985. Evaluation of a no-till tropical legume ley-farming strategy. In R.C. Muchow (ed.) "Agro-Research for the semi-arid tropics", Univ. Qld. Press, Australia pp. 450–472.

Mehanni, A. H. 1974. Short-term effect of some methods of improving soil structure in red-brown earth soils of the northern irrigation areas, Victoria. *Australian Journal of Experimental Agriculture and Animal Husbandry* **14**: 689–693.

Melville, I. R. 1978. Conservation tillage for N.T.? A look at minimum tillage systems in North America and Queensland. *N.T. Rural News-Magazine* **3**(4): 14–15.

Miller, J. C. and S. M. Bell (eds.). 1982. Crop production using cover crops and sod as living mulches. International Plant Protection Centre, Oregon State Univ., Corvallis, OR, p. 123.

Mittra, M. M., and J. W. L. Pieris. 1968. Paraquat as an aid to paddy cultivation. Proc. 9th British Weed Conf., pp. 668–674.

Mongi, H. O. and P. A. Huxley. 1979. Soils research in agroforestry. ICRAF, Nairobi, Kenya.

Moomaw, J. C., S. K. De Datta, D. E. Seaman, and P. Yogaratnam. 1968. New directions in weed control research for tropical rice. Proc. 9th Brit. Weed Control Conf. pp. 675–691.

Morel, R. and P. Quantin. 1972. Observation on the long-term fertility of cultivated soils at Grimari (Central African Republic) *Agronomie Tropicale* **27**:667–739.

Muller, J. and G. de Bilderling, 1953. Les methods culturals indigenes sur les sols equatoriaux de plateau. *I.N.E.A.C. Belg. Bull. Inform.* **2**:21–30.

Nangju, D. 1979. Effect of tillage methods on growth and yield of cowpea and soybean. In. R. Lal (ed.). *Soil Tillage and Crop Production*. IITA Proc. Series 2, IITA, Ibadan, Nigeria, pp. 93–108.

Nangju, D., H. C. Wien, and T. P. Singh. 1975. Some factors affecting soybean variability and emeyence in the lowland tropics. World Soybean Res. Conf. 3–8 Aug., 1975, Urbana-Champaign, Il.

Nicou, R. 1974a. Contribution to the study and improvement of the porosity of sandy-clay soils in the dry tropical zone: Agricultural consequences. *Agronomie Tropical* **29**:110–1127.

Nicou, R. 1974b. The problem of caking with the drying out of sandy and sandy clay soils in the arid tropical zone. *Agronomie Tropicale* **30**:325–343.

Nicou, R. 1977. Le travail du sol dans les terres exondées du Sénégal: motivations; contraintes. ISRA/CNRA, Bambey, Senegal.

Nicou, R. 1979. Tillage in the West African tropical zone. Proc. 8th ISTRO Conf., Hohenheim, Germany, pp. 429–434.

Nicou, R. and J. L. Chopart. 1979. Water management in sandy soils of Senegal. In R. Lal (ed.). *Soil Tillage and Crop Production*. IITA Proc. Series 2, IITA, Ibadan, Nigeria, pp. 248–257.

Northwood, P. J. and J. C. Macartney. 1971. The effect of different amounts of cultivation on the growth of maize on some soil types in Tanzania. *Tropical Agriculture* **48**:25–33.

Nur, I. A. and A. A. E. Gasim. 1974. Effect of certain tillage operations on groundnut performance in Sudan Gezira. *African Soils* **19**:29–37.

Nyoka, G. C. 1980. Studies on the germination, growth and control of weeds in upland rice fields under different fallow periods in Sierra Leone. Ph.D. Dissertation, Njala Univ. College, Sierra Leone.

Nyoka, G. C. 1983. Potential for no-tillage crop production in Sierra Leone. In I. O. Akobundu and A. E. Deutsch (eds.). *No-Tillage Crop Production In The Tropics*. IPPC Doc. 46-B-83, Oregon State Univ., Corvallis, OR, pp. 66–72.

Ofori, C. S. 1973. The effect of ploughing and fertilizer application on yield of cassava (Manihot esculenta). *Ghana J. Agric. Sci.* **6**:21–24.

Ofori, C. S. and S. Nandy. 1969. The effect of method of seed cultivation on yield and fertilizer response of maize grown on a forest ochrosol. *Ghana J. Agric. Sci.* **2**:19–24.

Ogborn, J. 1982. No-till cropping systems for small holder farmers. *Appropriate Technology* **9**(2):7–9.

Ogunremi, L. T., R. Lal, and O. Babalola, 1986. Rice response to tillage methods and soil compaction in an Ultisol. Soil Till. Res. (In Press).

Okigbo, B. N. 1973. Maize experiments on the Nsukka plain. V. Effect of pre-planting cultivations, mulching and weeding frequency on the yield and general performance of maize. *Agronomie Tropicale* **28**:54–74.

Okigbo, B. N. 1979. Effects of pre-planting cultivations and mulching on yield and performance of cassava (*Manihot esculenta*). In R. Lal (ed.). *Soil Tillage and Crop Production*. IITA Proc. Series 2, Ibadan, Nigeria, pp. 75–92.

Olaniyan, G. O. 1983. No-tillage production of maize, rice, and cowpea in Nigeria. In I. O. Akobundu and A. E. Deutsch (eds.). *No-Tillage Crop Production in the Tropics*. IPPC Document 46-B-83, Oregon State Univ., Corvallis, OR, pp. 127–131.

Opara-Nadi, O. A. and R. Lal. 1983. The effects of tillage methods on hydrological properties of a tropical Alfisol. *Proc. Germany Soc. Soil Sci.* Sept., 1983, Trier, Germany, pp. 4–10.

Ospina, V. and P. Nel. 1957. La agriculture moderna y la conservation de les recurses naturales. *Agron. Trop. (Paris)* **13**:1036–1047.

Panikar, S. N. and W. G. Walunjkar. 1976. Effect of mulches on soil micro-climate, growth, maturity and yield in cigar-wrapper tobacco. *Indian Journal of Agronomy* **21**:415–419.

Peake, D. C. I., R. K. Jones, and R. L. McCown. 1983. Dryland farming systems for the semi-arid tropics of Australia. A.I.A.S. Refresher Training Course, Brisbane, 24–28 Jan. 1983, pp. 258–266.

Peers, A. W. 1962. Establishment of grass leys by undergrowing to cereals. *E. Afr. Agric. For. J.* **27**:145–419.

Pereira, H. C. 1956. A rainfall test for structure of tropical soils. J. Soil Sci. **7**:68–74.

Pereira, H. C. and V.R.S. Beckley. 1952. Grass establishment on eroded soil in a semi-arid African reserve. Emp. J. Exp. Agric. **21**:1–15.

Pereira, H. C., E. M. Chenery, and W. R. Millas. 1954. The transient effects of grasses on the structure of tropical soils. *Emp. J. Exp. Agric.* **22**:148–160.

Pereira, H. C., P. H. Hosegood, and M. Dagg. 1967. Effects of tied ridges, terraces and grass leys on a lateritic soil in Kenya. *Expl. Agric.* **3**:89–98.

Pereira, H. C. and P. A. Jones. 1954. A tillage study in Kenya coffee Part 1. The effects of tillage practices on coffee yields. *Emp. J. Exp. Agric.* **22**:231–240.

Pereira, H. C., R. A. Wood, H. W. Brzostowski, and P. A. Hosegood. 1958. Water conservation by fallowing in semi-arid tropical East Africa. Emp. J. Expl. Agric. **26**:213–228.

Perez Escolar, R. and G. Ortiz Lugo. 1973. Vertical topsoiling, a practicable technique to manage soils with markedly compact subsoils. *Journal of Agriculture of the University of Puerto Rico* **57**:186–195.

Pla Sentis, I., A. Florentino, and T. Perez. 1979. Relation between soil physical properties and problems of soil management and conservation in agricultural soils of Venezuela. In R. Lal (ed.). *Soil Tillage and Crop Production*. IITA Proc. Series 2, IITA, Ibadan, Nigeria, pp. 184–196.

Poulain, J. F. and R. Tourte. 1970. Effects of deep preparation of dry soils on yields from millet and sorghum to which nitrogen fertilizers have been added. *Afr. Soils* **15**:533–537.

Prentice, A. N. 1946. Tie-ridging, with special reference to semi-arid areas. *East African Agric. Jour.* **12**:101–108.

Rijn, P. J. Van. 1983. Pests and their control in no-tillage in the tropics. In I. O. Akobundu and A. E. Deutsch (eds.). *No-Tillage Crop Production In The Tropics*. IPCC, Document 46-B-83, Oregon State Univ., Corvallis, OR, pp. 86–101.

Rodriguez, M. and R. Lal. 1979a. Comparison of zero and conventional tillage systems in an acidic soil. In R. Lal. (ed.). *Soil Tillage and Crop Production*. IITA Proc. Series 2, IITA, Ibadan, Nigeria, pp. 197–205.

Rodriguez, M. and R. Lal. 1979b. Tillage/fertility interactions in paddy rice. In R. Lal (ed.). *Soil Tillage and Crop Production*. IITA Proc. Series 3, Ibadan, Nigeria, pp. 349–356.

Sanchez, P. A. 1973a. Puddling tropical rice soils: I. Growth and nutritional aspects. *Soil Sci.* **115**:149–159.

Sanchez, P. A. 1973b. Puddling tropical rice soils. 2. Effects of water losses. *Soil Sci.* **115**:303–308.

Sanchez, P. A., D. E. Bandy, J. H. Villachica, and J. J. Nicholaides. 1982. Amazon soils: management for continuous crop production. *Science* **216**:821–827.

Sanchez, P. A. and A. M. Briones, 1973. Phosphorus availability of some Philippine rice soils as affected by soil and water management practices. *Agron. J.* **65**:226–228.

Sar, R. van Der, Tillage for dry annual crops in the humid tropics. Surinaamse Landbouw **24**:93–98 (1976).

Scheltema, W. 1974. Puddling against dry plowing for lowland rice culture in Surinam: effect on soil and plant, and interactions with irrigation and nitrogen dressing. Agricultural Research Reports, Wageningen 828.

Seth, A. K., C. H. Khaw, and J. M. Fua. 1971. Minimum and zero-tillage techniques and post planting weed control in rice. Third Asian Pacific Weed Sci. Conf. pp. 188–200.

Shenk, M. D. and J. L. Saunders. 1983a. Vegetation management systems for crop production in tropical regions of central America: The case of Costa Rica. In I. O. Akobundu and A. E. Deutsch (eds.). *No-Tillage Crop Production in Tropics* IPPC Doc. 46-B-83, Oregon State Univ., Corvallis, OR.

Shenk, M. D. and J. L. Saunders, 1983b. Insect population responses to vegetation managment systems in tropical maize production. In I. O. Akobundu and A. E. Deutsch (eds.). *No-Tillage Crop Production in the Tropics*. IPPC Doc. 46-B-83, Oregon State Univ., Corvallis, OR, pp. 73–85.

Sidiras, N., J. C., Henklain, and R. Derpsch. 1982. Comparison of three different tillage systems with respect to aggregate stability, the soil and water conservation and the yields of soybean and wheat on an Oxisol. Proc. 9th ISTRO Conf., Osijek, Yugoslavia: pp. 537–544.

Sidiras, N., R. Derpsch, and A. Mondardo. 1983. Effect of tillage systems on water capacity, available moisture, erosion and soybean yield in Parana, Brazil. In I. O. Akobundu and A. E. Deutsch (eds.). *No-Tillage Crop Production In The Tropics*. IPPC Document 46-B-83, Oregon State Univ., Corvallis, OR, pp. 154–165.

Srivastava, S. P., and R. Singh. 1972. A note on row tilling in cotton. *Indian Journal of Agronomy* **17**(2):119–120.

Stibbe, E. and D. Ariel. 1970. No-tillage as compared to tillage practices in dryland farming of a semi-arid climate. *Neth. J. Agric. Sci.* **18**:293–307.

Subramanian S., S. Loganathan, V. Ravikumar, and K. K. Krishnamoorthy. 1975. Effect of tillage and organic amendments on the physical properties of soil and yield of bajra. *Madras Agricultural Journal* **62**:106–109.

Takayi, S. K. 1970. Effects of land preparation on maize yields. *Ghana J. Agr. Sci.* **3**:151–154.

Thamburaj, S., K. G. Shanmugavelu, C. R. Muthukrishnan, and M. Vijayakumar, 1980. Studies on "no-tillage" in tapioca. Nat. Seminar Tuber Crops Prod. Techn., Tamil Nadu Agric. Univ.

Vergara, N. T. 1982. New directions in agro-forestry: the potential of tropical legume trees, E.P.I., East West Centre Honolulu, Hawaii.

Verma, S. M. 1973. A note on response of tapioca to variable tillage. *Indian Journal of Agronomy* **18**:97–99.

Vine, H. 1953. Experiments on the maintenance of soil fertility at Ibadan, Nigeria, 1922–1951. *Emp. J. Exp. Agric.* **21**:65–85.

Wade, M. K. and P. A. Sanchez. 1983. Mulching and green manure applications for continuous crop production in the Amazon Basin. *Agron. J.* **75**:39–44.

Walton, P. D. 1962. The effect of ridging on the cotton crop in the Eastern Province of Uganda, Emp. J. Expl. Agric. **30**:63–76.

Wang, M. M. 1975. Effects of soil pulverization by tractor wheels on rice growth. *Taiwan Agriculture Quarterly* **11**:85–92.

Weng, C. K., J. A. Rajaratnam, and L. I. Hock. 1979. Ground cover management and its effects on soil physical properties under oil palm cultivation in Malaysia. In R. Lal (ed.). *Soil Tillage and Crop Production*. IITA Proc. Series 2, IITA, Ibadan, Nigeria, pp. 235–245.

Whiteman, P.T.S. 1975. Moisture conservation by fallowing in Botswana. *Expl. Agric.* **11**: 305–314.

Wijewardene, R. 1980. *Conservation Farming*. IITA Sri Lanka Program, Colombo, p. 19.

Wilkinson, G. E. 1975. Effect of grass fallow rotations of Northern Nigeria. *Trop. Agric.* **52**: 97–103.

Wilkinson, G. E. and P. O. Aina. 1976. Infiltration of water into 2 Nigerian soils under secondary forest and subsequent arable cropping. *Geoderma* **15**:51–59.

Wilkinson, G. E. and P. O. Aina. 1977. Shifting tropical forest soils in Nigeria from bush to arable crops: the effect on the infiltration of water. *Geoderma* **15**:51–59.

Willcocks, T. J. 1979a. Tillage energy requirements for semi-arid crop production in Botswana. In R. Lal (ed.). *Soil Tillage and Crop Production*. IITA Proc. Series 2, IITA, Ibadan, Nigeria, pp. 303–323.

Willcocks, T. J. 1979b. Semi-arid tillage research in Botswana. Proc. "Appropriate Tillage Workshop" 16–20 Jan. 1979, IAR Zaria, Nigeria, Commonwealth Secretariat, U.K., pp. 127–146.

Willcocks, T. J. 1981. The tillage of clod forming sandy loam soils in the semi-arid climate of Botswana. *Soil Tillage Res.* **1**:323–350.

Williams, J., K-J. Day, R. F. Isbell, and S. J. Reddi. 1983. Soils and Climate. Conference Proc. "Agro-Research for Australia's Semi-arid Tropics" 21–25 March, 1983, Darwin, Australia.

Wilson, G. F. 1979. The potential of legume covers in no-tillage cropping in the tropics. In R. Lal (ed.). *Soil Tillage and Crop Production* IITA Proc. Series 2, IITA, Ibadan, Nigeria, pp. 109–123.

Wilson, G. F. and K. L. Akapa. 1983. Providing mulches for no-tillage cropping in the tropics. In I. O. Akobundu and A. E. Deutsch (eds.). *No-tillage Crop Production In The Tropics*. IPPC Document 46-B-83, Oregon State Univ., Corvallis, OR, pp. 51–65.

Wilson, G. F., R. Lal, and B. N. Okigbo. 1982. Effects of cover crops on soil structure and on yield of subsequent arable crops grown under strip tillage on an eroded Alfisol. *Soil Tillage Research* **2**:233–250.

Wood, T. A. 1950. *The Groundnut Affair*. The Bodley House, London.

Yu, S. L., J. S. He, and G. H. Zhang, 1981. Study on the effect of mulching groundnuts with plastic film on soil fertility and activity of microflora. *Zhongguo Youliao* No. 3:50–52.

11

PRINCIPLES OF WEED MANAGEMENT WITH SURFACE-TILLAGE SYSTEMS

G. B. TRIPLETT
Professor of Agronomy
Department of Agronomy
Mississippi State University
Mississippi State, Mississippi

A. D. WORSHAM
Professor of Weed Science
Department of Crop Science
North Carolina State University
Raleigh, North Carolina

INTRODUCTION

Farmers control weeds or their crops fail. This maxim has dominated agricultural practice for centuries, and the challenge was met with hands, hoes, plows, and cultivators until recently. Tilling the soil not only increased erosion but brought weed seed to the surface, which germinated and had to be killed by more tillage. Pioneer agriculturists tried to reduce the amount of tillage required to produce crops, but their systems often failed because weeds were not controlled. Herbicides added a new dimension to the war on weeds. First supplementing and then substituting for tillage, herbicides have permitted major reductions in cultivation and tillage. No-tillage production eliminates practically all soil disturbance; yet, the basic requirement of weed control remains unchanged.

Tillage is treated in this chapter as a continuum of conventional plowing, seedbed preparation, and postemergence cultivation. No-tillage, with spray–plant–harvest operations, represents the other extreme. Surface, minimum, and conservation tillage employ intermediate amounts and kinds of tillage (Moncrief, 1984).

Tilling the soil disturbs growing plants. Roots, crowns, and rhizomes of established plants are dislodged and broken. Above ground parts of plants are buried or severed from soil contact, moisture, and/or light. Most weeds are killed outright by dessication, and perennial species that survive represent some of the most persistent noxious weeds. Tillage as a weed control method has limitations. To avoid injury the crop must not be disturbed, so weeds in the row too tall to be covered with soil often survive. Vegetatively regenerative parts regrow and rhizomes, stolons, and tubers, when transported to other areas of the field or to other fields, spread the infestation. Weed seed buried by previous tillage may be unearthed and germinate. Cultivation prunes shallow crop roots and damages the crop in other ways, such as breaking limbs or stems. Bringing soil into contact with stems of many crops increases disease.

Surface-tillage production of crops has been a goal of agriculturists for decades. The primary factor limiting the use of less tillage was the inability to control weeds at planting time and those that developed later. Until the early 1950s, tillage was the only method available to prepare a seedbed, temporarily free it of weeds, and control weeds that developed later.

Prospects for controlling weeds by alternate means, however, improved during the 1950s with the advent of a host of new herbicides. These discoveries probably led Harper (1957) to write, "for efficient longlasting weed control, ploughing should be avoided, surface tillage reduced to a minimum and any weed seeds which are formed should be left on the surface to be killed by spraying when they do germinate." The discovery and subsequent development of a new class of nonselective herbicides in the United Kingdom and marketing of the contact herbicide paraquat* in the United States (ca 1960) opened new opportunities for surface-tillage production. New crop production techniques soon were developed and adopted in many areas of the United States. An estimated 87–90 million acres of U.S. cropland were in some form of surface tillage in 1983 and another 10–12 million acres were planted no-tillage (Magleby et al., 1984).

Modern herbicides make no-tillage crop production possible, but even with the many compounds available, weeds and weed control remain the dominant concern. A survey in 25 leading corn-producing† states of the United States in 1980 led agronomists in three states to list lack of herbicide effectiveness and an

*Common and chemical names of herbicides are listed in Appendix A.

†Common and botanical names of crops and weeds are listed in Table 11.4. (see pages 342–343)

increase in perennial weeds as major reasons that no-tillage corn production likely would not increase in their states by 1990. Respondents in all 25 states listed perennial weed control as a problem currently encountered in no-tillage corn and in 16 states perennial weeds were listed the most important problem. Poor weed control was listed by respondents in 24 of the 25 states as a serious problem and was predicted to worsen if no-tillage corn acreage increased. Insects and poorly drained, cold soils were the next most-listed factors limiting the expansion of no-tillage corn acreage in 12 states (Worsham, 1980).

A widely held view among scientists is that effective weed control is the most important single problem limiting acceptance of surface- and no-tillage management systems, and farm acceptance will be expanded as the herbicides now being developed are incorporated into weed-management systems. For example, control of some perennial weeds with the nonselective, systemic herbicide, glyphosate, and of perennial grass weeds in broadleaf crops with new, postemergence "grass" herbicides is now possible. The remainder of this chapter identifies principles and provides examples of developing weed-management systems to fit specific situations. Herbicides often do not perform as well under no-tillage systems as they do with clean-tillage systems. Soil incorporation and tillage–herbicide combinations for use with no-tillage systems differ from those for plow-systems.

WEED MANAGEMENT PROGRAMS

The objective of integrated weed management (IWM) is to encourage crop growth by combining several methods to lessen weed competition. Just as integrated pest management (IPM) has provided better, less costly, and safer insect and disease control in a new environment (Chapter 12), IWM is equally effective to control weeds. Through this approach, adequate weed suppression can be obtained with a long-term program combining several methods. Traditional methods remain applicable with some major differences. These methods are discussed as they are employed in plow-tillage systems and surface- and no-tillage systems. No-tillage systems have the same requirements for economic and effective weed control as do plow-systems. The major difference is that more burden is placed on chemicals to deter weed growth. In no-tillage systems, herbicides are relied on for preplant, preemergence, and postemergence control of all weeds. Tillage after planting is rarely an option.

The essential components of weed management of these cropping systems consist of (1) weed prevention, (2) control of existing vegetation at planting, (3) residual weed control (herbicides in the soil to kill germinating weed seeds), and (4) postemergence weed control (herbicides applied postemergence to control late weed development). The major techniques, "tools," employed include (1) weed seed prevention, (2) crop rotation, (3) crop competition, (4) mechanical tillage, and (5) biological, predator control, aside from herbicides (Lewis and Worsham, 1981; McWhorter and Chandler, 1982).

Use can be made of crop rotations, crop competition, and biological methods to integrate these components into a total weed management program for any crop. A delicate balance must be reached to be effective. Weed pressure can be decreased by practices that (1) reduce the weed seed reservoir; (2) prevent new weed seeds from being brought onto the farm with contaminated seeds, feed, hay, machinery or manure; and (3) control of perennials that spread vegetatively.

Reducing the Weed Seed Reservoir

The reservoir of seeds of various weed species in the soil can exceed billions per hectare. Many of these seeds remain viable for several decades while remaining dormant until conditions are favorable for germination. The reservoir is replenished each year as uncontrolled weedy plants are allowed to mature and produce seeds. High populations of germinating weed seeds are hazardous to crop production from two aspects. First, 95% control of a large population of weeds may leave a sufficiently large number of weedy plants to noticeably reduce crop productivity. Second, weed seedlings may "compete" with each other in uptake of herbicides applied or the larger weeds may intercept herbicides before reaching smaller ones in postemergence applications, thereby, seedling populations of the smaller may not absorb an adequate herbicide dose for control. Management systems that reduce the weed seed population make weed control programs more effective. To do this, nearly all weed plants must be prevented from forming seed. Single plants of various weed species often produce up to 100,000 seeds or more.

Schweizer and Zimdahl (1984a, 1984b) reported that weed seed numbers in the soil were reduced by more than 95% over a 6-year period through use of effective herbicides to prevent weed seed production. Results were similar in continuous corn and a rotation that included corn. When herbicide applications were stopped, the weed seed reservoir increased. A program that prevents weed seed production is useful in overall weed control strategy but will not lead to total eradication. Weed populations are dynamic and shift under different types of pressure. Species not controlled rapidly increase in numbers.

Preventing weed seed production in no-tillage crops is especially important, since tremendous numbers of seed can quickly build up in the upper few centimeters of soil. Weed seed near the soil surface are not as long lived as deeply buried seeds, and with effective control systems, seed reservoirs can be reduced to a minimum within a few years in no-tillage systems.

Aside from natural means of weed seed transport by water, wind and birds, weeds may be introduced with contaminated seed, and transported by tillage and harvesting equipment moving from infested areas. Hard seed of weed species ingested by animals are spread in droppings and manure. Each should be recognized as a potential source of infestation and measures should be taken to minimize them.

Perennial weeds that spread vegetatively are a serious problem and often require special control measures. Infestations of certain perennials restrict adoption of reduced-tillage systems. Localized perennial infestations should be identified and eradicated before they become a serious menace. Once identified, crops can be selected that permit use of effective herbicides to eradicate troublesome perennials. Prevention of serious weed problems requires careful appraisal of weeds present and prompt adjustment of weed control programs.

Crop Rotation

Some weed species increase under cultural practices unique to the management of the crop. This is becoming increasingly evident and is an important factor in herbicide–weed–crop associations. Crop rotations must not be overlooked as an important weed-management tool. Weaknesses in herbicide programs for specific weeds are much easier to overcome in some crops than in others. For example, weeds such as lambsquarter are more easily and economically managed in corn than in soybeans, peanuts, or cotton. Large-seeded broadleaf weeds such as cocklebur, morningglory, and sicklepod can be controlled at three different times during the life cycle of corn, whereas only postemergence applications can be used effectively in soybeans. Timing is critical and crop tolerance may be marginal at those crop growth stages when weed control is most effective (Lewis and Worsham, 1981). Deep-rooted broadleaf perennials such as trumpetcreeper, bigroot morningglory, and horsenettle can be managed in corn but not in soybeans.

Rotating crops helps prevent the build-up of problem weeds. Equally, if not more important, the herbicides also normally will be rotated with the crops. Perennial crops such as some hay crops, fruits, permanent pastures, and rangelands are not rotated as frequently as annual crops, but some of these can be intermixed in long rotation sequences (Aldrich, 1984).

Rotations are similar in surface- and no-tillage systems. Exceptions exist where a heavy residue mulch or a killed living mulch may interfere with planting or introduce other undesirable factors. For example, a widely held view of crop specialists has been that peanuts cannot be planted and grown successfully without tillage before planting to bury plant residues as a means of reducing disease and insect problems. Recent experimental work in at least four southeastern states has been successful in planting peanuts into various kinds of mulches and residues (Worsham, 1985). In double- and triple-cropping, no-tillage is preferred since crops in the sequence can be planted sooner with less loss of land use, soil moisture, time, and labor.

Crop Competition

As plant density per unit area increases, individual plants, whether weeds or crops, compete with each other for space, light, water, and nutrients. Response to competition by individual plants includes etiolated growth and decreased yield. If only crop plants are present, yield per unit area increases as population density increases to the optimum for the specific crop.

Time of competition in the development of the crop is important. Weeds emerging with the crop and remaining until maturity severely reduce yields. If weeds are controlled until the crop is 4–6 weeks old, competition from the crop suppresses weed seedlings emerging after this time, and these usually have little effect on crop productivity. When crops are planted into weed-free fields, weed and crop seedlings emerge at the same time. Competition between newly emerged seedlings does not become critical for several weeks. Thus, application of postemergence herbicides 2–4 weeks after planting to remove weed seedlings is also effective in eliminating weed-competition effects on the crop.

A grower can increase crop competitiveness appreciably by planning well to encourage it. This is possibly the most overlooked weed-management tool. Many plants have been chosen as crops because of their vigor and ability to compete well: that competitiveness is increased by using a combination of production practices to maximize vigor. Shading of weeds by the crop is an important factor. High-quality seed of vigorous cultivars, proper fertilization and liming, effective disease and insect control, narrow row spacing, and timely planting are all important in giving the crop an advantage over weeds. Cultivars may also vary in their competitiveness through strong rooting habits and morphological characteristics that provide dense shade. The sooner the crop canopy is closed, the better the weed control with or without herbicides (Aldrich, 1984; Lewis and Worsham, 1981; Klingman and Ashton, 1982). Corn, soybeans, and wheat are likely best suited for no-tillage because of large seed and robust seedling growth, as well as the array of effective herbicides available for these crops.

Many weeds interfere with crop growth through allelopathic effects.* Some crops are allelopathic against weeds, but cultivars vary in their allelopathic effects on some weeds (Putnam and DeFrank, 1983; Radosevich and Holt, 1984; Rice, 1984).

Use of production practices to promote fast emergence, rapid growth, and vigorous crops to shade weeds is common to all cropping systems. Surface- and no-tillage systems are at a disadvantage in more northern latitudes with warm temperature crops planted in killed cover crops, heavy infestations of weeds or fields with large amounts of previous crop residue. Emergence and early growth are slowed during the first few weeks after planting owing to slower warming of the soil in spring where a complete mulch cover is present. Both crop and weed seedlings may be shaded during emergence if the mulch is excessive. Other factors influencing germination and early growth rates may be a temporary nitrogen deficiency (see Chapter 5) and phytotoxic by-products of plant residues and microorganism decomposers (Putnam and DeFrank, 1983).

*Effects of plants or microorganisms on each other mediated through chemicals released by the organisms (Aldrich, 1984; Klingman and Ashton, 1982; Radosevich and Holt, 1984; Rice, 1984).

Mechanical

Mechanical weed management includes primary and secondary tillage, postemergence cultivation, and timely mowing. Primary tillage is very effective in uprooting shallow-rooted perennials and causing depletion of their root-energy reserves. Secondary tillage breaks rhizomes of perennials into small pieces, further dessicates them, often making herbicides more effective.

Postemergence cultivations uproot and bury seedling weeds. Furthermore, they increase soil aeration and water infiltration. However, cultivations compact soil and destroy aggregates, which encourages crusting and erosion (see Chapter 2). Shallow cultivation controls emerging seedlings and incorporates surface-applied herbicides to a depth of 1–2 cm, increasing their effectiveness when rainfall is delayed following application. On the negative side, improper cultivation 5 cm or deeper prunes crop roots and reduces yield. Cultivation also increases soil moisture loss by evaporation, decreases organic matter content, and destroys soil structure (Klingman and Ashton, 1982).

The major differences between no-tillage and plow-tillage systems occurs during primary and secondary cultivations. In plow-tillage systems, tillage removes weeds and cover to provide a clean weed-free seedbed and controls weeds with tillage after the crop emerges. Limited postemergence tillage is possible in some reduced tillage systems except with no-tillage culture. Exceptions include tillage with sweep cultivators in double-crop soybeans following small grain harvest where little or no straw residue remains. Ground-driven rotary cultivators can be used if moderate residue is present. Accordingly, heavy pressure is put on the herbicide for complete control, whether preplant, preemergence, or postemergence.

Predators

There are several outstanding examples of controlling weeds with other organisms. These have, in the past, included release of phytophagus insects and, more recently, use of fungal plant pathogens in a "bioherbicide" or "mycoherbicide" approach. The former has worked best in large areas infested dominantly with one weed species, the latter on selected weed species in row crops and orchards (Klingman and Ashton, 1982). Crop rotation, crop competition, and crop allelopathy are also forms of biological control. These methods plus defoliation by grazing or mowing should be equally effective in plow- or no-tillage cropping systems.

Chemical

Weed management in crops was revolutionized with the discovery of chemicals that selectively control weeds, first reported as 2,4-D and 2,4,5-T, members of the phenoxy group of herbicides (Hamner and Tukey, 1944). Today more than 180 different chemicals are available and registered for use. These uses may include as many as 30 individual herbicide treatments for each major crop. At least a score of new chemicals are under development at any given time. They may be applied preplant, preemergence, at emergence, early-postemergence, late-postemergence, directed postemergence, semidirected postemergence, and pre- and postharvest. Weed-management systems for use with surface- and no-tillage place great reliance on the chemical component. The herbicide (or herbicides) must kill all existing vegetation at planting time (whether a living cover crop or weeds) and retain enough residual preemergence activity to provide pre- and postemergence control as necessary.

Lower herbicide rates or band treatments may be used in some instances to temporarily retard growth of existing vegetation (weed or crop) to permit establishment of an interplanted crop. Examples include planting of small-seeded legumes into grass pastures, grasses into forage legumes, or corn into coastal bermudagrass, tall fescue, or other forage grasses. The success of reduced tillage systems requires keen knowledge of the crop, the weed, and the environment for complex managerial decisions to be effective.

WEED–CROP ECOLOGY IN SURFACE-TILLAGE

Problem weeds are simply defined as those not adequately controlled by available techniques or that require difficult and/or expensive methods. The list changes with time, geographical location, weed, and crop grown. For example, quackgrass, a perennial, has been a problem weed in crops in the Northeast and Midwest United States for many years. Today, quackgrass is still a major problem in small grains and several forage legumes, a minor problem in soybeans, and no longer a problem in corn production because of selective herbicides available for use with these crops.

When both tillage and herbicides are used for weed control, the list of problem weeds is shortened. As either practice is reduced, the number of weeds causing problems often increases because of inadequate control (Witt, 1984). Most surely, weeds that are troublesome where both tillage and herbicides are used will become more so as tillage is lessened. Growers must recognize that changes in tillage practices change the weed spectrum and the relative importance of different weed species. Fields must be carefully monitored to identify weed problems as they develop and adjust practices to deal with these. A list of weeds that are particularly troublesome with reduced tillage, in various crops and geographical areas, is presented in Table 11.1. With continued herbicide development, this list will diminish.

Eliminating tillage causes shifts in weed species present (Triplett and Lytle, 1972). Perennials, such as poison ivy, horsenettle, trumpet creeper, and tree seedlings, some of which are readily controlled by tillage, become established and persist in untilled fields. Weeds botanically related to the crop and others that escape control often increase in number to become dominant problems. A classic example of this developed with the use of atrazine to control weeds in

TABLE 11.1. Problem Weeds in No-Tillage and Surface-Tillage Production of Different Crops in the United States

Corn	Soybeans	Sorghum	Wheat	Cotton
No-tillage should not be attempted in fields that are severely infested with these or similar weeds:				
johnsongrass	johnsongrass[b]	johnsongrass	quackgrass	silverleaf nightshade
bermudagrass	bermudagrass[b]	bermudagrass	jointed goatgrass	johnsongrass[b]
wild proso millet	hemp dogbane	hemp dogbane	Russian knapweed	bermudagrass[b]
shattercane	redvine	Russian knapweed		redvine
bluestem	briars	wild proso millet		hog potato
switchgrass	wild proso millet[b]	shattercane		spurred anoda
trumpetcreeper	climbing milkweed	switchgrass		spurges
broomsedge	leafy spurge	bluestem		nutsedge
dallisgrass	sicklepod	broadleaf signalgrass		wollyleaf bursage
purple nutsedge	purple nutsedge	Texas panicum		Texas blueweed
itchgrass	trumpetcreeper	itchgrass		groundcherry
	forage grasses[b]			field bindweed
	quackgrass[b]			horsenettle
In addition to the weeds listed above, these weeds are also difficult to control with reduced tillage:				
briars	milkweed	jointed goatgrass	wild oats	velvetleaf
hemp dogbane	alfalfa	fall panicum	cheat	unicorn plant
climbing milkweed	horsenettle	horsenettle	Canada thistle	cocklebur
fall panicum	sunflower	sandbur	bindweed	lanceleaf sage
horsenettle	burcucumber		annual bromegrasses	Venice mallow
burcucumber	broadleaf signalgrass[b]			croton
broadleaf signalgrass	switchgrass[b]			

[a] Common and botanical names of crops and problem weeds are listed in Table 11.4.
[b] Recently developed postemergence grass herbicides effectively control these weeds in certain broadleaf crops.

SOURCE: Triplett et al. (1983).

corn. At first, atrazine controlled most annual weeds found in corn fields. Fall panicum, not a problem weed, tolerates atrazine and increased dramatically in continuous corn. Without midsummer cultivation, fall panicum pressure rendered atrazine inadequate as a sole herbicide in corn. Within weed species, biotypes tolerant of certain herbicides have appeared. Pigweed and lambsquarter resistant to atrazine have so been identified and have become problem weeds in parts of the United States and Canada (Bandeen et al., 1982). Fortunately, many species can be controlled with other herbicides.

A recent, encouraging development in weed ecology is the discovery that many annual broadleaf weeds are suppressed if mulches, especially small grain cover crops, are left on the soil surface (Liebl and Worsham, 1983; Putnam and DeFrank, 1983; Shilling et al., 1985). This beneficial effect, largely due to allelopathic interactions, can help suppress difficult-to-control annual broadleaf weeds in many broadleaf crops and possibly reduce the need for postemergence herbicide applications with surface- or no-tillage.

HERBICIDE TECHNOLOGY

Since herbicides and their effective use are central to the development of surface-tillage systems, a short review of herbicides and their group characteristics is in order. Selectivity is the most important attribute of a modern herbicide. A herbicide applied at a sensitive time in the crop growth cycle kills weeds while permitting the crop to survive uninjured. Previously used herbicides, such as sodium chlorate, were nonselective, killing noxious perennial weeds, a very useful function, but at the same time killing all crops. Selective herbicides kill weeds in the crop row without injury to the crop or need for cultivation. Herbicides available today are so divergent in their mode of action and selectivity that the only common thread shared is the biological activity they stimulate. Selectivity is largely a function of concentration of the herbicide at active site(s) within the plant.

Modern herbicides are primarily organic compounds that belong to one of several major chemical families. Some chemicals within families differ only by slight modifications in structure of the basic molecule. Slight alterations of molecular structure often change selectivity, thereby changing the spectrum of weeds controlled and crop sensitivity. The common thread in biological activity is the complementary relationship of herbicide chemistry to that of the plant. Herbicides, as formulated, range from high to low aqueous solubility and mammalian toxicity. While most are less toxic to mammals than aspirin or table salt, a few, such as paraquat and dinoseb, are very toxic.

Effects on Plants

The effects of most herbicides on plants consist of a disruption of metabolic processes. Paraquat causes phytotoxic by-products within the plant that destroy tissue and dessicate and defoliate top growth. This kills most seedling plants, but established perennials often recover. Other herbicides stop cell division, some distort growth processes, and some disrupt enzyme systems. Many herbicides inhibit photosynthesis, and some interfere with natural growth-regulating chemicals in plants. Various herbicides enter the plant through the foliage, through roots, while some have multiple entry sites. Once inside the tissue, some herbicides move little if any from the site of entry, while others are translocated to other parts of the plant. Some move from the site of entry only through the transpiration stream (xylem), while others move in both the xylem and phloem.

Plant growth stage influences herbicide effectiveness. Some herbicides that kill seedlings do not kill mature plants, while others control susceptible plants of all ages. In most cases, small and actively growing weeds are most susceptible to herbicides. Knowledge of these and other factors is essential that herbicides be made effective.

Herbicide Selectivity

Selectivity to escape herbicide injury, is gained by timing, placement and the crop and chemical involved. Paraquat defoliates nearly all plants, but has no soil activity. Before the crop emerges, paraquat can be applied to growing weeds with complete safety. Glyphosate, with is relatively nonselective when applied to foliage, controls a broader spectrum of weed and crop species including perennials, but also can be used safely before emergence. This type of selectivity is confined to weeds and crops emerging after herbicide application. Species tolerance by physiological or biochemical action constitutes a second form of selectivity. Here, the crop either metabolizes the herbicide, immobilizes it by binding to a natural plant constituent, or is immune to its effects while susceptible weeds are sensitive. Within the current generation of herbicides, species selectivity has not developed to the point where all weeds are killed and all crops escape.

Selectivity may also be achieved by herbicide placement. Directed applications to small weeds at the base of more mature crop stems may be made safely with herbicides that would severely damage the crop if applied to the leaves and young stems. Selectivity by placement is obtained with glyphosate in recirculating sprayers and various wiping devices. Recirculating sprayers employ a horizontal spray stream above the crop canopy that is directed at a pan to catch unused spray. Weeds taller than the crop interrupt the spray stream and are treated, while the crop escapes. Other forms of wiping devices consist of a rope or carpet wet with herbicide solution and passed above the crop to wipe taller weeds. Both treat weeds extending above the crop canopy without treating the crop.

Rates of application (dosage) determine herbicide selectivity due to physiological dependence on a concentration differential at the active site of crop and weed. Rates required for vegetation control range from 2 to 4 kg/ha with some older herbicides (such as atrazine) but a few grams per hectare are sufficient for some materials now on the market and being developed (Anonymous, E. I. du Pont Co.). Each herbicide has a threshold rate of application below which there is no measurable effect on a target organism. Each plant species has a threshold level that also may vary with maturity and that provides a basis for regulating crop–weed selectivity. Various symptoms appear on target organisms as the rate is increased. The morphology of the plant may be changed, growth may be slowed or increased. Effects may persist for a short period; and the organism later resumes normal growth. As the rate is increased further, effects become more pronounced until death of the target plant occurs. Frequently, sublethal doses provide crop plants with a competitive advantage over weeds. Reduced weed vigor plus crop competition often provide an adequate level of control (Elliott, 1960).

Available herbicides vary in selectivity of crops and weeds, application requirements, crop safety, and effectiveness with small and large or annual and perennial weeds. Postemergence treatments in most crops consist of early-postemergence, overtop sprays, strictly directed sprays (directing the spray on small weeds under the crop, keeping spray off the crop foliage), and semidirected sprays (directing the spray toward the base of the crop plant with some of the lower crop leaves and stems contacted).

The crop must tolerate rates of sprays that control weeds present. Examples include atrazine and oil for small annual weeds in corn and sorghum; cyanazine in corn not beyond the 4-leaf stage; 2,4-D and dicamba for broadleaf annual and perennial weeds in corn and sorghum; sethoxydim and fluazifop for annual and perennial grass weeds in soybeans and cotton; bentazon and acifluorfen for small broadleaf weeds in soybeans, and DSMA, MSMA, or fluometuron for small broadleaf and grass weeds in cotton.

Selectivity in crops with little tolerance is gained by directing the spray to the base of the crop. This is accomplished by mounting spray nozzles on a rigid shank, a sliding or rolling support, and/or having shields to cover the crop. To be used effectively, a height difference between crop and weed is necessary. Examples include DSMA, MSMA, fluometuron, diuron, cyanazine, linuron, dinoseb, and oxyfluorfen in cotton; linuron, metribuzin, dinoseb, and paraquat for grass and broadleaf weeds in soybeans; and ametryn and linuron in corn. These and similar herbicides act mainly through contact activity and defoliate small weeds (including the crop if the foliage is sprayed). Directing sprays may be more difficult in no-tillage fields, if tall crop stubble (such as in double-crop soybeans, where the small grain stubble is high) or if tall weeds are present. Unless the crop is shielded with some type of fender, splashing of the chemical onto the crop could occur. Examples of semidirected sprays are 2,4-D and dicamba on larger corn. At this time, weeds need to be smaller than the corn for effective results. A listing of weed species controlled, timing, method of application, and crop safety considerations are found on the label of each herbicide.

Herbicides in Soil

The principles noted above apply to both root- and foliar-active herbicides. Soil activity is required to control seedlings as they emerge. Applied to soil, most herbicides decay exponentially, and the active herbicide must be maintained above the threshold level until the crop is well established. This may be up to 4–8 weeks, depending on the crop and weed species involved (Klingman and Ashton, 1982). Rates of herbicide decay vary widely between chemicals. With some a short persistence may be adequate. Other herbicides persist longer, and control weeds longer, but may injure sensitive crops that follow in the rotation. Herbicide labels warn the user of these hazards and opportunities.

Soil characteristics have little effect on foliar-active herbicides. Soil-applied herbicides initiate processes that reduce the amount of active herbicide present. To varying degrees, some herbicides are volatilized from the soil surface and/or are destroyed by sunlight (photodecomposition). In some situations, these processes are so rapid that herbicides must be incorporated immediately or injected into the soil to be effective. Other herbicides are adsorbed to varying degrees by soil clay particles and organic matter. Adsorption is so rapid that rates of these herbicides must be adjusted to compensate for high and low soil organic matter and clay contents. Although the herbicide remains present in the soil, it may be bound so tightly to clays or organic particles as to be unavailable to plants and becomes biologically inactive.

Paraquat and glyphosate are thus bound to soil colloidal matter and immediately inactivated on soil contact. Others, such as triazines and ureas, remain active in the soil for several weeks. The dinitroaniline herbicides are active only in soil and are effective only before weeds emerge; emerged weeds present at application must be destroyed by other means. Many triazine and urea herbicides possess both soil and foliar activity and have a wide latitude of effective timing.

As organic compounds, some herbicides are subject to biological decomposition, serving as energy sources for various soil microorganisms. This in turn limits persistence in the soil. With repeated use, the inactivation rate of these herbicides increases because of increases in microflora inocculum. Certain thiocarbamates are thus affected, as is 2,4-D, and the increased breakdown rate renders them effective for a limited period of time.

Leaching contributes to disappearance of soil-applied herbicides. The movement by leaching of some chemicals accounts for most of the removal of a herbicide such as TCA from the soil. However, leaching is beneficial to movement and placement of the chemical in the weed seed zone and may account for selectivity or lack of it.

Tillage Changes

Important soil changes take place with continuous no-tillage. Soil organic matter content increases and its distribution in the soil profile is changed.

Plow-tillage buries organic debris, whereas that which accumulates on the surface of untilled soil (see Chapter 2) may reduce herbicide effectiveness. High soil organic matter content increases absorption of most herbicides, requiring increased rates of application. Because of these and other changes, the dynamic herbicide performance in untilled soil cannot be predicted well from performance evaluations conducted on tilled soils. Therefore, herbicide combinations must be evaluated for the particular cropping system in use. Changes in herbicide performance with changes in tillage apply primarily to soil activity.

Climatic Effects

Weather influences both foliar effectiveness and the persistence of herbicides in the soil. Herbicides active with foliar application should not be applied if heavy rainfall is imminent, since washoff reduces efficacy. Soil-applied herbicides that do not require incorporation often require 1–2 cm of rainfall for activation. This is important with some herbicides, those used to inhibit seedling emergence. Weeds escape unless the herbicide is activated at early seedling emergence. Alachlor, metolachlor, oryzalin, and others fit this category. Triazine and urea herbicides, on the other hand, kill weed seedlings after emergence. Delayed activation of these compounds with low soil moisture for 2–3 weeks does not reduce their effectiveness markedly and low rainfall serves to extend their persistence. Temperature also exerts a major effect on herbicide persistence; herbicide activity in soil is extended with cool temperatures.

Acidity

Soil pH influences persistence and crop tolerance to some herbicides, especially the triazines. The pH of soil surface layers may be reduced to 4.5 or less with repeated surface applications of acid-forming nitrogen fertilizers. Below pH 4 or 5 the triazine molecule is hydrolized rapidly and the herbicide is inactivated. Above pH 7.5, metribuzin (a triazine used for soybean weed control) becomes more active, and crop safety is reduced. Decrease of the metribuzin application rate helps maintain crop safety at high soil pH levels.

Herbicide Application

To be effective, herbicides must be applied properly, and several important differences exist between tilled and untilled systems. The goal with soil-active herbicides is uniform distribution of the pesticide over the surface. With foliage applications, complete coverage of the target species is needed for nontranslocated herbicides, such as paraquat. If the spray application does not penetrate the vegetative canopy, weeds in the understory will not be controlled. Herbicides that are readily translocated require contact with only a part of the foliage to be effective, such as with glyphosate applied with wiping devices or recirculating sprayers. Granular (applied dry) and sprayable formulations of several herbicides used in reduced-tillage systems are available. Granular applications sometimes are made in bands over tilled, weed-free rows or after emerged weeds have been killed by a contact herbicide. Presence of established weeds requires liquid formulations.

Proper herbicide application is even more important in no-tillage systems because postemergence mechanical cultivation is rarely an option to correct application mistakes. Basic liquid delivery systems consist of a tank to hold the dilute spray solution, a pump to move it, and an atomizer system to deliver the pesticide to the target. Many herbicides are applied as suspended solids or emulsions, necessitating agitation in the spray tank. Some sprayers are equipped for mechanical agitation but hydraulic stirring through excess pump capacity is common. The pesticide must be mixed into the tank with agitation while the tank is being filled. If spraying and agitation are interrupted for several hours, agitation must be sufficient to remix the pesticide for resuspension of solid particles that have settled.

Atomizers or nozzles break the solution into small droplets and distribute them over the area to be treated. Patterns of hydraulic nozzles include fans and hollow cones. Droplet size is determined by nozzle characteristics and pressure. Small droplets are produced at high pressures and provide better coverage but are subject to drift by wind currents. Controlled droplet applicators (CDAs) produce droplets of uniform size by directing the spray solution onto a spinning disk. To date, these have not proved more effective for reduced tillage than hydraulic nozzles (Reichard and Triplett, 1983).

Application rate is a function of pesticide concentration in the spray tank and the spray volume per unit area. Spray volume per area is a function of atomizer flow rate, spacing, and speed of travel. The operator must recognize these interrelationships and make calibration adjustments to obtain the desired active ingredient (ai) per unit area. Flow rates of specific atomizers vary with the square root of line pressure. Thus, only minor flow rate modifications are convenient by adjusting pressure on farm spray equipment. However, atomizer sizes include a range of flow rates and these may be selected to provide the spray volumes and coverage desired. Nozzle spacing usually is fixed, making ground speed and concentration of the spray solution the most easily manipulated variables. Volume applied is inversely proportional to ground speed.

The spray volume selected represents a compromise between several factors. High volumes, of 150–200 liters per hectare, are recommended for optimum performance of some herbicides, which creates a logistics problem of water supply to the field. Volumes of 20–50 liters per hectare may be adequate for soil-active herbicides and may be preferred for some foliar-active herbicides such as glyphosate and the postemergence grass herbicides. Tall weeds or cover crops often are present in no-tillage systems. Spray booms must be high enough for the nozzle pattern to overlap above the canopy to get needed coverage.

Liquid fertilizers are sometimes used as carriers for herbicides. Spray volumes of 200–400 liters per hectare often are needed to apply required amounts of nutrients. Combining fertilizer and pesticide in a single application is convenient and often is done by custom applicators. Compatibility of pesticides and liquid fertilizer formulations must be determined. Fluid fertilizer suspensions containing clay may inactivate some herbicides, such as paraquat and glyphosate.

Surfactants as spray additives are required for best performance of most foliar-active herbicides. Surfactants lower surface tension of the solution and increase spread of individual spray droplets on the foliage of target species, enhancing penetration of the pesticide through the cuticle. Some pesticide formulations include surfactants, others may require additions. Several types of surfactants are available, including nonionic polar molecules and petroleum oils containing emulsifiers. The surfactant must be matched with the pesticide to ensure performance.

Personnel and crop safety are important concerns. Herbicide movement to nontarget areas is not great with proper application techniques; however, some are toxic to mammals and drift of phenoxy herbicides and dicamba cause injury to sensitive crops. This can be minimized by careful application and timing. Nonvolatile formulations reduce vapor drift of phenoxy herbicides and dicamba. Use of nozzles that produce coarse droplets, low line pressure, and/or a spray drift reduction additive reduces drift of spray particles. Movement of soil-applied herbicides can be by runoff or leaching. Greatest movement in runoff arises from heavy rainfall immediately following application (Triplett et al., 1978); with some herbicides, this may be 5% of that applied. Transport is usually less than 1% of that applied if runoff does not occur within 7–10 days after application. Much of the transported herbicide may be adsorbed to soil particles or organic matter; if so, the activity of the chemical in nontarget areas is reduced. Surface and no-tillage systems reduce runoff and soil erosion, thus lessening the movement of all pesticides from treated fields. To avoid damage to current or following crops, applications with excessive overlap of the spray swath and wind drift of spray particles should be avoided.

Protective clothing and face shields should be used to avoid contact of the pesticide with eyes, nasal passages, or skin. Spills should be promptly flushed with water. Programs to train and license applicators do much to improve safety of pesticide use.

Adoption of reduced- and no-tillage practices has important effects on herbicide application, use, and effectiveness (Hayes, 1982, Soil Cons. Soc. of Amer., 1983; Turner, 1983; Young, 1982). Dependence on herbicides to destroy existing vegetation becomes complete as tillage operations are eliminated. This requirement is normally met by adding a herbicide with foliar activity, such as paraquat or glyphosate, to the residual herbicide(s) selected to control weeds that emerge later. An evaluation of vegetation present at the site must be made before selecting the materials and rates. Herbicides requiring incorporation cannot be employed with no-tillage. This includes most of the dinitroanilines and all of the thiocarbamate herbicides. Recent research, however, is exploring means of increasing the effectiveness of volatile thiocarbamate herbicides with no-tillage by impregnation onto fertilizer granules and by injection (Richardson, 1983).

HERBICIDE SYSTEMS

The above discussion of herbicide technology, while useful in outlining parameters, does not provide information on use characteristics of specific products; nor does it match herbicides, or combinations, to specific situations. Qualities of some herbicides, and combinations for weed control in reduced-tillage crop production are listed in Table 11.2. Most of this information is included on the pesticide label; however, some points may be obscure. Important information not included here is a detailed listing of major weed species controlled by a particular herbicide. This information appears on the pesticide label along with suggestions for tank mix combinations. A list of species not controlled may or may not be included on the label or in the accompanying literature.

Characteristics

Advisory information in Table 11.2 includes:

Crop use. If one crop fails, other named crops may be planted safely the same season following the use of this herbicide. Herbicides registered for use in one geographic area may not be registered for use on that same crop in other areas.

Time of application (to crop). Some herbicides, such as paraquat can only be used before crop emergence. Others may be applied safely to emerged crop plants or as over-the-top sprays, directed sprays or semi-directed sprays.

Soil activity (time). Wide variations shown here reflect herbicide rate, soil, and climatic effects. Six to eight weeks residual life is usually required for season-long weed control. Extended residual activity can injure sensitive crops that follow.

Species controlled. General categories of broadleaf and grass weeds are given. The effect on annuals or established perennials within these categories is also important.

Movement in plants. Herbicides with movement in the xylem (transpiration stream) enter roots and move to the foliage. When applied to foliage, these products will not move down to kill roots. Other herbicides are translocated through phloem both to root and stem growing points. Some herbicides move in either direction.

Foliar activity. Some of the herbicides act to kill top growth of emerged vegetation, while others are ineffective.

TABLE 11.2. Characteristics of Herbicides Widely Used in Reduced Tillage Crop Production

Quality	Atrazine	2,4-D	Paraquat	Alachlor
Crop Use	Corn, sorghum	Corn, sorghum, small grains	Most	Corn, soybeans sorghum (with seed safener)
Time of application	Pre- and early postemergence	Pre- and post-emergence	Preemergence	Pre- and post-emergence
Kill emerged plants	Yes (if small)	Yes (broadleafs)	Yes	No
Soil activity (time)	6–20 weeks	2–10 days	None	3–6 weeks
Species controlled				
Broadleaf spp.	Most seedlings	Many annuals and perennials	Most seedlings[a]	A few
Grass spp.	Most seedlings, some established perennials	Seedlings, at high rates	Most seedlings	Most seedlings
Movement in plants	In xylem	Through phloem	Very slight	None
Foliar activity	Good	Excellent	Excellent	None
Crop Safety				
Preemergence	Excellent	Good	Excellent	Good
Postemergence	Excellent	Good	Directed only	Good
Adjuvant effects	Yes	Yes	Yes	No
Use rates (kg/ha)	2–4	0.25–1.0	0.25–0.5	2–4
Drift hazard	Little	Moderate	Some	Little
Washoff hazard	No	Yes, 2 hr	Yes, 3 hr	No
Minimum temperature requirement	No	5–10°C	5–10°C	No
Antagonism with other herbicides	No	Yes	Yes	No

Crop safety. Some herbicides injure crops if improperly applied. Others have a relatively wide latitude of crop safety with respect to timing and rates.

Adjuvant effects. For foliar-active herbicides adding crop oil, or some other wetting agent, increases activity. Little benefit can be expected in soil activity or persistence.

Use rates. This is conditioned by soil type and crop and/or weed species and growth stage.

Drift hazard. Most herbicides do not drift to cause damage to nontarget species when applied properly. For the few that do volatilize and/or have spray droplets that drift to damage sensitive crops, special application precautions

TABLE 11.2. *(Continued)*

Linuron	Glyphosate	Bentazon	Fluazifop
Soybeans, corn, wheat	Most	Soybeans, corn	Soybeans, cotton
Preemergence postdirected	Preemergence	Postemergence	Postemergence
Yes	Yes	Yes	Yes
4–10 weeks	None	Slight–none	Slight
Many seedlings	Most species	Seedlings annuals	None
Small seedlings	Most species	None	Annual, perennial
In xylem	In phloem	None	In phloem
Yes	Yes	Yes	Yes
Rate and soil sensitive	Excellent	N/A	N/A
Only directed	Only directed	Good	Excellent
Yes	Yes with low carrier volume	Yes	Yes
0.5–2.0	1.0–4.0	0.5–1.0	0.2–0.4
Little	Little	None	Little
No	Yes (5 hr)	Yes (8 hr)	Slight
No	10°C	Yes[b]	Yes[b]
No	Yes	Yes	Yes

(continued)

must be followed. Application should be when wind speed is low and the direction away from sensitive crops.

Wash-off hazard. Rainfall soon after application removes the herbicide from the plant and reduces its effectiveness.

Temperature requirements. Many herbicides are active only through foliage, and plants must be metabolically active at the time of application. Low temperatures slow growth and reduce herbicide activity.

Antagonism with other herbicides. In some combinations, effectiveness of one or both herbicides can be reduced or enhanced. Paraquat, for example, defoliates plants rapidly. Combining paraquat with herbicides such as

TABLE 11.2. *(Continued)*

Quality	Simazine	Pendimethalin	Dicamba
Crop Use	Corn	Soybeans, corn	Corn, small grains
Time of application	Preemergence	Preemergence	Pre- and postemergence
Kill emerged plants	Slight activity	No	Yes (broadleafs)
Soil activity (time)	10–40 weeks	8–20 weeks	1–3 weeks
Species controlled			
Broadleaf spp.	Most seedlings	Seedlings of some species	Wide range, annual and perennial
Grass spp.	Most seedlings	Seedlings, broad spectrum	None
Movement in plants	In xylem	None	In phloem
Foliar activity	Slight	None	Yes
Crop Safety			
Preemergence	Excellent	Good	Good
Postemergence	Good	Good	Good
Adjuvant effects	Slight	No	Yes
Use rates, kg/ha	1–4	0.5–2	0.25–1
Drift hazard	None	None	Significant (broadleaf crops)
Washoff hazard	N/A	N/A	Yes, 2 hr
Minimum temperature requirement	No	No	5–10°C
Antagonism with other herbicides	No	No	Yes

glyphosate, 2,4-D, bentazon, or fluazifop, that are active primarily through foliar absorption, reduces their effectiveness. The paraquat effect is additive when combined with most soil-active herbicides such as atrazine, cyanazine, linuron, and metribuzin, which also exert some degree of foliar activity.

Components of the System

Existing Vegetation Control

Complete control of existing vegetation at planting is essential before crop emergence in no-tillage systems, except in cases where one crop is interplanted into another (Anonymous, 1983). This control is accomplished best with a quick-acting, contact herbicide, such as paraquat, or a slower-acting,

TABLE 11.2. *(Continued)*

Metolachlor	Metribuzin	Sethoxydim	Acifluorfen
Corn soybeans, sorghum (with seed softener)	Soybeans, wheat	Soybeans, cotton	Soybeans
Preemergence	Preemergence, post directed	Postemergence	Postemergence
No	Yes	Yes	Yes
4–8 weeks	3–8 weeks	Slight	Slight
A few	Most seedlings	None	Most seedlings
Most seedlings	Many seedlings	Annual, perennial	Very small, annual
None	In xylem	In phloem	Little
None	Yes	Yes	Yes
Good	Fair–good	N/A	N/A
Good	Good (directed)	Excellent	Good
No	Yes	Yes	Yes
1–3	0.4–0.7	0.2–0.4	0.4–0.6
Little	Little	Little	Some
No	No	Yes, 2 hr	Yes, 6 hr
No	No	Yes[b]	Yes[b]
No	No	Yes	Yes

[a] Defoliates large plants, which may recover.
[b] Avoid cool temperatures when weeds are not actively growing.

translocated herbicide, such as glyphosate. In some instances of sparse populations of very small annual weeds, residual herbicides with contact activity, such as cyanazine, atrazine + crop oil, linuron, or metribuzin, may be used satisfactorily at planting without a contact herbicide.

Analysis of the weed spectrum and stage of growth before and at planting is essential for the grower to determine the herbicide and rate required to control the weeds most effectively and economically (Table 11.3). Different situations frequently dictate different treatments. For example, no-tillage corn plantings may be made into perennial grass or legume sods, annual cover crops (grasses or legumes), annual broadleaf and grass weeds, and a few perennial broadleaf and grass weeds. Surface-tillage soybeans have not been recommended even when low infestations of perennial weeds are present. However, the availability of new postemergence herbicides (sethoxydim, fluazifop) now makes possi-

TABLE 11.3. Examples of Situations in Surface-Tillage Cropping and Herbicides to Use for Vegetation Control at Planting

Situation at Planting	Herbicide	Rate (kg/ha)
A few perennial weeds or legumes present	Glyphosate	2–4.5
Annual broadleaf weeds and grasses over 15 cm tall, small grain cover crops	Glyphosate	1.7
Annual broadleaf weeds and grasses less than 15 cm tall	Glyphosate	1.0
Cutleaf evening primrose, wild lettuce, horseweed	Glyphosate	1.7
Annual broadleaf and grass weeds 2–8 cm tall[a]	Paraquat	0.3
Rye (before jointing)	Paraquat	0.3 + 0.3[b]
Rye (after jointing)	Paraquat	0.3
Annual broadleaf weeds 10–15 cm tall wheat, oats, barley cover crops	Paraquat	0.6
Interplanting crops or overseeding pastures	Paraquat	0.3–0.6

[a] Pennsylvania smartweed, common ragweed, and lambsquarters are some of the annual weeds which become resistant to paraquat as they grow beyond the 8 cm stage.
[b] One application about 1 week prior to planting and a second at planting.

ble the control of perennial grass weeds in soybeans and some other broadleaf crops.

Residual Control

Herbicides used for residual control of annual weeds in surface tillage cropping systems are essentially the same as those used in plow-tillage systems where similar weed species and populations are present. One exception is herbicides that must be soil incorporated (most dinitroanilines and thiocarbamates) and cannot be used with no-tillage or where large amounts of crop residue remain on the soil surface. Many preemergence herbicide labels and accompanying literature give directions for shallow soil incorporation with moderate amounts of surface mulch. This allows use of these herbicides while maintaining enough cover to reduce soil erosion.

Postemergence Control

Controlling weeds with postemergence herbicides in reduced-tillage crops differs little from methods and chemicals used in plow-tillage systems. An array of herbicides that are applied postemergence to crops and weeds is available for use with most agronomic crops. In most reduced-tillage systems there generally is more reliance on postemergence herbicides. They are invaluable tools in controlling escaping weeds or those tolerant to preemergence applications.

Formulating Herbicide Systems

A major goal of the reduced-tillage grower is to match herbicide characteristics with weed species and crops grown to implement season-long vegetation control, crop safety, and no residue in the soil to harm the next crop. Principles and herbicide characteristics noted above are illustrated by the rationale of several different strategies used in middle and eastern United States.

EXAMPLE 1. Corn is to be planted following an alfalfa–grass meadow in the upper midwest and northeast. Besides alfalfa and orchardgrass, weedy broadleaf perennial species such as dandelion and curly dock have invaded the sward. Annuals have been suppressed by perennial plants and are a minor consideration. Disking alone will not control the established sod effectively but serves to stimulate weed seed germination.

Atrazine + Glyphosate. This combination controls all species listed, but foliage kill is less rapid than with herbicide mixtures containing paraquat. Thus, application one to several weeks before planting may be desirable to prevent excessive soil moisture depletion by the sward. No follow-up application should be required.

Atrazine + Crop Oil + 2,4-D (or dicamba). When applied at high rates to vegetative plants (before heading), atrazine will control cool-season perennial grasses such as orchardgrass, quackgrass, timothy, and Kentucky bluegrass. Oil increases rate of foliar kill with atrazine but does not interfere with the action of 2,4-D or dicamba and broadleaf species are controlled. Atrazine + oil applied early postemergence provides good crop safety for emerged corn and controls many weeds missed by the initial spray application.

Atrazine + Paraquat. This is not the best choice for this weed situation. This combination will kill orchardgrass but merely defoliates broadleaf perennial species, which recover to compete with the crop. These weeds can be controlled with an application of 2,4-D or dicamba after broadleaf species regrow. Tank mixing 2,4-D and, to a lesser extent, dicamba with paraquat reduces their effectiveness, so a separate application is required.

Therefore, several effective options exist for control of a mixed sod where corn is to be planted (Triplett, 1985). A final factor that also must be considered in selection of a system is herbicide cost.

EXAMPLE 2. Corn following an annual crop with or without a cover crop. Here the weed spectrum normally is comprised of summer and/or winter annuals. Some of these may be well established at time of herbicide application while seeds of other weed species in the soil may germinate later. After emerged plants are controlled, either by tillage or herbicides, emphasis shifts to soil residual control or postemergence herbicides. Several different herbicide combinations and strategies may be considered:

Atrazine + Paraquat. This combination controls a wide spectrum of

TABLE 11.4. Common and Botanical Names of Crops and Weeds

Common name	Botanical Designation
	Weeds
Bermudagrass	*Cynodon dactylon* (L.) Pers.
Bigroot morningglory	*Ipomoea pandurata* (L.) F.G.W. Meyer
Bluestem	*Andropogon* spp.
Briars	*Rubus* spp.
Broadleaf signalgrass	*Brachiaria platyphylla* (Griseb.) Nash
Broomsedge	*Andropogon virginicus* L.
Burcucumber	*Sicyos angulatus* L.
Bursage, woolyleaf	*Ambrosia grayi* (A. Nels.) Shinners
Canada thistle	*Cirsium arvense* (L.) Scop.
Climbing milkweed	*Sarcostemma cynanchoides* Dcne.
Common milkweed	*Asclepias syriaca* L.
Cheat	*Bromus secalinus* L.
Cocklebur	*Xanthium strumarium* L.
Croton	*Croton* spp.
Curly dock	*Rumex crispus* L.
Dallisgrass	*Paspalum dilatatum* Poir
Dandelion	*Taraxacum officinale* Weber in Wiggers
Fall panicum	*Panicum dichotomiflorum* Michx.
Field bindweed	*Convolvulus arvensis* L.
Foxtail	*Setaria* spp.
Groundcherry	*Physalis* spp.
Goatgrass, jointed	*Aegilops cylindrica* Host
Hemp dogbane	*Apocynum cannabinum* L.
Hogpotato	*Hoffmanseggia glauca* (Ortega) Eifert
Horsenettle	*Solanum carolinense* L.
Horseweed	*Conyza canadensis* (L.) Cronq.
Itchgrass	*Rottboellia exaltata* (L.) F.
Johnsongrass	*Sorghum halepense* (L.) Pers.
Kentucky bluegrass	*Poa pratensis* L.
Lambsquarter	*Chenopodium album* L.
Leafy spurge	*Euphorbia escula* L.
Lanceleaf sage	*Salvia reflexa* Hornem
Morningglory	*Ipomoea* spp.
Nutsedge, purple	*Cyperus rotundus* L.
Nutsedge, yellow	*Cyperus esculentus* L.
Pigweed	*Amaranthus* spp.
Poison ivy	*Rhus radicans* L.
Quackgrass	*Agropyron repens* (L.) Beauv.
Redvine	*Brunnichia ovata* (Walt.) Shinners
Russian knapweed	*Centaurea repens* L.
Sandbur	*Cenchrus* spp.
Shattercane	*Sorghum bicolor* (L.) Moench.
Sicklepod	*Cassia obtusifolia* L.

TABLE 11.4. *(Continued)*

Common name	Botanical Designation
Silverleaf nightshade	*Solanum elaeagnifolium* Cav.
Smartweed	*Polygonum* spp.
Spurges	*Euphorbia* spp.
Spurred anoda	*Anoda cristata* (L.) Schlecht.
Sunflower	*Helianthus annuus* L.
Timothy	*Phleum pratense* L.
Switchgrass	*Panicum virgatum* L.
Texas blueweed	*Helianthus ciliaris* D.C.
Texas panicum	*Panicum texanum* Buckl.
Trumpetcreeper	*Campsis radicans* (L.) Seem. ex Bureau
Unicorn plant	*Proboscidea louisianica* (Mill.) Thellung
Venice mallow	*Hibiscus trionum* L.
Velvetleaf	*Abutilon theophrasti* Medik.
Wild oat	*Avena fatua* L.
Wild proso millet	*Panicum miliaceum* L.
	Crops
Alfalfa	*Medicago sativa* L.
Corn (maize)	*Zea mays* L.
Cotton	*Gossypium hirsutum* L.
Orchardgrass	*Dactylis glomerata* L.
Peanut (ground nut)	*Arachis hypogaea* L.
Sorghum	*Sorghum bicolor* (L.) Moench
Soybean	*Glycine max* (L.) Merr.

broadleaf and grass species but is ineffective on later-germinating fall panicum. Use of this combination for two or three seasons will shift the weed population to fall panicum dominance. During the first year, when fall panicum is not a dominant species, atrazine + paraquat can be used effectively.

Atrazine + Paraquat + Alachlor (or) Metolachlor (or) Pendimethalin. Inclusion of one of the three latter herbicides increases preemergence activity of the mixture against seedling grasses such as fall panicum and seedling johnsongrass. Care must be taken with pendimethalin to be sure the planter slot is closed and seed are covered to avoid corn injury. The short duration residual activity of alachlor may not provide full-season control and metolachlor gives slightly longer control. With these mixtures, glyphosate can be substituted for paraquat to increase control of perennials or large annual grass plants.

Atrazine + Simazine + Paraquat. Inclusion of simazine in the mixture greatly increases length of residual activity and controls fall panicum. At higher rates residual activity of simazine may persist and injure sensitive crops that follow.

Early postemergence herbicides represent an option only if fall panicum or other difficult-to-control grasses are not present. None of those listed are effective against established fall panicum; however, cyanazine may be used when corn has from one to four leaves. Broadleaf species may be controlled with 2,4-D, dicamba, or atrazine + oil. Atrazine + oil also controls many seedling grasses, including foxtails. If a differential height between corn and weeds develops, a postemergence directed application of linuron or ametryn can be used to control emerged annual broadleaf or grass weeds. A postemergence directed application of linuron or ametryn along with any earlier weed control treatment usually is necessary in southeastern United States, where broadleaf signalgrass or Texas panicum are present.

EXAMPLE 3. Soybeans following winter wheat or barley. When soybeans are planted as a second crop following small grain harvest, competition from the growing grain crop suppresses summer annual weeds and shifts species composition. Weeds present are etiolated seedlings that develop rapidly when crop competition is removed following small grain harvest. The weed control objective is to eliminate plants present and to prevent reinfestation until the new crop canopy closure. Both residential and postemergence herbicides may be used in this system (Lewis, 1985).

Paraquat + Linuron + Alachlor (or) Metolachlor (or) Oryzalin. This combination controls most annual weeds in the stubble but may not control horeseweed and smartweed. Residual activity usually does not control large seeded broadleaf species such as cocklebur, morningglory, and velvetleaf. Substitution of glyphosate for paraquat improves control of well-established plants. Postemergence application of bentazon or acifluorfen controls most annual broadleaf species, and fluazifop or sethoxydim controls most grass species, including johnsongrass and fall panicum, with good crop safety. Several herbicides and combinations of herbicides can be used as postemergence directed sprays to control broadleaf weeds; however, the soybeans must be taller than the weeds.

Paraquat + Metribuzin + Alachlor (or) Metolachlor (or) Oryzalin. The metribuzin in this combination generally improves residual control of broadleaf weed species and increases foliar activity. Use of metribuzin preemergence is prohibited on coarse (sandy) soils with less than 2% organic matter due to risk of crop injury.

Small grains produced in arid areas represent a major use of reduced-tillage systems (Chapter 7). Wicks deals with fallow crop systems for small grains and sorghum and these are not covered here since weed populations are specific to dry areas. The basic approach of matching herbicides to weeds present, however, applies to crops produced in all surface-tillage environments. The examples above are intended to illustrate the rationale of general principles. For specific situations growers should contact specialists with experience in local conditions.

SUMMARY

Basic weed-management principles are applicable both to conventional and surface-tillage crop culture. As tillage is reduced, reliance on chemical weed control increases and the successful producer must develop detailed understanding of how herbicides function, their strengths and limitations, if they are to develop effective systems. Herbicide performance cannot be extrapolated safely from conventional to surface-tillage systems. Each system must be an integrated entity. New herbicides as they become available must be evaluated and fitted into existing systems or used to formulate new systems. Expansion of surface- and no-tillage to different crops and into new areas will continue as new herbicides and opportunities for weed control develop.

LITERATURE CITED

Aldrich, R. J. 1984. *Weed-Crop Ecology*. Breton Publishers, North Scituate, MA.

Anonymous. 1983. Control vegetation for successful no-till corn. *Conservation Tillage Guide*. Successful Farming, Des Moines, IA, p. 14.

Anonymous. undated. Glean weed-killer technical bulletin. Publication No. E-46095. E. I. duPont de Nemours and Company.

Bandeen, J. D., G. K. Stephenson and E. R. Cowett. 1982. Discovery and distribution of herbicide-resistant weeds in North America. In Homer M. LeBaron and Jonathan Gressel (ed.). *Herbicide Resistance in Plants*, Wiley-Interscience, New York, pp. 9–30.

Elliott, J. G. 1960. The evaluation of herbicidal activity on a mixed sward. Proc. 8th Int. Grassl. Congr. Reading, U.K., pp. 267–271.

Hammer, C. L. and H. B. Tukey, 1944. The herbicidal action of 2,4-dichlorophenoxyacetic acid and 2,4,5-trichlorophenoxyacetic acid on bindweed. *Science* **100**:154–155.

Harper, J. L. 1957. Ecological aspects of weed control. *Outlook Agric.* **1**(6):197.

Hayes, W. A. 1982. *Minimum Tillage Farming*. No-Till Farmer, Inc., Brookfield, WI.

Klingman, Glenn C. and F. M. Ashton. 1982. *Weed Science: Principles and Practices,* 2nd ed. Wiley-Interscience, New York.

Lewis, W. M. 1985. Weed control in reduced tillage soybean production. In A. F. Wiese (ed.). *Weed Control in Limited-Tillage Systems.* Weed Sci. Soc. Amer. Monogr. Ser. No. 2, Champaign, IL.

Lewis, W. M. and A. D. Worsham. 1981. Weed management in no-till. In W. M. Lewis (ed.). No-till crop production systems in North Carolina—corn, soybeans, sorghum, and forages. N. C. Agric. Ext. Serv. Bull. AG-273, pp. 8–11.

Liebl, Rex A. and A. D. Worsham. 1983. Inhibition of pitted morningglory (*Ipomoea lacunosa* L.) and certain other weed species by phytotoxic components of wheat (*Triticum aestivum* L.) *Straw. Jour. Chem. Ecol.* **9**(8):1027–1043.

Magleby, R., D. Gadsby, D. Colacicco, and J. T. Thigpen. 1984. Conservation tillage—who uses it now. Conference Proceedings, Nat. Conf. on Cons. Tillage - Strategies for the Future. Cons. Till. Info. Ctr., Fort Wayne, IN, pp. 73–74.

McWhorter, C. G. and J. M. Chandler. 1982. Conventional weed control technology. In R. Charudattan and H. Lynn Walker (eds.). *Biological Control of Weeds with Plant Pathogens.* Wiley-Interscience, New York, pp. 5–27.

Moncrief, J. 1984. Where we're headed with conservation tillage. *Solutions Magazine* **28**(3): 17–22.

Putnam, A. R. and J. DeFrank. 1983. Use of phytotoxic plant residues for selective weed control. *Crop Prot.* **2**(2):173–181.

Radosevich, S. R. and J. S. Holt. 1984. Plant growth and interference. In *Weed Ecology Implications for Vegetation Management.* Wiley-Interscience, New York, pp. 118–121.

Reichard, D. L. and G. B. Triplett, 1983. Paraquat efficacy as influenced by atomizer type. *Weed Sci.* **31**:779–782.

Rice, E. L. 1984. *Allelopathy.* 2nd ed. Academic Press, Orlando FL.

Richardson, L. 1983. New tools adapt PPI herbicides to reduced-tillage farming. In *Tillage for the Times.* Agrichemical Age.

Schweizer, E. E. and R. L. Zimdahl. 1984a. Weed seed decline in irrigated soil after six years of continuous corn (*Zea mays*) and herbicides. *Weed Sci.* **32**:76–83.

Schweizer, E. E. and R. E. Zimdahl. 1984b. Weed seed decline in irrigated soil after rotation of crops and herbicides. *Weed Sci.* **32**:84–89.

Shilling, Donn G., R. A. Liebl, and A. D. Worsham. 1985. Rye (*Secale cereale* L.) and wheat (*Triticum aestivum* L.) mulch: the suppression of certain broadleaved weeds and the isolation and identification of phytotoxins. In A. C. Thompson, (ed.). *The Chemistry of Allelopathy.* Amer. Chem. Soc. Symposium Series No. 268, Amer. Chem. Soc., Washington, DC, pp. 247–271.

Soil Cons. Soc. of Amer. 1983. Conservation tillage a special issue. *Jour. Soil Water Cons.* **38**(3).

Triplett, G. B., Jr., 1985. Principles of weed control for reduced tillage corn production. In A. F. Wiese (ed.). *Weed Control in Limited-Tillage Systems.* Weed Sci. Soc. America Monograph Series No. 2. Champaign, IL.

Triplett, Glover B., Jr., J. R. Abernathy, C. R. Fenster, W. Flinchum, D. L. Linscott, E. L. Robinson, L. Standifer, and J. D. Walker. 1983. *Weed Control for Reduced Tillage Systems.* Minnesota Agric. Ext. Serv. Pub. AD-FO-2279.

Triplett, G. B., Jr., B. J. Conner, and W. M. Edwards. 1978. Transport of atrazine and simazine in runoff from conventional and no-tillage corn. *J. Environmental Qual.* **7**:77–84.

Triplett, G. B., Jr. and G. D. Lytle. 1972. Control and ecology of weeds in continuous corn grown without tillage. *Weed Sci.* **20**:453–457.

Turner, J. H. (ed.). 1983. *Fundamentals of No-Till Farming.* Amer. Assoc. for Vocational Instr. Materials, Driftmier Engineering Center, Athens, GA.

Witt, W. W. 1984. Response of weeds and herbicides under no-tillage conditions. In Ronald E. Phillips and Shirley H. Phillips (eds.). *No-Tillage Agriculture Principles and Practices.* Van Nostrand Reinhold Co., New York, pp. 152–170.

Worsham, A. D. 1980. No-till corn—its outlook for the '80's. *Proc. Ann. Corn and Sorghum Res. Conf., Amer. Seed Trade Assoc.* **35**:146–163.

Worsham, A. D. 1985. No-till tobacco (*Nicotiana tabacum*) and Peanuts (*Arachis hypogaea*). In Allen F. Wiese (ed.). *Weed Control in Limited-Tillage Systems.* Weed Sci. Soc. Amer. Monogr. Ser. No. 2, Champaign, IL.

Young, H. M., Jr. 1982. *No-Tillage Farming.* No-Till Farmer, Inc., Brookfield, WI.

12

MANAGEMENT OF VERTEBRATE AND INVERTEBRATE PESTS

J. N. ALL
Professor of Entomology
Department of Entomology
University of Georgia
Athens, Georgia
G. J. MUSICK
Professor and Head
Department of Entomology
University of Arkansas
Fayetteville, Arkansas

INTRODUCTION

As new agricultural systems evolve, apprehensions arise concerning the association between tillage operations and pest outbreaks. Conservation tillage systems are not new; some of these systems were used in early times when high technology in the form of sophisticated equipment and effective agrichemicals was not available. Today, conservation tillage systems are common to many regions of the world. These systems have been adopted by single-family subsistence farmers who must minimize production inputs by practical necessity. Low yields are typical in these situations and are due to a variety of factors, including the depredation of insects.

It is important to recognize that the current agronomic practices associated with conservation-tillage agriculture involve new methodologies. Generally, there is a reliance on cost-effective modern technological innovations. Pest problems are viewed within the context of crop management, with solutions having an intrinsic requirement of economic efficiency. (See Chapter 14 for details on the economic considerations.) Thus, in all conservation-tillage systems management of vertebrate and invertebrate pests is a major concern.

Foremost among several improvements that have ensured the success of conservation-tillage systems has been the development of effective herbicides that provide a nearly weed-free environment for crop growth and allow the accumulation of plant residues at or near the soil surface. In conservation tillage the lack of soil disturbance and the presence of surface crop residue provide a contrast in habitat for pests, compared to the bare soil environment of plow tillage. The potential of damaging pest infestations developing in the more diverse environment of conservation tillage has been of concern to entomologists during the past decade, especially since most research information on the ecology and management of major agricultural pests has been developed in plow-tillage systems. The unique habitat associated with conservation-tillage systems can have positive, negative, or neutral influence on the biological potential of pests. Only recently has sufficient information begun to accumulate on pests associated with conservation tillage systems. Broad generalizations are not possible; each pest/crop situation is different and must be considered independently.

Continuous and Intermittent Conservation Tillage Systems

When considering the hazards of conservation-tillage systems on potentially damaging pests, it is not only important to specify the tillage method, but the concomitant cropping system as well. Several types of conservation-tillage systems involve varying degrees of soil disturbance associated with seedbed preparation: plow-tillage, surface-tillage, no-tillage (see Chapter 2 for definitions). The intensity of conservation tillage operations may be continuous season after season or intermittent. These different intensities of tillage and other agronomic operations can influence pest populations differently. For example, continuous no-tillage planting of corn (*Zea mays* L.) has been practiced in some areas of the United States for over 20 years. Unfortunately, the undisturbed plant residues in continuous no-tillage corn can serve as a haven for certain pests, such as cutworms (*Agrotis* spp.) and increase damaging populations.

A less intense practice, "double cropping," is popular in the southern United States. Generally, conservation-tillage operations are alternated with a plow-tillage operation every one or two years. Typically in this situation, a no-tillage planting of a field crop, such as corn, sorghum (*Sorghum bicolor* L. Moench), or soybean (*Glycene max* Merrill), is followed by a fall plowing to prepare a seedbed for small grains. Although the periodical plowing operation may reduce those pest problems associated with an increase in surface residues, the double-cropping approach may increase the hazards of some pest outbreaks as the crop is often planted later than normal in order to obtain maturity of the first crop (i.e., small grain) in the planting sequence. Certain pests [i.e., fall armyworm *(Spodoptera frugiperda* (J. E. Smith)]; corn earworm [*Heliothis zea* (Boddie)]; etc.} pose a significant threat to late plantings in both "double-cropping" and single-cropping systems.

Management Tactics for Insects in Conservation Tillage

Producers, whether involved in high- or low-intensity agriculture, have similar objectives in managing their pest problems. Proficient farmers recognize that management of pests cannot be left to chance. They are aware of the devastating potential of the various pests and of the severity of yield losses that result from failure to maintain insect control. These producers are quick to adopt management tactics that avoid pest problems. This philosophy is termed *preventive control.*

A second philosophy, *suppressive control*, is concerned with the application of pesticides for control of developing pest outbreaks. Insecticide use is usually mediated by cost–benefit analysis involving chemical application expense, crop value, and expected loss from the pest outbreak.

PREVENTIVE CONTROL

Typically, preventive control tactics (i.e., biological, biochemical, cultural, physical, mechanical, etc.) require detailed knowledge of pest biology/ecology and population dynamics. Certain preventive control measures directly and adversely affect the pest. Crop rotation, burning of crop residues, use of pest-resistant crops, intermittent tillage, and manipulation of biological control agents have direct lethal effects on pests and indirect effects such as stressing the pest, making it more susceptible to attack by natural enemies. Other preventive tactics are used because they help avoid pest problems, for example, manipulation of planting date(s) and utilization of pest-resistant crop(s). Finally, some preventive tactics foster crop tolerance or recovery from pest damage. Breeding crops tolerant of pest infestation has been successful. Also, practices such as irrigation, fertilization, and subsoiling may enhance plant recovery or tolerance of pest infestations. Table 12.1 lists some preventive control options and their relative potential for use in conservation-tillage systems.

TABLE 12.1. Preventive Control Strategies for Major Insect Pests of Field Crops Utilized in Conservation-Tillage Systems

Resistant crops

These are crop(s) varieties that possess heritable characteristics that negatively influence insect populations at the physiological or behavioral level so that the plant avoids debilitating infestation by specific pests. Insect preference for oviposition or for

(continued)

feeding may be reduced or eliminated by use of resistant crops. Pest development and growth or reproduction may also be adversely affected.

Tolerant crops

Plant varieties that possess heritable properties that enable normal growth and yields while under attack by the pest are considered tolerant to the pest.

Planting/harvest date

Altering times of planting or harvesting dates for crops is an age-old method of evading damaging insect populations. In utilizing planting/harvest date(s) as a control tactic, two considerations are important. First, the choices usually are localized (i.e., a recommendation in one region often is not applicable in another). Second, a farmer should consider the entire production system when manipulating planting/harvest date for pest control. Agronomic disadvantages or increased susceptibility of the crop to other pests must be carefully weighed against the advantages of using these procedures to control insects.

Crop rotation

Of long standing this tactic is based on the premise that many major pests have a restricted feeding habit to one (monophagous) or a few closely related species (oligophagous) of plants. However, this approach may not be effective with insects that infest many different crop species (polyphagous). Crop rotation involves periodic alternating of two or more crops within an area in such a manner that resident pest populations are unable to survive when one of the crops is in place. Thus, it is a highly effective management method for pests that tend to build damaging population levels over time and are somewhat sedentary in habit.

Prohibitive crop selection

This is similar to crop rotation, but is concerned with a management decision to use or forego use of a crop on a permanent or semipermanent basis because of its high or low risk to insect infestations. For example, in many regions where double cropping is practiced, farmers may use corn, sorghum, or soybeans as a second crop following small grains. Since many times these systems require later than normal planting dates, the crops are subjected to greater than normal insect pressure. Corn is often decimated by several pests, while soybeans have less relative insect pressure in double cropping. Selection of soybeans over corn, if agronomically feasible, is advised in these situations. Other economic considerations are outlined in Chapter 14.

Intermittent tillage

Many pest insects have one or more life stages associated with the soil or with surface trash in fields. Often these stages are quiescent (i.e., pupal stage) or otherwise vulnerable to control with plow-tillage operations. These insects may build up in certain conservation-tillage systems such as continuous no-tillage cropping. Also, these pests may be lurking in sod or meadowlands that are being planted for the first time in several years. Use of plow-tillage operations when opening new crop lands or a periodic-tillage operation in continuous no-tillage fields may be necessary when the pest hazard is high.

TABLE 12.1. *(Continued)*

Burning and other sanitation practices

Burning of crop refuse in conservation-tillage systems is practiced in many areas. This practice is debatable from several standpoints, not the least of which is the production of aerial pollutants. However, it probably has value in pest control by reducing plant debris that could harbor pest insects. Little entomological research information is available on thc incineration effects of burning, but it appears unlikely that the temperatures in a quick running fire over conservation-tillage land would reach lethal temperature levels for sufficient duration to affect soil-inhabiting pests. Periodic flooding may be feasible in some conservation-tillage fields, and this would be completely lethal to most pests trapped in the deluge. Various chopping/shredding practices, such as in silage operations or postharvest shredding of crop remnants, have control value for some pests. Grain drying often results in direct lethal temperatures for pests, and these operations reduce grain moisture to levels that insects cannot survive.

Insecticides

The application of insecticides to protect crops has been a standard insect control tactic since the introduction of the organochlorine insecticides several decades ago. Many of the chemicals have changed over the years, but the philosophy of applying insecticides to ensure the prevention of pest outbreaks is probably as popular today as ever. These compounds are used in situations where high resident insect populations have been detected, or in conditions that have a high potential hazard for pest infestations. Current knowledge indicates an insecticide can be used effectively in conservation-tillage crops in the same way as for plow-tillage systems. When used intelligently, they are a potent tool for the pest manager. However, utilization of prophylactic insecticide applications needs to be mediated by knowledge of their limitations. Insecticides are expensive, and unless their use is justified as the only alternative over other preventive methods (most of which are less expensive), they will reduce net profit. Problems with pest resistance to prophylactic insecticides is well documented. Lack of specificity, toxicity to nontarget animals, and other environmental insults are often problems cited against broad-scale use of insecticides to protect crops.

Biological control

The use of natural enemies to control insect pests is a topic that is easy to talk about but difficult to manipulate in practical pest-management systems. Techniques of biological control that are available for conservation-tillage crops include conservation (avoiding practices that are detrimental to natural enemies) and enhancement (utilizing procedures that favor the biological potential of natural enemies). Mounting evidence indicates that the increased diversity of the habitat associated with many conservation-tillage systems favors some biological control agents.

Cultural practices

These are practices, such as irrigation, fertilization, planting date(s), subsoiling, etc., that favor optimum crop growth, and in many situations enable plants to tolerate higher insect infestations. Many times these procedures entail appreciable cost input, and thus they rarely are utilized as a pest control tactic alone. They should have strong agronomic value as a primary requisite to this consideration.

SUPPRESSIVE CONTROL

Use of pesticides to control damaging pest outbreaks is often the only control tactic many farmers consider. Although pesticides are a viable tactic for suppression of pest outbreaks, they must be used in concert with other management tactics. Not only are pesticides expensive (resulting in reduced net profit), they may cause an unnecessary insult to the environment. Therefore, implementation of effective preventive control tactics should be of highest priority when selecting an insect-pest-management program for a specific conservation-tillage system. When pesticides are required, they should be applied in the most efficient manner possible. Although most of the research on efficacy of pesticides has been conducted in plow-tillage systems, recent studies indicate that, for many pests, insecticides are equally efficacious across tillage systems.

A prerequisite for implementation of any management strategy is to estimate the magnitude of the pest population and the damage potential associated with specific population levels. Additionally, the impact of pest damage on crop yield must be established so that an "action threshold" can be developed and used as a guideline for implementing the management strategy. An action threshold is defined as the pest population density at which control tactics should be initiated to prevent insect numbers from reaching an economic injury level (Stern et al., 1959) as mediated by the effects of the control tactic(s) on beneficial organisms, the environment, aesthetics, and sociological stresses to people in the nearby ecosystems (Martin et al., 1980). Insecticides are used when insect populations and concomitant damage exceed the action threshold. Limited research has been done on action thresholds for pests associated with conservation-tillage systems. Observations indicate that action thresholds developed for plow-tillage systems will be similar for these tillage systems.

Integrated Pest-Management Strategies

A modern agricultural control program typically incorporates one or more preventive tactics into the management of an individual pest or pest complex. The suppressive insect control tactic is utilized when preventive efforts have failed to keep the pest populations below some predefined action threshold. Use of various preventive and suppressive tactics in as compatible a manner as possible while keeping the pest population at levels below those causing economic damage to the crop is termed *integrated pest management* (IPM). IPM programs have been developed for specific pests on individual crops for relatively localized regions. Although most of the IPM programs have been developed for plow-tillage systems, current knowledge suggests that strategies based on plow tillage are readily adaptable to conservation-tillage systems.

This work is published with the approval of the Directors, Georgia and Arkansas Agricultural Experiment Stations.

Research has demonstrated that the greatest differences in pest populations between plow tillage and conservation tillage occur in the seed and/or seedling stage. As crop growth proceeds through the postseedling stages, pest outbreaks associated with conservation-tillage systems generally parallel those found in the plow-tillage system.

SEED AND SEEDLING PESTS

Generally, conservation-tillage practices minimize the deleterious impact of intensive tillage (i.e., plow tillage) on some pests, especially soil insects. Gregory and Musick (1976) reported that in northern United States some pests have become more serious because herbicides have eliminated their preferred host(s). Also, cool spring temperatures cause slower germination of the seed increasing the time of exposure to pest attack. In addition, the status of some noninsect pests (i.e., slugs and mice) has been increased as a result of the reduction in tillage.

Seed and seedling pests are among the most difficult to control. Suppressive applications of insecticide often have limited efficacy and are difficult to time properly. In general, early detection of damaging infestations on seeds or seedlings is difficult, and very limited information on action thresholds is available, especially for soil-inhabiting pests. Consequently, there is heavy reliance on preventive approaches to pest control.

Care must be exercised in associating the failure of seeds to germinate or emerge with pest attack. In cool, wet spring conditions, seeds may rot in the soil as a result of poor seed–soil contact (i.e., failure of opening made by the planter to close) and colder soil temperatures associated with heavy mulch covers and for some soil types. This has been a significant problem in soils with "hard pans" near the surface. Before any definitive statements on causal agent(s) are possible, it is imperative that the seed be examined closely.

The following discussion covers examples of seed and seedling pests of major concern in crops grown using conservation-tillage practices. The biology/ecology of the pest as it is related to crop damage and control will be emphasized. In addition the current IPM program will be outlined. No attempt will be made to identify specific pesticides, as new and more effective chemicals are continually being developed. Current pesticide recommendations should be obtained from publications of the Cooperative Extension Service or from an extension entomologist.

Armyworm [*Pseudaletia unipuncta (Haworth)*]

The armyworm is among the most consistent and devastating pests of corn grown under conservation-tillage systems, especially no-tillage (Musick and Petty, 1973; Wrenn, 1975).

Biology/Ecology Relative to Conservation Tillage

All crops belonging to the grass family are susceptible to attack. Most species of small grains, especially rye (*Secale cereale* L.), and other grasses are attractive for oviposition, and thus many conservation-tillage/crop situations are vulnerable. These include double-cropping systems where corn is planted following harvest of small grains. When spring applications of herbicides are used to kill the small grains, armyworm larvae readily move to corn. In meadow situations, the same pattern exists; the meadow is killed with herbicides and the larvae move to corn.

Musick and Beasley (1978) reported that in Ohio no-tillage cornfields, 32% of the corn planted into various types of meadowlands had serious damage (Fig. 12.1). Only 11% of the fields in continuous no-tillage corn were seriously attacked; all of these fields had a rye cover crop previous to planting of the corn. Armyworm damage was present in 86% of all of the surveyed fields during the 3-year study. In fields with heavy damage, the duration of the meadow prior to the planting of conservation-tillage corn had no effect on the magnitude of damage. Harrison et al. (1980) reported that for sweet corn in Maryland, the armyworm was more damaging to no-tillage plantings than to plow-tillage plantings.

FIGURE 12.1. Armyworm damage is significantly increased in no-tillage corn. Larvae (Inset) consume the entire plant literally overnight. (Inset, courtesy of R.R. Kriner, Hendersonville, N.C.)

Pest Management in Conservation Tillage

The seasonality of infestations is an important consideration in developing a control program for armyworms in conservation-tillage systems. Early season attack can markedly reduce yields. Musick (1973) reported that if severe pruning of corn occurs prior to early June, replanting should be considered. However, if severe damage occurs after early June, higher yields could be obtained using properly timed suppressive insecticidal applications and allowing the corn to grow out of the damage. Replanting should be a last resort in this latter case. Each field situation must be monitored (i.e., scouted) before decisions on control are made. Generally, insecticidal controls are initiated when 25–30% of the plants show damage and two or more larvae are present per plant. Banded or broadcast applications of various insecticides have been effective.

Plow-tillage and surface-tillage systems give excellent control of the armyworm through early season destruction of host plants attractive for oviposition. This should be considered an alterntive to conservation tillage in areas where the armyworm hazard is considered high.

Although biological control agents {red-tailed tachina fly [*Winthemia quadripustalata* (Fabr.)]; braconid wasps (*Apanteles* spp.); egg parasites (*Telenomus minimus* Ashmead)} are effective, they have not kept armyworm populations in check in no-tillage situations. In addition, armyworm mortality caused by these parasites often occurs too late in the season to prevent significant yield losses to early-planted corn.

Cutworms

Regardless of tillage system, several species of cutworms have caused significant damage to crops across the United States. In general, cutworm problems in conservation-tillage systems have paralleled those in plow tillage, except for the black cutworm [*Agrotis ipsilon* (Hufnagel)], in corn (Musick and Beasley, 1978) and in cotton (*Gossypium hirsutum* L.) (Dumas, 1983). Although incidence of attack has increased, little change in management strategies is required among tillage systems. In some states cutworms are considered to be among the most important pests of conservation-tillage corn.

Biology/Ecology Relative to Conservation Tillage

Musick and Beasley (1978) reported that 46% of continuous no-tillage cornfields examined for 3 years in Ohio had cutworm feeding damage of economic importance (Fig. 12.2). In experiments comparing no-tillage and plow-tillage corn, Musick and Petty (1973) observed that the black cutworm attacked 15% of seedlings in no-tillage areas and only 1% of the seedlings in adjacent plow-tillage portions of the same cornfields. Harrison et al. (1980) concluded that cutworm was second in frequency among pests attacking no-tillage sweet corn in Maryland, and Dumas (1983) reported significantly higher black cut-

FIGURE 12.2. Black cutworm damage is decreased in no-tillage corn. Larvae sever plant at or below the surface of the soil.

worm damage to cotton planted in a conservation-tillage system as compared to plow tillage.

Busching and Turpin (1976) found that during early spring moths are attracted to annual and perennial broadleaf weeds for oviposition. Moths appeared to prefer low, dense weed forms. Also, cutworm larval mortality increased as tillage of the soil increased.

Gaylor et al. (1983, unpublished data) found that significant variegated cutworm [*Peridroma saucia* (Huebner)] populations developed in cotton grown in conservation-tillage systems including crimson clover (*Trifolium incarnatum* L.) as a cover crop.

The only substantiated report of increased cutworm activity on crops grown in conservation-tillage systems is for black and variegated cutworms. However, the following cutworm species may pose a threat: pale western cutworm (*Agrotis orthogonia* Morrison); army cutworm [*Euxoa auxiliaris* (Grote)], and dingy cutworm (*Feltia ducens* Walker).

Pest Management in Conservation Tillage

Although the incidence of cutworm problems in conservation-tillage systems may be greater, management remains similar among tillage systems. For preventive control, agronomic measures that reduce early-season weed growth usually minimize cutworm problems.

Suppressive controls can be applied when cutworm damage approaches some predetermined action threshold if cutworms are still present in the fields.

The action threshold varies among states, and Cooperative Extension Service recommendations should be consulted prior to treatment.

Southern Corn Billbug [*Sphenophorus callosus* (Oliver)]

A major pest of corn in the southeastern United States, the southern corn billbug produces significantly higher infestations in no-tillage than plow-tillage systems (All et al., 1984).

Biology/Ecology Relative to Conservation Tillage

Adult southern corn billbugs feed in the pith/meristem tissues of corn and produce symptoms varying from mild foliage perforations to severe stunting and death of seedlings (Metcalf, 1917) (Fig. 12.3). A debilitating type of damage is produced that is influenced by the interaction of insect, plant, and environment. Factors such as the stage of plant growth when feeding occurs and the length of time billbugs have access to a feeding site are important determinants of the degree of damage that will occur (Durant, 1982). Also, factors that aid vigorous seedling growth will influence tolerance to billbug feeding. Fast-growing seedlings can "outgrow" moderate billbug damage. Larval survival is low on older growth stages (Wright et al., 1983). Yellow

FIGURE 12.3. Southern corn billbug infestations are increased in no-tillage corn systems. Adults (Inset) feed in pith/meristem tissue of young corn plants. Figure shows infested no-tillage corn plots on the right treated by sub-soiling plus soil insecticide and on the left without these treatments. (Inset, courtesy of R. R. Kriner, Hendersonville, N.C.)

nutsedge (*Cyperus esculentus* L.) and large crabgrass [*Digitaria sanguinalis* (L.) SCOP] are favorable hosts for larvae (Metcalf, 1917).

Southern corn billbug adults are active at the time of corn planting, and highest populations in plow-tillage systems are associated with unplowed weed areas near fields (Metcalf, 1917). In continuous no-tillage systems, spring populations of billbugs in corn are associated with weeds and corn debris from the previous season. In double-cropping systems where corn is planted following a small grain, the spring generation of billbugs occurs in the small grain and infests corn after it is planted (All et al., 1984; All, unpublished data). Thus, billbug populations are enhanced in no-tillage cropping systems because late-season weeds or small grains serve as larval hosts and/or cover for overwintering populations.

Pest Management in Conservation Tillage

Various preventive methods are effective when used in conservation-tillage systems. Cornfields with a history of southern corn billbug problems should not be planted using no-tillage practices unless billbug populations have been reduced by other control tactics, such as insecticides. Since the southern corn billbug does not attack legumes, crop rotations (for example, soybeans and corn) could be considered. An application at planting time of certain insecticides reduces billbug infestations in no-tillage corn as effectively as in plow tillage (All and Jellum, 1977). The use of subsoiling in fields with hardpan layers materially aids plant recovery from billbug feeding, especially under drought conditions (Fig. 12.3). When billbug infestations are high, preventive insecticidal applications used in conjunction with subsoiling in no-tillage systems act synergistically to improve corn growth and yield (All et al., 1984).

Corn Root Aphid [*Anuraphis maidiradicis* (Forbes)] and Cornfield Ant [*Lasius allenus* (Forster)]

Gregory (1974) considered these aphids as the most important pests of no-tillage corn in Kentucky; the severity of damage by this pest complex seems to be particularly high in that region of the United States.

Biology/Ecology Relative to Conservation Tillage

The corn root aphid and attendant ants are one of the most interesting pest complexes in nature. Ants move the aphids to suitable hosts to feed. The aphid secretes honeydew during feeding, which the ants collect. Corn root aphids feed on a variety of grasses, but appear to prefer corn. The ants will often move the aphids considerable distances in the spring to establish them on corn. Aphid feeding on the roots of corn results in a yellowing of the plant and in reduced growth with subsequent loss in yield. Young plants may wither and die, especially under drought conditions. Gregory (1974) reported stand losses exceeding 50% in some fields. Most infestations were associated with corn planted into grass sods or in areas that had not been planted in field crops for several years. He indicated that plow-tillage corn also was severely infested in similar circumstances, and thus it is unclear if the no-tillage habitat per se favors outbreaks.

Pest Management in Conservation Tillage

As with many insect problems associated with conservation-tillage systems, growers in regions with historically serious corn root aphid problems should be cautious when using no-tillage systems for the first time. This is especially true when fields have laid fallow for one or more years. In these situations, a thorough examination of the field for ant mounds is recommended. One preventive control method is directed at the ants: early spring, deep-tillage operations can be used to disrupt and destroy the ant nests in fields with high ant populations. Use of plow tillage in ant/aphid infested regions might be wise for 1 or 2 years before initiating a no-tillage program. Selection of a nonhost crop, such as soybeans, also may be advisable for high-risk areas. Highly effective insecticides have been identified and are available for application to no-tillage fields on a suppressive basis.

Stalk Borer [*Papaipema nebris* (Guenee)]

The incidence of attack by this pest has been much greater in conservation-tillage systems than in plow-tillage systems (Musick, 1970).

Biology/Ecology Relative to Conservation Tillage

Moths deposit eggs in late August or September on curly dock (*Rumex crispus* L.), broadleaf dock (*Rumex obtusifolius* L.), giant ragweed (*Ambrosia trifida* L.), pigweed (*Amaranthus retroflexus* L.), and many grasses. Field observations indicate that they readily oviposit on volunteer wheat (*Triticum vulgarie* Vill.) and on fall-planted rye. The eggs hatch the subsequent spring. In general, stalk borer larvae will attack any plant large and tender enough for them to tunnel into. Giant ragweed appears to be preferred, although stalk borer has been found in a wide variety of plants used in conservation tillage. When preplant herbicides are applied in conservation-tillage cornfields the following spring, the young larvae move from the dead weed host to newly emerged seedlings (Fig. 12.4). Stalk borer larvae have a "restless" habit and readily move from one host plant to another, resulting in much greater injury to crops than would occur if development was completed within a single host.

Pest Management in Conservation Tillage

Weed control programs during late summer aid in preventing stalk borer infestations in conservation-tillage systems. Suppressive control with insecticides has been ineffective, as the larva develops within the corn stem (Musick and Beasley, 1978).

FIGURE 12.4. Stalk borer larva on young corn plant. They tunnel into the stalk and kill the plants (arrow points to larva).

Corn Rootworms, Western (*Diabrotica virgifera* LeConte); Northern [*D. longicornis* (Say)]; and Southern (*D. undecimpunctata howardi* Barber)

Rootworms are among the most destructive pests of corn. Although cultural practices may influence population levels in conservation-tillage systems, rootworms continue to pose problems (Gregory and Musick, 1976).

Biology/Ecology Relative to Conservation Tillage

The western and northern corn rootworms are of major concern as they are obligatory pests of corn. Tillage studies by Musick and Collins (1971), Chaing et al. (1971), and Kirk et al. (1968) have provided good information relative to their biology. In 1968, Kirk et al. showed that western corn rootworm adults preferred oviposition sites composed of large soil particles, open cracks, and moist soil. Chaing et al. (1971) concluded that placement of corn rows midway between the rows of the preceding year might reduce larval survival in both the western and northern species. Musick and Collins (1971), working independently, came to this same conclusion for the northern corn rootworm. In continuous no-tillage corn on a 40-in. row spacing, approximately three times more eggs were deposited in the soil as compared to corn in plow tillage. However, it required about four times more eggs in no-tillage to develop similar root damage compared to the plow-tillage system. There was a linear relationship between larval population and placement of the corn row in relation to the previous year's row. For each 10 in. from the old row, larval populations were reduced approximately one-third (Musick and Collins, 1971).

The southern corn rootworm has a very wide host range. Unlike the other species, which overwinter (diapause) in the egg stage, this pest overwinters as an adult. Adults become active in the spring and oviposit in the soil next to their host. Since this corn rootworm does not overwinter in northern climates, outbreaks in corn are less prevalent than in the southern United States. Widespread problems have not occurred in conservation-tillage systems.

Pest Management in Conservation Tillage

Because of the obligatory relationship of the northern and western corn rootworms with corn, crop rotation is an effective preventive control. However, given the economic incentive to produce corn in many areas, crop rotation often is not a realistic solution. Data are lacking on action thresholds for conservation-tillage systems. Therefore, if these pests are present, insecticidal applications will usually be required. Action thresholds are not available for rootworm larvae in any tillage system because of the unreliability of soil-sampling techniques for eggs and larvae. However, several states recommend scouting for adult beetles in July and August to ascertain the need for a soil insecticide to control the larvae during the subsequent spring. This approach to management is independent of tillage system. Although soil insecticides used in plow-tillage systems have been effective in conservation-tillage systems, Musick and Collins (1971) showed that the relative efficacy of a soil insecticide changed with tillage systems. Those soil insecticides that were susceptible to photodegradation appeared to perform more poorly.

For the southern corn rootworm, these management strategies will have limited effectiveness because the eggs are laid in the fields after corn has emerged. In plow-tillage systems, applications of a soil insecticide can be made during cultivation. Insecticide applications after planting are also possible in conservation-tillage systems, but little information is available on the effectiveness of these treatments for controlling the southern corn rootworm.

Wireworms (*Melanotus* spp., *Aeolus* spp., *Conoderus* spp., and *Ludius* spp.)

Many species of wireworms attack crops used in conservation-tillage systems. Gregory (1974) considered wireworms the second most important pest of no-tillage corn in Kentucky. In England, Edwards (1975) reported that losses from these pests could be two or three times greater in "direct-drilled" wheat as compared to plow tillage.

Biology/Ecology Relative to Conservation Tillage

Wireworms attack the crop as soon as it is planted and may gouge out the seed, preventing germination. Young seedlings also are attacked. Feeding in the pith/meristem of seedlings by wireworms causes the unfurling leaves to have a shredded appearance. The plant's growth potential is reduced or seedlings are killed outright. Wireworm infestations are characterized by patchy, sparse areas located at random in a field. It has been a rule of thumb for many years that in fields having a history of wireworm problems the problem will recur unless populations are controlled, as dispersal by both larvae and adults is limited (Thomas, 1940).

Entomologists have studied wireworms for over a century, but many aspects of their biology are still a mystery. Much of their activity is obscured in the soil. However, two characteristics seem common among wireworm species: their preference for cooler soil conditions with adequate moisture and their association with soils higher in organic matter. Many problems have occurred in crops planted in land that has been fallow for one or more years (Thomas, 1940). Musick and Beasley (1978) indicated that most wireworm problems in no-tillage corn were associated with plantings into former grassland areas, which is where Gregory (1974) and Edwards (1975) observed most wireworm infestations.

Pest Management in Conservation Tillage

Recognition of a potential wireworm hazard is an important step in implementation of pest-management programs. Plow-tillage fields with a recurring wireworm problem will undoubtedly continue to have damage when conservation-tillage systems are utilized. Fields that have been fallowed or have been in grasses for several years are of significant risk and should be examined closely. In recent years, baits and pitfall traps have aided in the detection of wireworm populations in fields.

Insecticides have been effective in controlling potentially damaging populations. Treatment of the seed with an insecticide or placement of an insecticide in the seed furrow or banded over the row at time of planting are popular control measures in infested fields. A variety of insecticides have been used over the years; many were chlorinated hydrocarbons which have been subsequently banned from use. However, some other classes of insecticides are now available. Preplanting and at-planting applications of insecticides are recommended for wireworm control in plow-tillage systems and appear to be effective in conservation tillage systems.

Sugarcane Beetle [*Euetheola rugiceps* (LeConte)]

Severe infestations in corn by the sugarcane beetle have been observed in conservation-tillage systems (All, unpublished data).

Biology/Ecology Relative to Conservation Tillage

The biology of the sugarcane beetle is not well known. Infestations of this pest in plow-tillage systems are often associated with freshly tilled sod fields or fields with high organic matter. Adults damage the young seedlings in much the same manner as described for the southern corn billbug. Young seedlings wither and die as if drought stricken. Examination of dying plants reveals large gougelike wounds in the pith tissue, usually at or slightly below the soil surface. Heavy beetle infestations have been observed in crops using either no-tillage or surface-tillage systems. In general, heaviest damage was located adjacent to wooded areas and/or pastures. This is consistent with the syndrome described for many plow-tillage situations; outbreaks are associated with grassland areas adjacent to cornfields. However, in double-crop systems utilizing conservation-tillage methods, small grains may serve as larval hosts from which adults emerge and attack corn.

Pest Management in Conservation Tillage

At present, the hazard of the sugarcane beetle as a major pest in conservation-tillage systems is not clearly established. However, caution should be exercised when any crop is planted following a pasture or other sod situations where high larval populations were observed. Deep plowing is recommended for recurring sugarcane beetle infestations. Little recent information is available concerning the efficacy of insecticides for the sugarcane beetle.

Seedcorn Maggots [*Hylemya platura* (Meigen)]

This insect has been especially damaging when substances of high organic matter (i.e., animal manure, rye, etc.) are used to maintain surface mulches in no-tillage systems (Musick and Beasley, 1978).

Biology/Ecology Relative to Conservation Tillage

Eggs of the seedcorn maggot are deposited in areas where an abundance of decaying vegetable matter is present. In conservation-tillage systems, applications of animal manure may lay unincorporated on the soil surface. Also, fall-planted cover crops have been used to maintain surface mulches, especially in no-tillage corn systems. These conditions provide an ideal environment for oviposition and subsequent development of maggots. In addition, seed germination may be delayed in conservation-tillage systems because of lower soil temperatures associated with heavy surface mulches. With this delay in seed germination and seedling emergence the seed is exposed to seedcorn maggot infestation for a longer time (Musick and Beasley, 1978).

Pest Management in Conservation Tillage

Delayed planting to ensure adequate temperature for seed germination and only limited use of surface mulches are valuable preventive controls. This strategy should minimize ovipositional attractiveness and exposure of the seed to larvae (Fig. 12.5). Since seedcorn maggots occur sporadically, no effective sampling procedure or action thresholds have been worked out. In situations conducive to maggot problems, an insecticidal seed treatment (planter box) should be used.

Seedcorn Beetle Complex [*Stenolophys lecontei* (Chandoir) and *Clivinia impressifrons* LeConte]

Occasional seedcorn beetle problems have been encountered in no-tillage corn systems (Musick and Beasley, 1978).

Biology/Ecology Relative to Conservation Tillage

The biology of this insect is poorly understood. Although the beetles are primarily predaceous, they have been known to destroy corn seeds. Problems in conservation-tillage systems are usually encountered when seed is of low vigor and/or environmental conditions greatly prolong the germination period (Musick and Beasley, 1978).

FIGURE 12.5. Seedcorn maggot larva on corn seed. Problems are enhanced when high organic matter field situations exist.

Pest Management in Conservation Tillage

Using seed of good vigor (i.e., high germination percentage based on a cold test) and delaying planting until the soil warms enough to ensure rapid germination are good preventive control tactics. No sampling methodology or action thresholds have been developed due to the sporadic occurrence of these pests. Seed treatment and in-furrow applications of insecticides are recommended in situations favoring this pest.

Lesser Cornstalk Borer [*Elasmopalpus lignosellus* (Zeller)]

This is a significant pest of all major field crops used in conservation-tillage systems in the southern United States. This insect can produce various amounts of damage in different types of tillage systems. Hazards of lesser cornstalk borer are often high in plow-tillage and surface-tillage (surface disking) systems, whereas no-tillage cropping is considered an effective preventive control for this pest (All, 1978, 1979a; All et al., 1982; All and Gallaher, 1976; Fig. 12.6).

Biology/Ecology Relative to Conservation Tillage

The lesser cornstalk borer has a semisubterranean biology. Larvae attack single-stem crops such as corn, sorghum, or soybeans in the young seedling

FIGURE 12.6. Infestations of the lesser cornstalk borer are reduced in no-tillage systems. Larvae attack plants at soil surface. No-tillage fields have less damage.

stage, and their tunneling in the stem girdles the moisture-conductive tissues or destroys the growing buds of these plants. Damaging infestations are associated with late planting, which is common in double cropping and with low rainfall situations (Tippins, 1982).

One type of conservation-tillage system is double cropping, using surface tillage after the first crop is harvested to prepare a seedbed for a second crop. Usually in such a system a disk harrow is passed once or twice over freshly harvested small grain fields to partially bury the residues, followed quickly with planting of a second crop such as corn, sorghum, or soybeans. Devastating infestations have occurred with these production systems, especially if droughty weather occurs for several weeks after planting. Conversely, in identical environmental situations, damage will be insignificant if the second field crop is planted using the no-tillage system (All, 1978). This phenomenon can be explained through knowledge of the feeding behavior of this pest. Lesser cornstalk borers are polyphagous, and larvae are often present in small grains or weeds when fields are plowed (All and Gallaher, 1977; All et al., 1979). The disking operations either bury or chop-up plant residues leaving them exposed for rapid desiccation. These tillage operations seem to have little detrimental effect on the resident lesser cornstalk borer populations; thus, larvae are present in the field to attack the second crop as soon as it germinates.

Research indicates that the lesser cornstalk borer is a semisaprophitic insect that is capable of feeding on plant residues in no-tillage systems (Cheshire and All, 1979a, 1979b). Thus, although resident populations are present in no-tillage situations, they may, in effect, be deterred from feeding on the germinating field crop because of the abundant food source in the surrounding environment (All, 1980b).

Pest Management in Conservation Tillage

Use of no-tillage practices, whenever feasible, in double-cropping systems using corn, sorghum, or soybeans following harvest of small grains would avoid damaging populations of the lesser cornstalk borer. Maintaining as much mulch as possible during the corn seedling stage is recommended. If conventional plow-tillage or surface-tillage operations are necessary, planting of the crop should be delayed for at least 2 weeks to allow resident populations of the lesser cornstalk borer to complete development and leave the area or to succumb to starvation. In experiments comparing the susceptibility of corn, sorghum, and soybeans to lesser cornstalk borer in double-cropping systems, it was found that soybeans had consistently lower infestations of this pest (Rogers and All, 1982). Thus, soybeans should have high priority in double-cropping systems when hazards of damage from lesser cornstalk borer are great.

Finally, insecticides may be used on a prophylactic or suppressive basis to control lesser cornstalk borer infestations (All, 1979b; All et al., 1979; Gardner and All, 1982). However, acceptable control is difficult to achieve, especially under dry conditions (Tippins, 1982).

Miscellaneous Seed and Seedling Pests

A myriad of other insects are known to attack crops utilized in conservation-tillage systems. Prior to the accumulation of definitive information on these insects, it may be assumed that their pest status will be at least comparable to that in plow-tillage systems. With the following insects some information is available to indicate that there will be increased problems in certain conservation-tillage systems.

Grape colaspis [*Colaspis brunnea* (Fabricius)] is a soil pest that may occur in conservation tillage on a sporadic basis. This pest has historically produced infestations in cornfields following clover (*Trifolium* spp.) or other legumes. Currently, there is strong interest in planting no-tillage crops, such as corn, in fields following nitrogen-producing legumes to save on fertilization costs. Thus, the status of the grape colaspis may increase in the future if this method of corn production is used on a large scale. Infestations of this pest are difficult to anticipate and effective sampling methodology is not available. However, control can be achieved with insecticide-treated seed or prophylactic application of insecticides.

White grubs (*Scarabaeidae*) are a multispecies complex of beetles whose larval stages feed on the roots of grasses. When high populations occur, serious damage has been observed in turf and in crops planted immediately following sod situations. Although serious problems have not often materialized in conservation tillage, sporadic infestations have been reported (Gregory, 1974). Conversely, Rivers et al. (1977) found significantly more white grubs in the soil around plants in plow-tillage systems as compared to surface-tillage systems. Musick and Petty (1973) reported that corn planted in killed sod with high grub populations failed to show significant root damage after 30 days. Selection of a nonhost crop such as soybeans is recommended if white grubs are expected to be a problem. Insecticides are available for controlling infestations.

The mint root borer [*Fumibotys fumalis* (Guenee)] is a serious pest of peppermint (*Mentha piperita* L.) in the Pacific Northwest. The borer infests the rhizomes of peppermint, and infestations have been associated with the practice of maintaining continuous stands of the crop for several years. Pike and Glazer (1982) reported that periodic rotary strip-tillage operations reduced borer populations by as much as 81% and recommended this practice as an alternative to the no-tillage practice currently used by farmers.

Slugs

Slugs are among the most destructive noninsect pests of crops grown under conservation-tillage regimes, especially no-tillage. Several species (*Limax* spp., *Derocerus* spp., *Milax* spp., *Agriolimax* spp.) have been encountered. The most serious slug problems have occurred in the northeastern quarter of the United States, with Maryland, Michigan, Ohio, Pennsylvania, New York,

New Jersey, and the New England states having the most consistent problems (Musick and Petty, 1973).

Biology/Ecology Relative to Conservation Tillage

Slugs are prevalent in high moisture situations, and no-tillage systems often provide a favorable environment for their survival. Eggs are laid in masses in damp situations and usually hatch in about a month. Warm wet springs are conducive to maximum damage. In continuous no-tillage row crops with little or no mulch either there is no damage or damage occurs in small confined areas within the field. In contrast, crops in no-tillage fields with heavy supplemental mulches (i.e., manure) or heavy crop residues (i.e., meadows, rye, old corn stubble, etc.) have been destroyed completely (Fig. 12.7). In side-by-side comparisons of plow-tillage corn with no-tillage corn, slugs destroyed up to 60% of the seedlings in no-tillage as compared to less than 5% in the plow-tillage area (Musick and Beasley, 1978).

Pest Management in Conservation Tillage

Slugs can be efficiently managed through reduction of heavy mulches or crop residues. Destruction of heavy mulch accumulations with intermittent-tillage operations should be considered in no-tillage regions where slug problems have occurred, especially if a cool wet spring is expected. Since mulches are necessary in some situations, delaying planting for warmer temperatures to ensure rapid seed germination and subsequent crop growth would be beneficial.

FIGURE 12.7. Slug damage is especially evident in the northeastern United States where heavy mulches occur. Lodging of young corn plants occurs as a result of root pruning by slugs.

Although some pesticides are toxic to slugs, no satisfactory efficacious pesticides are currently available. In situations where slugs are considered to be a problem, use of plow-tillage, if feasible, or rotation to a nonsusceptible crop (i.e., forage, meadows, etc.) is recommended.

Rodents

Several species of field mice have caused significant stand reductions in no-tillage systems. Generally, tillage operations (i.e., plow-tillage and surface-tillage) which significantly disrupt established "runs" have effectively discouraged these animals (Beasley and McKibben, 1976).

Biology/Ecology Relative to Conservation Tillage

Musick and Beasley (1978) provided an extensive account of the relative importance of rodents in conservation-tillage systems. Several common species of field mice may be present in no-tillage corn, especially where heavy stands of grasses form a cover within the field. When these grasses are killed with herbicides, sufficient mulch remains to provide rodents shelter and protection from predators. However, many of their food sources are destroyed. Consequently, the mice feed on the seed.

Prairie voles [*Microtus ochrogaster* (Wagner)], southern bog lemmings (*Synaptomys cooperi* Baird), deer mice [*Peromyscus maniculatus* (Wagner)],and house mice (*Mus musculus* L.) have caused problems in conservation-tillage fields. Populations of deer mice and house mice are generally too low to cause economic damage, except in small fields surrounded by densely vegetated fence rows and in fields adjacent to feed lots, barns, and storage buildings. The voles and lemmings cause scattered damage throughout cornfields. The prairie vole appears to be the most common and destructive rodent in no-tillage corn systems. Southern bog lemmings occur infrequently. Beasley and McKibben (1976) observed that damage by rodents in conservation-tillage systems was highest during brood producing periods.

In no-tillage corn, mice often burrow into the slot opened by the planter and continue down the row, taking one seed after another for a distance of several feet. Infestations are usually not apparent until the corn seedlings begin to emerge and stand reductions are evident. Occasionally, rodents will dig along side the emerging seedlings, cut through the roots, and consume the kernel. The injured seedling topples over and usually is killed. Seed/seedling damage by rodents commences the night immediately after planting and subsides after about 3 weeks (Musick and Beasley, 1978).

Pest Management in Conservation Tillage

Rodents make surface runways through the crop residues. In densely populated fields, a complex network of these runways exists. Grass cuttings are left along the runways, and at designated intervals (called latrins) the mice defecate. These latrins may be used to obtain an estimate of the population density of rodents in conservation-tillage fields prior to planting (Musick and Beasley, 1978).

Two characteristics are important regarding rodent control programs: hazards are great if (1) rodent populations are increasing or are in a reproductive phase and (2) a closed crown cover is present in the area. Under these high hazard situations, plow-tillage should be considered as an alternative to conservation-tillage cropping. Another control method is close cutting and removal of vegetation prior to planting no-tillage crops. Disking a sod field prior to planting should be helpful. Although some rodenticides are labeled, little is known concerning their value as a rodenticide or as a repellent for management of rodents in conservation-tillage systems (Musick and Beasley, 1978).

Birds

Unfortunately, there is little quantitative information on bird damage, but problems occur periodically in certain conservation-tillage systems. Flocks of pigeons (*Columba livia* Gmelin), blackbirds (*Euphagus* spp.), crows (*Corvus brachyrhynchos* Brehm), and starlings (*Sturnus vulgaris* L.) have been observed many times in no-tillage cornfields. In direct comparisons with plow-tillage systems, damage was concentrated in no-tillage (All, unpublished data).

Biology/Ecology Relative to Conservation Tillage

Birds are often attracted into freshly harvested small grain fields, presumably to feed on the seed lost from combining or on insects associated with this material. These flocks can often produce tremendous stand loss in no-tillage crops by destroying the planted seed or seedlings. This is especially apparent in situations where there has been poor seed coverage by no-tillage planters. Birds will peck into the planting slot and remove seed or will pull up young seedlings to gain access to the kernel.

Pest Management in Conservation Tillage

It is generally acknowledged that annual bird damage in most agricultural crops is tremendous, yet little control technology is available to contend with the problem. Noise-making devices, available at high cost, are only moderately effective. Scarecrows do not work. Repellent chemicals such as Mesurol® have been advocated as seed treatments or crop sprays, but the cost is high and effectiveness is inconsistent. Use of poisons is not legal in most regions.

POSTSEEDLING PESTS

These pests attack crops following seedling establishment and infest the plant through its fruiting and maturation stages. Defoliating insects such as the fall armyworm, grasshoppers, soybean looper, green cloverworm, etc.; stalk-infesting insects like the European corn borer and southwestern cornstalk borer; and pests that attack the fruiting structures of plants, including the corn earworm and maize weevil complex are examples of postseedling pests. In general, the hazard from infestations by these insects is similar in all tillage systems. Concern is increased in certain conservation-tillage systems, especially when they are used in conjunction with other cropping practices also having high pest hazards. Information currently available indicates that the survey methods, action thresholds, and control procedures developed for these pests in plow-tillage systems are readily adaptable to conservation-tillage systems.

Corn Earworm [*Heliothis zea* (Boddie)]

This major pest of corn also causes serious losses in other field crops when conservation-tillage production systems are used, including sorghum, soybeans, small grains, etc.

Biology/Ecology Relative to Conservation Tillage

No major increase in injury by the corn earworm in corn has been observed in conservation-tillage systems, except when crops are planted later than normal (All and Gallaher, 1976). Roach (1981a) reported that in comparisons of conservation-tillage and plow-tillage systems in cotton and tobacco (*Nicotiana tabacum* L.), *Heliothis* spp. populations were similar. However, he found that greater numbers of moths emerged from the conservation-tillage plots (Roach, 1981b). Serious damage has occurred on corn and sorghum planted as the second crop in a double-cropping system (All, 1980a). The corn earworm has multiple generations during the season in much of the United States. However, the insect does not survive the winter in many northern states, and infestations originate from northward migrations. Heavy infestations of this pest have long been associated with later than normal crop plantings (Fig. 12.8). When corn or sorghum is planted late, as a second crop following small grains, the whorl stage and plant fruiting structures are subject to increased earworm attack. Although damage to whorl-stage corn produces insignificant yield reductions, corn ear infestations can result in 100% loss of yield. Also, sorghum seed destruction can be greater than 50% in double-crop plantings.

Concern has been expressed that several insects, like the corn earworm, will be favored in conservation-tillage systems because pupation sites are not destroyed by tillage and ground cover provides protection from natural enemies (Hoards, 1970; Roach, 1981b; Watson et al., 1974). This has not occurred, probably for a variety of reasons. First, as will be discussed later, higher populations of natural enemies occur in conservation tillage, suggesting that pupating earworms are actually afforded less protection. Also, these insect pests feed on a wide variety of crop plants and have a strong migratory

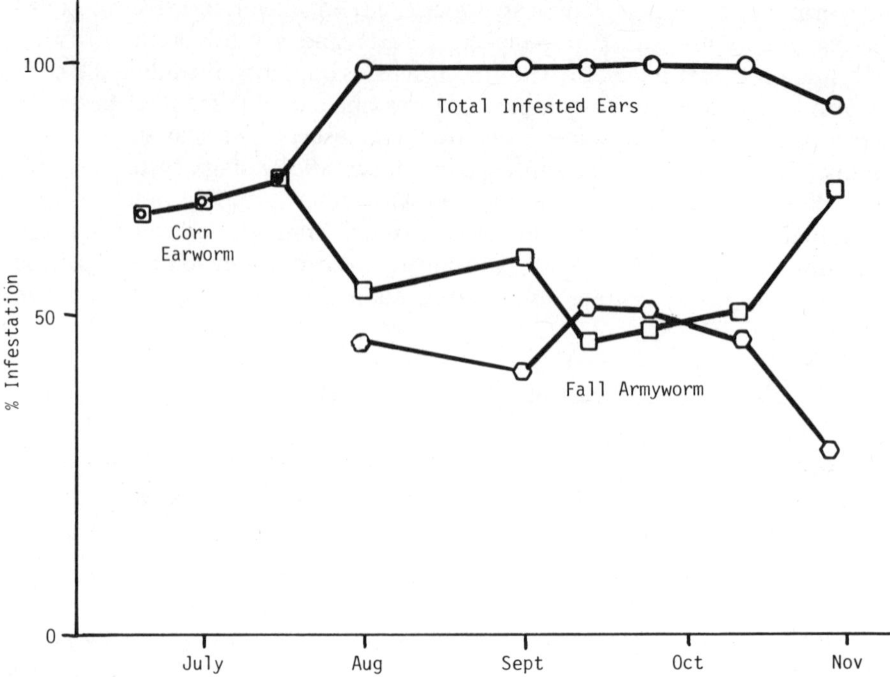

FIGURE 12.8. Seasonality of corn ear infestations by the corn earworm and fall armyworm in Georgia, 1980–1981.

behavior. This suggests that dispersal patterns from conservation-tillage fields would be similar to plow-tillage fields.

Pest Management in Conservation Tillage

In conservation-tillage systems, the recommendation to plant early is essentially the same as for the plow-tillage system. Corn and soybean cultivars are available with a degree of resistance to the corn earworm, but resistant varieties of other crops used in conservation-tillage systems are not widely available. Crop selection would be a wise pest-management consideration in certain conservation-tillage systems such as double cropping, where late planting dates often cannot be avoided. In comparisons of corn, sorghum, and soybeans, soybeans received substantially less damage from late season earworm populations compared to sorghum and corn (All and Rogers, 1983). A wide variety of effective insecticides are available for suppressive control of earworm infestations (All, 1979c). Action thresholds have been developed for many conservation-tillage crops. The use of insecticides is not always economical due to an unfavorable cost/benefit relationship as a result of low crop value and/or the requirement for multiple applications.

Grasshoppers: Redlegged Grasshopper [*Melanoplus femurrubrum* (DeGeer)], Differential Grasshopper [*M. differentialis* (Thomas), Migratory Grasshopper [*M. sanguinipes* (Fabricius)] and so on.

Recent studies indicate that grasshopper species may be a significant problem in certain conservation-tillage systems (Sloderbeck and Edwards, 1979; D. A. Crossley, Inst. of Ecology, University of Georgia, Athens, personal communication, Aug. 1983).

Biology/Ecology Relative to Conservation Tillage

Grasshoppers are the paradigm of grazing insects. Crop decimations by these pests are well known throughout the world. Reports of grasshopper infestations in conservation-tillage systems have been associated with double-cropping systems where corn, sorghum, or soybeans follow small grains in a rotation. This is logical because many of the most destructive species of grasshoppers have the following behavioral characteristics: (1) grasshoppers prefer small grains for feeding [Sloderbeck and Edwards (1979) reported that nymphs (immature stage) of the redlegged grasshopper developed on small grains and became pests on subsequent plantings of no-tillage soybeans]; (2) they are generally polyphagous feeders and may switch crops in multiple-cropping systems utilizing conservation tillage; and (3) they lay groups of eggs in small packets in undisturbed soil (plow-tillage has long been recognized as an effective method of destroying these egg packets). Although there have been no major outbreaks of grasshoppers in large-scale conservation-tillage operations, the potential for problems should be recognized in order to avoid the devastations from grasshoppers that have been recorded periodically throughout history.

Pest Management in Conservation Tillage

In double-cropping systems, close examination of fields during the germination and seedling phase of the second crop following small grains should be made, especially in areas with periodic grasshopper outbreaks. Action thresholds vary with crops, grasshopper species, and region. Information developed for plow-tillage crops in a particular region should be applicable to conservation-tillage systems. To reduce the number of overwintering eggs, intermittent-tillage operations may be required in some regions. To suppress grasshopper populations which tend to build up in weedy habitats and migrate into crops, good early season weed control in the areas adjacent to crop fields and in small grains is recommended. Use of nonpreferred crops should be considered in high-hazard grasshopper situations. Resistant varieties of sorghum should be selected over nonresistant corn or soybeans if this rotation is compatible with the farm production program. Grasshoppers can be controlled with a variety of insecticides. These are typically applied as suppressive treatments when economically justified.

European Corn Borer [*Ostrinia nubilalis* (Hubner)]

This insect is a major pest of plow and surface-tillage corn; no change in its status has been observed in conservation-tillage systems (All and Gallaher, 1976).

Biology/Ecology Relative to Conservation Tillage

European corn borers overwinter as mature diapausing larvae in cornstalk residue. Plow-tillage operations significantly reduce overwintering populations through destruction of the overwintering site. Since conservation-tillage systems increase amounts of stalk residue on the soil surface, it is speculated that elimination of plowing will result in increased borer problems (Hoards, 1970). Although no definitive studies have either supported or refuted this claim, evidence suggests that the environmental conditions during early season moth flights have a greater effect on population than does the tillage system (Musick and Beasley, 1978).

Moths that emerge in the spring oviposit on corn that is in the most advanced stage of development. In the eastern and midwestern United States, it is desirable to delay planting in conservation-tillage systems so that soil temperature can increase and ensure rapid seed germination, etc. This late-planted corn should be less attractive to the spring moths. However, first-generation moths emerging from corn in July are attracted to corn that is pollinating. Hence, corn in conservation-tillage plantings may be more attractive for oviposition later in the season.

Pest Management in Conservation Tillage

Resistant varieties of corn show various levels of antixenosis to both first- and second-generation corn borer larvae. A resistant variety should be used in conservation-tillage plantings. Although corn planted in late spring escapes early season corn borer populations, this corn is more attractive to moths emerging in July. Action thresholds vary by type of production (seed fields versus grain production fields) and state. Several effective insecticides are available.

Sorghum Midge [*Contarinia sorghicola* (Coquillett)]

This insect is a constant threat to sorghum in many areas of the world. Serious infestations have been encountered in conservation-tillage systems, but in direct comparisons to plow-tillage systems damage is usually similar (All, unpublished data).

Biology/Ecology Relative to Conservation Tillage

In conservation tillage, sorghum midge outbreaks have been observed in double-cropping situations where sorghum was planted late. Also, heavy infestations have occurred in no-tillage systems where johnsongrass [*Sorghum halepense* (L.) PERS] control was poor and moderate to high weed levels were present when sorghum was fruiting. Johnsongrass is attacked by the sorghum midge and has been implicated as an early season host of overwintering populations. Since sorghum midge adults are short lived, it is critical that an early blooming host be present as a host for overwintering populations. In many areas, the flowering of johnsongrass coincides with peak midge emergence in the spring, and the weed is an important factor in promoting the development of damaging midge infestations in sorghum (Roth and Pitre, 1975). The sorghum midge has multiple generations during the season, and a preponderance of females is produced in late-season generations. These factors probably contribute to the progressively greater damage that is often observed in later plantings (Young and Teetes, 1977).

Pest Management in Conservation Tillage

Control of johnsongrass and early planting of sorghum are tactics of proven effectiveness in plow-tillage systems and should be utilized in conservation-tillage sorghum where possible. Unfortunately, johnsongrass control is especially difficult in conservation-tillage systems since herbicides effective on johnsongrass are equally effective on other crops belonging to the grass family. Periodic rotation to a legume crop, such as soybeans, should be coupled with an all-out johnsongrass eradication program. Various herbicides effectively control johnsongrass, including some "over-the-top" herbicides that may be applied throughout the season in legumes (Antognini, 1981). Early planting may not be agronomically feasible in certain conservation-tillage systems (i.e., double cropping). In these situations, the selection of short season and uniform flowering varieties should be helpful. Sorghum midge adults are vulnerable to a variety of contact insecticides. A control program should be considered in areas with a history of midge outbreaks. Since all life stages, except the adults, are protected in the glumes, timing of the insecticidal application is critical. Spray applications of insecticides must be made during the period of sorghum flowering since adult midges are active at that time (Huddleston et al., 1972).

Fall Armyworm [*Spodoptera frugiperda* (J. E. Smith)]

This pest is similar to the corn earworm in many respects. It has a strong migratory habit, and in many areas of the United States (Fig. 12.8) damage increases greatly as the growing season progresses. Devastating infestations have occurred in both corn and sorghum. But, like the corn earworm, high population levels were influenced more by late planting than by tillage system.

Biology/Ecology Relative to Conservation Tillage

The fall armyworm is a tropical insect with an extremely reduced overwintering capability; overwintering is limited to the most southern regions of the United

States. This pest is highly mobile and regularly migrates to the northcentral and eastern portions of the United States, attacking a variety of crops but preferring grasses, especially corn. In corn, damage can occur in all stages up to and including maturity. The fall armyworm, like other armyworm species, is noted for its apparent ability to produce devastation of the crop seemingly overnight (Figure 12.9). Moths lay batches of eggs on foliage, and during outbreaks larvae can quickly devour seedlings, leaving only a stub. If infestations occur when corn is silking, up to six larvae can be found in an ear, often

FIGURE 12.9. (a) Fall armyworm infestations in late planted corn are devastating in any type tillage system. (b) Fall armyworm can totally destroy a no-tillage corn field.

reducing the ear to pulp (Sparks, 1979). During many years in the South, this pest is considered a limiting factor to the efficient production of corn in double-cropping systems. In direct comparisons in no-tillage and plow-tillage systems associated with double-cropping practices, it was found that infestations were initiated sooner on young corn seedlings in plow tillage. Very little damage was observed in no-tillage systems while the seedlings were growing in the small grain stubble. However, as the seedlings grew and were exposed above the mulch, infestations occurred rapidly and damage to older corn seedlings was equal to that observed in the plow-tillage system (All, 1980b).

Pest Management in Conservation Tillage

Preventive control tactics discussed for the corn earworm are effective against the fall armyworm. If agronomically feasible, late plantings should be avoided. In double-cropping conservation-tillage systems where fall armyworm hazards are high, crop selection is important. Soybeans and other legumes are less preferred by fall armyworm than grass crops. Sorghum is less vulnerable than corn. Since damaging infestations can occur rapidly, early detection of fall armyworm populations is especially important. A reliable, yet inexpensive, method of determining the onset of fall armyworm infestations is placement of red surveyor's flags in susceptible crops. Moths readily oviposit on these flags, and infestations can be detected before serious feeding damage commences (Thomson and All, 1982). Action thresholds have been established for the fall armyworm. Insecticides that are effective in plow-tillage systems are equally efficacious in conservation-tillage systems (All, 1980a).

Miscellaneous Postseedling Pests

Several other insect pests attack postseedling stages of crops and are potentially damaging in conservation-tillage plantings. Limited research data are available on the relative impact of these pests in conservation-tillage systems. However, knowledge of the biology of these pests in plow-tillage systems suggests that they may be important. The southwestern corn borer [*Diatraea grandiosella* (Dyar)] and the southern cornstalk borer [*Diatraea crambidoides* (Grote)] have been observed at low levels in continuous no-tillage cornfields (All and Gallaher, 1976; Gregory and Musick, 1976). Their damage and biological characteristics are similar to those of the European corn borer. Both insects overwinter near the crown of cornstalks and in other refuse that is not buried in continuous no-tillage corn. The potential for build up of populations in crop refuse has concerned entomologists, and the possibility of damaging outbreaks of these pests should be recognized by farmers utilizing conservation tillage.

Various defoliating pests such as the soybean looper [*Pseudoplusia includens* (Walker)], the green cloverworm [*Plathypena scabra* (Fabricius)], the velvetbean caterpillar (*Anticarsia gemmatalis* Huebner), and various species of armyworms occur on conservation-tillage crops in virtually identical situations as in plow tillage. Increased concern should be exercised in late plantings and appropriate pest management techniques applied when action thresholds are reached (Raney, 1974).

Mexican bean beetle (*Epilachna varivestis* Mulsant) populations were reduced in soybeans planted in no-tillage systems compared directly to plow tillage (Sloderbeck and Edwards, 1979). The reason for this was unknown, but the researchers speculated that adult beetles preferred the monoculture environment of conventional tillage to the more diverse no-tillage habitat.

The chinch bug [*Blissus leucopterus leucopterus* (Say)] has been observed intermittently in double-cropping systems of no-tillage corn planted after harvest of small grains for either silage or grain. Chinch bugs have a long history of producing devastating infestations in corn following population buildups in small grains. The chinch bugs migrate to corn when the small grains are harvested or are no longer palatable. These conditions readily occur in double-cropping systems in which conservation tillage is used.

Several species of aphids, spittlebugs, leafhoppers, stink bugs, plant bugs, and other sucking insects have biologies that make them adaptable to many plant hosts and have the potential of building populations on one crop, such as small grains or forage legumes, and transferring to a second crop planted by conservation-tillage methods. No major problems have occurred to date; however, these situations should be closely monitored when new conservation-tillage practices are started in an area.

Tarnished plant bugs (*Lygus lineolaris* Palisot de Beauvois) and bollworm/budworm (*Heliothis* spp.) populations were found to be economically damaging to cotton grown in conservation-tillage systems using a legume cover crop when the cotton was moisture stressed (Gaylor et al., unpublished data).

Hessian fly (*Mayetiola destructor* Say) numbers are increased in conservation-tillage systems utilizing rotations of spring and winter wheat [*Triticum vulgare Vill.* (aestivum L.)] in northern Idaho and eastern Washington (Dr. L. O'Keeffe, University of Idaho, personal communication, June, 1983). Crop damage directly related to conservation tillage has not been significant to date in the Northwest or in other grain production areas, but farmers should be concerned about Hessian fly outbreaks in continuous conservation-tillage systems utilizing small grains. Hessian flies undergo pupation and overwinter in ground stubble, and plow-tillage operations have long been advocated as a sanitation operation to destroy this life stage. Thus, if a potential exists for high emergence of flies in conservation-tillage fields, farmers should take particular care to utilize the monitoring and control programs available for Hessian flies in their region.

Maize weevils (*Sitophilus zeamais* Motschulsky) are frequently observed in no-tillage corn ears at harvest. Stored grain problems arise when infested grain is taken into storage from fields. Little is known on the field biology of the maize weevil. Recent research indicates that these pests are capable of overwintering among grain debris left in fields (Dix and All, 1983). Conservation-tillage systems may provide population reservoirs for this pest.

Ring-legged earwig [*Euborellia annulipes* (Lucas)] has been encountered in no-tillage corn in damaging populations (All and Gallaher, 1976). Groups of 1–10 adults and nymphs were found in the tips of ears when the grain moisture content was about 20% and appeared to be associated with previous corn earworm feeding. These insects produced little apparent damage, but the high populations found in some no-tillage cornfields was dramatic.

Rodents and Birds

Rodents may also attack corn ears in conservation-tillage systems. This has been demonstrated where dense ground cover was present and short hybrids were used, thereby placing the ears closer to the ground. Observations in continuous no-tillage corn indicate that damage to grain occurs during the milk and dough stages of corn development. Ear feeding occurs throughout the field, but is spotty in character. Typically 30% of the grain on individual ears is consumed, but total destruction of an ear is not uncommon (All, unpublished data).

Bird problems also occur in harvest-stage crops. Damage to corn is common, and bird losses in sorghum can be tremendous in conservation-tillage systems. However, observation indicates that bird problems in harvest-stage crops are similar in plow-tillage and conservation-tillage systems (All, unpublished data).

Crop varieties with some level of resistance to bird damage are available in some regions. If resistant varieties are not available, information on hybrids that are particularly susceptible to bird damage is often available, and these should be avoided. When good harvesting practices minimize seed losses, birds will be less attracted into freshly planted no-tillage fields.

Insect Vectors of Plant Virus Diseases

The two major virus diseases of corn, maize chlorotic dwarf and maize dwarf mosaic, can be serious problems in conservation-tillage systems (All, 1983). The epidemiology of these diseases involves the interaction of insects, the virus pathogens, and the overwintering weed host of the pathogens, johnsongrass. This represents a unique multifaceted challenge for pest management (All, 1983; All et al., 1981).

Biology/Ecology Relative to Conservation Tillage

Maize chlorotic dwarf and maize dwarf mosaic both produce debilitating effects on corn growth. The most striking symptom of both diseases is stunting of plants resulting, most often, in severe yield reductions. Maize chlorotic dwarf virus is transmitted by leafhoppers, particularly the black-faced leafhopper (*Graminella nigrifrons* Forbes). Maize dwarf mosaic virus is vectored by several species of aphids. Both diseases overwinter in rhizomes of johnsongrass, which is the only perennial host of these viruses.

Several factors are important determinants of the ultimate severity of both diseases on corn in conservation-tillage systems. First, the presence and abundance of the vectors at various times in the season influence the rate of disease spread within fields. Another factor is the presence of the overwintering host of the pathogens, johnsongrass. When this weed occurs at low or moderate levels in fields, it can have a tremendous influence on establishment and spread of the disease in corn. An additional highly important aspect is the time when the infection occurs in corn. In general, the younger a plant is when inoculated with a virus(es), the greater the severity of the disease(s) and the loss in yield.

Johnsongrass is often present in conservation-tillage fields at the time corn is planted. Consequently, vectors can acquire the pathogens from the weed and transmit it among corn plants during the early growth stages. In plow-tillage systems, even if johnsongrass is poorly controlled, it is "set back" 1 or 2 weeks, providing more time for corn to develop to larger and more-tolerant growth stages. There is a trend for earlier inoculation and increased severity of these virus diseases in no-tillage as compared to plow-tillage systems (All, unpublished data; All et al., 1977).

Pest Management in Conservation Tillage

A variety of management strategies for maize chlorotic dwarf and maize dwarf mosaic can be considered. The objective is to disrupt one or more links in the virus × vector × johnsongrass × corn interaction. Field corn hybrids with moderate to high tolerance for the diseases are available. Some systemic insecticides, applied at planting, effectively control the leafhopper vectors of maize chlorotic dwarf, resulting in reduced disease loss (All and Alverson, 1979; All et al., 1976, 1977). These systemic insecticides, although effective against aphids, are not effective in reducing transmission of maize dwarf mosaic (Kuhn et al., 1975). Early planting of corn results in reduced disease incidence because large vector populations are avoided. Also, irrigation and optimum fertilization practices are useful in aiding tolerance of corn to the diseases. Johnsongrass control with glyphosate resulted in only slight reductions in the disease in conservation-tillage systems as there is usually incomplete eradication of the weed. It is apparent that the threshold for reducing johnsongrass populations low enough to minimize the incidence of the virus diseases is lower than the threshold to eliminate it as a weed pest. Thus, corn production with conservation-tillage systems in areas that have both a history of the virus diseases and high johnsongrass populations is questionable. Crop rotation with a noncereal crop, like soybeans, may be advisable in these situations. This tactic has special merit since over-the-top herbicides can be used throughout the season in soybean fields to eradicate johnsongrass, thus allowing corn production during the subsequent year (All, 1983).

The optimum management strategy for maize chlorotic dwarf and maize dwarf mosaic in conservation-tillage systems would be an integrated program utilizing all of the tactics outlined. However, cost/benefit relationships indicate that use of disease-resistant hybrids and early planting would be the most efficient management strategy for these diseases. Planting-time applications of systemic insecticides in combination with these tactics may be justified when disease levels are high or when the toxicants will have additional efficacy against other pests (All, 1983).

Biological Control in Conservation Tillage

The environment near the soil surface in conservation-tillage systems provides a habitat that supports higher numbers and greater diversity of arthropods than in plow-tillage systems. Many of these arthropods, particularly spiders and beetles in the familes Carabidae and Staphylinidae, are predatory on pest insects (Blumberg and Crossley, 1982; House and All, 1981; House and Stinner, 1983; McPherson et al., 1982). It is particularly important that these predators be present in higher numbers in conservation-tillage fields at the time crops are germinating and becoming established. Their control of some seed and seedling pests can be substantial. In plow-tillage areas, these predators must move in from adjacent nonagricultural habitats. This takes time, and young crop seedlings may be more vulnerable to pest infestation. The actual role of predatory arthropods in controlling pests in conservation-tillage systems is not completely understood. The process involves a complex interaction between many abiotic and biotic factors in the unique environment of a particular conservation-tillage system. However, as more knowledge is developed, it is becoming clear that these predatory arthropods aid in preventing outbreaks of pests in crops grown using conservation-tillage systems.

Increased moisture, reduced temperature, and reduced lighting occur within the mulch residues and on the soil surface of conservation-tillage systems. These conditions are more favorable for the development of disease epizootics in pest populations (Burges and Hussey, 1971). Several insect pathogens, including fungi, entomophilic nematodes, viruses, and bacteria, may be enhanced in conservation-tillage habitats. Higher populations of entomophilic Rhabditoid nematodes were observed in no-tillage as compared to plow-tillage sorghum (Saunders and All, unpublished data). Additional research is needed on the influence of conservation-tillage systems on pathogens of insects.

Two techniques for enhancement of biological control involve a manipulation of the environment: conservation—avoiding practices that destroy natural enemies, and enhancement—practices that increase the attractiveness of an area for colonization by natural enemies or management that increases their longevity or reproductiveness (Stehr, 1975). The environment created by conservation-tillage systems serves both of these functions for many natural enemies of insect pests. By avoiding disruption of habitats with plow-tillage operations, a conservation of natural enemies is achieved, and enhancement of these organisms is more likely within the diverse and ecologically favorable environments of conservation-tillage systems.

CONCLUSIONS

It is apparent that insect pest management in conservation tillage systems is a mixture of the new and the old. It is new in the sense that in most situations throughout the world conservation tillage involves new technologies that establish agroecosystems that heretofore have not been encountered by entomologists. Current knowledge indicates that some pests may behave differently in these new systems. It is old in the sense that management strategies involve long-standing principles of applied entomology. That is, new agricultural production systems have continually evolved with similar attributes of entomological uncertainty. It has been the entomologist's challenge to develop management programs based on the biological idiosyncrasies of the new insect × crop × environment interaction. When anxiety and ignorance are dispelled, the "bugs" can be worked out and appropriate pest-management strategies can be developed. In conservation tillage system cropping, it is apparent that many of the pest-management strategies that have been developed for specific pests in plow-tillage systems are readily adaptable to the new systems. In certain situations, such as with cutworms, wireworms, aphid–ant complexes, slugs, rodents, and birds, new pest-management strategies must be developed.

The impact of conservation tillage systems on pest problems associated with the various crops is dependent on crop, insect, and tillage environment. Some general conclusions concerning vulnerability of major crops used in conservation-tillage systems are as follows:

1. *Corn.* For seed and seedling pests, as tillage is reduced, greater problems can be expected from armyworms, slugs, black cutworm, stalk borers, and southern corn billbug. Other pest problems may be more serious, but are sporadic enough in occurrence to make conclusions tenuous. Postseedling pests are not expected to cause any additional problems over those observed in plow-tillage systems, except when planting is delayed to accommodate double cropping. Although pest-management strategies will remain similar to those employed in the plow-tillage system, certain pests (stalk borers, southern corn billbug, and slugs) have not been managed effectively in conservation-tillage systems. Preventive tactics should be adopted where possible and suppressive control used when populations approach or exceed a predetermined action threshold.

2. *Soybean.* Conservation-tillage systems have been used in conjunction with double-cropping practices. Generally, pest problems have been influenced more by late planting date than by the tillage practices.

3. *Wheat.* Definitive studies for the midwestern wheat belt of Texas, Oklahoma, Kansas, and the Dakotas are lacking. For the northwestern United States, no measurable differences among most pests have been observed, with the exception of the Hessian fly, which seems to increase in conservation-tillage systems where spring wheat follows winter wheat.

4. *Forages.* Grasshoppers, leafhoppers, armyworms, and slugs have been devastating to young forage seedlings. Most severe problems tend to occur during late summer and early fall, especially in dry years.

5. *Other Crops.* Although some information is available on the relative importance of some pests in conservation-tillage systems for cotton, tobacco, vegetable, and peppermint production, additional research must be completed before definitive statements can be made.

Some of the principles of insect pest management strategies associated with conservation-tillage systems may be summarized as follows:

1. There are many types of conservation-tillage systems, each of which develops a unique agroecosystem that may have positive, negative, or neutral effects on pests and their natural enemies. Hence, specific entomological knowledge must be developed for the particular tillage system, including the concomitant cultural practices used in the system. For example, it is apparent that certain double-cropping no-tillage systems have increased pest hazards, but this hazard is associated with later-than-normal planting dates rather than the specific tillage system.

2. Special caution should be exercised when establishing conservation-tillage crops in areas that have laid fallow or have been in sod for several years. A variety of pests may be lurking in these fields.

3. Greatest hazard for pest infestation occurs in the seed and seedling stages from subterranean and semisubterranean pests. Fields should be examined prior to planting and as frequently as feasible for 3 or 4 weeks after crop establishment if pest infestations are to be accurately identified and managed.

4. With a few exceptions (i.e., lesser cornstalk borer), if a pest hazard existed in a plow-tillage situation in an area, a similar problem should be anticipated when conservation-tillage systems are employed.

5. Preventive control tactics for pests are typically the most economical pest-management strategies. However, they must be considered before the crop is planted.

6. Suppressive control of insect outbreaks with insecticides is expensive and should be used only with a knowledge of pest numbers and an awareness of current or potential damage to yield. Damage should exceed some action threshold before insecticides are applied.

7. In developing an integrated-pest-management strategy for conservation tillage crops, the entire pest/beneficial complex (insects, diseases, and weeds) must be considered. In addition, an understanding must be developed

on how preventive or suppressive measures directed at one pest problem may affect other components in the crop-protection system. Although not easy, this "total" integrated crop management program can be approached as knowledge on conservation tillage is advanced.

ACKNOWLEDGMENTS

The authors wish to express their sincere appreciation to Dr. L. E. O'Keeffe, University of Idaho, Moscow, Idaho; Dr. E. C. Berry, USDA-ARS, Corn Insects Research Laboratory, Iowa State University, Ankeny, Iowa; and Dr. J. M. Cheshire, University of Georgia, Georgia Experiment Station, Griffin, Georgia for their assistance in technical review of this manuscript. We also want to express our appreciation to Mrs. Marie Lavallard, Retired Editor, University of Arkansas, Department of Agricultural Publications, for her editorial review.

LITERATURE CITED

All, J. N. 1978. Insect relationships in no-tillage cropping. Proc. First An. S. E. No-Till Systems Conf. *Univ. Ga. Spec. Publ.* **5**:17–19.

All, J. N. 1979a. Insect relationships in no-till cropping. *Agrichemical Age* **23**:22–23.

All, J. N. 1979b. Consistency of lesser cornstalk borer control with Lorsban in various corn cropping systems. *Down to Earth* **36**:33–36.

All, J. N. 1979c. Sweetcorn, control of mixed infestation of corn earworm and fall armyworm. *Insecticide and Acaricide Tests* **4**:98.

All, J. N. 1980a. Reducing the lag from research synthesis to practical implementation of pest management strategies for the fall armyworm. *J. Fla. Entomol.* **63**:357–361.

All, J. N. 1980b. Pest management decisions in no-tillage agriculture. In R. G. Gallaher (ed.). *Energy Relations in Minimum Tillage Systems.* Proc. 3rd No-tillage Systems. Univ. Fla. Press, pp. 1–6.

All, J. N. 1983. Integrating techniques of vector and weed host suppression into control programs for maize virus diseases. Proc. II Intl. Maize Virus Dis. Colloquium and Workshop II. pp. 243–247.

All, J. N., and D. R. Alverson. 1979. Field corn, blackfaced leafhopper and maize chlorotic dwarf disease control. *Insecticide and Acaricide Tests* **4**:205.

All, J. N., and R. N. Gallaher. 1976. Insect infestations in no-tillage corn cropping systems. *Ga. Agric. Res.* **17**:17–19.

All, J. N. and R. N. Gallaher. 1977. Detrimental impact of no-tillage corn cropping systems involving hybrids, insecticides, and irrigation on lesser cornstalk borer infestations. *J. Econ. Entomol.* **70**:361–365.

All, J. N., R. N. Gallaher, and M. D. Jellum. 1979. Influence of planting date, preplanting weed control, irrigation, and conservation-tillage practices on efficacy of planting time insecticide applications for control of lesser cornstalk borer in field corn. *J. Econ. Entomol.* **72**:265–268.

All, J. N., W. A. Gardner, E. F. Suber, and B. Rogers. 1982. Lesser cornstalk borer as a pest of corn and sorghum. In A review of information on the lesser cornstalk borer, *Elasmopalpus lignosellus* (Zeller). Univ. Ga. Spec. Publ. 17, pp. 33–46.

All, J. N., R. S. Hussey, and D. G. Cummins. 1984. Southern corn billbug (Coleoptera: Curculionidae) and plant-parasitic nematodes: influence of no-tillage, coulter-in-row chiseling, and insecticides on severity of damage to corn. *J. Econ. Entomol.* **77**:178–182.

All, J. N., and M. D. Jellum. 1977. Efficacy of insecticide-nematocides on *Sphenophorous callosus* and phytophagous nematodes in field corn. *J. Ga. Entomol. Soc.* **12**:291–297.

All, J. N., C. W. Kuhn, R. N. Gallaher, M. D. Jellum, and R. S. Hussey. 1977. Influence of no-tillage-cropping, carbofuran, and hybrid resistance on dynamics of maize chlorotic dwarf and maize dwarf mosaic diseases of corn. *J. Econ. Entomol.* **70**:221–225.

All, J. N., C. W. Kuhn, and M. D. Jellum. 1976. The changing status of corn virus diseases: potential value of a systemic insecticide. *Ga. Agric. Res.* **17**:4–6.

All, J. N., C. W. Kuhn, and M. D. Jellum. 1981. Control strategies for vectors of virus and viruslike pathogens of maize and sorghum. In D. T. Gordon, J. K. Knoke, and G. E. Scott (eds.). *Virus and viruslike diseases of maize in the United States.* Southern Coop. Ser. Bull. 247, pp. 121–127.

All, J. N., and B. Rogers. 1983. Insect management in no-till. Proc. 5th S.E. No-Till Systems Conf. Florence, SC, Clemson, Univ. Circ. pp. 12–15.

Antognini, J. 1981. Selective, over the top grass control: the next revolution in herbicides. *Agrichemical Age* **25**:20–25.

Beasley, L.E., and G. E. McKibben. 1976. Mouse control in no-tillage corn. Ill. Agric. Expt. Stn. DSAC 4:27-30. Dixon Springs Agric. Ctr., Simpson, IL.

Blumberg, A. Y., and D. A. Crossley, Jr. 1982. Comparison of soil surface arthropod populations in conventional tillage, no-tillage and old field systems. *Agro-Ecosystems* **8**:247–253.

Burges, H.D., and N. W. Hussey (eds.). 1971. *Microbial Control of Insects and Mites.* Academic Press, New York.

Busching, M. K., and F. T. Turpin. 1976. Ovipositional preferences of the black cutworm moth among various crop plants, weeds, and plant debris. *J. Econ. Entomol.* **69**:587–590.

Chaing, H. C., D. Rasmussen, and R. Gorder. 1971. Survival of corn rootworm larvae under minimum tillage conditions. *J. Econ. Entomol.* **64**:1576–1577.

Cheshire, J. M., Jr., and J. N. All. 1979a. Monitoring lesser cornstalk borer larval movement in no-tillage and conventional tillage corn systems. *Ga. Agric. Res.* **21**:10–14.

Cheshire, J. M., Jr., and J. N. All. 1979b. Feeding behavior of lesser cornstalk borer larvae in simulations of no-tillage, mulched conventional tillage and conventional tillage corn cropping systems. *Environ. Entomol.* **8**:261–264.

Dix, D., and J. N. All. 1983. Field dynamics of the maize weevil (*Sitophilus zeamais*) in Georgia field corn. Proc. Entomol. Soc. Amer. Ann. Mtg.

Dumas, W. T. 1983. Specific technology for conservation-tillage cotton. Conservation Tillage Conf., Auburn Univ., Aurburn, AL.

Durant, J. 1982. Influence of the southern corn billbug (Coleoptera: Curculionidae) population density and plant growth stage infested on injury to corn. *J. Econ. Entomol.* **75**:892–894.

Edwards, C. A. 1975. Effects of direct drilling on the soil fauna. *Outlook Agric.* **8**:243–244.

Gardner, W. A., and J. N. All. 1982. Chemical control of the lesser cornstalk borer in grain sorghum. *J. Ga. Entomol. Soc.* **17**:167–171.

Gregory, W. W. 1974. No-tillage corn insect pests of Kentucky. . . A five year study. In *Proceedings No-Tillage Research Conference.* Univ. Kentucky, Lexington, KY, pp. 46–58.

Gregory, W. W., and G. J. Musick, 1976. Insect management in reduced tillage systems. *Bull. Entomol. Soc. Amer.* **22**:302–304.

Harrison, F. P., R. A. Bean, and O. J. Qawiyy. 1980. No-till culture of sweet corn in Maryland with reference to insect pests. *J. Econ. Entomol.* **73**:363–365.

Hoards, D. 1970. Reduced tillage systems cause insect problems. *Crops and Soils* **115**:359.

House, G. J., and J. N. All. 1981. Carabid beetles in soybean agroecosystems. *Environ. Entomol.* **9**:194–196.

House, G. J., and B. R. Stinner. 1983. Arthropods in no-tillage soybean agroecosytems: community compositon and ecosystem interactions. *Environ. Manage.* **7**:23–28.

Huddleston, E. W., D. Ashdown, B. Maunder, C. R. Ward, G. Wilde, and C. E. Forehand. 1972. Biology and control of the sorghum midge. 1. Chemical and cultural control studies in west Texas. *J. Econ. Entomol.* **65**:851–855.

Kirk, V. M., C. O. Calkins, and F. J. Post. 1968. Ovipositional preferences of western corn rootworms for various soil surface conditions. *J. Econ. Entomol.* **61**:1322–1324.

Kuhn, C. W., M. D. Jellum, and J. N. All. 1975. Effect of carbofuran treatment on corn yield, maize chlorotic dwarf, and maize dwarf mosaic virus diseases, and leafhopper populations. *Phytopathology* **65**:1017–1020.

Martin, P. B., B. R. Wiseman, and R. E. Lynch. 1980. Action thresholds for fall armyworm on grain sorghum and coastal bermudagrass. *Fla. Entomologist* **63**:375–405.

McPherson, R. M., J. C. Smith, and W. A. Allen. 1982. Incidence of arthropod predator in different soybean cropping systems. *Environ. Entomol.* **11**:685–689.

Metcalf, Z. P. 1917. Biological investigation of *Sphenophorus callosus* Oliver. N.C. Agric. Exp. Stn. Bull. 13.

Musick, G. J. 1970. Insect problems associated with no-tillage. *Proc. N.E. No-Tillage Conf.* **1**:44–59.

Musick, G. J. 1973. Control of armyworm in no-tillage corn. *Ohio Rpt.* **58**:42–45.

Musick, G. J., and L. E. Beasley, 1978. Effect of the crop residue management system on pest problems in field corn (*Zea mays* L.) production. In *Crop Residue Management Systems.* Amer. Soc. Agron., Madison, WI, pp.173–186.

Musick, G. J., and D. L. Collins, 1971. Northern corn rootworm affected by tillage. *Ohio Rpt.* **56**:88–91.

Musick, G. J., and H. B. Petty. 1973. Insect control in conservation tillage systems. In *Conservation Tillage: The Proceedings of a National Conference.* Soil Conserv. Soc. Amer., Ankheny IA.

Pike, K. S., and M. Glazer. 1982. Strip rotary tillage: a management method for reducing *Fumibotys fumalis* (Lepidoptera: Pyralidae) in peppermint. *J. Econ. Entomol.* **75**: 1136–1139.

Raney, H. 1974. Insects in no-till soybeans. In Proc. No-Tillage Res. Conf. Univ. Ky., Lexington, KY, pp. 59–65.

Rivers, R. L., K. S. Pike, and Z. B. Mayo. 1977. Influence of insecticides and corn tillage systems on larval control of *Phyllophaga anxia. J. Econ. Entomol.* **70**:794–796.

Roach, S. H. 1981a. Reduced vs conventional tillage practices in cotton and tobacco; a comparison of insect populations and yields in northeast South Carolina, 1977–1979. *J. Econ. Entomol.* **79**:688–695.

Roach, S. H. 1981b. Emergence of overwintered *Heliothis* spp. moths from three different tillage systems. *Environ. Entomol.* **10**:817–818.

Rogers, B., and J. N. All. 1982. Impact of no-tillage systems of corn, sorghum, and soybeans involving insecticides on lesser cornstalk borer infestations. *Proc. S.E. Branch Entomol. Soc. Amer.* **56**:10.

Roth, J. P., and H. N. Pitre. 1975. Seasonal incidence and host plant relationships of the sorghum midge in Mississippi. *Ann. Entomol. Soc. Amer.* **68**:654–658.

Sloderbeck, P. E., and C. R. Edwards. 1979. Effects of soybean cropping practices on Mexican bean beetle and redlegged grasshopper populations. *J. Econ. Entomol.* **72**:850–853.

Sparks, A. N. 1979. A review of the biology of the fall armyworm. *Fla. Entomol.* **62**:82–86.

Stehr, F. W. 1975. Parasitoids and predators in pest management. In R. L. Metcalf and W. H. Luckmann (eds.). *Introduction to Insect Pest Management.* Wiley, New York, pp. 147–188.

Stern, V. M., R. F. Smith, R. van den Bosch, and K. S. Hagen. 1959. The integration of chemical and biological control of the spotted alfalfa aphid. Part I. The integrated control concept. *Hilgardia* **29**:81–101.

Thomas, C. A. 1940. The biology and control of wireworms. A review of the literature. Penn. Agric. Exp. Stn. Bull. 392.

Thomson, M. S., and J. N. All. 1982. Oviposition by the fall armyworm onto stake flags and the influence of flag color and height. *J. Ga. Entomol. Soc.* **17**:206–210.

Tippins, H. H. (ed.). 1982. A review of information on the lesser cornstalk borer *Elasmopalpus lignosellus* (Zeller). Univ. Ga. Spec. Publ. 17.

Watson, T. F., K. K. Barnes, J. E. Slosser, and D. G. Fullerton. 1974. Influence of plowdown dates and cultural practices on spring moth emergence of the pink bollworm. *J. Econ. Entomol.* **67**:207–210.

Wrenn, E. 1975. Armyworms launch heavy attack on many corn fields in Virginia. S.E. Farm Press July 2, 1975. pp. 5, 28.

Wright, R. J., J. W. van Duyn, and J. R. Bradley, Jr. 1983. Seasonal phenology and biology of the southern corn billbug in eastern North Carolina. *J. Ga. Entomol. Soc.* **18**:376–385.

Young, W. R., and G. L. Teetes. 1977. Sorghum entomology. *Ann. Rev. Entomology* **22**: 193–218.

13

EFFECT OF SURFACE TILLAGE ON PLANT DISEASES

M. G. BOOSALIS
Professor
Department of Plant Pathology
University of Nebraska
Lincoln, Nebraska

B. L. DOUPNIK
Professor
Department of Plant Pathology
University of Nebraska
South Central Station
Clay Center, Nebraska

J. E. WATKINS
Professor
Department of Plant Pathology
University of Nebraska
Lincoln, Nebraska

INTRODUCTION

Traditionally, the most common tillage method was either fall or spring moldboard plowing leaving residue completely buried. The destruction of crop residue by burying is an effective means of destroying plant pathogens and, thus, an important means of plant disease control. Today, crop producers are shifting to surface-tillage systems to offset the rapidly rising costs of fuels, labor, and soil erosion (Phillips et al., 1980).

The effects of surface tillage on plant diseases is not fully understood. This forces plant pathologists to make predictions of potential benefits or problems that might arise due to the presence of residue on the soil surface. Alteration of the microclimate due to surface residues may retard, enhance, or have little effect on plant disease development. The degree of influence on plant diseases by residue generally relates to the amount of residue remaining after planting. A tillage system that leaves 20% cover is less likely to influence disease development than one leaving 90% residue cover after planting. Unfortunately, there is little information on the effect of different amounts of surface residues in relation to plant diseases. Most surface-tillage systems presented in this chapter leave over 40% residue cover.

Surface residues may affect plant diseases in several ways. They may provide a habitat for overwintering (survival), growth, and multiplication of plant pathogens—particularly fungal and bacterial pathogens. Some pathogens within diseased tissues continue to grow and multiply on crop residue after the crop is harvested. In other cases, pathogens from soil, weeds, and other hosts colonize residues from disease-free crops. Such colonized residues also provide excellent overwintering habitats for many pathogens.

There are many plant pathogens that overwinter best in surface residues because they are protected from the environment and other microorganisms. Surface tillage increases the chances of epidemics caused by such pathogens. Fortunately, such residues from surface-tillage systems in the American Great Plains have not caused widespread disease epidemics (Boosalis et al., 1981). This is due, largely, to the prevailing hot, dry, windy weather during the summer months, which is unfavorable for the development of diseases, particularly foliar diseases, the relatively short time that such practices have been in use, and the fact that such practices are only recently widely used. It generally takes considerable time for disease problems to become widespread and severe. Severe disease outbreaks related to surface residues are apt to be more common in areas with high rainfall and cooler temperatures.

RESIDUES AS HABITATS OF SURVIVAL, GROWTH, AND REPRODUCTION OF PLANT PATHOGENS

Surface tillage favors the survival of many fungal and bacterial pathogens. Here are a few examples.

The new physiologic race "T" of the *Helminthosporium maydis* Nisikado causing southern corn leaf blight, overwinters on corn residues, particularly on ear-husks (Burns and Shurtleff, 1972; Futrell and Scott, 1971; Gudauskas et al., 1971; Littrell and Sumner, 1971; Ullstrup, 1971). Also, the primary source of initial inoculum of *Helminthosporium turcicum* Pass., the incitant of northern leaf blight of corn, is from corn residue (Boosalis et al., 1967; Nelson and Sheifele, 1970). High populations of the fungal pathogens, *Colletotrichum graminicola* (Ces.) C. W. Wils. (Naylor and Leonard, 1977; Phillips et al., 1980; Vizvary and Warren, 1982) and *Phyllosticta maydis* Arny and Nelson (Arny et al., 1970), which cause anthracnose and yellow leaf blight of corn, respectively, survive in corn residues. *C. graminicola* survived and sporulated on about 30% of the infested corn stalks buried 1.3 cm in the soil until planting time (from December, 1981 to August, 1982), but the pathogen was destroyed in all of the infested stalks buried 10–15-cm depth. On the other hand, the pathogen survived and sporulated in all infested corn stalks left on the surface of the ground from December, 1980 to August, 1981 and continued to survive and sporulate on 52% of these stalks through May of 1982. This suggests that *C. graminicola* would be eliminated from corn stalks buried for one growing season, but it would require two growing seasons to eliminate the pathogen from stalks on the soil surface (Lipps, 1983). Martinson (1981) reported the severity of eyespot of corn (caused by *Kabatiella zeae* Narita & Hiratsuka) grown under different tillage systems and crop rotation schemes was the highest in a monoculture system utilizing no-tillage (Table 13.1). The fungal pathogen, *Pyrenophora trichostoma* (Fr.) Fckl, inducing tanspot of wheat, has spread rapidly throughout the midwest, extending from Canada (Tekauz, 1976) to Oklahoma. The disease is also prevalent and destructive in Australia

TABLE 13.1. Severity of Eyespot Disease on Corn Grown with Different Tillage Systems and Rotation Schemes

Crop[a]						Eyespot Lesions per Leaf[b]		
1976	1977	1978	1979	1980	Tillage	1978	1979	1980
Corn	Soybean	Corn	Soybean	Corn	Plow	17	—	21
					Chisel plow	16	—	26
					Till plant	55	—	9
					No-till	20	—	34
Corn	Corn	Soybean	Corn	Soybean	Plow	—	22	—
					Chisel plow	—	85	—
					Till plant	—	52	—
					No-till	—	75	—
Corn	Corn	Corn	Corn	Corn	Plow	22	52	34
					Chisel plow	217	177	103
					Till plant	200	197	119
					No-till	300	362	129

[a] Tillage treatments for 1978, 1979, and 1980 crops. All plots were plowed for the 1976 crop and chisel plowed for the 1977 crop.

[b] Average number of eyespot lesions on the ear leaf on August 31 in 1978 and 1979, and July 31 in 1980.

SOURCE: Martinson (1981).

especially in Queensland (Rees and Platz, 1979). Reduced and no-tillage, which provide surface residues for the survival of this pathogen, have contributed to the wide distribution and establishment of this disease (Hosford, 1976; Watkins *et al.*, 1978). Incorporation of residue harboring *P. trichostoma* into the soil prevents formation of the overwintering sexual stage of the pathogen, which is the source of initial inoculum. Plowing under wheat residue sharply decreased populations of *Cephalosporium gramineum*, Nisikado and Ikata, the incitant of Cephalosporium stripe of wheat (Wiese and Ravenscroft, 1975). The pathogen survives saprophytically in high populations on undisturbed straw on the soil surface producing conidiospores for at least 3 years (Wiese and Ravenscroft, 1975). Root rot of wheat caused by *Cochiobolus sativus* Ito and Kurib. was reduced by deep tillage of residue (Tanasevych and Korobil, 1978).

Surface tillage may lessen the incidence of some diseases or, for others, have no effect on their development. Minimum tillage decreased the incidence of take-all of wheat caused by *Gaeumannomyces graminis* var. *tritici* Walker (Brooks and Dawson, 1968; Hood, 1965); decreased or had no effect on eyespot of wheat induced by *Pseudoscercosporella herpotrichoides* (Fron.) Dei. (Hood, 1965); and had no influence on sharp eyespot and brown rot of wheat caused by *Rhizoctonia solani* Kühn and *Fusarium* spp., respectively (Hood, 1965).

Surface residues play a key role in the survival and spread of many bacterial plant pathogens (Schuster and Coyne, 1974). The widespread occurrence of Holcus leaf spot of corn caused by *Pseudomonas syringae* Van Hall is attributable to vast amounts of surface residues harboring the pathogen (Weihing and Vidaver, 1967). Goss's bacterial wilt and blight caused by *Corynebacterium michiganense* sp. *nebraskense* was first found in Nebraska in 1969 (Schuster, 1975). Since then, the disease has spread not only throughout the state but to most states in the western corn belt. Surface tillage most likely has played a key role in establishing this disease, since the pathogen survives primarily on surface corn residue.

There are many other fungal and bacterial pathogens that survive in residues (Boosalis et al., 1981; Cook et al., 1978; Menzies, 1963; Sumner et al., 1981). Some of the fungal diseases are septoria glume blotch (Scharen, 1971) and septoria leaf blotch of wheat and barley (Lutey and Frezer, 1960; Scharen, 1971), southern leaf blight of tomato (Worley et al., 1966), brown spot of corn (Burns and Shurtleff, 1973), and verticillium wilt of cotton (Brinkerhoff, 1969). Bacterial diseases include bacterial blight of soybean (Daft and Leben, 1973), halo blight of dry edible beans (Natti, 1967), bacterial blight of cotton (Brinkerhoff and Fink, 1964), tomato canker (Farley, 1971), black chaff of cereals (Boosalis, 1952), and many others.

There is little information on how surface tillage affects nematode and virus pathogens. Plowing reduced populations of migratory parasitic nematodes compared with no-tillage in continuously cropped wheat (Corbett and Webb, 1970). Similarly, no-tillage sustained greater populations of parasitic nematodes on corn, both in Iowa (Thomas, 1978). No-tillage fields had higher levels of nematode root injury and lower yields.

The role of crop debris (residue) in relation to survival of viruses in soil is succinctly presented by Zeyen (1979). Most viruses associated with residues are destroyed in the process of residue decomposition (Gray and Williams, 1971; Zeyen, 1979). However, some viruses such as tobacco mosaic, tobacco necrosis remain virulent in soil residues from one season to the next (Broadbent *et al.*, 1954; Temmink *et al.*, 1970). It was also shown that soil residues are the source of tobacco necrosis for the fungal zoospore vector that introduces the pathogen into the roots. Thus, residues may be a source of initial inoculum for some viruses and, in this way, relate to surface tillage.

Improper weed control in surface-tillage systems may sustain alternate hosts of plant viruses. These plants may provide virus inoculum and perhaps insect vectors to initiate epidemics of virus diseases. Also, surface tillage may decrease the variety of annual weed species and increase the variety of biennial and perennial weed species with the greatest increase in perennial weeds (Burnside, 1981). Such changes in weed species could also change the kinds and the incidence of crop diseases associated with weeds.

Yield reduction of corn was not significantly increased from maize chlorotic dwarf virus infection in no-tillage compared to conventional tillage despite the higher population of Johnsongrass [*Sorghum halepense* (L.) Per.] (a reservoir of the virus) and greater leafhopper population in no-tillage (All et al., 1977). However, the incidence of maize dwarf mosaic and maize chlorotic dwarf was greater in no-tillage corn, presumably because infected Johnsongrass provided inoculum for early infection. In contrast, preplant plowing destroys Johnsongrass and eliminates the source of early viral inoculum for corn.

The foregoing role of residues as habitats of survival of plant pathogens in monocropping systems is applicable to other cropping systems such as multiple and sequential cropping, intercropping, and monoculture. Definitions of these terms are given by Sumner et al. (1981).

Recent research has focused on the use of legumes as a soil cover and source of N for subsequent no-till crops (Martin and Touchton, 1983). Benefits from legumes as a mulch and N source in surface-tillage is both economical and applicable. There is little information on plant diseases in surface-tillage systems utilizing legumes (Power et al., 1983). A greater incidence of fungal root diseases could develop on legumes planted in a cool, wet soil created by surface-tillage systems. Research on these and other legume diseases associated with surface residues is needed to devise effective control measures in surface tillage.

STIMULANTS AND TOXINS FROM PLANT RESIDUES

Plant residue decomposition in soil affects plant diseases in several ways. Decomposition products of residues may be toxic to plants or microorganisms or both, or have no biological activity (Garrett, 1956; McCalla and Haskins, 1964; Patrick et al., 1963; Waksman, 1952). Some of the phytotoxic components of decomposing barley residue are benzoic, phenylacetic, 3-

phenylpropionic, and 4-phenylbutyric acids (Linderman and Toussoun, 1968). Treating cotton seedling roots with phenylpropionic acid or benzoic acid from extract of barley residue increased the incidence of root rot caused by *Thielaviopsis basicola* (Berk. & Br.) Ferr. (Linderman and Toussoun, 1968). It appears that barley toxins induce changes in root exudates increasing the virulence of the pathogen. Boyd and Phillips (1973) reported that phytotoxins from decomposing residues not only act directly on roots, but also predispose roots to infection by various weak pathogens. Another example is tobacco plants treated with extracts from decomposing rye and timothy residues increased the incidence of root rot caused by *T. basicola*. Moreover, the disease was equally severe on roots of resistant and susceptible varieties treated with extracts of residue. Again, as is often the case, phytotoxins from residues predispose plants to root diseases.

Organic substances from decomposing residues also may inhibit soilborne diseases. The low populations of *Fusarium oxysporum* in forest soils may be partly due to the presence of fungal stimulants from decomposing residues such as pine needles, which stimulate germination of resting fungal propagules in a hostile environment (Patrick and Koch, 1958; Toussoun et al., 1969). Germination in such an unfavorable environment for the growth of the pathogen could eliminate it or substantially reduce the population.

Boosalis et al. (1981) state that the distance and rate and diffusion of plant residue decomposition products into the soil solution may limit their effectiveness in activating pathogenic fungus propagules. The effective distance of diffusion of residue byproducts in their gaseous state, however, may be greater. Microsclerotia of *Verticillium dahlieae* Kleb. and sclerotia of *Sclerotium rolfsii* Sacc. were stimulated to germinate and grow when exposed to volatile aldehydes from decomposing alfalfa hay followed by lysis and net reduction in populations of these two fungal pathogens (Linderman and Gilbert, 1975). Sulfur-containing volatiles emitted from decomposing crucifer residues are inhibitory to *Aphanomyces euteiches* Drechs. and may reduce the incidence of root rot of peas caused by this pathogen (Lewis and Papavizas, 1975).

Residues fully incorporated (plowing) or partially incorporated (surface tillage or subtillage) in soil decompose faster than residues left undisturbed (no-tillage) on this soil surface. Furthermore, with plowing and surface tillage, there is greater opportunity for decomposing residues to come in contact with roots and predispose them to root diseases. Thus, it follows that the incidence of root diseases resulting from residue phytotoxins is less in no-tillage than in plowed and subtillage systems. However, this may not always be the case, since there are many factors that may affect the production of phytotoxins from decomposing residues. For example, the microenvironment of plowed soils may destroy some phytotoxins derived from decomposing residue. Such a microenvironment could also arrest the development of those microorganisms in residue that produce phytotoxin. This may help to explain why populations of *Penicillium urticae* Bainer, which produces the phytotoxin patulin in residues, were lower in plow and no-tillage than in subtillage plots (Behmer and McCalla, 1963; Dr. J. R. Ellis, Microbiologist, USDA-ARS, Lincoln, Nebraska, personal communication).

Some saprophytic organisms such as *Gliocladium roseum* and *Azotobacter chroococcum* associated with cereal straw residue prevent seed germination and retard seedling root development of cereals (Harper and Lynch, 1979; Lynch and Pryn, 1977). These microorganisms induce this type of injury by competing with the seed for available oxygen and by producing phytotoxic metabolites. Results from laboratory tests conducted by Lynch et al. (1981) showed that treating cereal seed with a formulation containing calcium peroxide and lime alleviated microbial inhibition of cereal establishment associated with straw residue.

CHEMICAL AND PHYSICAL ENVIRONMENT OF SURFACE-TILLAGE SYSTEMS IN RELATION TO PLANT DISEASES

Tillage can affect the concentration and distribution of nutrients which, in turn, can affect plant diseases. Baeumer and Bakermans (1973) present an excellent summary of the literature on nutrients in tilled and untilled soils. The amount of available K and P near the soil surface is higher in untilled than tilled soil, but the reverse was found in deeper layers. In contrast, Mg and Ca was lower near the soil surface and higher in lower layers of untilled soil. Generally, untilled soils have a lower concentration of soluble N and a lower pH. Untilled soils sustain higher populations of denitrifying bacteria and the potential for losing NO_3 is therefore greater. Also, the greater number of microorganisms in the surface 7.5–10 cm of untilled soils tie up greater amounts of N. Doran (1980a, 1980b) reported that increases in microbial populations result from higher water and organic contents at the surface of untilled soils. The water availability is a major factor regulating numbers and activities of microorganisms in the surface of untilled soils. In deeper soil (10–17 cm), populations of aerobic organisms are 25–50% lower with untilled than in plowed soils. Moreover, facultative anaerobes and denitrifying bacteria are more abundant in untilled soils and represent a greater proportion of the total microbial populations than in plowed soils. Although it is not yet clear how nutritional and microbial differences in tilled and untilled soils affect soilborne pathogens, these differences, nevertheless, are likely to affect disease outbreaks.

A review of the literature on nutrition in relation to plant diseases (Huber, 1981) and in relation to pathogens associated with residues (Boosalis et al., 1981) has been published. The survival of fungal pathogens residing in residues may be affected by nutrients released from decomposing residues. These nutrients may stimulate germination of fungal propagules and subsequent mycelial growth and sporulation and, thus, extend the survival of the pathogen. Nutrients from residues in the early stages of decomposition induced the best germination of chlamydospores of *Fusarium solani* (Mart.) Appel & Wr. f. sp.

phaseoli (Burk.) Snyd. & Hans. in soil in close proximity to residue fragments (Schroth and Hendrix 1962; Toussoun et al., 1963). In this way, residues are important in the survival and growth of fungal pathogens.

The amount of available nutrients, particularly N, in soil or residues may affect the survival of some soilborne fungal pathogens. Survival of the wheat root rot pathogens *Gaeumannomyces graminis* Sacc. and *Fusarium graminearum* Schwabe is enhanced when N supply in infested straw is not limiting. The reason for this is that N is required for growth of the pathogen into fresh cellulose substrates and without a fresh source of cellulose to colonize the pathogen is killed (Garrett, 1972, 1976). Conversely, the longevity of *Helminthosporium sativum* P. K. & B. in straw in soil is shortened if soluble N content is high (Garrett, 1966). Evidently, *H. sativum* is a poor competitor against other microorganisms under these conditions.

These and other examples clearly show that survival, growth, and reproduction of fungal pathogens associated with residues are profoundly affected by nutrients. The source of these nutrients may come primarily from decomposing residues and from fertilizer applications.

The physical environment of surface-tilled soil differs from that of plowed soil. Untilled soils are generally wetter, cooler, and more compact than plowed soils. The top 7.5–10 cm of untilled soils may contain 20–30% more water than do plowed soils at the same depth. Tillage operations incorporate residues into the soil, and this not only increases soil porosity, but also creates greater potential for water loss through surface evaporation. Another characteristic of untilled soil is that it generally has less resistance to rainwater infiltration than tilled soil.

Allmaras et al. (1973) state that soil temperatures under surface residues are governed by many factors, some of which are solar radiation, emission of long-wave radiation, soil and air heat flow, and evaporation. At a depth of 2.5 cm soil with 2240 kg/ha of wheat mulch, the soil surface may be 6°C cooler than bare soil in the spring. Other studies have shown that untilled soils have a substantially lower temperature near the ground surface than tilled soils. Lower soil temperatures in the spring and early summer with reduced tillage may retard seed germination and seedling development of many crops. This situation predisposes seedlings to damping-off and seedling and root rot diseases incited by pathogens favored by cool, wet soils. Thus, the importance of spatial distribution of residues away from the emerging plants such as with surface tillage is obvious. With the till planter, seed is planted in 10–15-cm ridges formed during cultivation. The ridges warm up and dry out quicker in the spring.

On the other hand, an increase in soil moisture and a decrease in soil temperature resulting from no-tillage or reduced tillage can reduce the incidence of some plant diseases. In this regard, a unique 3-year reduced-tillage rotation system known as "ecofallow" (ecofarming) significantly reduced the incidence and severity of stalk rot of corn and sorghum. Ecofallow is defined as a system of controlling weeds and conserving soil moisture in a crop rotation with minimum disturbance of crop residue and soil (Smika and Wicks, 1968). This conservation tillage system is based on initial research conducted by Phillips (1964). His work established a system to replace part or nearly all of the mechanical cultivation by using herbicides in a wheat–sorghum–fallow rotation. Subsequently, the ecofallow system evolved out of a study initiated by Smika and Wicks (1968) to determine the feasibility of reduced tillage in a winter wheat–grain sorghum–fallow rotation for the semiarid Central Great Plains of the United States. Most of the benefits of the ecofallow system are associated with the fallow period between winter wheat and grain sorghum (or corn) planting. These benefits arise primarily from the large amount of wheat residue maintained weed-free by herbicides and left undisturbed on the soil surface. The undisturbed wheat residue traps snow and reduces wind velocities at the soil surface (this, alone, can add 2.5–7.5 cm of soil moisture); stabilizes soil temperature and evaporation; reduces wind and water erosion, increases water infiltration and runoff and water sedimentation. The net gain in soil moisture storage (Smika and Wicks, 1968) has averaged about 5 cm. It is noteworthy that it takes about 22.5 cm of water to produce the first bushel of grain sorghum and that each additional 2 cm of water conserved adds about 680 kg/ha. Thus, 5 cm of additional water stored under ecofallow adds, on the average, about 1360 kg of grain sorghum.

Since plant diseases often increase (Boosalis and Doupnik, 1976; Boosalis et al., 1981) in other surface-tillage systems, the long-term effect of ecofallow on plant diseases is important. In developing the ecofallow system, it was noted that no foliar disease problem developed on either grain sorghum or wheat but fewer grain sorghum plants were lodged in the ecofallow plots than in the conventionally tilled plots. Subsequent field studies for 7 years examined the effect of ecofallow on foliar diseases of wheat and grain sorghum and on stalk rot of grain sorghum. The ecofallow system dramatically reduces stalk rot of sorghum and corn, a moisture and temperature stress disease caused primarily by *Fusarium moniliforme* Sheld. in Nebraska (Doupnik et al., 1975). The average decrease in stalk rot was 41% and 72% and an average increase in grain sorghum yield was 44% and 41% with minimum and no-tillage, respectively, over conventional tillage (Table 13.2). Although studies on similar rotation systems using corn in place of grain sorghum have not been extensive, similar affects on stalk rot incidence and yield were noted.

Since stalk rot of sorghum and corn is a stress disease, the increase of soil moisture storage and reduced soil temperature fluctuations under ecofallow are, undoubtedly, important factors in reducing its incidence. The average soil temperature with ecofallow is several degrees lower between the grain sorghum rows up until the time the plant canopy is fully developed in mid-July; thereafter differences in soil temperature compared to the conventional tilled plots are much less. The average temperature in the row, however, was slightly lower with ecofallow compared to conventional tillage throughout the growing season. Furthermore, the daily high soil temperature was much lower between rows in the ecofallow plots throughout the growing season. The daily soil

TABLE 13.2. The Effect of Ecofallow (Surface Tillage) on Stalk Rot and Yield of Grain Sorghum Grown in a Winter Wheat–grain Sorghum–fallow Rotation

Year	Tillage Treatment	Stalk rot[a] (%)	Yield[a] (kg/ha)
1972[b]	Conventional	45	1570
	Minimum	32	2072
	No-tillage	15	2324
1973	Conventional	51	3077
	Minimum	33	5401
	No-tillage	14	5275
1974	Conventional	20	3517
	Minimum	5	4270
	No-tillage	3	3894
3-yr avg.	Conventional	39a[c]	2721a[c]
	Minimum	23b	3914b
	No-tillage	11c	3831c

[a]Mean of five replications. For the stalk rot data, 100 stalks were examined in each plot.
[b]Yields were suppressed in 1972 due to a severe hail storm.
[c]Values not followed by the same letters are significantly different, $P = 0.01$, according to Duncan's multiple range test.

SOURCE: Doupnik et al. (1975).

temperature fluctuations between the rows (determined by subtracting the daily lows from the daily highs) were substantially lower under ecofallow compared to conventional-tillage soils throughout the growing season (Doupnik et al., 1975).

Increased soil moisture and the lower and more constant soil temperatures associated with ecofallow are, undoubtedly, major factors in reducing stalk rot. Under these reduced-stress conditions, the plants are less vulnerable to fungi inducing stalk rots.

The ecofallow system differs from most surface-tillage practices in that one crop is planted directly into the residue of another crop rather than into the residue of the same crop. Alternating crops effectively break disease cycles which involve pathogens that overwinter in residue.

The ecofallow system has been extensively used in western Nebraska and Kansas and eastern Colorado and acreage is increasing each year. Until recently, no new disease problems were associated with this cropping system. However, tanspot of wheat caused by *Pyrenophora trichostoma* and *Cephalosporium* stripe of wheat caused by *Cephalosporium gramineum* are now well established in the Great Plains states. Both of these pathogens overwinter in high populations in and on surface wheat residues. If these two wheat diseases become epidemic, it may be necessary to modify the crop rotation systems in ecofallow and develop cultivars resistant to the diseases. However, there appear to be no insurmountable problems to prevent the use of the ecofallow system in semiarid regions of the world where wheat and grain sorghum and corn can be grown. In addition, the ongoing concern about energy conservation, soil conservation, and ecology throughout the world makes ecofallow an appealing cropping practice alternative for nonirrigated cropping or where irrigation water availability is declining.

Martinson showed that in corn–soybean rotation plots utilizing four tillage systems, including conventional tillage, the incidence of stalk rot of corn was lower in the surface-tillage plots (personal communication).

Residue management is employed in the winter wheat dry areas of Washington and Oregon to reduce root and crown rot caused by *Fusarium roseum* (LK) emend. Snyd. & Hans (Papendick and Cook, 1974). A residue-dust mulch is prepared by the rod-weeder during the summer-fallow period to reduce evaporation losses and help retain water in the soil profile that accumulated during the rainy season (Papendick et al., 1973). This type of management of wheat residue increases the water stored during summer-fallow and increases the total amount of water available to a subsequent winter crop by reducing the chances of Fusarium root and crown rot.

ROOT DEVELOPMENT IN RELATION TO DISEASES IN SURFACE TILLAGE

As previously stated, surface residues may alter soil moisture, temperature, aeration, density, organic content, nutrition, and types and population levels of microorganisms. Such changes may, in turn, affect the growth pattern and configuration of roots (Allmaras and Nelson, 1971; Barley and Greacen, 1967; Chaudhary and Prihar, 1974; Nelson and Allmaras, 1969; Taylor, 1967). How changes in the growth pattern and configuration of roots affect disease and drought resistance is not yet known. Roots generally have a lower density and weight, and are shallower in untilled soils. Studies on the effect of crop residues on corn root development indicate greater horizontal root development (Allmaras *et al.*, 1973; Chaudhary and Prihar, 1974) with more adventitious roots initiated at high soil temperatures, while root elongation and branching was favored at lower optimum temperatures. Roots near the surface of the soil may be more prone to diseases because populations of plant pathogens are higher in the top 4–8 cm of soil. Nevertheless, a more vigorous root system produced near the soil surface in surface-tillage systems may be highly resistant to soilborne pathogens. In surface-tillage systems, improved chemical and physical environment, higher population of organisms antagonistic to pathogens, and enhanced nutrient availability to the roots may yield robust roots resistant to pathogens.

Ferguson and Boatwright (1968) reported, in a greenhouse study, that surface residues may influence the location of the crown node of winter wheat. The location of crown node is important to winter survival of the plant. Forming the crown node well below the soil surface reduces winter injury by protecting it from exposure to wind and providing a warmer habitat. In such an environment, the crown node can produce a more vigorous adventitious root system that can penetrate deeper into the soil under drought conditions in the fall. Weaver (1926) noted that a well-developed root system, at the start of winter, can better withstand stress. As the surface wheat stubble increased, the crown node formed closer to the surface of the soil and, in some cases, above the soil surface (Ferguson and Boatwright, 1968). Winter-hardy varieties formed crown nodes deeper in the soil than non-winter-hardy varieties of wheat. As the soil temperature decreased or the light available to the wheat seedling decreased, the crown node formed closer to the soil surface. There was a positive variety–temperature–light intensity interaction, and each variety reacted differently to a specific set of conditions. Nilsson (1969) noted a wheat varietal difference in root development and summarized the literature on the subject as well as the disease aspects relating to crown and root development.

Wilhelm et al. (1982) studied root development for winter wheat grown in a wheat–fallow rotation with fallow tillage treatments of plow, subtillage, and no-tillage. The no-tillage induced the greatest root weight (46 mg/dm^3) and the least root weight was with the subtillage treatment (26 mg/dm^3). By June, the upper 30 cm of soil contained 62% of the total root mass with maximum root depth greater than 120 cm for all three tillage treatments. The increased root weight evidently was due to increased numbers of roots. Also, the root system for the three tillage systems were equally fibrous. It is noteworthy that the grain yield was the same for all the tillage treatments. It is not known whether the greater root weight resulting from the no-tillage treatment relates to fungal or bacterial root diseases or to other factors. However, a more dense root system may provide more infection sites for pathogens requiring injured roots for penetration. On the other hand, a more dense root system may be able to better withstand attacks by pathogens that infect roots by direct penetration.

Changes in soil conditions caused by tillage may affect the population of vesicular-arbuscular mycorrhizal (VAM) fungi in soil. Increased aeration of soil increases the level of infection of roots by mycorrhizal fungi (Saif, 1981). Generally, levels of soil aeration are reduced when the soil is no longer tilled. However, the reduced aeration of untilled soil favors the accumulation of soil organic matter. Organic matter stimulates the growth of hyphae of mycorrhizal fungi (St. John et al., 1983) and mycorrhizal fungal hyphae are commonly in particles of decomposing organic matter (Koske et al., 1975; Mosse, 1959).

Khan (1975) found spore densities in the top 8 cm of no-tillage wheat soils and undisturbed pasture sod soils almost double those in the top 8 cm of plowed soils. However, he also found that these spore density relationships almost completely reversed in the 8–16-cm depth zones of these soils, and no differences in spore density existed between tillage treatments when these are combined into a single 0–16-cm value. Schenck et al. (1982) observed higher spore densities and root colonization levels under minimum tillage than under conventional tillage for sorghum, soybean, corn, oats, and vetch. They also found greater VAM fungal species diversity in the minimum tillage soils and that subsoiling (= subtill) had little effect on VAM incidence. The research of Yocom et al. (1985) support the suggestion by Schenk et al. that minimum tillage (no-tillage) has less effect on VAM fungi than does conventional tillage. Yocom et al. (1985) noted that VAM fungal populations in mechanically tilled soils decreased significantly more than in no-tillage soils, and root colonization levels of 1-month-old field-grown winter wheat declined to 8–11% in the mechanically tilled soils by October of the second crop cycle. Continuation of such early decline in VAM inoculum potential and root colonization in the mechanically tilled soils could eventually reduce winter wheat yield.

PESTICIDES IN RELATION TO PLANT DISEASES

Boosalis et al. (1981) stated that the success of reduced-tillage systems is dependent on the use of pesticides, particularly herbicides and, to a lesser extent, on insecticides and fungicides. Several comprehensive review articles on the effect of herbicides on plant diseases have been published (Altman and Cambell, 1977; Katan and Eshel, 1973).

Some herbicides and insecticides may increase or decrease the incidence and severity of plant diseaseses. Such pesticides may increase plant diseases by arresting the growth of or by killing organisms that are antagonistic to the pathogen, by increasing root exudation which may stimulate growth of the pathogen in the rhizosphere leading to infection, and by increasing the susceptibility of the host. On the other hand, pesticides may have precisely the opposite effect. More information is needed on how pesticides used in surface-tillage systems affect plant pathogens and disease development.

Although much more work also needs to be done before effects of herbicides on mycorrhizal fungi are fully understood, it has been shown that herbicides reduce the levels of mycorrhizal infection in plant roots and cause a decline in numbers of spores of these fungi in soil (Ocamp and Hayman, 1980; Pope and Holt, 1981).

PLANT DISEASE CONTROL WITH SURFACE-TILLAGE

Because of the benefits from reduced tillage, it is impractical to reject this cultural practice solely for the purpose of reducing potential plant disease threats. From a positive standpoint, many diseases associated with conservation tillage can be minimized in a variety of ways that do not significantly increase production costs.

The virulence and viability of bacterial and fungal pathogens decrease with time. A crop rotation sequence of 2 or 3 years induces sufficient deterioration of residues to reduce pathogen populations. Crop rotation is especially important for controlling diseases with surface tillage. Planting a crop back into its own residue from the previous season (monoculture) without a fallow period between crops is more likely to increase certain plant diseases than a crop rotation system where a crop is planted into the residue of a nonrelated crop.

Another way to reduce diseases associated with reduced tillage is to rotate tillage systems. Inclusion of tillage rotation with crop rotation is an excellent method of disease management. This could be done in a manner to allow retention of 20–30% of the surface residue; thus, providing the benefits of surface tillage while reducing the potential of disease outbreak.

Some plant pathogenic bacteria, fungi, viruses, and nematodes are capable of readily producing new strains that overcome host resistance. This necessitates a continuous research program for resistant varieties and varieties better adapted to surface-tillage systems. Development of crop varieties that are well adapted to surface tillage may offer unique ways of controlling plant diseases. For example, the survival of pathogens in crop residue may be affected by the genotypes of both the host and the strain of the pathogen. Nelson and Sheifele (1970) reported that races of *Helminthosporium turcicum* differ in their ability to survive in residue from different corn genotypes. This phenomenon of survival ought to be exploited in developing crop varieties that are inhospitable to the survival of pathogens. Crop varieties with increased seedling vigor are less likely to suffer stand losses due to damping-off pathogens that are favored by cool, moist soils associated with surface residues. Also, planting certified seed treated with fungicides ensures good germination and seedling vigor in a soil environment conducive to seed decay and seedling blight.

The application of foliar fungicides is an effective and practical means of controlling some diseases associated with surface tillage. Foliar fungicides are most effective when applied as a preventative. Thus, the producer should assess the cost–benefit aspects of this method before application.

The control of weeds in and around fields often reduces the primary inoculum source of certain pathogens associated with surface-tillage. Such a practice may also reduce the opportunity for the development of new strains of the pathogen.

Soilborne fungal diseases associated with surface tillage may be decreased by the kind, the amount, and the time of fertilizer application. Results from research conducted on the relationship of fertilization and soilborne fungal diseases may be applicable to diseases associated with surface tillage (Huber, 1972, 1981). In this regard, spring application of ammonium sulfate controlled take-all of wheat, whereas, fall application of N for spring-seeded wheat increased take-all (Huber, 1972). The source of N may affect the susceptibility of winter wheat to root rot (Huber et al., 1968). Huber and Watson (1972) reported that the ammonium form of N stabilized by the addition of nitrapyrin [2-chloro-6-(trichloro-methyl)pyridine] reduced the severity of take-all. Fall application of nitrapyrin with anhydrous ammonium reduced the amount of stalk rot of corn by 60–90% in Indiana (Warren et al., 1975).

It may be possible to kill or arrest development of pathogens on and in the residue. This might be achieved by spraying the residues with fungicides or other chemicals soon after the crop is harvested. Also, various kinds of nutrients may be applied to partially decompose the residue at the expense of the pathogen.

Integrated-pest-management methods employing several methods of control such as biological, chemical, and cultural, as well as the use of resistant or tolerant varieties, undoubtedly will be further refined and contribute to alleviating many diseases associated with surface tillage. Other disease management strategies integrating various plant disease control methods are extensively discussed by Apple (1977).

ACKNOWLEDGMENTS

We wish to thank Dr. Daniel Yocom and Dr. Charles Martinson who provided us with information on mycorrhizae and eye spot of corn, respectively. We also wish to thank Drs. Anne Vidaver, Leslie Lane, and James Partridge for their critical reading of this chapter.

Research in our laboratory was supported, in part, by funds from the North Central Regional Project NC-125. This work was conducted under Nebraska Agricultural Stations Projects 21-005 and 48-004.

LITERATURE CITED

All, J. N., C. W. Kuhn, R. N. Gallaher, M. D. Jellum, and R. S. Hussey. 1977. Influence of no-tillage-cropping, carbofuran, and hybrid resistance on dynamics of maize chlorotic dwarf and maize dwarf mosaic diseases of corn. *J. Econ. Ent.* **70**:221–225.

Allmaras, R. R., A. L. Black, and R. W. Rickman. 1973. Tillage, soil environment and root growth. In Conservation Tillage the Proceedings of a National Conference. Soil Conservation Soc. Am. Ankeny, IA.

Allmaras, R. R., and W. W. Nelson. 1971. Corn (*Zea mays* L.) root configuration as influenced by row-interrow variants of tillage and straw mulch management. *Soil Sci. Soc. Amer. Proc.* **35**:974–980.

Altman, Jack and C. Lee Cambell. 1977. Effect of herbicides on plant diseases. *Ann. Rev. Phytopathology* **15**:361–385.

Apple, J. 1977. The theory of disease management. In James G. Horsfall and Ellis B. Cowling (eds.). *Plant Disease An Advanced Treatise*. Vol. 1, How Disease is Managed. Academic Press, San Francisco, London, pp. 7–99.

Arny, D. C., G. L. Worf, R. W. Ahrens, and M. F. Lindsey. 1970. Yellow leaf blight of maize in Wisconsin: Its history and the reactions of inbreds and crosses to the inciting fungus (*Phyllosticta* sp.). *Plant Dis. Rep.* **54**:281–285.

Baeumer, K. and W. A. P. Bakermans. 1973. Zero-tillage. *Adv. Agron.* **25**:77–123.

Barley, K.P., E. L. Greacen. 1967. Mechanical resistance as a soil factor influencing the growth of roots and underground shoots. *Adv. Agron.* **19**:1–43.

Behmer, D. E., and T. M. McCalla. 1963. The inhibition of seedling growth by crop residue in soil inoculated with *Penicillium urticae* Bainer. *Plant and Soil* **18**:199–206.

Boosalis, M. G. 1952. The epidemiology of *Xanthomonas translucens* (J. J. and R.) Dowson on cereals and grasses. *Phytopathology* **42**:387–395.

Boosalis, M. G. and B. Doupnik, Jr. 1976. Management of crop diseases in reduced tillage systems. ESA Symposium: Crop production with reduced tillage systems. *Bull. Entomol. Soc. Am.* **22**:300–302.

Boosalis, M. G., B. Doupnik, and G. N. Odvody. 1981. Conservation tillage in relation to plant diseases. *Handbook of Pest Management in Agriculture*. CRC Press, Boca Raton, FL, Vol. 1, pp. 445–474.

Boosalis, M. G., D. R. Sumner, and A. S. Rao. 1967. Overwintering of conidia of *Helminthosporium turcicum* on corn residue and in soil in Nebraska. *Phytopathology* **57**: 990–996.

Boyd, H. W. and D. V. Phillips. 1973. Toxicity of crop residue to peanut seed and *Sclerotium rolfsii*. *Phytopathology* **63**:70–71.

Brinkerhoff, L. A. 1969. The influence of temperature, aeration, and soil microflora on microsclerotial development of *Verticillium albo-atrum* in abscised cotton leaves. *Phytopathology* **59**:805–808.

Brinkerhoff, L. A., and G. B. Fink. 1964. Survival and infectivity of *Xanthomonas malvacearum* in cotton plant debris and soil. *Phytopathology* **54**:1198–1201.

Broadbent, L., W. H. Read and F. T. Last. 1954. The epidemiology of tomato mosaic X. Persistence of TMV-infected debris in soil, and the effects of soil partial sterilization. *Ann. Appl. Biol.* **55**:471–483.

Brooks, D. H., and M. G. Dawson. 1968. Influence of direct-drilling of winter wheat on incidence of take-all and eyespot. *Ann. Appl. Biol.* **61**:57–64.

Burns, E. E. and M. C. Shurtleff. 1972. Factors affecting overwintering and epidemiology of *Helminthosporium maydis* in Illinois corn. *Phytopathology* **62**:749 (Abstr.).

Burns, E. E. and M. C. Shurtleff. 1973. Observations of *Physoderma maydis* in Illinois: effects of tillage practices in field corn. *Plant Dis. Rep.* **57**:630–633.

Burnside, O. C. 1981. Changing weed problems with conservation tillage. Conference on Crop Production with Conservation in the 80's. ASAE Publication. 7-81, St. Joseph, MI, pp. 161–174.

Chadhary, M. G., and S. S. Prihar. 1974. Root development and growth response of corn following mulching, cultivation, or interrow compaction. *Agron. J.* **66**:350–355.

Cook, R. J., M. G. Boosalis, and B. Doupnik. 1978. Influence of crop residues on plant diseases. In *Crop Residue Management Systems*. Madison, WI. Am. Soc. Agron. Spec. Publ. 31, pp. 147–163.

Corbett, D. C. M., and R. M. Webb. 1970. Plant and soil nematode population changes in wheat grown continuously in ploughed and unploughed soil. *Ann. Appl. Biol.* **65**:327–335.

Daft, G. G., and C. Leben. 1973. Bacterial blight of soybeans: field-overwintered *Pseudomonas glycinea* as possible primary inoculum. *Plant Dis. Rep.* **57**:156–157.

Doran, J. W. 1980a. Microbial changes associated with residue management with reduced tillage. *Soil Soc. Am. J.* **44**:518–524.

Doran, J. W. 1980b. Soil microbial and biochemical changes associated with reduced tillage. *Soil Soc. Am. J.* **44**:765–771.

Doupnik, B., Jr., M. G. Boosalis, G. Wicks, and D. Smika. 1975. Ecofallow reduces stalk rot of grain sorghum. *Phytopathology* **65**:1021–1022.

Farley, J. D. 1971. Recovery of *Corynebacterium michiganense* from overwintered tomato stems by the excised-petiole inoculation method. *Plant Dis. Rep.* **55**:654–656.

Ferguson, H. and G. O. Boatwright. 1968. Effects of environmental factors on the development of the crown node and adventitious roots of winter wheat *(Triticum aestivum). Agron. J.* **60**:258–260.

Futrell, M. C. and G. E. Scott. 1971. Overwintering and early season development of *Helminthosporium maydis* in Mississippi. *Plant Dis. Rep.* **55**:954–956.

Garrett, S. D. 1956. *Biology of Root-Infecting Fungi*. Cambridge University Press, Cambridge, England.

Garrett, S. D. 1966. Cellulose-decomposing ability of some cereal foot-rot fungi in relation to their saprophytic survival. *Trans. Br. Mycol. Soc.* **49**:57–68.

Garrett, S. D. 1972. Factors affecting saprophytic survival of six species of cereal foot-rot fungi. *Trans. Br. Mycol. Soc.* **59**:445–452.

Garrett, S. D. 1976. Influence of nitrogen on cellulolysis rate and saprophytic survival in soil of some cereal-root fungi. *Soil Biol. Biochem.* **8**:229–234.

Gray, T.R.G. and S. T. Williams. 1971. *Soil Micro-Organisms*. Hafner Publishing Company, NY.

Gudauskas, R. T., D. H. Teem, W.-N. To, and T. L. Whatley. 1971. Overwintering of *Helminthosporium maydis* in Alabama. *Plant Dis. Rep.* **55**:947–951.

Harper, S.H.T., and J. M. Lynch. 1979. Effects of *Azotobacter chroococcum* on barley seed germination and seedling development. *J. Gen. Microbiol.* **112**:45–51.

Hood, A.E.M. 1965. Plowless farming using "Gramoxone." *Outlook Agric.* **4**(6):286–294.

Hosford, R. M., Jr. 1976. Fungal leaf spot diseases of wheat in North Dakota. N.D. Agric. Exp. Sta. Bull. 500. N.D. State University, Fargo, ND.

Huber, D. M. 1972. Spring versus fall nitrogen fertilization and take-all of spring wheat. *Phytopathology* **62**:434–436.

Huber, D. M. 1981. The use of fertilizers and organic amendments in the control of plant disease. p. 357–394. In *Handbook of Pest Management in Agriculture*, CRC Press, Boca Raton, FL, Vol. 1, pp. 357–394.

Huber, D. M., C. G. Painter, H. C. McKay, and D. L. Peterson. 1968. Effect of nitrogen fertilization on take-all of winter wheat. *Phytopathology* **58**:1470–1472.

Huber, D. M. and R. D. Watson. 1972. Nitrogen form and plant disease. *Down to Earth* **27**:14–15.

Kahn, A. G. 1975. The effect of vesicular-arbuscular mycorrhizal associations on growth of cereals. II. Effects on wheat growth. *Ann. Appl. Biol.* **80**:27–36.

Katan, J. and Y. Eshel. 1973. Interactions between herbicides and plant pathogens. *Residue Rev.* **45**:145–177.

Koske, R. E., J. C. Sutton, and B. R. Sheppard. 1975. Ecology of *Endogone* in Lake Huron sand dunes. *Can. J. Botany* **53**:87–93.

Lewis, J. A., and G. C. Papavizas. 1975. Survival and multiplication of soil-borne plant pathogens as affected by plant tissue amendments. In G. W. Bruehl (ed.) *Biology and Control of Soil-Borne Pathogens*. Am. Phytopathol. Soc., St. Paul, MN, pp. 84–89.

Linderman, R. G. and R. G. Gilbert. 1975. Influence of volatiles of plant origin on soilborne plant pathogens. *In* G. W. Bruehl (ed.). *Biology and Control of Soilborne Plant Pathogens*. Am. Phytopathol. Soc., St. Paul, MN.

Linderman, R. G., and T. A. Toussoun. 1968. Predisposition to *Thielaviopsis* root rot of cotton by phytotoxins from decomposing barley residues. *Phytopathology* **58**:1571–1574.

Lipps, P. E. 1983. Anthracnose fungus survives in surface corn residues. Ohio Report. Research and Development in Agriculture, Home Economics and Natural Resources, Ohio State University, OARDC, Wooster, OH, Vol. 68, pp. 38–40.

Littrell, R. H. and D. R. Sumner. 1971. Overwintering of *Helminthosporium maydis* in Georgia. *Plant Dis. Rep.* **55**:951–953.

Lutey, Richard W., and Kane D. Fezer. 1960. The role of infected straw in the epiphytology of Septoria leaf blotch of barley. *Phytopathology* **50**:910–913.

Lynch, J. M., and S. J. Pryn. 1977. Interactions between a soil fungus and barley seed. *J. Gen. Microbiol.* **103**:193–196.

Lynch, James M., S. H.T. Harper, and M. Sladdin. 1981. Alleviation, by a formulation containing calcium peroxide and lime, of microbial inhibition of cereal seedling establishment. *Current Microbiol.* **5**:27–30.

Martin, G. W., and J. T. Touchton. 1983. Legumes as a cover crop and source of nitrogen. *J. Soil Water Conser.* **38**:214–216.

Martinson, C. A. 1981. Corn diseases and cultural practices. Iowa State Univ. Coop. Ext. Ser. CE1602j, Ames, IA.

McCalla, T. M. and F. A. Haskins. 1964. Phytotoxic substances from soil microorganisms and crop residues. *Bacteriol. Rev.* **28**:181–207.

Menzies, J. D. 1963. Survival of microbial plant pathogens. *Bot. Rev.* **29**:79–122.

Mosse, B. 1959. Observations on the extramatrical mycelium of a vesicular-arbuscular endophyte. *Trans. Br. Mycol. Soc.* **42**:439–448.

Natti, J. J. 1967. Overwintering survival of *Pseudomonas phaseolicola* in New York. *Phytopathology* **57**:343 (Abstr.).

Naylor, V. D. and K. J. Leonard. 1977. Survival of *Colletotrichum graminicola* in infected corn stalks in North Carolina. *Plant Dis. Rep.* **61**:382–383.

Nelson, R. R. and G. L. Sheifele. 1970. Factors affecting the overwintering of *Trichometasphaeria turcica* on maize. *Phytopathology* **60**:369–370.

Nelson, W. W., and R. R. Allmaras. 1969. An improved monolith method for excavating and describing roots. *Agron. J.* **61**:751–754.

Nilsson, H. E. 1969. Studies of root and foot rot diseases of cereals and grasses. *Ann. Agr. College Swed.* (Lantbrukshögskolans annaler) **35**(NR3):275–807.

Ocamp, J. A. and D. S. Hayman. 1980. Effects of pesticides on mycorrhiza in field-grown barley, maize and potatoes. *Trans. Br. Mycol. Soc.* **74**:413–416.

Papendick, R. I. and R. J. Cook. 1974. Plant water stress and development of Fusarium foot rot in wheat subjected to different cultural practices. *Phytopathology* **64**:358–363.

Papendick, R. I., M. J. Linstrom, and V. L. Cochran. 1973. Soil mulch effects on seedbed temperature and water during fallow in eastern Washington. *Soil Sci. Am. Proc.* **37**:307–314.

Patrick, Z. A., and L. W. Koch. 1958. Inhibition of respiration, germination and growth by substances arising during the decomposition of certain plant residues in soil. *Can. J. Bot.* **36**:621–647.

Patrick, Z. A., T. A. Toussoun, and W. Snyder. 1963. Phytotoxic substances in arable soils associated with decomposition of plant residues. *Phytopathology* **53**:152–161.

Phillips, R. E., R. L. Blevins, G. W. Thomas, W. W. Frye, and S. H. Phillips. 1980. No-tillage agriculture. *Science* **208**:1108–1113.

Phillips, W. M., 1964. A new technique of controlling weeds in sorghum in a wheat-sorghum-fallow rotation in the Great Plains. *Weeds* **12**:42–44.

Pope, P. E. and H. A. Holt. 1981. Paraquat influences development and efficacy of the mycorrhizal fungus *Glomus fasciculatus*. *Can. J. Bot.* **59**:518–521.

Power, J. F., R. F. Follett, and G. E. Carlson. 1983. Legumes in conservation tillage systems: a research prospective. *J. Soil Water Conser.* **38**:217–218.

Rees, R. G. and G. J. Platz. 1979. The occurrence and control of yellow spot of wheat in north-eastern Australia. *Aust. J. Exp. Agric. Anim. Husb.* **19**:369–372.

Saif, S. R. 1981. The influence of soil aeration on the efficiency of vesicular-arbuscular mycorrhizae. I. Effect of soil oxygen on the growth and mineral uptake of *Eupatorium adoratum* L. inoculated with *Glomus macrocarpus*. New Phytol. 88:649–659.

Scharen, A. L. 1971. Environmental influences on development of glume blotch in wheat. *Phytopathology* **54**:300–303.

Schenk, N. C., G. S. Smith, D. J. Mitchell, and R. N. Gallaher. 1982. Minimum tillage effects on the incidence of beneficial mycorrhizal fungi on agronomic crops. *Florida Scientist* **45** (Suppl.): 8 (Abstr).

Schroth, M. N., and F. F. Hendrix, Jr. 1962. Influence of nonsusceptible plants on the survival of *Fusarium solani* f. *phaseoli* in soil. *Phytopathology* **52**:906–909.

Schuster, M. L. 1975. Leaf freckles and wilt of corn incited by *Corynebacterium nebraskense* Schuster, Hoff, Mandel, Lazar. Nebr. Agric. Exp. Sta. Res. Bull. 270.

Schuster, M. L., and D. P. Coyne. 1974. Survival mechanisms of phytopathogenic bacteria. *Ann. Rev. Phytopathol.* **12**:199–221.

Smika, D. E., and G. A. Wicks. 1968. Soil water storage during fallow in the Central Great Plains as influenced by tillage and herbicide treatments. *Soil Sci. Soc. Am. Proc.* **32**:591–595.

Smith, T. F. 1978. A note on the effect of soil tillage on the frequency and vertical distribution of spores of vesicular-arbuscular endophytes. *Aust. J. Soil Res.* **16**:359–361.

St. John, T. V., D. C. Coleman, and C. P. P. Reid. 1983. Association of vesicular-arbuscular mycorrhizal hyphae with soil organic particles. *Ecology* **64**:957–959.

Sumner, D. R., B. Doupnik, Jr., and M. G. Boosalis. 1981. Effects of reduced tillage and multiple cropping on plant diseases. *Ann. Rev. Phytopathol.* **19**:167–187.

Tanasevych, I. I. and I. A. Korobil. 1978. Effect of preceding crops and method of soil tillage on the rate of winter wheat infections with root rot in western forest-steppe in the Ukrainian SSR. *Zakkyst Roslyn.* **25**:3–7 (From 1980 *Rev. Plant Pathol.* **58**:141).

Taylor, H. M. 1967. Effects of tillage-induced soil environmental changes on root growth. Conf. Proc.: Tillage for greater Crop production. Am. Soc. Agr. Eng., St. Joseph, MI. ASAE Publ. Proc. **168**:15–18.

Tekauz, A. 1976. Distribution, severity, and relative importance of leaf spot of wheat in western Canada in 1974. *Candian Plant Disease Survey* **56**:36–40. Contribution No. 661, Res. Sta., Agric. Can., 25 DaFoe Road, Winnepeg, Manitoba R3T 2 M9.

Temmink, J.H.M., R. N. Campbell, and P. R. Smith. 1970. Specificity and site of *in vitro* acquisition of tobacco necrosis virus by zoospores of *Olpidium brassicae*. *J. Gen. Virol.* **9**:201–213.

Thomas, S. H. 1978. Population densities of nematodes under seven tillage regimes. *J. Nematol.* **10**:24–27.

Toussoun, T. A., W. Menzinger, and R. S. Smith, Jr. 1969. Role of conifer liter in ecology of *Fusarium*: stimulation of germination in soil. *Phytopathology* **59**:1396–1399.

Toussoun, T. A., Z. A. Patrick, and W. C. Snyder. 1963. Influence of crop residue decomposition products on the germination of *Fusarium solani* f. *phaseoli* chlamydospores in soil. *Nature* **197**:1314–1316.

Ullstrup, A. J. 1971. Overwintering of race T of *Helminthosporium maydis* in midwestern United States. *Plant Dis. Rep.* **55**:563–565.

Vizvary, M. A., and H. L. Warren. 1982. Survival of *Colletotrichum graminicola* in soil. *Phytopathology* **72**:522–525.

Waksman, S. A. 1952. *Soil Microbiology*, Wiley, New York.

Warren, H. L., D. M. Huber, D. W. Nelson, and O. W. Mann. 1975. Stalk rot incidence and yield of corn as affected by inhibiting nitrification of fall-applied ammonium. *Agron. J.* **67**: 655–660.

Watkins, J. E., G. N. Odvody, M. G. Boosalis, and J. E. Partridge. 1978. An epidemic of tanspot of wheat in Nebraska. *Plant Dis. Rep.* **62**:132–134.

Weaver, J. E. 1926. *Root Development in Field Crops*. McGraw-Hill, New York.

Weihing, J. L., and A. K. Vidaver. 1967. Report on holcus leaf spot *(Pseudomonas syringae)* epidemic on corn. *Plant Dis. Rep.* **51**:396–397.

Wiese, M. V., and A. V. Ravenscroft. 1975. *Cephalosporium gramineum* populations in soil under winter wheat cultivation. *Phytopathology* **65**:1129–1133.

Wilhelm, W. W., L. N. Mielke, and C. R. Fenster. 1982. Root development of winter wheat as related to tillage practice in western Nebraska. *Agron. J.* **74**:85–88.

Worley, R. E., D. J. Morton, and S. A. Harmon. 1966. Reduction of southern blight on tomato by deep plowing. *Plant Dis. Rep.* **50**:443–444.

Yocom, D.H., H.J. Larsen, and M.G. Boosalis. 1985. The effects of tillage treatments and a fallow season on VA mycorrhizae of winter wheat. Proc. 6th North American Conference on Mycorrhizae. College of Forestry Oregon State University. Corvallis, Oregon, Published by Forest Research Laboratory, 1985, 297.

Zeyen, R. J. 1979. Viruses. In S. V. Krupa and Y. R. Dommergues (eds.). *Ecology of Root Pathogens*. Developments in Agricultural and Managed Forest Ecology, 5. Elsevier Scientific Publishing Co., Amsterdam-Oxford-New York, pp. 179–205.

14

THE ECONOMICS OF CONSERVATION TILLAGE

P. CROSSON
Senior Fellow
Resources for the Future
Washington, D.C.

M. HANTHORN
Agricultural Economist
Economic Research Service
U.S. Department of Agriculture
Washington, D.C.

M. DUFFY
Professor
Department of Economics
Iowa State University
Ames, Iowa

INTRODUCTION

Increased use of conservation tillage by U.S. farmers in the last 15 years was due primarily to its economic advantages relative to conventional tillage methods. Government programs to promote the technology were relatively unimportant. Although the U.S. Department of Agriculture (USDA) now gives high priority to conservation tillage as an erosion-control measure, budgetary constraints severely limit cost-sharing and other public measures to encourage adoption. In the future, as in the past, farmers' decisions to adopt reduced tillage will be made primarily on their calculations of its economic worth.

This chapter analyzes the principal factors entering into those calculations. To give perspective, we show amounts of land in reduced tillage in 1982 for the nation, for major regions, and for major crops. We draw on a literature review to consider the principal differences between conservation tillage and conventional tillage with respect to quantities of inputs and crop yields. This permits a general assessment of economic differences between the technologies and the conditions under which reduced tillage has advantage. We then present data from a survey of farmers by the USDA which provides insights into the economics of reduced and plow tillages under field conditions. No comparable survey has been undertaken before. We believe the results presented to be of major importance in understanding the economics of the new tillage systems.

We conclude with an overall assessment of the economics of reduced tillage and draw a judgment of what this portends for future adoption of the technology.

LAND IN CONSERVATION TILLAGE

Three conservation tillage acreage estimates for 1982 show distribution among crops, Table 14.1. The *No-Till Farmer* (NTF) data were collected from state agronomists of the U.S. Soil Conservation Service (SCS). Conservation Tillage Information Center (CTIC) data were also collected by state and are consensus estimates of personnel in state soil conservation agencies and state offices of the federal SCS and Cooperative Extension Service (CES). USDA data were collected from farmers through the 1982 Crop and Livestock Pesticide Use survey.

The NTF estimates of land in other tillage systems are substantially higher than those of either the CTIC or the USDA. This probably is due to differences in definition. By CTIC definitions a system must leave at least 20% of the residue from the previous crop on the soil surface after planting to qualify as conservation tillage. NTF does not mention crop residue in its definition. However, it defines conventional tillage as any system which mixes or inverts 100% of the topsoil, implying that systems which leave less than 20% of the crop residue on the soil surface could qualify as conservation tillage. See definitions, Chapter 1.

Apart from this, the estimates obtained through these three surveys are remarkably similar. This is particularly impressive when one considers that the NTF and the CTIC data were obtained from government officials and the USDA data were collected from a sample of farmers.

Table 14.2 lists the CTIC and NTF estimates of land in conservation tillage in 10 agricultural regions of the United States. Except for "other" conservation tillage in the Southeast and Mississippi Delta, the two sources differ little in the regional percent distribution.

TABLE 14.1. Estimates of Land in Conservation Tillage in 1982

Source	No-Tillage (000 ha)	Other Conservation Tillage (000 ha)	Total (000 ha)
NTF[a]			
Total	4,685	40,611	45,296
Land in corn	1,029	n.a.[d]	n.a.
Land in soybeans	1,815	n.a.	n.a.
Land in small grains	488	n.a.	n.a.
CTIC[b]			
Total	4,147	34,025	38,172
Land in corn	1,526	10,999	12,525
Land in soybeans	1,711	8,454	10,165
Land in small grains	504	11,090	11,594
USDA[c]			
Total	3,595	34,141	37,735
Land in corn	1,777	12,158	13,936
Land in soybeans	1,199	9,594	10,794
Land in small grains	591	10,871	11,462

[a] *No-Till Farmer*, March 1983. No-Till Farmer, Inc., Waukesha, WI.
[b] *1982 National Survey Conservation Tillage Practices*, 1983. Conservation Tillage Information Center, Ft. Wayne, IN.
[c] *1982 Crop and Livestock Pesticide Use Survey*. 1982. U.S. Department of Agriculture, Economic Research Service.
[d] n.a. = not available.

TABLE 14.2. Estimates of Land in Conservation Tillage in 1982, by Region

Region	CTIC[a] (000 ha)		NTF[b] (000 ha)	
	No-Tillage	Other Conservation Tillage	No-Tillage	Other Conservation Tillage
Northeast	455	677	537	839
Appalachia	1,195	1,535	1,232	2,195
Southeast	329	853	438	2,345
Lake States	113	2,227	213	2,130
Cornbelt	1,254	13,801	1,452	14,774
Mississippi Delta	96	786	100	2,519
Northern Plains	459	8,852	459	9,584
Southern Plains	64	2,253	30	2,149
Mountain States	130	2,637	174	2,964
Pacific	53	404	52	1,000
U.S. Total	4,148	34,025	4,687	40,499

[a] SOURCE: CTIC (1982).
[b] SOURCE: NTF (March 1983).

The CTIC (1983) distinguishes between five kinds of conservation tillage, as follows:

No-tillage. Seedbed preparation and planting are completed in one operation, leaving a minimum of 90% of the previous crop residue on the soil surface.

Ridge till (no-till on ridges). Seedbed preparation and planting are completed in one operation on ridges. A minimum of 66% of the crop residue is left on the soil surface.

Strip till (unridged). Seedbed preparation and planting are completed in one operation, with tillage limited to a narrow band centered on the growing row. A minimum of 50% of the previous crop residue is left on the soil surface.

Mulch till. Seedbed preparation loosens and/or mixes the soil, incorporating a portion of the previous crop residue into the soil. Tillage tools include chisels, wide sweeps, disks, and harrows. A minimum of 33% of the previous crop residue is left on the soil surface.

Reduced tillage. The reduction of tillage trips over the field as a result of vegetative chemical control, combined tillage operations, or multifunction tillage tools. A minimum of 20% of the previous crop residue is left on the soil surface.

Table 14.3 presents the amount of land in each of these tillage systems devoted to corn, soybean, and small grain. From these surveys (Tables 14.1, 14.2, and 14.3) one may conclude:

1. Some 25–30% of the nation's cropland was in some kind of conservation tillage in 1982.
2. Only 2.5–3.0% was in no-tillage, the rest being in a variety of other systems.
3. Fifty-seven percent of the land classed as conservation tillage was in reduced tillage, the system closest to conventional tillage by the CTIC definition, and which leaves the least amount of crop residue on the soil surface.
4. Twenty-eight percent of the conservation tilled land was in mulch tillage and 11% in no-tillage. Strip tillage and ridge tillage were insignificant with 2% each.
5. About 90% of the land in some kind of conservation tillage was in corn, small grains, or soybeans. Corn had about 33%, small grains 30%, and soybeans a little over 25%.
6. Twelve percent of the no-tillage land was in small grains and 43.3% in soybeans. For mulch-till and reduced-till land, however, the distribution among the three crops was similar.
7. The Cornbelt had 30% of land in no-tillage and Appalachia had 29%. No other region had more than 11% no-tillage land.

TABLE 14.3. Land Use by Corn, Soybean, and Small Grains in Various Conservation Tillage Systems in 1982

	Tillage System					
Land	No-Till	Strip Till	Mulch Till	Ridge Till	Reduced Till	All Conservation Till
Total hectares (000)	4,147	749	10,813	775.	21,684	38,168
Percent	10.9	2.0	28.3	2.0	56.8	100
Corn						
Hectares (000)	1,526	203	3,719	287	6,789	12,524
Percent by tillage	12.1	1.6	29.7	2.3	54.2	100
Percent by crop	36.8	27.1	34.4	37.1	31.3	32.8
Soybeans						
Hectares (000)	1,710	104	3,005	149	5,196	10,164
Percent by tillage	16.8	1.0	29.6	1.5	51.1	100
Percent by crop	41.3	13.9	27.8	19.3	24.0	26.7
Small Grains						
Hectares (000)	504	278	3,351	266	7,196	11,595
Percent by tillage	4.3	2.4	28.9	2.3	62.1	100
Percent by crop	12.1	37.2	31.0	34.3	33.2	30.4

SOURCE: CTIC (1983).

8. The Cornbelt had 35–40% of land in all forms of conservation tillage and the Northern Plains 20–25%. No other region had more than 8%.
9. In large measure, the economics of conservation tillage at present is the economics of producing corn, soybeans, and wheat with reduced-tillage and mulch-tillage systems, particularly in the Cornbelt and Northern Plains.

INPUT REQUIREMENTS AND YIELD

All forms of conservation tillage involve less disturbance of the soil and leave more crop residue on the soil surface than plow-tillage. These differences have implications of the amounts of labor, fuel, pesticides, fertilizers, machinery, and managerial skills required per acre by the two systems in preharvest operations. Input requirements for harvesting are about the same (Young, 1982).

Labor

In an earlier survey of the literature (Crosson, 1981) it was shown that labor requirements with conservation tillage systems were only 33–80% of those with plow-systems, depending on the crop and system. Most of the studies reviewed put labor needs with reduced tillage at about half those with plow-tillage, no-tillage being the lowest. Subsequent studies have confirmed these findings (Colvin, et al., 1983; Ervin, et al., 1982; Jolly et al., 1982; Pfost, 1982).

Conservation tillage requires less labor because the farmer makes fewer passes over the field in preparing the seedbed and cultivating to control weeds. The value to the farmer of the time saved depends on the value he or she places on alternative uses of time, such as other farm work, off-farm employment, or increased leisure. Clearly the value will be different for different farmers, but for most, if not all, it will be positive.

Machinery

Machinery costs of conservation tillage systems are variable, but consistently lower than those for conventional systems (Crosson, 1981). Subsequent studies confirm this (Christensen and Norris, 1983; Colvin et al., 1983; Hamlett et al., 1981: Hayes, 1982, citing studies done at Purdue University; Jolly et al., 1982; *Successful Farming*, 1983; Young, 1982). There are three reasons why machinery costs are lower with conservation tillage: (1) less machinery and equipment are needed; (2) smaller, less expensive machinery will do the job; (3) fewer trips across the field reduce machinery wear.

In an Iowa State University study by agronomists and engineers (*Successful Farming*, 1983), costs and kinds of machinery needed to farm 243 ha with plow-tillage and three kinds of reduced tillage were compiled. Average machinery prices for 1976–1980 were used, and presumably the data reflect Iowa soils planted to corn, soybeans, or a corn–soybean rotation. The three kinds of reduced tillage were a full-width system (CTIC's definition), a strip-tillage system, and a no-tillage system. The plow system required 13 pieces of equipment with annualized costs per hectare of $128.32; full-width tillage, 11 pieces and $114.71; strip tillage, 8 pieces and $95.14; and no-tillage, 8 pieces and $95.59.

Young (1982, pp. 129–130) finds similar differences in equipment and tractor horsepower when comparing plow-tillage and no-tillage on a hypothetical 1000 acre farm (half in soybeans and half in corn) in "the central areas of the United States."

The other studies cited above do not provide the detail on equipment needs and costs given in the Iowa or Purdue reports, but all are consistent with finding less, and less expensive, machinery and equipment with conservation tillage than with plow-tillage.

There is a caveat to this conclusion, however. The comparisons are for situations in which farmers use either some form of conventional tillage or some variant of conservation tillage. Some farmers who adopt conservation tillage may want to retain the capacity to plow also, for example, to achieve better weed control by plowing every few years. Some of their soils may be poorly drained, hence not well adapted to conservation tillage, particularly no-tillage (more on this below). Farmers in these kinds of situations may conclude it would be worthwhile to shift to conservation tillage on part of their land, or on all of their land part of the time, but savings in equipment costs would not be a major factor in the decision.

Fuel

Fewer trips across the field and smaller horsepower tractors result in fuel economies with reduced tillage compared to plow-tillage. The largest savings are with no-tillage systems (Crosson, 1981). No-tillage fuel requirements typically are one-quarter to one-half those for plow-tillage Cristensen and Norris, 1983; Crosson, 1981). Data from central Iowa (Colvin et al., 1983) showed fuel requirements with a till-plant system (CTIC's strip-till system on ridges) 58% of those with plow-tillage. Other Iowa data (Christensen and Norris, 1983) show that with combinations of three crop rotations, six tillage systems (one conventional and five conservation, including no-tillage) and three soil types (36 combinations) fuel requirements for conservation tillage were less than for conventional tillage for all comparable combinations but six. A corn–soybean rotation with fall plowing used less fuel on light, medium, and heavy soils than did a fall chisel plow system (comparable to CTIC's reduced tillage) with the same rotation and same soils. The fall-plow system also had a fuel advantage over the chisel-plow system on all three soils in a corn–meadow rotation.

Fertilizer

There is much ambiguity in the discussion of per acre fertilizer requirements of reduced-tillage systems relative to plow systems (Crosson, 1981). A commonly expressed view is "that under given soil conditions conservation tillage often requires more nitrogen fertilizer than conventional tillage because the cooler and more moist soils slow mineralization of [organic] nitrogen and promote denitrification." Work published subsequent to Crosson's review supports this view (Buchholz and Hanson, 1982; Christensen and Norris, 1983; Hayes, 1982; Siemens, 1982; *Successful Farming*, 1983; Young, 1982).

Ambiguity arises because the concept of fertilizer "requirements" is usually not clearly defined. It reasonably can be assumed that farmers will apply fertilizer to the point where cost of an additional unit equals the value of the additional output achieved. By this criterion the "required" amount of fertilizer depends on the yield response to varying amounts and kinds of fertilizers and crops and prices of both. Much of the discussion of fertilizer use in different tillage systems is not cast in these terms. When it is, the evidence indicates that for no-tillage the yield response may fully or more than compensate for the additional nitrogen required. Researchers in Kentucky assert that although no-tillage requires more nitrogen than plow-tillage because of losses and slower mineralization, this is offset by higher yields with no-tillage (Phillips et al., 1980). Data from four experimental farms in Maryland in 1981 show that at very low levels of nitrogen application corn yields with plow-tillage generally are higher than with no-tillage (Bandel, 1982). However, at 135 kg N/ha (only a little less than the national average used on corn) no-tillage corn yields exceeded those for plow-tillage by from 3.5 to 42%. Average yields of the two systems on the four farms averaged 130.0–192.9 kg/ha. (See Chapter 5.)

Bandel (personal communication) points out that these results were obtained at the end of nine continuous years of experimental use of conventional and no-till systems on the same land. In the beginning, the yield advantage of no-tillage, at higher levels of N fertilization, was less marked, or nonexistent. According to Bandel this is likely in the first years after adopting no-tillage on soils that have long been row cropped with a plow. Organic matter in such soils is typically very low with consequent weak structure, poor tilth, and a tendency to seal after rain. Plowing loosens the soil and so overcomes this tendency, and fertilizer can compensate for the low natural nutrient content. With no-tillage, plowing is not an option, so crop yields initially often compare unfavorably with those of plow-tillage. With time, no-tillage increases soil organic matter with consequent improvements in soil structure and tilth. Eventually yields may equal or exceed those with plow-tillage.

Citing agronomists and soil scientists in Iowa, Illinois, Indiana, Minnesota, and Wisconsin, *Successful Farming* (1983) reports that availability of potassium fertilizer may be less with reduced-tillage systems than with plow-tillage. Evidence, however, is not as strong as for nitrogen. Plant-available phosphorus seems little affected by the type of tillage system. Siemens (1982) asserts that "research in the central Cornbelt shows that requirements of phosphorus and potassium fertilizer should not vary greatly among various tillage systems."

The evidence is clear that with reduced tillage all three major nutrients accumulate in the top 3 in. of soil, while plow-tillage mixes them more evenly through the plow layer. Close-to-the-surface accumulation of P and K does not appear to affect yields significantly. However, the surface application of nitrogen increases the acidity of the top few inches of soil, which sometimes reduces crop uptake of nutrients and the effectiveness of triazine herbicides (Hinish, 1980; *Successful Farming*, 1983). Soil acidity can be corrected by liming, but this constitutes an added cost of nitrogen fertilizer with reduced tillage.

Evidence on differences in nitrogen fertilizer requirements is clearest when comparison is between plow-tillage and no-tillage. For other kinds of conservation tillage the evidence is more vague and differences are smaller. In a comparison of energy requirements of plow-tillage with those of chisel and disk systems, Siemens (1982) assumed identical amounts of nitrogen. Similarly, for comparison of production costs of plow-tillage with three kinds of conservation tillage, including no-tillage, Jolly et al. (1982) assumed identical fertilizer requirements.

On balance, the literature suggests that no-tillage systems often require more nitrogen fertilizer than conventional tillage, although yields in time are correspondingly higher. With other kinds of tillage, requirements for all three major nutrients are about the same. For P and K this is true also for no-tillage.

Pesticides*

Conservation tillage systems, particularly no-tillage, generally depend more on herbicides and less on cultivation to control weeds than plow-tillage. This substitution of herbicides for cultivation in weed control tends to increase the amount of herbicide used per acre. More herbicide may be needed also because surface residues intercept the spray, and because cooler and more moist microclimates alter effectiveness and foster biological breakdown of the herbicide.

It often is noted in the literature that control of some weeds by herbicides permits the emergence of others previously suppressed by the target weed. Perennials in particular have become more troublesome in some areas and are difficult to control adequately with presently available herbicides. In areas heavily infested with such weeds, farmers face increased costs of weed control, and perhaps a yield penalty, if they opt for some form of reduced tillage.

There is evidence that continuous use of herbicides over a long period may induce genetic resistance in weeds to those herbicides (Hileman, 1982; Hinkle, 1983; Marx, 1983). Should this become significant, it could force farmers to increase rates or shift to more expensive materials to achieve adequate weed control. Although herbicide requirements may increase for plow-tillage as well as for reduced tillage, the latter would be particularly affected because of its greater reliance on herbicides.

The main thrust of the material reviewed is that because of more crop residue on the soil surface conservation tillage is likely to require more herbicide than conventional tillage. There is evidence, some anecdotal and some scientific, that this is a well-established fact. In the summer of 1981, W. B. Shafer, an employee of Stauffer Chemical Company, made a 2-week trip through Iowa and Illinois talking about tillage practices with farmers, dealers, and technical and sales representatives. He reports (Shafter, 1981) that most dealers were recommending and most farmers were using more herbicides per acre with conservation tillage than with conventional tillage.

In a 5-year experiment with a corn–soybean rotation on 1 acre plots in central Iowa with conventional tillage and three kinds of conservation tillage (a full-width reduced-tillage system, a strip-till system, and a slot till-ridge system) herbicide costs were the same for the plow and full-width systems, but 25% and 51% higher, respectively, for the other two tillage systems (Jolly et al., 1982). Siemens (1982) estimated a chisel plow system required 14% more herbicide per hectare than a plow system; a disk system 29% more and a no-tillage system 43% more.

*Other chapters in this volume deal in depth with the use of pesticides to manage weeds, insects, and disease in conservation tillage system. The discussion here focuses only on those aspects that affect the quantities of these materials used relative to conventional tillage.

Colvin et al. (1983) in a 3-year field experiment compared a till-plant system and a plow-tillage system in a corn–soybean rotation in central Iowa. In 1979 and 1981 the same amounts of herbicide per hectare applied in both systems controlled weeds satisfactorily. In 1980, the soybean year, they encountered serious weed problems in the till-plant plots and yields were significantly reduced even though those plots received almost three times as much herbicide as the plow-tillage.

In his earlier review, Crosson (1981) concluded that "insect and disease problems are more likely with conservation tillage because the crop residues left on the soil surface provide a favorable habitat for various insects and diseases." Because no-tillage systems leave the most residue, they are the most affected. This view continues to find support (Keaster and Fairchild, 1982; Pfost, 1982), but it does not go unchallenged. (See Chapters 12 and 13.) In Georgia (All, 1979; All and Gallaher, 1977) some corn insects were encouraged by no-tillage, others were suppressed or unaffected by it. All (1979) emphasizes that the no-tillage systems studied were not continuous, a tillage operation occurring every 1–2 years in a multicrop sequence. Since tillage suppresses some insects, the neutral or beneficial effects of no-tillage found might not apply to situations of continuous no-tillage.

Although insect and disease problems may be more serious with conservation tillage, it does not necessarily follow that more insecticides or fungicides are needed with these systems, except in some situations. The amounts of insecticides applied to soybeans and wheat are small and when needed in reduced tillage, the additional cost is small. The use of pest-resistent or tolerant crop varieties is much more effective in controlling diseases than fungicides (University of Illinois, 1979). Crop rotation often provides effective control, since many pests are host specific (*Successful Farming*, 1983).

After cotton, corn receives the most insecticides in the United States. If conservation tillage significantly increased the amount of insecticide needed on corn, the effect on production costs would be important. A University of Illinois study (1979) reported the effects on corn of different tillage practices on populations of 12 kinds of insects and mice. Spring plowing and fall plowing either suppressed insect and rodent populations or had no effect. Surface tillage encouraged the European corn borer, perhaps encouraged the corn rootworm (the most important insect pest of corn in Illinois), and suppressed four others and mice. It had unknown effects on the rest. No-tillage unequivocably encouraged all of the pests, including mice.

The study thus suggests that more insecticides and rodenticides are needed with no-tillage corn and, where the corn rootworm is the main target, more might be required also with a reduced-tillage system. Farmers who plant continuous corn would face this situation more acutely. Rotating corn with soybeans controls the rootworm since it is host specific. Such a rotation would require less insecticide than continuous corn.

Finally, it is noteworthy that in three or four studies examined which compare quantities of inputs required by plow- and reduced-tillage systems

(Colvin et al., 1983; Ervin et al., 1982; Jolly et al., 1982; Siemens, 1982) only that by Ervin et al. listed insecticide costs, and found them to be 22% higher with a minimum of tillage and 62% higher with no-tillage. We conclude that in some situations reduced tillage requires more insecticides than plow-tillage, but in others it does not.

Management

Crosson (1981) found the literature virtually unanimous in asserting that reduced-tillage systems require more skillful management than plow systems. This is especially true of no-tillage. Subsequent work points to the same conclusion (Pfost, 1982; Walsh, 1982). The increased management requirement stems fundamentally from problems caused by less tillage and more crop residue. Fewer tillage operations means fewer opportunities to correct previous mistakes in seedbed preparation. Increased surface residue makes it more difficult to place seeds at the right depth and in firm contact with soil, and, as already noted, residue frequently promotes weed, insect, and disease problems. The farmer therefore needs more information about the properties of a wider variety of pesticides and more skill in applying them. He or she should also understand how crop rotations and disease- and insect-resistant varieties substitute for or supplement pesticides. Conservation tillage also requires changes in machinery and may require more care in operation and maintenance, especially with no-tillage systems where soil obstructions and large amounts of crop residue exist.

Increase in management skill is not quantifiable, like an increase in pounds of herbicides or fertilizers required. It is directly related to production costs, however, and therefore is a key component of an evaluation of the economics of conservation tillage. The farmer who attempts reduced tillage, especially no-tillage, without the requisite skills will encounter high costs per unit of output. He or she will either give up the attempt or invest money and time in learning needed skills.

The literature provides few assessments of management costs of conservation tillage, but experience indicates that farmers who attempt to adopt these systems without proper preparation may find the costs high. Weed infestations may cause severe losses in no-tillage fields where herbicides have been mismanaged. We believe, however, that the costs of acquiring the needed skills are unlikely to pose a serious obstacle to the future spread of conservation tillage. The history of U.S. agriculture since World War II indicates that U.S. farmers quickly learn to manage new technology when it is in their economic interest. The rate at which they adopt reduced tillage is determined primarily by the success of farm machinery manufacturers in developing suitable planting and tillage equipment; of pesticide manufacturers in developing more effective herbicides; and of the scientific community in expanding knowledge of how less tillage and more crop residue affect insects, weeds, and other soil biota. By comparison, the costs of acquiring the necessary management skills likely will be minor.

Summary

This review shows that, in general, conservation tillage systems require less labor and fuel than plow-tillage; have lower machinery costs; often require more nitrogen fertilizer, but not more potassium and phosphorus; usually require more herbicides (always with no-tillage); may or may not require more insecticides and fungicides; and are more exacting in management skills, especially with no-tillage. Several recent studies provide insights into how these differences in input quantities affect input costs. Cost ratios computed from the studies indicate that reduced tillage costs are generally 5–10% less than costs of conventional plow-tillage (Table 14.4). Considering the diversity of sources and areas covered by these studies with corn and soybean, the agreement among them is impressive.

Harman (1982) studied the relative costs of conventional and no-tillage systems in an irrigated wheat–fallow irrigated–sorghum rotation and in an irrigated wheat–fallow dryland–sorghum rotation in the semiarid High Plains region of Texas. Disking comprised the conventional tillage system. The no-tillage systems relied on atrazine, 2, 4-D, and paraquat to control weeds. Cost

TABLE 14.4. Ratios of Conservation Tillage Costs to Conventional Plow-Tillage Costs[a]

Source	Corn	Soybeans
Crosson (1981)[b]	0.93	0.91
Hudson [1980, as given in Christensen and Norris (1983)][c]		
No-tillage	0.87	0.86
Ervin et al. (1982)[d]		
Minimum tillage	0.95	0.93
No-tillage	0.96	0.99
Jolly et al. (1982)[e]		
Full-width system	0.94	0.93
Strip tillage	0.91	0.89
Slot tillage	0.93	0.91
Colvin et al. (1983)[f]		
Till-plant		
1979	0.88	
1980		0.91
1981	0.90	

[a] Excludes land.
[b] Does not distinguish among kinds of conservation tillage.
[c] Tennessee, 1980.
[d] Missouri, 1982, on Mexico soil with 3% slope.
[e] Central Iowa, 1976–1980, Clarion-Nicollet–Webster soil association.
[f] Central Iowa, machinery investment costs not included.

and yield data for the two systems were collected over a 7-year period. They showed that in both rotations per hectare profits (1982 prices) on average were about $40 higher with no-tillage. Lower labor, fuel, and machinery costs of the no-tillage system more than offset higher costs of chemicals. In addition, yields with no-tillage were almost 15% higher because of water conservation.

The marked economic advantage of no-tillage revealed by the study suggests that most farmers in the region using a wheat–fallow–sorghum rotation would have adopted no-tillage in place of the conventional system. However, according to Allen Wiese, an agronomist at the Department of Agriculture's experiment station in the region (Bushland, Texas), this has not occurred (personal communication). The reasons for the failure to adopt are not known.

The Texas High Plains study illustrates that although costs are key elements in the economics of tillage systems, differences in net revenue fundamentally guide a farmer's choice among alternative systems. Given the cost differences, differences in net revenue depend on how yields vary among systems. Yield, therefore, must be considered.

YIELD

Long Term

Erosion removes soil nutrients, reduces soil available water holding capacity, and, if continued indefinitely constricts the crop rooting zone. Sooner or later, with technology fixed, continued erosion in excess of new soil formation will cause crop yields to decline, the amount of the reduction depending on the original condition and depth of the topsoil and subsoil, and the amount of erosion per unit of time.

The evidence is clear that on erosive soils conservation tillage decisively reduces erosion compared to conventional plow-tillage (Crosson, 1981; Hayes, 1982; Moldenhauer, et al., 1983; Pfost, 1982; Young, 1982). The reduction is greatest with no-tillage. It follows that over the long-term conservation tillage has a yield advantage over plow-tillage on these soils.

Generally, the advantage will be stronger (1) the greater the advantage of conservation tillage in reducing erosion; (2) the shallower the soil; (3) the less favorable the subsoil is for yield; (4) the higher the prices of fertilizer and other inputs that substitute for soil; (5) the longer the period over which the yield differences matter to the farmer; (6) the higher the farmer expects future crop prices to be relative to current prices; and (7) the lower the rate of discount the farmer applies to future costs and revenues.

The calculation of the long-term yield advantage of surface-tillage relative to plow-tillage is complex. The outcomes will vary widely from place to place and time to time, depending on soil conditions, changes in agricultural technologies, farmers' preferences between present and future income, and their expectations of interest rates and prices of inputs and crops.

Studies of how these various factors interact to affect farmers' judgments of long-term advantages of conservation tillage suggest that the effect is small. Work done by soil scientists at the University of Minnesota indicates that continuation of recent rates of erosion for 100 years would reduce average corn yields 5–10% if there were no advances in yield-increasing technology (Larson et al., 1983; Pierce et al., 1983). The losses would be considerably greater on land with slopes of 12% or more, but relatively little cropland in the United States is this steep. Even a very modest rate of development of new technology to improve yields—say 1% per year—would far outweigh the prospective erosion-induced productivity loss. Consequently, most farmers probably view the long-term threat of erosion as not very important.

The University of Minnesota scientists did not examine the economics of erosion–yield relationships, but those who have find generally that they do not favor investment by farmers in erosion control measures. Seitz, et al. (1979) found that under price and technological conditions prevailing in 1974, corn–soybean farmers in a typical Cornbelt watershed would be insensitive to the long-term yield effects of erosion unless their planning horizon was 40 years and they did not discount future costs and revenues. If they discounted at 5%, their planning horizon would have to be 60 years before they would have incentive to invest in erosion control. These findings are particularly significant because corn and soybean prices in 1974 were much higher than they have been since then. If continuation of those prices over 40–60 years would give little incentive to control erosion, the incentive given by more recent lower prices would be even weaker.

The Palouse area of the Pacific Northwest has erosion rates among the highest in the nation. Walker (1982) studied the economics of growing wheat with conventional tillage and conservation tillage in that area. He found that the annual per hectare costs of the two technologies were almost identical but that wheat yields with conservation tillage were 3% lower because of problems with seedbed preparation and weed control. Consequently, a farmer considering only immediate profitability would not adopt conservation tillage.

Over the long term, however, the erosion-reducing characteristics of conservation tillage gave it a yield advantage. On more shallow soils this was sufficient to make conservation tillage more profitable than the conventional system, given certain assumptions about future crop prices and interest rates. On deeper soils, or with lower crop prices and higher interest rates, the conventional system still was more profitable.

Soils in the Palouse region are not typical of those in the rest of the country. They are deeper, but they also are on more sloping land and suffer substantially higher rates of erosion. Consequently, Walker's specific numerical results cannot safely be extrapolated to other regions. His study is particularly valuable, however, because it vividly illustrates the great complexity and subjectivity of calculations farmers must make in judging the significance for profits of conservation tillage's long-term yield advantage.

Short term

Uncertainty about the economics of long-term yield differences between conservation tillage and conventional tillage would be unimportant if conservation tillage had a clear short-term yield advantage. This is not the case, however. Christensen and Norris (1983), in a very useful exercise, reviewed studies of comparative yields in 10 states. Of 63 comparisons, 53 were for corn, 6 for soybeans, 3 for sorghum, and 1 for sunflowers. Forty-six of the comparisons were for no-tillage and 17 for other kinds of reduced tillage.

Thirty-nine of the comparisons showed a yield advantage for reduced tillage, twenty-one showed a disadvantage and three showed no difference. Thirty-three of the corn comparisons showed an average yield advantage of eight bushels per acre with reduced tillage, nineteen showed a yield disadvantage of twenty-three bushels per acre (fourteen bushels excluding three extreme observations), and one showed no difference.

Thirteen of twenty-one comparisons in Indiana showed a yield disadvantage for reduced tillage and eight showed a yield advantage. Eleven comparisons were from Kentucky, all for no-tillage. Ten showed a yield advantage and one was neutral. Twelve comparisons from Ohio also were all no-tillage. Seven showed a yield advantage and five a disadvantage.

These data are the most complete we have found showing yield comparisons for plow-tillage and conservation-tillage systems, but they cannot be taken as representative of the country as a whole. No-tillage is vastly overrepresented, considering the relatively small proportion of conservation-tilled land in no-tillage (see Table 14.1). Corn is greatly overrepresented, and wheat is not represented at all. The distribution of the comparisons by state also is unrepresentative with three states, Indiana, Kentucky, and Ohio accounting for 32 of the 46 comparisons of no-tillage yields and 44 of the total of 63 comparisons. According to the CTIC survey, these states had 25% of the no-tillage land in 1982 and 11% of all conservation-tilled land.

Although they lack representativeness, the data compiled by Christensen and Norris leave little doubt that yields with reduced tillage compare favorably with those of plow-tillage under some soil and climate conditions and unfavorably under others. In this the data are consistent with those reviewed by Crosson (1981).

Despite the variability in yield response, some generalizations are possible.

1. The surface residue left with conservation tillage reduces both evaporation and runoff of water. In areas where inadequate rainfall is the principal factor limiting plant growth, the moisture-conserving feature of a surface mulch is a distinct plus and likely accounts for the high percentage of land in conservation tillage in the Northern Plains (Table 14.2).

2. Surface residue, and associated increase in soil moisture, slows soil warming in spring, delaying seed germination and seedling emergence. Where

the growing season is already short, as at high latitudes, this characteristic of conservation tillage provides a yield disadvantage. The relatively small percentage of land in conservation tillage in the Lake States (Michigan, Minnesota, and Wisconsin, Table 14.2) is consistent with this. So is the diminishing percentage of land in conservation tillage as one moves north in the Northern Plains: 41% of cropland in Kansas; 44% in Nebraska; 20% in South Dakota; and 4% in North Dakota (CTIC, 1983).

3. Conservation tillage generally suffers a yield disadvantage on poorly drained soils. This proposition commands a stronger consensus than any other about comparative yields of different tillage systems (Crosson, 1981). It is noteworthy that in the compilation by Christensen and Norris (1983) in all five instances in Ohio no-tillage yields were less than plow-tillage yields on poorly drained soils. In only one case did no-tillage show advantage on a poorly drained soil, 1 bushel of corn in a corn–oats–meadow rotation. Cosper (1979) concluded that soil wetness is the most important single soil factor restricting adoption of conservation tillage in the Cornbelt and that it becomes less restrictive as one moves from east to west. The CTIC data showing percentages of cropland in conservation tillage are consistent with this: Ohio, 31%; Indiana, 35%; Illinois, 47%; and Iowa, 53%.

Evidence reviewed in Crosson (1981) suggests that disease organisms and weeds, favored by moist environments, are reasons for poor yield on poorly drained soils; cold, moist soil conditions also slow mineralization of organic nitrogen and facilitate denitrification and breakdown of herbicides by soil bacteria.

4. Reduced tillage systems suffer a yield penalty wherever weeds are not adequately controlled by herbicides. Perennial weeds in particular may become troublesome because they are less vulnerable than annuals to herbicides due to below-ground regeneration.

5. Conservation tillage saves time between harvesting one crop and planting of another, and so is more favorable to double cropping than plow-tillage. Taking two crops per year from the land instead of only one increases the economic yield of the land. This advantage is most marked in the southeast, where stretching of the already long growing season favors double cropping. It is practiced also in parts of the southern Cornbelt.

With conservation tillage it may be economical to put steep, erosive land into row crops that otherwise could economically be put only in some lower-valued use. Furthermore, pasture, meadow, and woodland can be improved effectively using conservation-tillage systems. This potential yield advantage could become important if the demand for cropland grows much beyond that of 1981 and 1982, when the cropland base was fully used. If that happens, the use of conservation tillage on currently marginal land may help contain the rise in cropland prices that otherwise would occur. However, crop prices would have to be substantially higher than in 1983 and the rate of development of land-saving technology modest by the standard of the last several decades, before much of this marginal land would be converted to crop production.

Summary

The erosion-reducing effect of conservation tillage provides a long-term yield advantage over conventional plow-tillage. Under prevailing conditions of soils, rates of erosion, crop and fertilizer prices, interest rates, and rate of technological advance, the advantage is small except on steep and shallow soils not generally characteristic of U.S. cropland.

On relatively well-drained soils in areas below the northern tier of states where weeds can be effectively controlled, yields with reduced tillage equal or exceed those with conventional tillage. Under these conditions the cost advantage of conservation tillage translates into a net revenue advantage and farmers will adopt the technology. Where these conditions are not met, conservation tillage likely suffers a yield penalty, which discourages adoption.

THE USDA STUDY

The discussion to this point has dealt exclusively with studies investigating input use and costs and crop productivity among tillage practices under experimental conditions. Such studies are valuable, but are inherently limited because they do not reflect actual production conditions in which farmers operate. Analysis of farm-level production data recently collected through joint U.S. Department of Agriculture (USDA) surveys sheds additional light on how different tillage practices affect input use and costs and crop productivity for U.S. corn and soybean farmers.

Data Sources and Description

In 1980, USDA's Corn and Soybean Pesticide Use Survey (CSPUS) was taken in conjunction with its Objective Yield Survey (OYS). Fields chosen for the OYS were randomly selected by crop using an area frame sampling technique (USDA, 1975b). Farmers working corn and soybean fields were personally enumerated, providing information about crop yields, tillage practices, and fertilizer and pesticide use. Enumerators also took plant population counts in randomly selected spots in each surveyed field, from which corn and soybean seeding rates were estimated.

Compilation was made of the responses for 1193 nonirrigated cornfields harvested for grain in the 10 major producing states, which account for 77% of the total U.S. corn crop in 1980. A total of 759 surveyed cornfields selected for the study were conventionally tilled, while no-tillage and reduced-tillage fields totaled 17 and 417, respectively.

Responses for 1464 soybean fields located in the 17 major producing states were also analyzed. These states produced 91% of the total U.S. soybean crop in 1980. For the analysis, surveyed soybean fields were stratified regionally to account for differences in soil characteristics, weather, length of growing season, and relative homogeneity of production input use and practices.

Soybean regions include the Midwest, Midsouth, and Southeast. The Midwest represents fields located in Illinois, Indiana, Iowa, Kansas, Kentucky (excluding the extreme western counties), Minnesota, Missouri (excluding the bootheel), Nebraska, Ohio, and South Dakota. The Midsouth includes Alabama, Arkansas, western Kentucky, Louisiana, Mississippi, the bootheel of Missouri, and Tennessee. The Southeast comprises Georgia, North Carolina, and South Carolina. Conventionally (plow) tilled soybean fields total 576 in the Midwest, 418 in the Midsouth, and 145 in the Southeast. Soybean fields planted to reduced tillage total 212 in the Midwest, 43 in the Midsouth, and 26 in the Southeast. No-tillage soybean fields total 15 in the Midwest, 21 in the Midsouth, and 8 in the Southeast.

Per-acre corn and soybean production data obtained through CSPUS and the OYS were compared in a partial budget framework that included harvested bushels of nonirrigated corn and all soybeans; total application rates for nitrogen, phosphorus, potassium, herbicides, and insecticides; seeding rates; and number of mechanical cultivations to control weeds.

Crop price and input cost data were not collected in these surveys. We used average state, regional, and national prices obtained from several USDA, state-level, and private publications. For instance, 1980 USDA season average prices for corn and soybeans at the state level were assigned to observations for each respective state. Per-pound USDA regional costs for seed and state-level costs for N, P, and K were assigned to observations according to location. National average prices were used for specific herbicide and insecticide formulations. Per-acre cultivation costs for each region are based on budget estimates of costs to conduct this activity.

Data were not collected for labor, fuel, and machinery expenditures incurred during field preparation. However, these are inputs about which much is known. The literature reviewed above is virtually unanimous in showing that conservation-tillage systems, especially no-tillage, require less field preparation than conventional plow systems. Current uncertainties relate primarily to differences in fertilizer, pesticide, and seed applications, and the subsequent effect on crop productivity.

Since data were not collected for field preparation input use and costs, we used USDA per-acre cost of production estimates for labor, fuel, and machinery, which were adjusted for specific tillage systems based on cost differences estimated by Ervin et al. (1982). This study was selected as a basis for comparison over several other studies because it reflects recent conditions, presents a cost comparison for three tillage systems similar to ours, and provides estimated field preparation cost differences for both corn and soybeans.

Farmers were asked in CSPUS whether they used no-tillage, reduced tillage, or conventional (plow) tillage during 1980. These practices were not specifically defined in the survey instrument; therefore, some farmers may have inadvertently given inconsistent responses with regard to using reduced- and conventional-tillage practices.

A comparison of corn and soybean acreage planted to various tillage practices shows that U.S. farmers are increasingly adopting conservation-tillage strategies. Data collected through CSPUS in 1980 and through USDA's 1982 Crop and Livestock Pesticide Usage Survey reveal that the proportion of corn acreage planted to no-tillage increased from 1% to 4% during this 2-year period (Table 14.5). Reduced-tillage soybean acreage also increased from 26% to 39% in the Midwest, 9% to 18% in the Midsouth, and 15% to 40% in the Southeast.

Following is a summary of the 1980 USDA tillage study in which a comparison of input use, input costs, and crop yields for corn and soybean farmers using different tillage practices is presented.

Weed Control

For most corn and soybean farmers using different tillage practices, total per-hectare weed control costs by crop were not significantly different. In the Midwest, however, weed control costs for no-tillage soybean farmers were significantly higher than costs for either reduced- or conventional-tillage soybean farmers (Table 14.6). As corn and soybean farmers shift from conventional to conservation tillage, with no-tillage being the extreme conservation

TABLE 14.5. Type of Tillage Employed in the Production of Corn and Soybeans in the United States, 1980 and 1982

	Hectares Planted to Tillage Systems (000 ha)					
	No-Tillage		Reduced Tillage		Plow-Tillage	
Crop, Region	1980	1982	1980	1982	1980	1982
Corn (10 states) (1193)[a]	359	1,094	10,309	10,906	15,162	13,790
Soybeans (17 states)						
Midwest (803)[a]	321	483	4,175	6,534	11,562	9,798
Midsouth (482)[a]	307	407	690	1,279	6,670	5,446
Southeast (179)[a]	100	114	377	1,105	2,033	1,514

[a]Numbers of farms surveyed.

SOURCE: USDA (1982, survey described in text)

TABLE 14.6. Average Per-hectare Weed Control Input Use and Costs for Different Tillage Practices, 1980[a]

Crop, Tillage Practice	Herbicide		Mechanical Cultivation		Weed Control Cost (dollars)
	Use (kg a.i.)[b]	Cost (dollars)	Use (number)	Cost (dollars)	
Midwest corn					
No-tillage	3.92a	42.58a	0.35a	3.73a	46.31a
Reduced tillage	3.78a	31.37b	1.13b	11.91b	43.27a
Conventional tillage	3.36a	28.13b	1.13b	11.91b	40.04a
Midwest soybeans					
No-tillage	3.61a	75.58a	0.47a	4.94a	80.52a
Reduced tillage	2.08b	43.60b	1.19b	12.57b	56.17b
Conventional tillage	2.27b	44.26b	1.33b	14.08b	58.34b
Midsouth soybeans					
No-tillage	2.24a	58.49a	0.19a	2.10a	60.59a
Reduced tillage	1.95a	45.28ab	1.42b	15.64b	60.91a
Conventional tillage	1.62a	38.58b	1.79b	19.74b	58.32a
Southeast soybeans					
No-tillage	1.88a	46.61a	0.00a	0.00a	46.61a
Reduced tillage	1.63a	29.66a	1.12b	11.07b	40.73a
Conventional tillage	1.54a	26.85a	1.73b	17.19b	44.04a

[a]The Tukey–Kramer method (SAS Institute, Inc., 1982) was used to test for differences in means among tillage practices in each region. Regional means in each column followed by different letters are significantly different from each other at the 5% level.
[b]a.i. = active ingredient.

strategy, different mixes or additional herbicides are substituted for mechanical weed control.

No-tillage farmers generally do not break the soil surface in their fields prior to planting, therefore, they do not use herbicides that require incorporation. Rather, they are more likely to use selective broad spectrum materials that can be applied singly, tank mixed with other broad spectrum or burndown herbicides, or can be mixed and applied with fertilizer. The USDA data, however, suggest that in most instances a substitution in weed control inputs does not change per-hectare weed control costs significantly.

Differences in per-hectare quantities of herbicides applied were not statistically significant for corn and soybean farmers, except in the Midwest where no-tillage soybean farmers applied 3.6 kg active ingredient (ai) per hectare compared to about 2.1–2.3 kg ai per hectare for both reduced- and plow-tillage soybean farmers. In the Midwest, no-tillage soybean farmers applied additional herbicides in either early-season or postemergence applications compared to soybean farmers who tilled and cultivated their fields more intensively.

Although total weed control costs and applications of herbicides applied per hectare did not vary significantly among tillage practices for most corn and soybean farmers, we found that total per-hectare herbicide costs were more variable. Except for southeast soybeans, average per-hectare herbicide costs were significantly higher for no-tillage farmers compared to costs for virtually all other farmers tilling more intensively. Herbicide costs were significantly higher for no-tillage farmers because they either applied mixes of more expensive broad spectrum herbicides or additionally used a nonselective, burndown material like glyphosate or paraquat. But, since no-tillage farmers cultivated their fields significantly less than other farmers, the increase in herbicide costs is offset by a drop in cultivation costs (Table 14.6).

The research literature suggests that tillage practices make little difference in the per-hectare amounts of phosphorus and potassium farmers apply, but that conservation tillage may require more nitrogen. The USDA survey data indicate that corn farmers in the Midwest applied more of all three nutrients per hectare as amount of tillage declined (Table 14.7). However, none of the differences were statistically significant. The survey data for soybeans gave mixed results but none of the differences among tillage practices in amounts of nutrients applied were statistically significant. For all three nutrients per-

TABLE 14.7. Average Per-hectare Fertilizer Use and Costs for Different Tillage Practices, 1980[a]

Crop,	Nitrogen		Phosphorus		Potassium		Total
Tillage Practice	Use (kg)	Cost (dollars)	Use (kg)	Cost (dollars)	Use (kg)	Cost (dollars)	Cost (dollars)
Midwest corn							
No-tillage	161a	67.63a	86a	50.98a	117a	28.43a	147.04a
Reduced tillage	152a	63.92a	74a	43.82a	97a	23.44a	131.18ab
Conventional tillage	131a	54.88a	67a	39.94a	86a	20.95a	115.77b
Midwest soybeans							
No-tillage	7a	2.89a	27a	16.01a	33a	8.00a	26.90a
Reduced tillage	3a	1.26a	14a	8.15a	23a	5.56a	14.97a
Conventional tillage	4a	1.61a	15a	8.72a	22a	5.29a	15.62a
Midsouth soybeans							
No-tillage	6a	3.09a	19a	11.41a	22a	5.83a	20.33a
Reduced tillage	5a	2.27a	25a	14.94a	29a	7.58a	24.79a
Conventional tillage	5a	2.57a	26a	15.76a	36a	9.53a	27.86a
Southeast soybeans							
No-tillage	15a	7.14a	28a	15.66a	52a	12.67a	35.47a
Reduced tillage	13a	6.50a	42a	23.05a	74a	17.96a	47.51a
Conventional tillage	9a	4.27a	30a	16.82a	65a	15.68a	36.77a

[a]The Tukey–Kramer method (SAS Institute, Inc., 1982) was used to test for differences in means among tillage practices in each region. Regional means in each column followed by different letters are significantly different from each other at the 5% level.

hectare amounts applied to soybeans in the Southeast were higher than in the Midwest and Midsouth because soils in the southeast are less favorable.

The only statistically significant difference in total per-hectare fertilizer costs was between no-tillage and conventional tillage corn farmers in the Midwest.

Seed, Insect Control, and Yield

We found that in the Midwest no-tillage corn farmers planted significantly more seed per hectare than did corn farmers who conventionally tilled their fields (Table 14.8). However, none of the differences in seeding rates for soybeans were statistically significant.

No-tillage corn farmers applied significantly more pounds ai of insecticides per hectare than did corn farmers using either reduced or plow systems (Table 14.8). Insecticide use on corn generally fluctuates more than herbicide use, since insect infestations are more variable than weed problems from year to year and across tillage systems. In 1980, a higher proportion of the no-tillage corn acreage was treated with a foliar insecticide application, primarily to

TABLE 14.8. Average Per-hectare Selected Input Use and Costs and Crop Yields for Different Tillage Practices, 1980[a]

	Seed		Insecticide		Yield
Crop, Tillage Practice	Use (thousand)	Cost (dollars)	Use (kg a.i.)[b]	Cost (dollars)	(bushels)
Midwest corn					
No-tillage	53.3a	30.83a	1.01a	11.36a	245.3a
Reduced tillage	50.7ab	29.29ab	0.51b	7.81a	243.5a
Conventional tillage	47.6b	27.54b	0.50b	7.81a	235.6a
Midwest soybeans					
No-tillage	356.6a	23.88a	0.00a	0.00a	68.4a
Reduced tillage	313.0a	20.97a	0.02a	0.42a	81.2ab
Conventional tillage	315.6a	21.14a	0.02a	0.20a	85.8b
Midsouth soybeans					
No-tillage	310.4a	17.69a	0.06a	0.82a	36.5a
Reduced tillage	281.8a	16.06a	0.22a	1.80a	45.1a
Conventional tillage	312.0a	17.78a	0.19a	2.54a	47.4a
Southeast soybeans					
No-tillage	316.3a	18.35a	0.97a	14.62a	45.5a
Reduced tillage	340.9a	19.78a	0.39a	4.64a	42.0a
Conventional tillage	261.6a	15.17a	0.76a	11.63a	41.0a

[a]The Tukey–Kramer method (SAS Institute, Inc., 1982) was used to test for differences in means among tillage practices in each region. Regional Means in each column followed by different letters are significantly different from each other at the 5% level.

[b]a.i. = active ingredient.

control armyworm and cutworm infestations, than was acreage tilled more intensively. This suggests that during the 1980 growing season, insect problems requiring insecticide control after crop emergence were more prevalent on no-tillage cornfields than on cornfields tilled more intensively.

Although the 1980 average per-hectare insecticide cost for no-tillage corn farmers was 45% greater than the cost for other corn farmers, the difference was not statistically significant. Within each region tillage practice differences in amounts and costs of insecticides applied to soybeans were not statistically significant. Very few Midwest soybean farmers applied insecticides in 1980, while average per-hectare insecticide use and costs for all three tillage strategies were highest in the Southeast.

Variation in average per-hectare yields among tillage practices was not statistically significant for corn in the Midwest or soybeans in the Midsouth and Southeast. No-tillage soybean yields, however, were significantly lower than conventional plow-tillage yields in the Midwest, averaging about 17 bushels per hectare less. Separate analysis of Midwest weather conditions during 1980 determined that average soil moisture conditions early and late in the season were not as favorable in surveyed no-tillage soybean fields as in surveyed soybean fields tilled more intensively. For no-tillage fields, soil moisture levels were higher early in the season and substantially lower during the pod filling growth stage than moisture levels in more intensively tilled fields. The combined effect of slower germination and reduced pod development caused no-tillage soybean yields to average well below plow-tillage yields.

Costs and Returns

Per-hectare gross revenue and derived returns for Midwestern corn farmers using no-tillage and reduced-tillage practices were higher than dollar amounts for plow-tillage corn farmers. However, differences in the gross revenue figures among tillage strategies were not statistically significant (Table 14.9). Total per-hectare costs, however, for the sampled inputs (seed, fertilizer, pesticides, and mechanical cultivation) were significantly higher for both no-tillage and reduced tillage corn farmers compared to costs for farmers who conventionally tilled their fields. These higher costs were more than offset by a reduction in the assumed field preparation costs, resulting in slightly lower net costs per hectare for the conservation tillage systems.

In the Midwest, average per-hectare gross revenue for no-tillage soybean farmers was significantly lower than revenue for plow-tillage soybean farmers because, for reasons given above, no-tillage yields averaged over 17 bushels per hectare lower than plow-tillage yields. Sampled per-hectare input costs for no-tillage soybean farmers in the Midwest also were significantly higher than costs for other soybean farmers, but lower field preparation costs more than offset the rise in these costs. The net effect of substantially lower yields causes the per-hectare derived return to no-tillage soybeans in the Midwest to fall well below reduced and conventional tillage soybean returns.

TABLE 14.9. Average Per-hectare Gross Revenue, Costs ($), and Returns for Different Tillage Practices, 1980[a]

Crop, Tillage Practice	Gross Revenue	Cost for Sampled Inputs	Assumed Field Preparation Costs (dollars)	Total Affected Costs	Derived Return
Midwest corn					
No-tillage	789.93a	235.56a	145.26	380.82	409.11
Reduced tillage	784.52a	211.60a	178.56	390.16	394.36
Conventional tillage	758.71a	191.15b	202.69	393.84	364.87
Midwest soybeans					
No-tillage	517.49a	131.15a	108.51	239.66	277.83
Reduced tillage	614.14ab	92.45b	143.85	236.30	377.84
Conventional tillage	648.35b	95.20b	168.97	264.17	384.18
Midsouth soybeans					
No-tillage	286.89a	99.39a	117.87	217.26	69.63
Reduced tillage	354.37a	103.59a	158.48	262.07	92.30
Conventional tillage	372.97a	107.27a	186.81	294.08	78.89
Southeast soybeans					
No-tillage	350.72a	115.03a	106.65	221.68	129.03
Reduced tillage	323.69a	112.66a	143.95	256.61	67.09
Conventional tillage	315.91a	115.55a	169.71	285.26	30.65

[a]The Tukey–Kramer method (SAS Institute, Inc., 1982) was used to test for differences in means among tillage practices in each region. Regional means in each column followed by different letters are significantly different from each other at the 5% level.

Difference among tillage practices in per-hectare gross revenue and sampled input costs were not statistically significant for soybean farmers in either the Midsouth or Southeast. Estimated per-hectare derived returns among tillage systems followed different patterns in these regions, primarily because of diverging yields. Although differences were not statistically significant, average soybean gross returns for no-tillage farmers were lower than for other farmers in the Midsouth, while just the opposite was the case in the Southeast.

CONCLUSIONS

Conservation (reduced) tillage on soils highly susceptible to erosion has a long-term yield advantage over conventional (plow) tillage because it substantially reduces soil loss over time. On steep, shallow soils this provides an economic incentive to farmers to adopt conservation tillage. But most U.S. cropland now in production has neither steep slopes nor shallow soils. Consequently, we do not expect the long-term yield advantage of conservation tillage to give farmers strong incentive to adopt it unless the growth of crop demand induces movement of crops to more steeply sloped land with shallow soils.

The adoption of conservation tillage will depend primarily on farmers' perceptions of its short- and near-term economic benefits relative to plow-tillage. In this regard, the results of the USDA study indicate that only with soybean farmers in the Midwest were no-tillage gross returns significantly lower and costs of sampled inputs significantly higher than with plow systems. As noted, low no-tillage soybean yields in the Midwest in 1980 may have been attributable to unusual weather conditions that year. Consequently, we do not interpret the reported results as necessarily typical of no-tillage soybean yields in the Midwest.

Corn farmers in the Midwest using no-tillage and reduced tillage systems also had significantly higher costs of sampled inputs than farmers using plow systems. However, these higher costs were more than offset by lower field preparation costs. As a consequence, total affected costs for corn in the Midwest rose slightly with the intensity of tillage.

Given these findings, why is most corn and soybean acreage under conventional plow systems? Many farmers will not shift from these systems to conservation tillage unless there is a good prospect for substantial monetary gain brought about by either increased output or decreased costs. The USDA results suggest that corn and soybean farmers adopting conservation tillage can expect somewhat lower costs but no significant gains in gross revenue.

A simple comparison of per-hectare costs and returns, however, does not capture the full range of factors that farmers consider when choosing a tillage practice. Historically, most U.S. corn and soybean farmers have used a plow system to prepare their fields for planting. Familiarity with conventional practices, the perception of lower yields, increased weed problems, the need to purchase additional farm machinery, and a general lack of knowledge about conservation tillage systems are major reasons why more farmers have not adopted these systems.

An abrupt modification in tillage systems involves increased risk, particularly with no-tillage, until farmers gain sufficient experience with different crop production conditions. For some farmers, the prospect of doing about as well financially with conservation tillage is not enough to justify accepting higher levels of risk. But, corn and soybean farmers using conservation tillage derive other monetary benefits not addressed in the USDA study. By reducing field operations, conservation tillage farmers achieve greater flexibility early in the growing season and lower the risk of having to reschedule or repeat operations due to inclement weather. With lower variable costs, they can operate with less borrowed capital. And in some regions farmers using no-tillage can double-crop their fields each season.

Development of more effective postemergence herbicides and research to overcome the disadvantages of conservation tillage systems on poorly drained soils will encourage more U.S. farmers to adopt the systems. The rate of adoption, however, will also depend on whether relative input costs remain constant or change over time. Increasing energy and finance costs during the

1970s and early 1980s encouraged adoption of conservation tillage as a way to maintain or lower operating costs. Similar input price rises in the future may spur further adoption of conservation tillage, particularly reduced tillage practices which are less radical than no-tillage and more appropriate for use on most corn and soybean fields.

LITERATURE CITED

All, J. N. 1979. Insect relationships in no-till cropping, *Agrichemical Age* **23**:22–23.

All, J. N. and R. N. Gallaher. 1977. Detrimental impact of no-tillage corn cropping systems involving insecticides, hybrids, and irrigation on lesser cornstalk borer infestations. *J. Econ. Entomology* **70**:361–365.

Bandel, V. A. 1982. Fertilization of no-tillage corn. *The Agronomist* **19**:7–8.

Buchholz, D. D. and R. G. Hanson. 1982. Soil fertility considerations in conservation tillage systems. In *Conservation Tillage Seminar Proceedings*, College of Agriculture, Univ. of Missouri, Columbia, MO.

Christensen, L. A. and P. E. Norris. 1983. *A Comparison of Tillage Systems.* USDA, Econ. Res. Serv., Ag. Econ. Report No. 499, Washington, D.C.

Colvin, T., D. Erbach, S. Marley, and H. Erickson. 1983. Large-Scale Evaluation of a Till Plant System. Paper presented at summer meeting of Amer. Agric. Eng., Montana State Univ., Bozeman.

Conservation Tillage Information Center. 1983. 1982 National Survey Conservation Tillage Practices, Ft. Wayne, IN.

Cosper, H. R. 1979. *Soil Taxonomy as a Guide to Economic Feasibility of Soil Tillage Systems in Reducing Nonpoint Pollution.* U.S. Dept. of Agric., ESCS, Washington, D.C.

Crosson, P. 1981. *Conservation Tillage and Conventional Tillage: A Comparative Assessment.* Soil Conservation Society of America, Ankeny, Iowa.

Duffy, M. and M. Hanthorn. Returns to Soybean Pest Management Practices submitted for publication as an Agricultural Report, USDA, Econ. Res. Serv.

Ervin, D. E., P. D. Nolte, and D. D. Raitt. 1982. "The Economics of Conservation Tillage: Some Preliminary Observations in Missouri." In *Conservation Tillage Seminar Proceedings*. College of Agriculture, U. of Missouri, Columbia, MO.

Givan, W. 1980. "What it costs to operate machinery. *Progressive Farmer*, 56E.

Givan, W., C. Dorminey, and G. Westberry, 1979. 1980 Georgia Crop Enterprise Cost Analysis, Misc. Pub. No. 27, Coop. Ext. Ser., U. of Georgia.

Hanthorn, M. and M. Duffey. 1983. Returns to Corn Pest Management Practices, AER-501, USDA, Econ. Res. Serv.

Hayes, W. A. 1982. *Minimum Tillage Farming.* No-Till Farmer, Inc., Brookfield, WI.

Hein, N., R. Rudel, and R. Schottman, 1981. 1980 Custom Rates for Farm Services in Missouri, UMC Guide 302, University of Missouri–Columbia Extension Division.

Hicks, J. 1983. Plant Breeder with Pioneer Hi-Bred International, Inc., Greenville, Mississippi, personal communication, April 27.

Hileman, B. 1982. Herbicides in agriculture. *Environmental Science and Technology* **16**: 645A–659A.

Hinish, W. W. 1980. Soil fertility. *Crops and Soils* **32**:7–11.

Hinkle, M. 1983. Problems with conservation tillage. *J. Soil Water Conser.* **38**:201–206.

Jolly, R. W., W. M. Edward, and D. C. Erbach. 1982. An economic analysis of conservation tillage under Iowa conditions: A case study. In *Farm Agricultural Resources: Management Conference on Conservation Tillage*. Iowa St. U., Ames, IA.

Keaster, A. J. and M. L. Fairchild. 1982. Insect problems associated with reduced tillage systems. In *Conservation Tillage Seminar Proceedings*. College of Agriculture, U. of Missouri, Columbia, MO.

Kuper, K. 1983. Agronomist with Pioneer Hi-Bred International, Inc., Cedar Falls, Ia., personal communications, March 24 and April 15.

Larson, W. E., F. J. Pierce, and R. H. Dowdy. 1983. The threat of soil erosion to long-term crop production. *Science* **219**:458–465.

Marx, J. L. 1983. Plants' resistance to herbicide pinpointed. *Science* **220**:41–42.

Mason, B. Y. 1980. Agchemprice, Agsystems Research, El Paso, Texas, April 8.

May, D. and W. Walden. 1980. Arkansas Soybean Budgets, 1980, Special Report 80, Coop. Ext. Ser., U. of Arkansas.

Metcalfe, D. S. and D. M. Elkins. 1980. *Crop Production, Principles and Practices*, 4th ed. Macmillan, New York.

Moldenhauer, W. C., G. W. Langdale, W. Frye, D. K. McCool, R. I. Papendick, D. E. Smika, and D. W. Fryrear. 1983. Conservation tillage for erosion control. *J. Soil Water Conser.* **38**:144–151.

Pfost, D. L. 1982. Tillage systems: technical advantages and limitations. In *Conservation Tillage Seminar Proceedings*. College of Agriculture, Univ. of Missouri, Columbia, MO.

Phillips, R. E., R. L. Blevins, G. W. Thomas, W. W. Frye, and S. H. Phillips. 1980. No-tillage agriculture. *Science* **208**:1108–1133.

Pierce, F. J., W. E. Larson, R. H. Dowdy, and W. A. P. Graham. 1983. Productivity of soils: assessing long-term changes due to erosion. *J. Soil Water Conser.* **38**:39–44.

Pioneer Hi-Bred International, Inc., Pioneer Brand Soybean Seed.

Romander, L. 1982. Tillage for the times, *Agrichemical Age*, July 21–31.

SAS Institute, Inc. 1982. *SAS Users' Guide: Statistics*, Cary, North Carolina, pp. 139–99.

Seitz, W. D., C. R. Taylor, R. G. F. Spitze, C. Osteen,and M. C. Nelson. 1979. Economic impacts of soil erosion control. *Land Economics* **55**:28–42.

Shafer, W. B. 1981. Conservation tillage trends; implications for pesticides. Unpublished paper.

Siemens, J. C. 1975. Machinery related costs for tillage alternatives, U. of Ill., Urbana.

Siemens, J. C. 1982. Energy requirements for tillage-planting systems. Unpublished paper.

Soil conservation Society of America, 1973. Conservation Tillage, the Proceedings of a National Conference, Ankeny, Ia.

Successful Farming, 1983. Conservation Tillage Guide, Des Moines, Ia.

USDA. 1975a. Office of Planning and Evaluation, Minimum Tillage: A Preliminary Technology Assessment.

USDA. 1975b. Scope and Methods of the Statistical Reporting Service, Stat. Reporting Serv., Misc. Pub. No. 1308.

USDA. 1980. Commercial Fertilizers, Consumption for Year Ended June 30, 1980, Crop Reporting Board, Econ. and Stat. Serv., SpCr 7(11-80).

USDA. 1981a. Agricultural Statistics.

USDA. 1981b. Agricultural Prices, Annual Summary 1980, Crop Reporting Board, Eco. and Stat. Ser., Pr 1-3(81).

USDA. 1983. Crop Production 1982 National Summary, Crop Reporting Board CrPR 2-1(83) Washington, D.C.

United States Senate. 1982. Cost of Producing Selected Crops in the United States—Final 1979, 1980, and Preliminary for 1981. Committee on Agriculture, Nutrition, and Forestry.

University of Illinois. 1979. *Tillage System for Illinois.* Circular 1172. Cooperative Extension Service, U. of Illinois, Urbana, IL.

Walker, D. 1977. An Economic Analysis of Alternative Environmental and Resource Policies for Controlling Soil Loss and Sedimentation from Agriculture, Ph.D. dissertation, Iowa State University, Ames.

Walker, D. 1982. A damage function to evaluate erosion control economics. *American J. Ag. Econ.* **64**:690–698.

Walsh, R. 1982. My experience with conservation by limited tillage. In *Soil and Water Conservation: The Principle and the Practice.* Special Report 290, Ag. Exp. Sta., U. of Missouri, Columbia, MO.

Young, Jr., H. M. 1982. *No-Tillage Farming.* No-Till Farmer, Inc., Brookfield, WI.

15

TILLAGE MANAGEMENT FOR A PERMANENT AGRICULTURE

M. A. SPRAGUE
Professor Emeritus
Department of Soils and Crops
Rutgers University
New Brunswick, New Jersey

G. B. TRIPLETT
Professor of Agronomy
Department of Agronomy
Mississippi State University
Mississippi State, Mississippi

INTRODUCTION

A production enterprise must build and maintain its resource base to sustain itself. Wise management and improvement of that base are essential to compensate for the imbalance arising with its use. Agriculture has three major adjustable resources on which it depends—the soil, the climate, and the crop—and their interdependence is critically important to maintain proper balance for optimum performance. The weather (rainfall, temperature, light) is likely the most unpredictable, but can be chosen, predicted, adjusted, and controlled within limits. The soil can be evaluated and some corrective amendments made. The crop can be selected for adaptability and genetically altered, in the long range, to improve its performance. The latter requires much time and effort, and once a crop or cultivar is selected, it is fixed for the duration of the season. The management of these resources in reduced tillage systems discussed in this volume provides control mechanisms with which to maintain the balance required for a sustained agriculture.

The time is now and the opportunity is here to plan for a predictably more secure and stable world agriculture. For the farmer this means a year-to-year and generation-to-generation assurance that his or her land will remain productive. For the urban consumer it provides confidence that he or she can be fed. For all of society it beautifies the landscape and permits life to be more free from strife. A permanently productive agriculture is attainable today as never before in history through the wise application of our expanding knowledge of environmental resources and their use. Each generation in its turn must exercise good stewardship to maintain the resource base.

The volume, in succession, identifies and describes the many environmental features that make up that resource and interact to regulate crop growth with low-intensity tillage technology. It discusses the means by which tillage can be manipulated to maintain efficiency in the use of those resources on which crops depend for yield, quality, and profit. Separate considerations are given middle latitude grain crops in humid regions; humid region pastures and meadowlands; grain crops in the semiarid regions; range grasslands and varied opportunities in the tropics. Aside from the management of soil, water, and climatic inputs, other aspects impinging on growth and development are discussed, including crop adaptation, weed and other pest control, tillage and planting equipment, crop rotation, harvest sequence, and economics. Effectively integrated, through short-term and long-term no-tillage technology, on-going production systems are available to match a wide variety of crops, sites, and soil situations.

Crosson et. al. (Chapter 14) observe that increased use of reduced tillage systems by farmers is determined by their economic advantages. The inference is that their incentive for use is centered on near-term profitability on the better, but more difficult to manage, erosive soils where the economic advantage is greatest. On farms the focus is to use, yet retain and recharge, the resource base as a means to a continuing profitable return. The incentive long term is amplified as total demand (urban and rural) for less erosion-prone cropland increases, thus moving crop production onto the more erosive land. In the process the land base is shifted, and, with no-tillage, more erosive land can be safely brought into the food chain. A second major advantage is improved labor effectiveness, allowing one operator to farm more acres more efficiently through lower per land unit time requirements. No-tillage, in particular, so reduces the time equation required for seedbed preparation and subsequent cultural management that a single operator is able to plant and care for many times the area possible with a plow and cultivator.

TILLAGE SYSTEMS

In contemplating cultural management systems a plowed and prepared seedbed occupies the polar position of intensive tillage, while no-tillage is polar to the minimum amount of tillage required for crop growth. Between these two extremes of tillage intensity lie many variations and combinations of soil manipulation intended to meet specific requirements for the establishment and growth of each of many crop species under a wide range of soil and climatic conditions. Seedbed preparation is of initial interest when choosing a tillage system because of its importance to seedling establishment, which determines, in large part, the subsequent growth of a crop. No one system is suitable for all crops and situations; none are simple or equally successful on all accounts. Many compromising factors are unpredictable or uncontrollable within the time limits and resources available. Compensating management planning is essential on a continuing basis.

Plow-tillage inverts, pulverizes, and mixes the surface soil profile to incorporate plant and animal materials that have accumulated on the surface since last it was disturbed. It thus mixes and exposes a fresh, uniform, and uninhibiting soil surface. No-tillage involves no mechanical disturbance of the soil prior to planting. With some crops a temporary opening is needed to insert the seed into the soil through a surface mulch that otherwise remains undisturbed. Small-seeded crops often can be broadcast directly on the surface, and the natural forces of gravity, rain, and frost can be relied on to provide coverage and intimate soil contact. Between these two polar entities a wide variety of effective seedbeds can be provided, determined by the kind of weeds present; kind and quantity of materials on the surface; soil and water conditions; design of the equipment used; and size, morphology, and physiology of the seed. These manipulations of the soil and its surface cover are also exceedingly important to limit soil erosion, maintain soil structure, manage soil moisture, place seed properly, control competitive pests, and maximize crop growth. No-tillage and surface-tillage eliminate need for the plow.

DIVERSE SITUATIONS

Perhaps the most consistent theme throughout this volume is the great diversity among tillage systems to best serve different crops in their interacting environments. It becomes axiomatic that no two systems are alike, only similar in critical ways. Each is patterned to cope with controllable and uncontrollable inputs and have optimum influence on crop growth as each adjustment alters the balance of others in sequence. The challenge of this complex new technology is to regulate that balance sufficiently well to create an improved integrated environment within which the crop may flourish.

Individual crop species respond differently to environments, and in many ways intensely tilled and not tilled environments are not favorable. Crops differ greatly in their cardinal temperatures for growth and survival and respond differently to soil, water, mineral, and light regimes. Inputs adjust and supplement existing resources. Thus, in establishing a management system, the naturally occurring regimes, such as climate and soil, must be assessed and predicted and compensation made should the prediction err. Soil amendments, water, and tillage provide some means of control aided by an array of chemicals, delivery equipment, and energy sources to the extent they are effective and available. Rates, timing, and sequence are exceedingly important to success that the crop may progress unhindered through each stage of maturity. Finally, control lies with the farm manager who will choose the crop and variety and determine the cultural management and pest control based on his or her objectives, motivations, experience, capital, and ability. It is understandable that great diversity should exist among tillage systems.

LAND AVAILABILITY

When civilizations were small, claims on land for its soil and water resources were easily honored. The demand was minuscule by comparison with the seemingly inexhaustible supply. Methods of land use were primitively simple and production returns not dependable. As land use, crops, and their management improved, production increased, civilizations grew larger, and competition for land became more intense. New lands and climates were explored and claimed with increasing rapidity. The competition for land has never been more severe than at present. New stresses are pushing land use into marginal areas. Salinity, soil erosion, and desertification are expanding at a rapid rate, and much of the most productive land is lost to agriculture permanently for urban economic reasons. As productive land is lost, more efficient management of poorer sites must follow or hunger will prevail.

PRODUCTION ASSESSMENTS

Ancient writings report yield in terms of number of seed harvested per seed sown, reflecting the sacrifice of seed for planting in place of consumption. As fertile, arable land became limiting and seed more plentiful, yield per unit of land became an important measure. As additives increased production, yield per unit of input (irrigation water, mineral fertility, pesticides, tillage, equipment, power) became units of evaluation. The agricultural revolution of the past century shifted the focus of yield increase dramatically from human labor and natural resources to fossil fuels and supplementary inputs of growth control. Most of these are now purchased off the farm, so capital has thus become a critical measure of return. Improved labor effectiveness initially was gained through intensive tillage mechanization. More recently, specialized equipment and chemicals have permitted less tillage with improved labor-saving along with soil and water conservation. To the individual farm enterprise, return per unit of land is but one factor influencing economic advantage; capitalization per operator output has become of equal importance. The most dramatic change in agriculture during the past few decades has been the expanded use of debt capital to finance the new technologies.

THE SOIL MATRIX

Surface tillage, with the resultant surface mulch, in place of clean cultivation, has been recognized for several decades as a means of holding soil in place while it is being cropped. Its potential, however, with the advent of herbicides and no-tillage, has expanded far beyond the parameters of soil conservation alone. In semiarid regions surface tillage fallow was initially practiced for soil moisture retention, presumably by breaking capillary columns and reducing evaporation. Call and Sewell (1917) in Kansas, however, determined that the major value of tillage for moisture retention lay not in the tillage-produced dust-mulch but in the control of weedy plant growth as consumers of soil moisture. As research continued seasonal control of available water became important in humid regions as well. Tillage technology initially was largely for weed control; with herbicides as replacements for it, no-tillage has emerged as a focal center of cultural management carrying with it improved soil organic matter and texture and other influences on crop growth.

Plowing was advanced also as a means of deep placement of lime and fertilizer. This has lessened in importance, as no-tillage was found to encourage shallow rooting coincident with improved soil moisture management and soil aeration. Erosion control with a well-anchored mulch has expanded land use capability by moving annual grain crops to steeper slopes and untillable stony ground, thus expanding the productive land available to the food chain. Soil improvement is realized slowly, but there is abundant evidence of long-term increases in soil organic matter content, nitrogen and phosphorous use efficiency, increased water absorption, and more stable growing environments (Chapters 2, 4, and 5). Short-term advantages take the form of fewer tillage operations requiring less fuel consumption, labor, and wear on field equipment. As new engineering designs are introduced more versatile and fewer pieces of machinery are required to perform essential tasks better and more rapidly.

The authors of Chapters 2, 6, 7, 8, 9, and 10 clearly establish that no one tillage system is best for all soils, crops, and sites. For some soils and some crops, no-tillage may be the best choice to create an optimum environment for the crop while protecting the resource base. In other situations, a degree of tillage may be necessary to insure that specific needs of the crop are met and thus avert a decline in productivity. Only recently have we been able to evaluate tillage with weed control excluded to confirm these relationships. Chapters 2 and 10 develop schemes of tillage selection for the corn belt and tropics, respectively, based on soil and crop characteristics. An important effort for future research will be to test these schemes over a broad range of climates, soils, and crops, and to refine the systems in different localities. The full effect of tillage modification on the cropping environment cannot be arbitrarily predicted. Therefore new systems must be tested to assess their suitability.

MULCH EFFECTS ON TEMPERATURE AND MOISTURE

There is a clear message in Chapter 2, 6, 7, and 10 that soil and crop responses to surface- and no-tillage systems are remarkably similar in principle, but different in detail in the tropics, middle latitudes, humid, and semiarid climates. Temperature and moisture effects on soil weathering, organic matter, mineral uptake, microflora, and root growth follow familiar patterns. Surface mulches insulate to maintain more stable temperature and moisture regimes, the major differences being the seasonal sequences and extent of the extremes. Slower warming delays early planting following winter in the higher latitudes and altitudes, whereas slower cooling extends the growing season further into fall with winter small grains and perennial forages. In winter, less diurnal temperature fluctuation reduces winter crop losses from heaving and low temperature, where a surface mulch is present. In contrast, during summer insulation is favorable throughout most of the tropics and subtropics by avoidance of excessively high soil temperature. In both regions a mulch of plant residue reduces air circulation near the soil, maintains higher relative humidity in the microclimate, and thereby lessens soil moisture loss by evaporation. Chapter 7 describes how opportunity cropping increases the frequency of high-return crops in the rotation; thus, raising the total return from the land by better moisture management with no-tillage. A covering mulch also encourages some insects and diseases while encouraging earthworm activity.

A stubble mulch, with no-tillage winter wheat, traps and holds snow (Chapter 7) thus adding moisture to the soil profile in spring as well as insulating the plants against injurious cold temperatures. Experience (personal communication, E. H. Stobbe, University of Manitoba, Winnipeg, 1983) indicates this moves the area of winter wheat adaptability further north. To be fully effective the stubble must remain 8–10 inches (20–25 cm) high following harvest. A shift from spring to fall seeded wheat increases crop yield and economic return.

Some rocky soils, such as those in parts of New England, eastern Canada, southern Florida, and Yucatan are so stony as to be impossible to plow and equally forbidding to surface tillage of any type for weed control. Only a sharpened stick is effective for seed placement between the rocks, and weed control becomes impossible without herbicides. In many such areas, however, a deep, productive well-drained soil exists between the stones. (Sprague et al. 1979)

As with clean cultivation, the effects of reduced tillage in the farm enterprise are additive, overlapping, and accumulative, but not all are positive. Herbicides provide weed control and reduce competition in prescribed situations, while in other situations they are not dependable. A surface mulch and more fertilizer contribute to soil organic matter increases, but also increase acidity near the surface, which should be corrected. Reduced water runoff with a residue mulch is related to a greater total water penetration rate, adding to leaching losses, altered microclimates, lower evaporation rates, slowed drying, and reduced salinity problems in dry regions. Favorable soil moisture levels encourage shallow rooting habits, greater availability of surface-applied lime and phosphorus, and less water stress in some crop situations. Higher soil bulk density may interfere with root and tuber development in some tropical soils; soil aeration needs further study. In summary, water saved with no-tillage improves water use efficiency through better moisture management. A balance is reached in the environment which lessens extremes of temperature and moisture stress.

WEED CONTROL

Tillage uproots weeds, severing their contact with the soil, which leads to death by dessication or covering. Herbicides have allowed farm operators to relinquish this time-honored weed control practice in favor of better control with a more dynamic and dependable attack. The weed control objective, to eliminate competition, is greatly enhanced by the greater efficiency of herbicides. Technologies applied differ among species of crops and weeds, planting modes, soils, and climatic regions, but the objective remains the same, to encourage maximum crop growth by relieving competition without tillage. A tillage revolution, with associated benefits, is implemented through chemical growth control of competing species. Its introduction affects essentially all aspects of cultural management.

Selective herbicides first supplemented tillage in arable crop production where tillage was ineffective in the control of field bindweed. The technology has since been extended and expanded to function with a wide range of species and conditions of climate and soil to provide control of competition not possible with tillage alone. Special target weeds first received attention. Then developed a philosophy of complete riddance of all weedy plants prior to planting without tillage. In no-tillage the complete burden of weed control is shifted to the herbicide, and tillage is eliminated.

A single herbicide sometimes is not adequate; accordingly, combinations of materials are frequently needed. Capacities of one herbicide are matched with limitations of others for effective mixtures to complete the task. Suitable application techniques and correct timing are essential to cope with each problem. Shifts of weed species in a field must be recognized and programs altered to counter new problem situations. Crops also compete with weeds.

The performance of most herbicide combinations is conditioned by the tillage system in which they perform. Interpolation from one system to another is not predictable, particularly with soil-active materials. The grower–operator must gain sufficient knowledge of the herbicides he or she is using to understand their action characteristics and to use them safely with respect to both personnel and crops. He or she must have reliable sources of information and understandable descriptions of soil persistence, crop safety, effective rates, foliar and soil activity, timing, temperature, and application requirements. Compatability with other herbicides, carriers, crops, and weed species to be controlled is essential to implement effectiveness. This is not difficult to comprehend as instructions are listed on the pesticide labels in an orderly manner. Strategies should represent annuals, perennials, monocultures, multicropping, row crops, forages, and other crops, and weed situations as applicable. New materials currently being developed hold considerable promise with new and difficult weeds in many important situations.

DISEASE, INSECT AND OTHER PEST CONTROL

Tillage affects all biological systems in agriculture. The effects of tillage and no-tillage on insect and disease organisms are sufficiently well understood to permit predictions of benefits and problems likely to occur short term and long term (Chapters 12, 13). Pest-management strategies are mixtures of the old and the new in reduced tillage systems; old in the sense that firmly established principles are employed and new in the use of new methods in newly created environments. Just as weeds and crop plants respond to changes in microclimate in and near the soil surface, so predator life forms similarly respond. The degree of influence relates to the amount and kinds of residues in place at planting (both inocula and food sources), the temperature, the moisture, and the kinds of disease and insect forms present in relation to the host crop. Integrated pest management (IPM) employs specific control strategies adjusted to match the particular pest, site, and agricultural situation at hand.

Many pest-management strategies developed for plow-tillage systems have for the most part been effective with reduced tillage. In a few instances effective control measures have yet to be well developed for use in some areas. Most notable are grasshoppers, armyworms, and slugs in seedling forages; and stalk borers, southern billbug, and slugs in corn. Incomplete information is available also on some pests in cotton, tobacco, and certain vegetables. Generally, reduced tillage pest problems with double-crop soybeans are more influenced by late planting than tillage. As the parameters of reduced tillage expand to new agricultures, new strategies will be needed to protect crops in the coming decades.

MACHINERY AND EQUIPMENT

Precision tillage and planting equipment are needed to implement reduced tillage systems. Great strides have been made as needs are identified, and more improvements are on the drawing boards. The cost of equipment must be amortized in each situation by improved performance with respect to precision operation and time and fuel consumption, durability, and capital investment. Soil-working equipment design for plow-tillage evolved over the years as separate specialty tools for the most favorable sites and soil situations to be used in combinations and sequence. As surface- and no-tillage were introduced, some conventional tools were no longer needed nor adequate for the new tasks, particularly with respect to seed and fertilizer placements in seedbeds covered with crop residue and other impediments.

Planting and cultivation equipment were designed largely for use on level, smooth soil surfaces free of plant residue. Redesign required that such equipment operate consistently through light to heavy mulches of varying textures, at precision depths, on rough surfaces, in high-density, hard, or wet soils and, in the case of planters, to close an opening behind a seed for seed–soil contact. Beyond that, equipment should be able to operate on moderately steep slopes and in stony soils. Furthermore, it is essential that the engineer working with a variety of soil and mulch matrices provide for a rapid flow-through of materials without clogging. On stony soils provision must be made for rapid, smooth passage without breakage. Since tillage is relieved of pest control responsibilities in no-tillage systems, pesticide delivery systems are separated from tillage yet are no less important.

Equipment cost must include capital investment, repairs and maintenance, fuel, area covered, and amortization. Its function must conform to soil surface configurations, textures, and moisture contents and provide strength, weight, and durability to function dependably at high speeds without excessive corrosion, wear, or breakage. Planting equipment must be versatile to deliver seed, fertilizer, and pesticide materials over a broad range of sizes and textures. In no-tillage situations all these should be accomplished in a single pass over the field if possible.

A myriad of devices have been invented to accomplish these tasks, ranging from different size, shape, and placement of coulters to "hairpin" devices to prevent clogging and press wheels to firm soil around each seed (Erbach et al., 1984). A sharp coulter has emerged as *the* key opener tool; in surface- and no-tillage systems up to four types are commonly employed.

Much of the machinery available in many countries utilizes the basic tools with new variations in design, placement, and combinations to meet specific performance and power requirements. Modular design permits the availability of units for use on both large and small acreages. Developments are likely to continue to be innovative for many decades to meet projected needs in a wide variety of agricultures.

ENERGY AND TILLAGE

If only the tillage portion of the energy budget for production is considered, a move from plow-tillage provides a dramatic reduction in fuel use. Up to 40–50 liters of diesel fuel per hectare may be consumed by primary and secondary tillage. This can be reduced by as much as 80% (to about 10 liters per hectare) with no-tillage. The total energy requirement of a midwestern corn crop (fuel, nitrogen fertilizer, harvesting, shelling, drying) must also reflect the energy required for the manufacture of pesticides substituted for tillage. Estimates vary from 6 to 8 liters of diesel equivalent for each liter of pesticide produced. If the grower currently uses no herbicides, total energy savings would be relatively small; whereas, if the grower adds herbicides to a plow-tillage system, the total savings would be considerable.

Another dimension enters this equation. Engineering of plow-tillage systems and equipment have been so refined during the past decades that little opportunity remains to dramatically improve energy efficiency. Pesticides are still progressing in their development. When the first successful no-tillage was initiated in the late 1940s, 3–4 kg/ha or more of herbicide were required for weed control. A third of a century later, experimental herbicides are reportedly effective at rates of from 10 to 20 g/ha. If energy required for production of these new chemicals is near to those noted above, this represents a 10- to 100-fold increase in energy efficiency of herbicide production in but three decades.

GRASSLANDS

Forages (including woodland pasture) constitute the largest single segment of land use in world agriculture, aside from forestry, and much, if not all, is the poorest or least intensely managed. Some of the soils encompassed are nearly as productive as the best arable land but, for one of many reasons, have been allotted to the production of forage. It may have been due to their expanse, inaccessability, stones or other impediments, excessive or imperfect drainage, slope, or difficult climate. Most have been neglected with respect to fertilization, losing minerals annually with the removal of meat, milk, and wool and erosion. Some of the soils are thin and/or potentially unproductive due to limitations of precipitation, poor drainage, high or low temperature, salinity, or low natural fertility and so may not warrant large input investment. A sizable portion however does warrant attention with additions of inputs for improving production and quality of feed for livestock up to several hundred percent.

The largest single segment of the world's agriculture is a livestock enterprise. Most grasslands have received little or no management continuously for thousands of years. The potentially more productive areas have received some attention only during the past century or less, and returns have been very rewarding. Owing to their perenniality, location, or erodibility, these more productive forage lands are well suited to integrated no-tillage management and promise great percentage return.

NO-TILLAGE ACCEPTANCE

Development and adoption of new agricultural methods follow predictable patterns. Agricultural research workers and innovative growers generate and evaluate a new practice with available products. The science journals, extension specialists, and farm press describe the experience when it looks promising. Other researchers and more adventuresome growers follow to test the method in new ways and new environments and discuss the results in the classroom, at farmers' meetings, and through the grapevine. These initial phases frequently require a decade or more and represent but a small fraction of the potential production area. A few growers in time establish competence in its use in their particular operation; neighbors will adjust to it in their own. Industry plays a large role in overcoming the inertia of resistance to change with new products.

Considerable time and effort are required for introduction and adoption of a new practice. Hybrid corn required nearly 30 years from its discovery by Shull in 1909 until it occupied nearly 90% of the corn acreage in some areas. Caution exhibited by growers and suppliers is warranted; the investment of time and resources is great. The introduction of a many-faceted new practice, such as no-tillage, involving new products, equipment, and timing of land use, is far more complex than a mere substitution of seed of a new cultivar. Adoption of no-tillage cultural management, however, has followed a path similar to that of hybrid corn, only more slowly and carefully. It has now been shown that a knowledgeable manager who knows his or her land, the crop and its needs, weeds and herbicides and other pests can readily develop a useful no-tillage system specially suited to his or her own farm situation.

Major departures from traditional practices involve considerable risk and expenditure on the part of a farm business manager. Without a margin of profit no business venture can continue long, and no-tillage is no exception. New methods involve new equipment, herbicide–pesticide schedules, labor capabilities, and capital inputs. These are investments that must be amortized by returns from the operation. No-tillage improved water management and fertilizer use efficiency may increase grain yield and quality in the short or intermediate term. Improved labor and fuel efficiency may reduce costs per hectare in some instances, but also may be delayed until a few years' experience have implemented better planning. Returns from soil improvement and erosion control will likely be delayed in the long term. See Conclusions, Chapter 14.

Interest rates and costs of fertilizer, pesticide and land fluctuate. Short-term profit or loss depends on an operator–manager's good planning and few mistakes. This requires quality information, improved products, and guidance in their use. The crescendo increase of no-tillage acreage in some areas during recent years is testimony that the information and products with which a farmer has been provided have been effective and are improving. As they continue to improve and systems are refined in the years ahead the tillage revolution will likely continue to expand.

A FORWARD LOOK

Major changes in agriculture are nearly always slow, initially difficult, and require clear understanding. New complex adjustments resist change from familiar practices, and no-tillage is a radical departure from cultural methods common only a few decades ago. It follows that we are in the process of improving our ability to predict and regulate biological performance in newly balanced environments in order that all the biologs involved may act more in unison and in a complementary fashion. The pursuit of sustainable improved production lies in the balance of interacting inputs and responses.

Much of the discussion in this volume represents research findings with the field crops corn, soybean, wheat, and other small grains, sorghum, rice, cotton, and the forage crops. Standifer and Beste (1985) offer a state of the art assessment of reduced tillage systems with several vegetable crops. Included is a study involving vegetable production in rye stubble as a means to reduce wind erosion. Some vegetable seedlings are often severely injured by wind-blown sand particles after emergence in production fields. Covers of many types reduce the abrasive action of particle movement to negligible levels. Sweet corn, snap beans, lima beans, and direct-seeded tomatoes grown with a mulch cover are reported to respond well and to be commercially feasible. Such reduced tillage stubble mulches are as yet of marginal value and not feasible with watermelon and cucumber. A plastic film mulch is reported to be functional but economically feasible only with very high value crops to meet the greater cost involved.

New Cultivars

Boyer (1982) points out that, because plants are essentially immobile, genetic recombinations have resulted in an evolution of selected cultivars that take advantage of growth agents that move and change in the environment, namely, temperature, wind, water, light, and soil. In an analysis of eight major crops, he shows that a large genetic potential for yield is unrealized because of need for better adaptation of crop varieties to improved environments. Average farm yields were only 21.6% of record yields in 1977. The focus of natural selection for millennia has been on survival of which only a part is yield. In natural environments, and to a lesser extent in agriculture, plants are under varying kinds and amounts of stress including drought, shallow soil, cold, salinity, competition, pests, and others. No-tillage and surface-tillage ease these stresses and in many ways stabilize them with respect to extremes, compared with plow-tillage cultivation.

During the past 35 years, corn yield has more than tripled by changing the environment to better serve the available gene base. Most of the increase during the first 12–15 years is attributed to the use of hybrid vigor. Since 1950, although efforts have been made to obtain new germ plasm, increases have been largely attributed to increased use of plant nutrients and control of weed competition and other pests. In doing so, new environments have been created and their effectiveness established by plant response. Genetic selection for these improved environments has had too minor a role in agriculture.

Inadequate attention is being directed to the mechanisms of plant growth in stress environments and the generation of new genotypes to better fit them. Studies of specific mechanisms have shown that plants have evolved ways to respond to environments and change their adaptation, and that these can be transmitted genetically. Truly, adaptation to improved environments can make major contributions to agriculture, but the process at present is insufferably slow. A stable gene base is inadequate. Recent advances in biochemical techniques to study gene manipulations are encouraging that plant breeders will be able to expand their horizons to better fit the crop to the more dependably favorable environments of no-tillage. For example, a recent announcement (*Progressive Farmer*, November 1985, p. 32) indicates that, through genetic engineering in the laboratory, a strain of corn has been developed which is tolerant of a new class of herbicides called the imidazolinones (Imazaquin and related compounds). This member of the group controls a broad spectrum of plants which now severely restrict no-tillage corn production. The new gene should lead to a corn hybrid that could be directly planted in fields infested with johnsongrass without injury to the crop.

Computer Models

Through applied mathematics with computers it is becoming increasingly possible for teams of scientists in different disciplines to pool their knowledge in appropriate simulation models to scrutinize the impacts of different events alone and in concert. The result is better understanding of the contribution of each of the ingredients in the system in the presence of others. From this an intelligent research thrust toward both short-term and long-term objectives can be launched. With mathematical modeling it is not enough to know what a tillage system does, how it works, and what it producers, but why it works, to understand why it may not work sometimes, and to be able to change it so it will. No-tillage is ready and in need of this kind of scrutiny and appraisal to guide its future.

No-tillage is no longer a new technology. It has been explored and tested from the subarctic to the tropics, from the deserts to the rainforests, and all major crops have been represented. It now needs refinement. It is respected as a means to a continuing agriculture and is being used on many farms. Since it remains a juvenile in terms of total agricultural land, however, the work is not complete. It remains a challenge and presents great opportunity for further development. For example, preliminary tests in New Jersey with the self-seeding subterranean clover, *Trifolium suberraneum* L., have been promising as a live mulch weed management tool in continuous corn production (Personal communication, R. D. Ilnicki, C. Rosenzweig, Rutgers University, New Brunswick, N. J.). Certain varieties (1) have survived winter temperatures, (2) reseeded well in late May, (3) die in June, (4) germinate in early August, (5) produce high quality forage in fall and spring, and (6) contribute nitrogen to the soil and exhibit no apparent competition to growing corn. Patterns of herbicide use are now being developed which require care in selection of rates and timing of application, especially with respect to materials which persist in soil.

A farm enterprise will make good use of its own computer analysis to determine the means to profitability. Software programs are becoming available (Edwards, 1984), and more will follow to analyze the effectiveness of planting and cropping patterns, cover crops, and fertilizer placements in a myriad of soil types and situations, climates, and other variables as yet undetected. With a soil loss equation, a weather probability function, and a genetic potential estimate a production management projection may be possible for each farm, field, and operator. Such analyses build organized plans not on emotion, hunches, or unsupported decisions, but on facts of situations at hand and calculated opportunities in monitored environments. Tillage decisions and alternatives form one focal point of management planning to meet the challenges of the production marketplace. The final, fine-tune, adjustment decisions must be made by the farm operator in the field at the time of planting, spraying, and fertilization. Constant surveillance of crop performance is essential, especially at early stages of growth.

The risks and challenges of agriculture are ever changing. Timely adjustments to them determine the difference between profit and bankruptcy; abundance or famine. Soil improvement builds on the existing base as the long-term variable; soil loss from wind or water spurs the effort. For the more sophisticated, computer technology provides prompt answers for a multitude of decisions in a dynamic enterprise. To function dependably, reliable data must be available. The no-tillage revolution is well underway, but still has a long way to go.

SUMMARY

Cultural management was dependant on plows and cultivators for many centuries and settled into reasonably stable systems of crop production. These discussions represent the base on which a new agriculture is unfolding through the substitution of herbicides for tillage resulting in more favorable environments for crop growth.

LITERATURE CITED

Boyer, J. S. 1982. Plant productivity and environment. *Science* **218**:443–448.

Call, L. E. and M. C. Sewell. 1917. The soil mulch. *J. Amer. Soc. Agron.* **9**:49–61.

Edwards, M. 1984. Moving soil conservation ahead with computer programs. *Agrichemical Age* **28**: (May) 36.

Erbach, D. C., J. E. Morrison, and D. E. Wilkins. 1984. Conservation tillage puts spurs to equipment innovation. *Agrichemical Age* **28**: (March) 12–22.

Sprague, M. A., J. C. Anderson, H. L. Motto, and J. M. Lopez Diaz. 1979. Alternatives to a monoculture of henequen in Yucatan. II Studies with maize, sorghum and sesame. *Interciencia* **4**:84–91.

Standifier, L. C. and C. E. Beste. 1985. Weed control methods for vegetable production with limited tillage. In A. F. Wiese (ed.). *Weed Control in Limited Tillage Systems*. Weed Science Soc. Amer. Monogr. Ser. No. 2, Champaign, IL.

APPENDIX

COMMON AND CHEMICAL NOMENCLATURE OF HERBICIDES

Acifluorfen	5-[2-Chloro-4-(trifluoromethyl)phenoxy]-2-nitrobenzoic acid
Alachlor	2-Chloro-*N*-(2,6-diethylphenyl)-*N*-(methoxymethyl)acetamide
Ametryn	*N*-Ethyl-*N′*-(1-methylethyl)-6-(methylthio)-1,3,5-triazine-2,4-diamine
Atrazine	6-Chloro-*N*-ethyl-*N′*-(1-methylethyl)-1,3,5-triazine-2,4-diamine
Bentazon	3-(1-Methylethyl)-(1*H*)-2,1,3-benzothiadiazin-4(3*H*)-one 2,2-dioxide
Bromoxynil	3,5-Dibromo-4-hydroxybenzonitrile
Chlorsulfuron	2-Chloro-*N*-[[(4-methoxy-6-methyl-1,3,5-triazin-2-yl)amino]carbonyl]benzenesulfonamide
Clopyralid	3,6-Dichloro-2-pyridinecarboxylic acid
Cyanazine	2-[[4-Chloro-6-(ethylamino)-1,3,5-triazin-2-yl]amino]-2-methylpropanenitrile
Dalapon	2,2-Dichloropropanoic acid
Dicamba	3,6-Dichloro-2-methoxybenzoic acid
Dinoseb	2-(1-Methylpropyl)-4,6-dinitrophenol
Diuron	*N′*-(3,4-Dichlorophenyl)-*N*,*N*-dimethylurea
DSMA	Disodium salt of MAA
EPTC	*S*-Ethyl dipropyl carbamothioate
Fluazifop	(±)-2-[4-[[5-Trifluoromethyl)-2-pyridinyl]oxy]phenoxy]propanoic acid
Fluometuron	*N*,*N*-Dimethyl-*N′*-[3-(trifluoromethyl)phenyl]urea
Glyphosate	*N*-(Phosphonomethyl)glycine

Hexazinone	3-Cyclohexyl-6-(dimethylamino)-1-methyl-1,3,5-triazine-2,4(1*H*,3*H*)-dione
Imazaquin	2-[4,5-Dihydro-4-methyl-4-(1-methylethyl)-5-oxo-1-*H*-imidazol-2-yl]-3-quinolinecarboxylic acid
Karbutilate	3-[[(Dimethylamino)carbonyl]amino]phenyl(1,1-dimethylethyl)carbamate
Linuron	*N'*-(3,4-Dichlorophenyl)-*N*-methoxy-*N*-methylurea
Mefluidide	*N*-[2,4-Dimethyl-5-[[(trifluoromethyl)sulfonyl]amino]phenyl]acetamide
Metolachlor	2-Chloro-*N*-(2-ethyl-6-methylphenyl)-*N*-(2-methoxy-1-methylethyl)acetamide
Metribuzin	4-Amino-6-(1,1-dimethylethyl)-3-(methylthio)-1,2,4-triazin-5(4*H*)-one
MSMA	Monosodium salt of MAA
Oryzalin	4-(Dipropylamino)-3,5-dinitrobenzenesulfonamide
Oxyfluorfen	2-Chloro-1-(3-ethoxy-4-nitrophenoxy)-4-(trifluoromethyl)benzene
Paraquat	1,1'-Dimethyl-4,4'-bipyridinium ion
Pendimethalin	*N*-(1-Ethylpropyl)-3,4-dimethyl-2,6-dinitrobenzenamine
Picloram	4-Amino-3,5,6-trichloro-2-pyridinecarboxylic acid
Pronamide	3,5-Dichloro(*N*-1,1-dimethyl-2-propynyl)benzamide
Propazine	6-Chloro-*N*,*N'*-bis(1-methylethyl)-1,3,5-triazine-2,4-diamine
Sethoxydim	2-[1-(ethoxyimino)butyl]-5-[2-(ethylthio)propyl]-3-hydroxy-2-cyclohexen-1-one
Silvex	2-(2,4,5-Trichlorophenoxy)propanoic acid
Simazine	6-Chloro-*N*,*N'*-diethyl-1,3,5-triazine-2,4-diamine
Sodium chlorate	Sodium chlorate
TCA	Trichloroacetic acid
Tebuthiuron	*N*-[5-(1,1-Dimethylethyl)-1,3,4-thiadiazol-2-yl]-*N*,*N'*-dimethylurea
Terbutryn	*N*-(1,1-Dimethylethyl)-*N'*-ethyl-6-(methylthio)-1,3,5-triazine-2,4-diamine
Triclopyr	[(3,5,6-Trichloro-2-pyridinyl)oxy] acetic acid
2,4-D	(2,4-Dichlorophenoxy)acetic acid
2,4,5-T	(2,4,5-Trichlorophenoxy)acetic acid

SOURCE: Composite list of herbicides. 1985. In D.E. Davis (ed.). *Weed Science Terminology*, Vol. 33, Supplement 1, pp. 5–11. Reproduced with permission.

APPENDIX

B

FACTORS FOR CONVERTING NON-SI TO ACCEPTABLE UNITS

Multiply *Non-SI Units*	*By*	To Obtain *Acceptable Units*
Acre	4.05×10^3	Square meter, m^2
Acre	0.405	Hectare, ha (10^4 m^2)
Acre	405×10^{-3}	Square kilometer, km^2
Atmosphere (bar)	0.101	Megapascal, MPa
Bar	0.1	Megapascal, MPa
Cubic feet	0.028	Cubic Meter, m^3
Cubic feet	28.3	Liter, L
Cubic inch	164×10^{-5}	Cubic meter, m^3
Foot	0.305	Meter, m
Gallon	3.78	Liter, L
Gallon per acre	9.35	Liter per hectare, L ha^{-1}
Inch	25.4	Millimeter, mm
Micron	1.00	Micrometer, μm (10^{-6} m)
Mile	1.61	Kilometer, km (10^3 m)
Mile per hour	0.477	Meter per second, m s^{-1}

(continued)

Appendix B *(Continued)*

Multiply *Non-SI Units*	*By*	To Obtain *Acceptable Units*
Ounce	28.4	Gram, g
Ounce (fluid)	2.96×10^{-2}	Liter, L
Pint (liquid)	0.473	Liter, L
Pound	454	Gram, g
Pound per acre	1.12	Kilogram per hectare, kg ha^{-1}
Pound per bushel	12.87	Kilogram per cubic meter, kg m^{-3}
Pound per cubic foot	16.02	Kilogram per cubic meter, kg m^{-3}
Pound per cubic inch	2.77×10^{4}	Kilogram per cubic meter, kg m^{-3}
Pound per square foot	47.9	Pascal, Pa
Pound per square inch	6.90×10^{3}	Pascal, Pa
Quart (liquid)	0.946	Liter, L
Quintal (metric)	10^{2}	Kilogram, kg
Square centimeter per gram	0.1	Square meter per kilogram, m^2 kg^{-1}
Square feet	9.29×10^{-2}	Square meter, m^2
Square mile	2.59	Square kilometer, km^2
Temperature (°F − 32)	0.556	Temperature, °C
Temperature (°C + 273)	1	Temperature, K
Tonne (metric)	10^{3}	Kilogram, kg
Ton (2000 lb)	907	Kilogram, kg
Ton per acre	2.24×10^{3}	Megagram per hectare, Mg ha^{-1}

SI system (Le Systeme International d'Unites), National Bureau of Standards. 1977. International system of units. Spec. Pub. 30 *U.S. Government Printing Office, Washington, D.C.* Selected from *Publications Handbook and Style Manual*, 1984, pp. 50–51, with permission of the American Society of Agronomy, Crop Science Society of America, Soil Science Society of America.

INDEX